Radiotracer Methodology in the Biological, Environmental, and Physical Sciences

PRENTICE-HALL BIOLOGICAL SCIENCE SERIES

William D. McElroy and Carl P. Swanson, *Editors*

Radiotracer Methodology in the Biological, Environmental, and Physical Sciences

C. H. Wang
Director of Radiation Center
Professor of Chemistry
Oregon State University

David L. Willis
Professor of Biology
Oregon State University

Walter D. Loveland
Associate Professor of Chemistry
Oregon State University

Prentice-Hall, Inc., Englewood Cliffs, New Jersey

Library of Congress Cataloging in Publication Data

Wang, Chih Hsing (date)
Radiotracer methodology in the biological, environmental, and physical sciences.

Published in 1965 under title: Radiotracer methodology in biological science.
Icnludes bibliographies and index.
1. Radioactivity—Measurement. 2. Radioactive tracers. 3. Radiobiology—Research. I. Willis, David L., joint author. II. Loveland, Walter D., joint author. III. Title.
QC795.42.W36 1975 500.2'02'8 75-11714
ISBN 0-13-752212-6

Previously published under the title of
Radiotracer Methodology in Biological Science © 1965

10 9 8 7 6 5 4 3 2

Printed in the United States of America

PRENTICE-HALL INTERNATIONAL, INC., *London*
PRENTICE-HALL OF AUSTRALIA, PTY. LTD., *Sydney*
PRENTICE-HALL OF CANADA, LTD., *Toronto*
PRENTICE-HALL OF INDIA PRIVATE LIMITED, *New Delhi*
PRENTICE-HALL OF JAPAN, INC., *Tokyo*
PRENTICE-HALL OF SOUTHEAST ASIA (PTE.) LTD., *Singapore*

Contents

Preface xiii

Introduction 1

1 Atoms and Nuclides 5

A. Atomic Structure and Energy Relations *5*
B. The Nucleus *6*
C. Nomenclature *8*
D. Stable Nuclides *9*
E. Radionuclides (Radioisotopes) *10*
1. Naturally Occurring Radioisotopes *11*
2. Artificially Produced Radioisotopes *12*
Bibliography *12*

2 The Nature of Radioactive Decay 14

A. Radionuclides and Nuclear Stability *14*
B. Types of Radioactive Decay *18*
1. Decay by Negative Beta (Negatron) Emission *18*
2. Decay by Positive Beta (Positron) Emission *19*

3. Decay by Electron Capture (EC) *20*
4. Gamma Ray Decay *22*
5. Internal Conversion (IC) *22*
6. Decay by Alpha Particle Emission *22*
7. Extranuclear Effects of Radioactive Decay *22*
C. Rate of Radioactive Decay *24*
1. Radioactive Decay Kinetics *25*
2. Half-life *28*
3. Composite Decay *32*
4. Average Life *33*
D. The Standard Unit of Radioactivity—The Curie *34*
E. Specific Activity *37*
Bibliography *38*

3 Characteristics of Ionizing Radiation 39

A. Alpha Particles *39*
1. Energy *39*
2. Half-life and Energy Relationships *40*
3. Interaction With Matter *41*
4. Range *42*
5. Practical Considerations *45*
B. Beta Particles *46*
1. General Nature *46*
2. Energy of Beta Decay *47*
3. Interaction With Matter *51*
4. Range *53*
5. Practical Considerations *55*
C. Gamma Rays *57*
1. Nature and Source *57*
2. Interaction of Gamma Radiation With Matter *60*
3. Absorption Relations *63*
4. Practical Considerations *71*
D. Summary *74*
Bibliography *74*

4 Nuclear Instrumentation 75

A. Introduction *75*
B. The Preamplifier *77*
C. The Amplifier *79*
D. Pulse Height Selectors *81*

E. Scalers *82*
1. Binary Scalers *82*
2. Decade Scalers *83*
F. Count Rate Meters *84*
G. Pulse Height Analyzers *86*
1. The Single-Channel Analyzer *86*
2. Multichannel Pulse Height Analyzers *87*
H. Output Devices *89*
I. Coincidence Circuitry *90*
1. Circuitry Used *91*
2. Techniques and Problems *91*
Bibliography *94*

5 Measurements of Radioactivity: General Considerations and the Methods Based on Gas Ionization **96**

A. Types of Radioactivity Measurement *96*
B. General Detector Characteristics *97*
C. Basic Principles of Radiation Detection *97*
1. Gas Ionization *98*
2. Scintillation *98*
3. Semiconductor Detectors *98*
4. Nuclear Emulsions and Autoradiography *99*
D. Gas Ionization *99*
1. Without Gas Amplification *100*
2. With Gas Amplification *106*
3. Proportional Detectors *111*
4. Geiger-Müller Detectors *114*
5. Summary of Gas Ionization Detectors *119*
Bibliography *121*

6 Gamma Ray Counting Using Solid Scintillators **122**

A. Basic Facets of the Scintillation Phenomenon *122*
B. Solid Scintillation Detectors *123*
1. Mechanism of Solid Scintillation Detection *123*
2. Components of Solid Scintillation Detectors *129*
C. Integral Counting *134*
1. Effect of Photomultiplier Potential *135*
2. Effect of Amplifier Gain *136*
3. Detection Efficiency *137*
Bibliography *141*

7 **Gamma Ray Spectrometry Using Solid Scintillation Detectors** **142**

A. Fundamentals *142*
1. Low-Energy Gamma Rays (γ_1) *144*
2. Intermediate-Energy Gamma Rays (γ_2) *146*
3. High-Energy X-Rays (γ_3) *148*
B. Special Effects *148*
C. Components of a Solid Scintillation Spectrometer *156*
D. Energy Resolution *158*
E. Identification of Radionuclides *160*
F. Quantitative Analysis of Gamma Ray Spectra *163*
G. Single-channel Counting *166*
Bibliography *167*

8 **Semiconductor Radiation Detectors** **168**

A. Basic Nature of Semiconductors *168*
B. Basic Operating Principles of Semiconductor Radiation Detectors *171*
C. Basic Types of Semiconductor Radiation Detectors *173*
Bibliography *180*

9 **Measurement of Radioactivity by the Liquid (Internal-Sample) Scintillation Method** **181**

A. Mechanism of Liquid (Internal-Sample) Scintillation Detection *182*
1. Energy-Transfer Steps in the Fluor Solution *182*
2. Energy-Transfer Efficiency for Very Soft Beta Particles *185*
B. Evaluation of the Liquid Scintillation Method *186*
1. Advantages of Liquid Scintillation Counting *186*
2. Problems Inherent in Liquid Scintillation Counting *189*
C. Components of a Liquid Scintillation Counter *194*
1. The Detector Assembly *194*
2. The Electronics *199*
D. Special Types of Liquid Scintillation Detectors *210*
1. Large-Volume External-Sample Detectors *210*
2. Continuous-Flow Scintillation Detectors *212*
E. Operating Characteristics of Liquid Scintillation Counters *214*
1. Selection of Optimal Counter Settings *214*
2. Determination of Counting Efficiency *217*
F. Summary *224*
Bibliography *224*

10 **Measurement of Radioactivity by Emulsion and Track Detectors** **233**

A. Introduction *233*
B. General Principles of Autoradiography *235*
1. Resolution and Radioisotope Characteristics *235*
2. Film Emulsion and Sensitivity *236*
3. Determination of Exposure Time *237*
4. Sample Preparation and Artifacts *237*
C. Specific Autoradiographic Techniques *238*
1. Temporary Contact Method (Apposition) *238*
2. Permanent Contact Method *239*
D. Nuclear Track Detectors *242*
Bibliography *245*

11 **Preparation of Counting Samples** **248**

A. Factors Affecting Choice of Sample Form for Counting *249*
B. Conversion of Biological Samples to Suitable Counting Form *253*
1. Ashing Methods *253*
2. Combustion Methods *254*
3. Combustion of Samples for Liquid Scintillation Counting *256*
4. Miscellaneous Methods *260*
C. Assay of Samples in Various Counting Forms *261*
1. Assay of Gaseous Counting Samples *261*
3. Assay of Liquid Counting Samples *261*
3. Assay of Solid Counting Samples *262*
4. Radiochromatogram Scanning *266*
D. Preparation and Assay of Liquid Scintillation Counting Samples *268*
1. Basic Considerations in the Choice of Scintillation Solution *268*
2. Homogeneous Counting Systems *269*
3. Heterogeneous Counting Systems *276*
Bibliography *280*

12 **Nuclear Statistics** **294**

A. Introduction *294*
B. Basic Statistical Distributions for Radioactive Decay *294*
C. Parameters that Describe Statistical Distribution Functions *298*
D. General Rules for Calculations Involving Normally Distributed Quantities *300*

E. Nuclear Counting Statistics *302*
1. Average Counting Rate *302*
2. Sample and Background Counting *303*
3. Optimum Choice of Detector *305*
4. Weighted Averages *306*
5. Rejection of Abnormal Data *307*
6. Other Statistical Tests *307*
F. Application of Nuclear Counting Statistics to Radiotracer Experiments *308*
Bibliography *308*

13 Correction Factors in Radiotracer Assay **310**

A. Background *310*
B. Geometry *312*
C. Detector Efficiency *313*
D. Resolving Time Losses *314*
E. Backscattering *317*
F. Absorption *319*
1. Window and Air Absorption *319*
2. Self-Absorption *320*
G. Summary *323*
Bibliography *323*

14 Design and Execution of Radiotracer Experiments **326**

A. Unique Advantages of Radiotracer Experiments *326*
B. Preliminary Factors in Design of Radiotracer Experiments *327*
1. Basic Assumptions Underlying the Validity of Radiotracer Experiments *327*
2. Evaluation of the Feasibility of Radiotracer Experiments *331*
C. Basic Features of Experimental Design *333*
1. The Nature of the Experiment *333*
2. The Scale of Operation *333*
3. Detection Efficiency *335*
4. Specific Activity *335*
5. Example of Experimental Design *336*
6. Anticipated Experimental Findings *338*
D. Execution of Radiotracer Experiments *338*
E. Data Analysis *339*
1. Expression of Results *339*
2. Interpretation of Results *340*
Bibliography *342*

15 **Availability of Radioisotope-Labeled Compounds** 344

A. Primary Production of Radionuclides *344*
1. Introduction *344*
2. Energetics and Probability of Nuclear Reactions *346*
3. Practical Techniques *347*

B. Conversion of Primary Radioisotopes to Labeled Compounds *352*
1. Chemical Synthesis *353*
2. Biosynthesis *356*
3. Tritium Labeling *356*
4. Radiolysis of Labeled Compounds *359*

Bibliography *360*

16 **Nuclear Safety** 362

A. The Standard Units of Radiation Exposure and Dose *364*
1. The Roentgen *364*
2. The Rad *365*
3. The Q Factor, rem, and LET *365*

B. Hazard Factors in Handling Radioisotopes *366*
1. External Hazards *367*
2. Internal Hazards *368*

C. Radiation-Monitoring Instrumentation *371*
1. Area Monitoring *371*
2. Personnel Monitoring *372*

D. Decontamination *375*

E. Disposal of Radioactive Wastes, Licensing, and Transportation of Radioactive Materials *376*

F. Radioisotope Laboratory Safety Rules *377*

Bibliography *379*

17 **Radioanalytical Techniques** 381

A. Introduction *381*

B. Isotope Dilution Analysis (IDA) *381*
1. Direct IDA *381*
2. Inverse IDA *383*
3. General Comments *383*
4. Special IDA Techniques *385*

C. Radiometric Techniques *388*

D. Activation Analysis *389*
1. Basic Description of Method *389*
2. Advantages and Disadvantages of Activation Analysis *391*

3. Practical Considerations in Activation Analysis *394*
4. Applications of Activation Analysis *397*
Bibliography *398*

18 **Environmental Applications of Radiotracers** **399**

A. Introduction *399*
B. Experimental Methods *400*
1. Artificial Radiotracers *400*
2. Stable Activable Tracers *404*
3. Naturally Occurring Radionuclides *411*
C. Methods of Sampling *416*
D. Low-Level Techniques *420*
1. Introduction *420*
2. Blanks *420*
3. Low-Level Counting: General Principles *421*
4. Low-Level Counting: Details *423*
Bibliography *430*

19 **Tracer Applications in the Physical Sciences** **433**

A. Tracing of Physical Processes *433*
B. Chemical Applications of Tracers *434*
C. Isotope Effects *438*
Bibliography *445*

Problems **447**

Appendix Nuclear Reaction Calculations **461**

A. Energetics of Nuclear Reactions *461*
B. The Probability of a Nuclear Reaction *463*
Bibliography *467*

Index **469**

Preface

The radiotracer method is clearly one of the most powerful tools in scientific research. Despite the number of excellent volumes that have appeared during the past two decades describing radiotracer techniques, it appears that a need still exists for a systematic treatment that covers the necessary fundamental background as well as up-to-date developments, such as γ-ray spectroscopy using Ge(Li) detectors.

As many readers will recognize, some of the material presented here is taken from *Radiotracer Methodology in Biological Science* by C. H. Wang and D. L. Willis. The new book is intended as a revision and a replacement of the older text. The motivation for writing it was manyfold. First, in the years that have passed since the first printing of Wang and Willis, many new, important tracer methods and radiation-measuring instruments have been developed that deserve discussion. Second, we have found many places in which we wished to correct errors in the older text and to present material in a pedagogically more effective manner. Finally, we wished to provide what insight we could concerning the use of radiotracers in the new, burgeoning field of environmental science, plus up-to-date coverage of the use of tracers in the physical sciences.

The new book is, we believe, somewhat more sophisticated in content than the old edition, but we have strived to retain straightforward explanations of the basic material. The material in Chapters 1, 2, and 3 on atomic structure, radioactive decay, and the interaction of radiation with matter has been updated, and a slightly more physical approach to the subject material has been used. A new chapter, Chapter 4, on nuclear instrumentation has been added in reflection of the authors' belief that single-channel analyzers, scalers, and similar instruments cannot be treated by a competent experimentor as "black boxes" whose workings are a mystery. Chapter 5 on gas ionization detectors is only slightly changed from the older book. Chapters 6 and 7 on γ-ray counting and spectroscopy with NaI(Tl) detectors represent a considerable expansion and sophistication of previous discussions by us of this subject. This expanded coverage was motivated by the importance of these techniques in physical and environmental research as well as biological research. Chapter 8 is a new chapter containing information on semiconductor radiation detectors, and it was necessitated by the increasing use of Ge(Li) and Si(Li) detectors in modern research. Chapter 9 contains a much expanded treatment of liquid scintillation counting, with a comprehensive bibliography of pertinent reference material. Chapter 10 on autoradiography has been expanded slightly to cover the use of the new nuclear track detectors of potential importance in environmental research. Chapter 11 on the preparation of samples for counting has been greatly expanded to include more up-to-date information on liquid scintillation samples and new sections on the preparation of sources for α-, β-, and γ-counting by other techniques. Chapter 12 on nuclear statistics represents a new chapter that deals at some length with this important topic. Chapters 13 and 14 remain virtually unchanged from the older book. Chapter 15 contains new material on the production of radionuclides, with a further discussion of this subject in the Appendix. Chapter 16 on nuclear safety has been completely rewritten to reflect a newer and more cautious approach to radiation safety. A new chapter, Chapter 17, on radioanalytical methods, especially isotope dilution and activation analysis, has been added. Chapters 18 and 19 are new and cover in detail, examples of the various applications of radiotracers in the environmental and physical sciences.

The title, *Radiotracer Methodology in the Biological, Environmental, and Physical Sciences*, has been a challenge and a burden to us. We have tried to collect in a single volume all the basic information necessary to allow one conduct radiotracer experiments in the biological, environmental, and physical sciences. We hope the present volume might serve as a text for some and a reference book for other workers in the field. As might be expected in such a broad undertaking, perhaps we have slighted some areas and overemphasized others. We recognize that fact and yet hope that we have been able to discuss the basic principles of a given subject and, through the bibliographies, point the reader to more detailed treatment of the subject.

Many hands and minds have aided in the preparation of this work. Particularly appreciated are the contributions made by the following colleagues at Oregon State University: Professor R. A. Schmitt, who critically reviewed the entire manuscript and contributed to the development of many of the problems and laboratory exercises; and Mr. Thurman Cooper, who supervised our radiotracer student laboratories and offered many useful comments. In addition, the accumulated experience gained from daily contact with students and colleagues has helped to shape the eventual form and content of the present volume. We appreciate the detailed review of the manuscript by Dr. Robert Ballentine of Johns Hopkins University.

To all the above, named and nameless, the authors are deeply indebted. Any errors and shortcomings in this volume occurred in spite of their counsel and aid, certainly not because of it.

The authors also wish to express their gratitude to the United States Atomic Energy Commission, the National Science Foundation, and the National Institutes of Health. These agencies have provided generous support to the authors' research and instructional programs at Oregon State University. It is the personal experience gained from these supported programs that forms much of the basis for this volume.

Finally, the first two authors wish to make clear the great contribution made by the new third author. His freshness of approach and greater familiarity with some of the more contemporary techniques and instrumentation of nuclear chemistry have been of inestimable value. The burden of both revising the older volume and producing *de novo* several chapters on current technologies fell on his shoulders. Were it not for his earnest endeavors, this volume would remain only an idea instead of the reality it now is.

C. H. Wang
David L. Willis
Walter D. Loveland

Corvallis, Oregon

Introduction

The detonation of the first nuclear weapon on the New Mexico desert in July 1945 marked the beginning of the nuclear age. Although military development of nuclear weapons continues, peaceful uses of nuclear energy have advanced at a rapid pace since the end of World War II. These uses include not only the more impressive employment of nuclear reactors for power generation but also the development of an entirely new methodology concerned with the utilization of some of the by-products of nuclear fission reactions—radioisotopes.*

The use of radioisotopes in industrial processes, medical applications, and as a research tool can be generally classified into five major categories on the basis of the respective properties of the radioisotopes utilized. These categories are (a) uses based on the effect of ionizing radiation on matter; (b) uses based on the effect of matter on ionizing radiation—that is, the absorption of ionizing radiation by matter; (c) age-dating based on the characteristic radioactive decay of certain natural radioisotopes; (d) direct energy transformation, such as thermoelectric nuclear batteries for spacecraft; and (e) the use of radioisotopes to trace the fate of stable isotopes in physical or biological processes.

*Throughout this book the term "radioisotope" is used in place of the more technically correct term "radionuclide." We have chosen to use the former term because of the widespread acceptance of the technically incorrect usage. Furthermore, in tracer work, one is generally interested in emphasizing the similarity of behavior of the tracer and the thing whose behavior is being traced. Hence the emphasis on the word isotope, indicating similar chemical properties.

The first attribute of radioisotopes involves providing high-energy radiation. Typical γ-ray energies, for example, far exceed chemical bond energies and can induce chemical reactions or disrupt structures in biological systems. Thus γ-irradiation in varying doses is commonly employed at present in polymer manufacture, cancer therapy, and food sterilization. In this type of application, the important consideration is the delivery of a massive dose. The chemical identity of the isotope used is of little importance here, as long as the desirable amount and type of radiation energy are provided.

The second category of radioisotope technology takes advantage of the characteristic interaction of ionizing radiation with matter. Such interaction results in the absorption of a defined amount of the energy associated with nuclear radiation, determined by the type of matter and its thickness. The phenomenon has been employed as the basic principle in devising various types of industrial-thickness gauges. Most such thickness gauges involve the use of small amounts of β-emitters and, occasionally, some soft γ-emitters.

The basic principle underlying the age-dating method is the defined half-life that is a distinctive characteristic of each radioisotope. Use is made of the presence in the environment of a number of naturally occurring radioisotopes, such as uranium-238, potassium-40, carbon-14, and hydrogen-3. In the case of carbon-dating, the carbon-14 content of living matter is the same as that of atmospheric carbon dioxide. After the death of any living matter, such as a tree, the ^{14}C in its remains would no longer be able to exchange with atmospheric ^{14}C, and hence the ^{14}C content will continuously decline at a defined rate determined by the half-life of ^{14}C. If we analyze the amount of ^{14}C in a given weight of ancient timber, for example, we can calculate its age fairly accurately by comparing this determined value with the ^{14}C content in the atmosphere. For the development of this unique method, Libby received the Nobel Prize. Age-dating methods have proven very beneficial to the fields of geology and archaeology in recent years.

The heat produced by certain radioisotopic sources (^{238}Pu, ^{90}Sr, etc.) may be directly converted to electricity by means of an assembly of thermocouples. Such isotopic power generators require no battery storage system and are quite compact. Thus they are highly suited for applications where weight-to-power ratios must be kept low. Since 1961, several such devices (designated SNAP) have been successfully employed as the electrical power sources in orbiting satellites. Slightly different units have found use as power sources for unattended, automatic weather stations in the Arctic.

The use of radioisotopes as tracers has provided research workers with a powerful new tool. Not only can the fate of a given compound in either physical or biological processes be readily traced by the use of radioisotope-labeled compounds, but the ease and sensitivity of radioactivity assay permit the detection of an extremely minute amount of a given labeled compound in a sample as well. The magnifying power of a typical radiotracer experiment may be as high as 10^8-fold with respect to the amount that can be detected. The radiotracer method is readily applicable to problems in the biological, chemical, and engineering sciences.

The value of this powerful research tool can be illustrated by the accomplishments of Calvin and Benson at the University of California at Berkeley in the study of photosynthetic mechanisms. The fate of CO_2 in the photosynthetic process had long been a challenging problem, one that baffled many earlier research endeavors. Using $^{14}CO_2$ as a tracer, these workers were able to elucidate the complete fate of atmospheric CO_2 in plants, leading to the formation of sucrose and other carbohydrates.

The use of the radioisotope method as a research tool necessitates a good understanding of certain basic concepts involved in radiotracer experiments. One must recognize the properties of radioisotopes and their radiation. The detection of radioactivity also constitutes a basic facet of radiotracer methodology, since proper instruments and procedures must be selected for a given experiment. This is a particularly difficult task inasmuch as advances in nuclear instrumentation have occurred rapidly in recent years.

Good experimental design is also a prerequisite in insuring the reliability of results in any radiotracer experiment. Acquiring such a background in nuclear physics, nuclear chemistry, and electronics through formal course work, in addition to gaining the required depth and breadth in his own field, often works a hardship on the scientist. Moreover, an adequate background in radiotracer methodology is of paramount importance in the evaluation of the research findings of fellow scientists employing the radiotracer method. The intended purpose of this book, therefore, is to present a summary of the essential and pertinent information for the scientist. The content has been chosen on the basis of courses in radiotracer methodology taught at Oregon State University for many years to senior undergraduates, graduate students, and research workers.

This book is intended as a reference or source book in the field of radiotracer methodology. It attempts to set forth only the most basic concepts of radioactivity necessary to the practical use of radiotracers. Nevertheless, it ranges from a coverage of the older techniques, such as the use of the Geiger-Müller counter for β-assay, up to such current techniques as liquid scintillation counting of tritium-labeled compounds and the use of semiconductor radiation detectors in γ-ray spectrometry and neutron-activation analysis. Thus it is primarily intended as a brief but up-to-date introduction to the field of radiotracer methodology. It has been developed with a formal-course sequence in mind but should prove equally valuable to the established investigator who desires to familiarize himself with this field.

Because of the book's introductory character, it was deemed especially important to include a rather comprehensive bibliography. The interested reader can thus readily find the additional information concerning theory and technique that he may require, without the text itself taking on an encyclopedic nature. References are grouped by chapter. Other general texts concerning radiotracer methodology, in general, are cited at appropriate points. Since references to older techniques are thoroughly covered in these texts, it was felt that the bibliographies in the present book should stress the more recently developed and currently used radiotracer methods. To this end, a most comprehensive bibliography of liquid scintillation detection and sample preparation techniques, as well as of tritium-labeling methods, is included.

A set of problems selected from practical situations facing the user of radiotracers is appended. It is strongly suggested that the reader make an effort to work through these problems as part of the study of this text.

It is the authors' sincere hope that this book will serve as a key to open the door to the utilization of radiotracer techniques for many a hopeful or experienced physical, environmental, or biological scientist. Although brilliantly exploited over the past three decades, the methodology has only begun to shed light into some of the dark corners of our present ignorance.

1

Atoms and Nuclides

In order to use radioisotopes advantageously, it is highly important to understand their physical nature. Although detailed knowledge of nuclear physics may not be necessary, it is essential to comprehend at least some of its general concepts. Hence a brief, simple review of some basic information on atomic and nuclear structure should precede our discussion of radioisotopes. Readers well versed in the physical sciences may not need this review, but some of those whose formal training is largely in the biological sciences may find it valuable. (See the bibliography at the end of this chapter for a more detailed discussion of the physical aspects of radioisotopes covered in the first three chapters of this book.)

A. ATOMIC STRUCTURE AND ENERGY RELATIONS

It is well known today that an atom is characteristically composed of two major components—a positively charged nucleus and a surrounding cloud of negatively charged electrons. The electron cloud may include from 1 to over 100 electrons that move in orbits of varying distance from the nucleus. Atoms typically have radii of the order of 1 Å (10^{-8} cm); the dense nuclei have radii of the order of 0.0001 Å (10^{-12} cm); and the radius of an electron is of the same order of magnitude. Such dimensions are well below even the best resolving powers obtainable with the electron microscope. These figures show that the nucleus occupies only a minute fraction of the volume of the atom.

A simple calculation will show just how dense the nucleus of an atom really is. Density is defined as mass per unit volume, or

$$\text{Density} = \frac{\text{mass}}{\text{volume}}$$

Thus for the hydrogen nucleus with a mass of 1.66×10^{-24} g and a radius of 1.2×10^{-13} cm, we can easily calculate the density as 230,000 metric tons/mm^3! Just think of how much energy must be stored in an object with such a density and what strong forces must be operating to hold the object together.

The phenomenon of radioactivity is concerned with changes within the nuclei of atoms, whereas ordinary chemical phenomena involve interactions between the orbital electrons of atoms. Energy changes resulting from orbital electron interactions during a chemical reaction are relatively small. In comparison, the energies involved in nuclear transformations are large indeed. As an example, the complete combustion of 1 gram atom of carbon (graphite) to CO_2 would liberate approximately 94,000 calories; by contrast, it can be calculated that the energy released in the complete radioactive transformation of a gram atom of carbon-14 to nitrogen would amount to well over 3 billion calories!

Energy relations involving radioactivity are most commonly discussed on a per nucleus basis. Since the calorie is too large an energy unit to be suitable for such usage, another unit—the electron volt (eV)—is employed. The electron volt is equivalent to the kinetic energy that an electron would acquire while moving through a potential gradient of 1 volt. This is an exceedingly small energy unit; hence several larger units are more commonly used—namely, the kilo electron volt (keV), the million electron volt (MeV), and the billion electron volt (GeV). The electron volt may also be expressed in the more familiar energy unit of calories; 1 electron volt is equivalent to 3.85×10^{-20} calories.

With reference to these energy units, it can be stated that the energy range of chemical reactions involving orbital electrons is of the magnitude of 10 eV per atom, whereas the kinetic energy of individual gas molecules in ordinary room air is only a few hundredths of an electron volt. On the other hand, nuclear changes associated with radioactivity involve energies on the keV or MeV levels per nucleus.

B. THE NUCLEUS

Atomic nuclei are presently believed to be composed of two major components: protons and neutrons. The collective term for these is *nucleons.* Protons are positively charged particles, with a mass approximately 1836 times greater than that of an orbital electron. A proton must not be thought of as an immobile particle in the nucleus. The protons within the nucleus are in constant motion.

Since the atom is known to be neutral with respect to electrical charges, the number of protons must equal the number of orbital electrons in any given atom. This number of nuclear protons is equivalent to the *atomic number* of the element or, symbolically, the Z value. Proton numbers run from 1 for hydrogen up to 105 for hahnium, the most recently produced element. The term *element* thus specifies a type of atom having a defined nuclear proton number and hence the same number of electrons. For example, atomic nuclei with six protons and various numbers of neutrons are called *carbon;* if the proton number is 53, the element is known as *iodine.*

Neutrons are uncharged nucleons with masses approximating those of protons. The total number of neutrons in the nucleus is called the neutron number, N. The neutrons and protons in the nucleus are held together by one of the four fundamental forces in nature, the strong interaction or nuclear force. We shall discuss the strength of this force shortly. The total number of nucleons in the nucleus is an integer, A, called the *mass number* of the nucleus. Obviously,

$$A = N + Z$$

Please note that although A is an integer, the actual mass of the nucleus is generally not an integer.

To designate a specific nucleus, we use a shorthand notation. The chemical symbol for the element appears with a left-hand superscript indicating the mass number, A, a left-hand subscript indicating the proton number, Z, and a right-hand subscript indicating the neutron number, N. Thus we have

$${}^{A}_{Z}\text{Chemical Symbol}_{N}$$

or for the case of the nucleus made up of six protons and eight neutrons

$${}^{14}_{6}\text{C}_{8}$$

or more simply

$${}^{14}\text{C}$$

In some of the older literature (i.e., prior to 1960), another system was used that designated a nucleus by the notation

$${}^{N}_{Z}\text{Chemical Symbol}^{A}$$

Thus, ${}^{14}_{6}C_{8}$ was ${}^{8}_{6}C^{14}$ or simply C^{14}. Although this book and almost all new works use the modern system (i.e., ${}^{14}C$), it is still common in the *spoken* terminology of radiotracer work to talk of "carbon-14" rather than "14-carbon." In recognition of this idea, the name of the element is often spelled out, followed by a hyphen and the mass number of the species in question—that is, carbon-14.

C. NOMENCLATURE

A specific nuclear species—that is, a nucleus with a given number of neutrons and protons—is properly referred to as a *nuclide*. Over 1500 nuclides are known today. As a biological analogy, it can be said that the relation between element and nuclide corresponds roughly to the relation between the classification levels of genus and species, respectively.

Isobars are nuclides with the same mass number, A. In other words, isobars have the same number of particles (nucleons) in the nucleus. Thus ^{3}He and ^{3}H are isobars, for each has three particles per nucleus. *Isotones* are nuclei with the same neutron number, N. Thus ^{3}He and ^{2}H are isotones because $N = 1$ for both species. Similarly, ^{30}Si, ^{31}P, and ^{32}S are isotones because all have 16 neutrons per nucleus.

The term *isotopes* (from the Greek, "same place") refers to nuclei that contain the same number of protons, Z, but varying numbers of neutrons. Therefore ^{2}H and ^{3}H are isotopes of the element hydrogen, but ^{3}He and ^{3}H are not isotopes, for they contain different numbers of protons. Isotopes are nuclei of a given chemical element that have differing mass numbers. Since, in general, the chemical behavior of an atom depends only on the properties of the atomic electrons, and isotopes have the same electronic structure for their atoms, their chemistry is the same. The number of known isotopes for each element varies quite widely. It ranges from hydrogen with three isotopes to polonium and tin with 26 isotopes. In general, lighter elements tend to have fewer isotopes than those of higher atomic numbers.

TABLE 1-1

Neutron-proton Combinations for Hydrogen and Carbon

Element	Atomic Number (number of protons)	Neutron Number	Mass Number (number of nucleons)	Nuclide Stability
Hydrogen	1	0	1	Stable
(Deuterium)	1	1	2	Stable
(Tritium)	1	2	3	Unstable
	1	3	4	Nonexistent
Carbon	6	3	9	Extremely unstable
	6	4	10	Unstable
	6	5	11	Unstable
	6	6	12	Stable
	6	7	13	Stable
	6	8	14	Unstable
	6	9	15	Unstable
	6	10	16	Unstable
	6	11	17	Extremely unstable

It should be emphasized that it is not true that any number of neutrons could be associated with the fixed number of protons in a nucleus to produce an indefinite number of isotopes of an element with the same chemical properties. Only a relatively few combinations are stable enough to persist in nature. They are known as *stable isotopes.* Certain other combinations are possible, but, for reasons to be discussed later, they are unstable and do not persist indefinitely. Other conceivably possible combinations of neutrons and protons are essentially so unstable as to be nonexistent. Even if produced artificially, they apparently disintegrate in too short an interval to be detectable. Table 1-1 illustrates some stable, unstable, and nonexistent combinations for several elements.

D. STABLE NUCLIDES

The number of stable nuclides varies widely from element to element. The element uranium, for example, has no stable nuclear species. Others, such as sodium, have only one stable nuclide. From this point, the range extends up to a maximum of ten stable isotopes for tin. The total number of stable nuclides for all the elements lumped together is approximately 280. The natural elements found in the earth's crust or atmosphere are actually composed of mixtures of the stable isotopes of each element. For any given element, the relative abundance of stable isotopes is remarkably constant, regardless of the geographic locality from which samples come.

Table 1-2 shows the relative abundance of naturally occurring stable isotopes for several of the biologically important elements. In a number of cases, one stable isotope is much more abundant than the others and is considered the "normal" nuclide for that element. Some examples are ^{1}H, ^{12}C, ^{14}N, ^{16}O, ^{40}Ca, and ^{32}S, each of which accounts for more than 95 % of the isotopic abundance of its respective element.

At present, we do not fully understand all the reasons for the stability of certain combinations of neutrons and protons and the instability of others. It is apparent, however, that nuclear stability, in general, is related to the ratio of neutrons to protons, which is, in turn, related to the binding within the nuclei. The forces tending to nuclear stability are greatest in those nuclides where the nuclear particles are paired. This principle can be readily deduced from the following observations: (a) The greatest number of stable nuclides, about 166, are those that contain even numbers of both protons and neutrons in the nuclei. (b) Fewer stable nuclides are found when either the number of protons or the number of neutrons is odd—about 106. (c) Only six stable nuclides have an odd number of both protons and neutrons. This situation is restricted to elements with a mass number less than 14, rather exceptional cases.

TABLE 1-2

Relative Natural Abundance of the Stable Isotopes of Several Elements of Biological Importance*

Element	Nuclide	Natural Abundance (%)
Hydrogen	^{1}H	99.985
	^{2}H	0.015
Carbon	^{12}C	98.89
	^{13}C	1.11
Nitrogen	^{14}N	99.635
	^{15}N	0.365
Oxygen	^{16}O	99.759
	^{17}O	0.037
	^{18}O	0.204
Sodium	^{23}Na	100
Magnesium	^{24}Mg	78.60
	^{25}Mg	10.11
	^{26}Mg	11.29
Phosphorus	^{31}P	100
Sulfur	^{32}S	95.0
	^{33}S	0.76
	^{34}S	4.22
	^{36}S	0.014
Chlorine	^{35}Cl	75.53
	^{37}Cl	24.47
Potassium	^{39}K	93.22
	^{41}K	6.77
Calcium	^{40}Ca	96.97
	^{42}Ca	0.64
	^{43}Ca	0.145
	^{44}Ca	2.06
	^{46}Ca	0.0033
	^{48}Ca	0.185
Iron	^{54}Fe	5.84
	^{56}Fe	91.68
	^{57}Fe	2.17
	^{58}Fe	0.31

*Data from Lederer, Hollander, and Perlman (6).

E. RADIONUCLIDES (RADIOISOTOPES)

It will be recalled that although certain neutron-proton combinations in a given nucleus are possible, these nuclides may not be stable. Because of this basic instability, sooner or later they will undergo nuclear changes. These changes usually result in an adjustment of the neutron-to-proton ratio so that the nucleus reaches a position of greater stability. Such changes in the nucleus are explored in greater detail in Chapter 2.

The nuclear changes described involve spontaneous emission of particles and/or electromagnetic radiation from the nucleus. (This electromagnetic radiation is in the characteristic form of discrete energy quanta known as *gamma rays*, or *photons*, which are much like X rays but considerably shorter in wavelength.) This type of phenomenon is known as *radioactivity*, and nuclides that can undergo it are termed *radioactive nuclides* or radioactive isotopes. The term is frequently contracted to *radioisotopes*. The number of these radioisotopes known today exceeds 1250 and, therefore, far exceeds the number of known stable nuclides. The vast majority of radioisotopes are artificially produced. Only a small minority occur naturally.

1. Naturally Occurring Radioisotopes

At the time of the earth's creation, about 4.6×10^9 years ago, it is probable that many radioisotopes were existent that do not occur naturally today. These nuclides, such as ^{244}Pu, were so unstable that they have long since decayed to stable forms. Those originally formed radioisotopes that still appear today are nuclides that either decay very slowly or are members of *decay series*—that is, unstable nuclides that are produced by the decay of naturally occurring radionuclides and that, in turn, decay to other unstable nuclides, reaching stability only after a dozen or more decay steps. Such nuclides are usually of high atomic number.

The three following decay series have member nuclides occurring in the earth's crust: the uranium-radium series, in which ^{238}U decays through some 14 intermediates to stable ^{206}Pb; the actinium series, in which ^{235}U decays through a series of 11 intermediate nuclides to stable ^{207}Pb; and the thorium series, in which ^{232}Th decays through a series of 10 intermediates to stable ^{208}Pb. We may note in passing that the stable end nuclide of each of these natural decay chains is an isotope of lead, but a different isotope in each case. This is the explanation of the observation that the isotopic composition of naturally occurring lead varies with the locality from which samples are taken. Thus lead in a deposit derived largely from decay of the actinium series will differ in percentage composition of isotopes from a sample derived largely from decay of the thorium series.

Of the several originally formed natural radioisotopes of lower atomic number, only one is biologically significant—^{40}K. Because of the general solubility of potassium compounds, this nuclide is quite widely distributed in both the earth's crust and bodies of water. Even more significant, potassium is a common constituent of living tissue; hence a small but measurable amount of ^{40}K is present in all living tissue. An average 160-lb man, for example, contains about 1.7×10^{-5} g of ^{40}K in his body, or 0.012 % of all the potassium present. These factors make ^{40}K a significant contributor to the "background" radiation to which all living things are constantly exposed.

Other radioisotopes decay too rapidly to have remained since the time of the creation of matter. A few, however, are being continually produced in the earth's upper atmosphere due to cosmic-ray bombardment of atmospheric atoms. In the case of these nuclides, the rate of decay has reached an equilibrium with the rate of formation, so that their abundance is relatively constant in nature.

Tritium (^{3}H) is one of the nuclides being produced in this way. When ^{14}N nuclei are struck by neutrons, one pair of possible reaction products is tritium (^{3}H) and ^{12}C. The neutrons that strike the ^{14}N nuclei are themselves produced by the interaction of cosmic-ray particles with other nuclei in the upper atmosphere. Another pair of products of the reaction (^{14}N plus neutron) is ^{14}C and ^{1}H. Thus either ^{3}H or ^{14}C can be produced as a result of cosmic-ray interaction with ^{14}N. The ^{3}H so formed combines readily with oxygen atoms in the vicinity and slowly mixes with the molecular water on the surface of the earth. Similarly, the ^{14}C apparently combines with available oxygen atoms to form radioactive $^{14}CO_2$ through an intermediate ^{14}CO species. Both these radioactive molecules are readily incorporated into living organisms. As a result, during the life of an organism there is a defined concentration of carbon-14 in its tissue. The decrease in this concentration after death makes possible radiocarbon, or ^{14}C dating, as previously described (8, 9).

2. Artificially Produced Radioisotopes

The first artificial production of radioisotopes was not achieved until 1934, when Curie and Joliot transformed ^{27}Al to ^{30}P. In the intervening four decades, hundreds of artificial radionuclides have been formed. This process has amounted essentially to the transmutation of elements—the long-sought goal of the medieval alchemists.

The basic method for inducing nuclear transformations commonly involves the bombardment of stable nuclei with various particles, either charged or uncharged. This bombardment produces temporary imbalance in the target nuclei, usually resulting in the ejection of nuclear particles or the emission of electromagnetic radiation, and the resultant formation of a new species. Physicists have investigated many such nuclear reactions. The most important reaction types in the production of radioactive nuclides involve the use of various charged particles or neutrons. A detailed discussion of the production of radioactive nuclides by artificial means is given in Chapter 15.

BIBLIOGRAPHY

1. HARVEY, B. G. *Nuclear Chemistry*. Englewood Cliffs, N.J.: Prentice-Hall, A good, simple discussion of the basic principles of nuclear chemistry.
2. CHASE, G. D., and J. L. RABINOWITZ. *Principles of Radioisotope Methodology*. Minneapolis: Burgess, 1967. A very good, simple treatment that parallels much of the discussion in this book.

3. Harvey, B. G. *Introduction to Nuclear Physics and Chemistry*. 2nd ed. Englewood Cliffs, N.J.: Prentice-Hall, 1969. An excellent, more advanced, more "physical" treatment of the basic principles of nuclear chemistry.

4. Friedlander, G., J. W. Kennedy, and J. M. Miller. *Nuclear and Radiochemistry*. New York: Wiley, 1964. A "chemically oriented," somewhat out-of-date classic treatment of nuclear chemistry.

5. Marmier, P., and E. Sheldon. *Physics of Nuclei and Particles*. Vols. I and II. New York: Academic Press, 1970. An excellent, advanced treatment.

6. Lederer, C. M., J. M. Hollander, and I. Perlman. *Table of Isotopes*. 6th ed. New York: Wiley, 1968. A definitive, somewhat out-of-date compilation of the properties of nuclei with references to the original literature.

7. *Nuclear Data Sheets*. New York: Academic Press. The definitive, up-to-date information on nuclear decay schemes presented in journal format.

8. Libby, W. F. *Radiocarbon Dating*. 2nd ed. Chicago: University of Chicago Press, 1955. A definitive older work on carbon dating.

9. Lal, D., and H. Suess. "The Radioactivity of the Atmosphere and Hydrosphere," *Ann. Rev. Nucl. Sci.* **18**, 407(1968). An important review of current information on naturally occurring radioisotopes.

2

The Nature of Radioactive Decay

A. RADIONUCLIDES AND NUCLEAR STABILITY

It has been seen that radioactive nuclides undergo spontaneous nuclear changes, leading to a more stable condition. A condition of stability has been shown to be related to the neutron-to-proton ratio in the nuclei. For each element there is a specific neutron-to-proton ratio that makes for the greatest stability. In the elements of lowest atomic weight, this ratio approximates one neutron to one proton, but as we move up the scale to elements of higher atomic weight, the ratio approaches 1.5 neutrons to one proton for maximum stability. In Figure 2-1 nuclides of lower atomic weight have been plotted on a neutron-proton diagram. The striking feature of this plot is the tendency for all stable nuclides to group within a narrow band. The solid line drawn through this band represents the *line of stability*.

The position of an unstable nuclide with regard to this line of stability generally determines the type of radioactive decay that it will show in attempting to reach greater stability. Unstable nuclides to the right of the line of stability in Figure 2-1 have a higher ratio of neutrons to protons. Their decay pattern, therefore, represents an attempt to decrease the neutron content and/or increase the proton content. On the other hand, radioactive nuclides to the left of the line of stability have a higher ratio of protons to neutrons. Their decay pattern then is an attempt to reduce proton content and/or increase neutron content. Thus the types of decay to be described are directly related to the foregoing problem of nuclear instability.

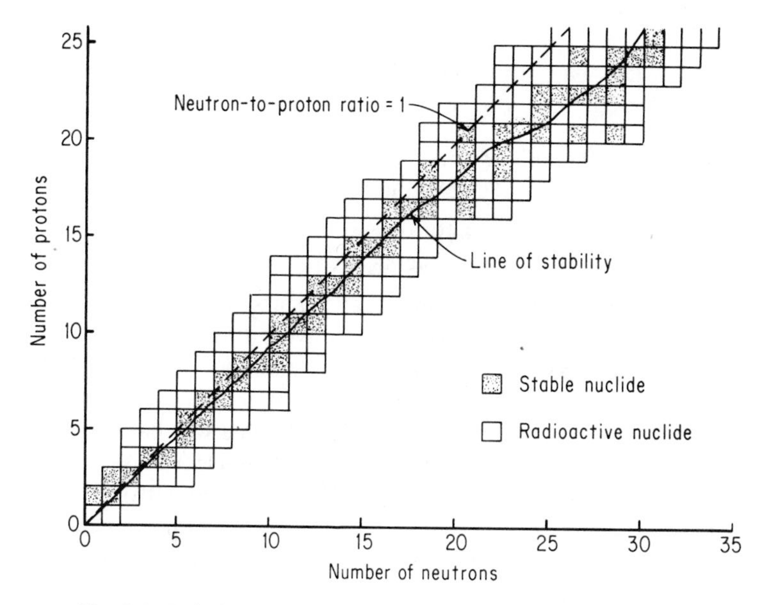

Fig. 2-1. Relation of nuclear composition to the line of stability.

Another, more quantitative way of explaining why radioactive decay takes place is in terms of the energy changes involved. Before beginning such a discussion, however, we must define a few new units and terms. First, let us consider nuclear masses. The mass of a nucleus is very small, being of the order of 10^{-23} g. Instead of using such small numbers, we can define two convenient units to discuss nuclear masses. The first is the *atomic mass unit* (amu), which equals 1.66053×10^{-24} g. Thus if the mass in grams of a ^{12}C nucleus is 1.992×10^{-23} g, the mass expressed in atomic mass units is

$$\frac{1.992 \times 10^{-23}\ \text{g}}{1.660 \times 10^{-24}\ \text{g/amu}}$$

or 12.00 amu. The second convenient unit of mass used is the *million electron volt* (MeV), which is an energy unit. (To get a feeling for the magnitude of the MeV unit, note that 1 MeV/molecule equals 23,045,000 cal/mole.) How can we express mass in terms of energy? Remember Einstein's equation

$$E = mc^2 \tag{2-1}$$

That is, we can associate with any mass (m) an energy (E) of magnitude mc^2, where c is the speed of light.

Let us work an example to show how a nuclear mass expressed in atomic mass units can be expressed in energy units. Suppose that we consider a nuclear mass of 1 amu or, in other words, 1.66053×10^{-24} g. We know that the speed of light c is 2.997925×10^{10} cm/sec. Thus, using Equation (2-1), we calculate

$$E = mc^2 = (1.66053)(10^{-24})(2.997925)^2(10^{10})^2$$
$$E = 1.492 \times 10^{-3} \text{ ergs}$$

Remembering that the unit of energy, 1 erg, is equal to 1.6022×10^{-6} MeV, we can convert E to units of MeV thus:

$$E = \frac{(1.492)(10^{-3}) \text{ ergs}}{(1.6022)(10^{-6}) \text{ ergs/MeV}} = 931.5 \text{ MeV}$$

In other words, we have just shown that a mass of 1 amu has an energy equivalence of 931.5 MeV. So in order to convert a nuclear mass from units of amu to units of MeV, we just multiply by 931.5. Since the changes in nuclear mass that occur in most processes are small, it is most convenient to express them in energy units (whose numerical value will be larger).

Now that we are familiar with the appropriate nuclear mass units we are ready to consider the concept of *nuclear binding energy* (BE). The binding energy of any nucleus is the energy liberated when a group of nucleons combine to form a nucleus. Thus the binding energy of ^{4}He is given by

$$\text{BE } (^4\text{He}) = 2\, M_{^1\text{H}} + 2\, M_n - M_{^4\text{He}} \tag{2-2}$$

where $M_{^1\text{H}}$ is the atomic mass of ^{1}H, M_n is the neutron mass, and $M_{^4\text{He}}$ is the atomic mass of ^{4}He.* Using values from a table of atomic masses, we see that

$$\text{BE } (^4\text{He}) = 2(1.00813) + 2(1.00896) - 4.00398 = 0.03030 \text{ amu}$$

Since 1 amu equals 931.5 MeV,

$$\text{BE } (^4\text{He}) = 28.22 \text{ MeV}$$

A convenient quantity is the *binding energy per nucleon* or, in other words, the total nuclear binding energy divided by the number of nucleons in the nucleus. Since there are four nucleons in ^{4}He, the binding energy per nucleon is 28.21 MeV/4 or 7.1 MeV. Calculating this quantity for each nucleus in the periodic table and plotting it versus the nuclear mass number, A, we arrive at Figure 2-2. Clearly, the nucleus with the highest binding energy per nucleon will be most tightly held together—that is, will be most stable. A rough analogy to this idea is the slogan "You get what you pay for." In other words, the more energy expended in binding the nucleons in the nucleus together, the better the binding—that is, the more stable the product.

Now we can start to answer anew the question of why nuclei decay. After all, the nuclear force holding the nuclei together is quite strong (as shown by the enormous nuclear density). Basically, a nucleus will decay or disinte-

*Atomic masses are used for convenience in these calculations instead of nuclear masses. Note that the extra electron masses cancel in the calculations.

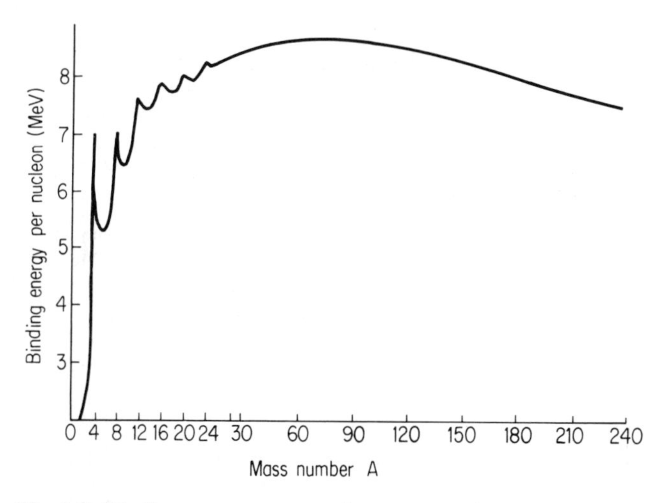

Fig. 2-2. Binding energy per nucleon versus the nuclear mass number, *A*.

grate only if it is energetically favorable for it to do so. So the driving force behind nuclear decay is not the "total" stability of a given nucleus but its stability relative to any possible products of its decay. All one must do to see if a nucleus can decay is to compute the energy release (called *the Q value*) of the decay. Consequently,

$$Q \equiv \text{(mass of the initial nucleus)} - \text{(mass of the decay products)} \qquad (2\text{-}3)$$

If Q is a positive number, decay can occur. If Q is a negative number, then decay cannot occur. In chemical terms, spontaneous nuclear decay must be exoergic.

For example, consider the stability of ^{235}U with respect to the emission of an α-particle (a ^{4}He nucleus). We have

$$^{235}\text{U} \longrightarrow {}^{231}\text{Th} + {}^{4}\text{He}$$

$$\begin{aligned} Q &= M_{^{235}\text{U}} - (M_{^{4}\text{He}} + M_{^{231}\text{Th}}) \qquad (2\text{-}4) \\ &= 235.043915 - (4.002603 + 232.036291) \\ &= 0.005021 \text{ amu} = 4.68 \text{ MeV} \end{aligned}$$

In other words, since Q is a positive number, the decay can occur.

In addition to determining whether or not a given decay will occur, the energetics also influence the *rate* at which the decay will take place. For example, in general, for a given element, the half-lives (see p. 28 for definition of half-life) of its radioactive isotopes are inversely related to the decay energy Q. Table 2-1 illustrates this relationship in the case of carbon and sodium.

TABLE 2-1

The Relation of Half-Life and Energy Release During Decay to Stability for Isotopes of Carbon and Sodium

Nuclide	Half-life ($t_{1/2}$)	Maximum Decay Energy (Q)
^{9}C	0.127 sec	16.6 MeV
^{10}C	19.4 sec	3.61 MeV
^{11}C	20.3 min	1.98 MeV
^{12}C	Stable	—
^{13}C	Stable	—
^{14}C	5730 yr	0.156 MeV
^{15}C	2.4 sec	9.77 MeV
^{16}C	0.74 sec	8.0 MeV
^{20}Na	0.39 sec	14.0 MeV
^{21}Na	23 sec	3.54 MeV
^{22}Na	2.60 yr	2.84 MeV
^{23}Na	Stable	—
^{24}Na	15.0 hr	5.5 MeV
^{25}Na	60 sec	3.83 MeV
^{26}Na	1.0 sec	8.5 MeV

B. TYPES OF RADIOACTIVE DECAY

A so-called decay scheme is a convenient means of concisely summarizing information concerning a specific type of radioactive decay. It is a conventionalized plot of energy against atomic number, although no scale is used. The symbol, mass number, and half-life of the nuclide appear on the uppermost horizontal line. Decay leading to emission of a negative particle is indicated by a diagonal arrow to the lower right; decay involving positive particle emission, or orbital electron capture, is shown by a similar arrow to the lower left. These arrows terminate on lower horizontal lines representing the energy levels of the daughter nuclei. If a daughter nucleus is produced in an excited state, the consequent γ-ray emission is denoted by undulating vertical arrows (no change in atomic number) leading to a lower energy state of the nucleus. Notation of the maximum kinetic energy in MeV of emitted particles and the energy in MeV of γ-radiation is made near the respective arrows. Where nuclei may follow more than one decay path, the percentage occurrence of each path is indicated. Typical decay schemes are shown for the various decay types in the following sections.

1. Decay by Negative Beta (Negatron) Emission

Under certain conditions nuclides with excess neutrons may reach stability by the conversion of a neutron to a proton, culminating in the ejection of a negative β-particle (electron) and a tiny particle called an antineutrino ($\bar{\nu}_e$). Note that the electron originates in the nucleus and is not to be confused with the orbital electrons. However, apart from this distinction on the basis of particle origin, the β^--particle is identical with the electron. It is, in fact, an energetic electron. Such a nuclear change results in a loss of one neutron and a gain of one proton, thereby shifting the nuclide toward the line of stability. This type of decay can be summarized as follows:

$$n \longrightarrow p + \beta^- + \bar{\nu}_e$$

Note that negatron emission results in an increase of one unit of atomic number (Z) but no change in mass number (A) for the nucleus involved. Any excess energy in the nucleus following beta emission is given off as one or more gamma rays, or photons.

Decay schemes of varying complexity involving negatron emission are shown in Figures 2-3, 2-4, and 2-5. Close observation of the decay schemes for ^{60}Co and ^{131}I will disclose that although two or more decay paths may be followed, the total disintegration energy to the stable ground state is (within experimental error) a constant value for each respective nuclide. For ^{60}Co, this value is 2.82 MeV; for ^{131}I, it is approximately 0.970 MeV.

2. Decay by Positive Beta (Positron) Emission

Where the number of protons in a nucleus is in excess, positron emission may occur in order to reach stability. Positrons are positively charged

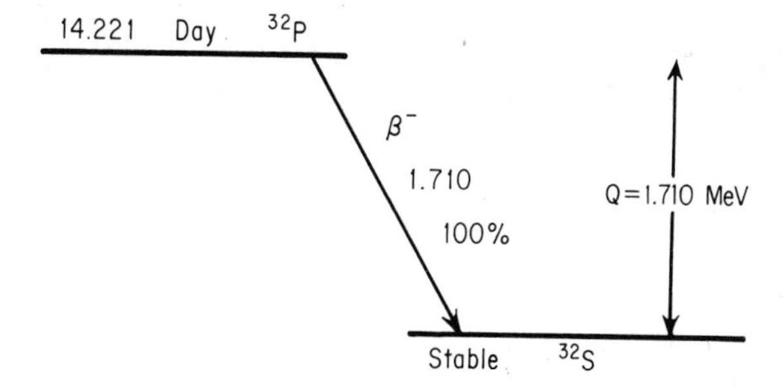

Fig. 2-3. Decay scheme of ^{32}P.

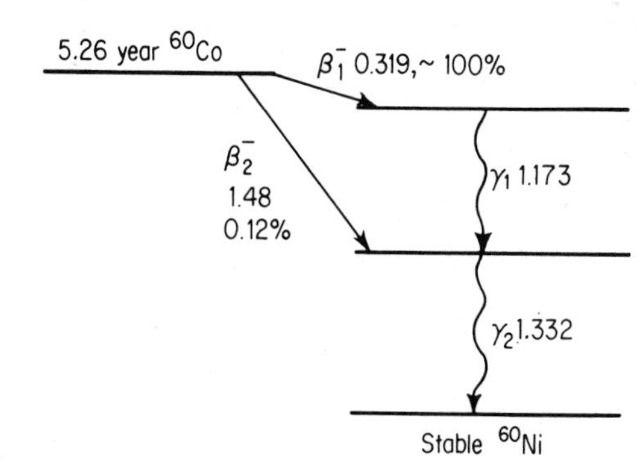

Fig. 2-4. Partial decay scheme of ^{60}Co.

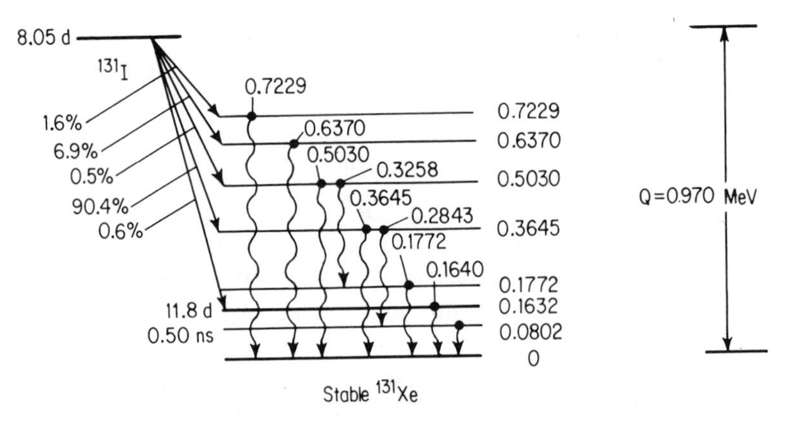

Fig. 2-5. Partial decay scheme of ^{131}I. The numbers to the right of the excited states of ^{131}Xe are their excitation energies.

β-particles. In this type of transformation, a nuclear proton is converted to a neutron, together with the ejection of a high-speed positron from the nucleus and a tiny particle called the neutrino (ν_e). This type of decay may be generally indicated as follows:

$$p \longrightarrow n + \beta^+ + \nu_e$$

Positron decay results in no change in mass number (A), but the nucleus involved decreases one unit in atomic number (Z). Again, any excess energy in the nucleus following positron ejection is emitted as gamma radiation. The decay scheme of ^{13}N, shown in Figure 2-6, illustrates simple positron decay.

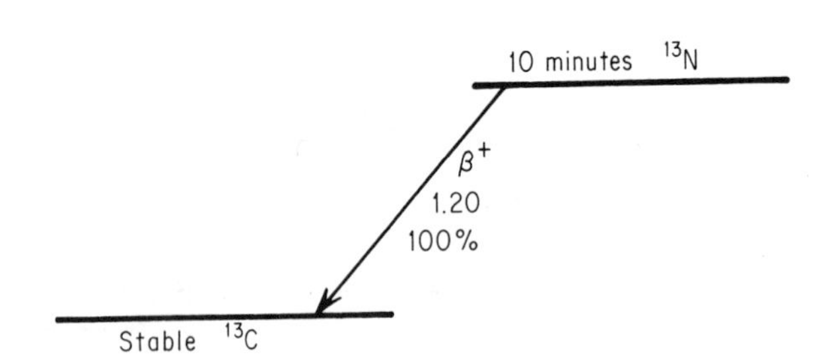

Fig. 2-6. Decay scheme of ^{13}N.

3. Decay by Electron Capture (EC)

The process of electron capture represents another means of reducing the number of protons in the nucleus. Thus it has essentially the same result as positron emission. In this type of radioactive decay, one of the inner orbital electrons interacts with the nucleus, combining with a nuclear proton to form a neutron and a small massless, chargeless particle, the electron neutrino (ν_e). Symbolically, we have

$$e^- + p \longrightarrow n + \nu_e$$

The overall result is again the loss of a proton and the gain of one neutron per nucleus. The only particle directly emitted in electron capture decay is the electron neutrino, which is very difficult to detect. X rays, however, are emitted as a consequence of the rearrangement of the orbital electrons following electron capture and can be detected. When the captured orbital electron comes from the K shell (as it does most of the time), this is also called K *capture*. An isotope of biological interest, ^{55}Fe, decays by electron capture, and its decay scheme is shown in Figure 2-7. For this radionuclide, the only detectable radiation is the X ray emitted as a result of orbital electron rearrangement.

Frequently, an unstable nuclide decays alternately by electron capture or positron emission. ^{22}Na displays this pattern, as indicated in Figure 2-8. Note that electron capture, like positron emission, results in no change in mass number (A), but it does result in a decrease of one unit of atomic number (Z) for the nucleus involved. Positron emission dominates as a decay mode over electron capture when the Z value is low or the decay energy is large.

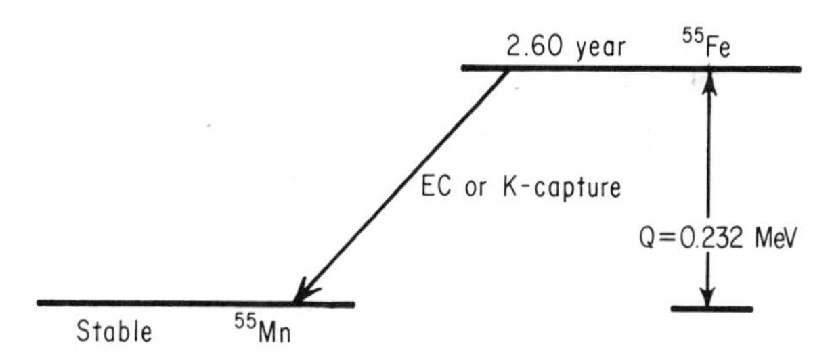

Fig. 2-7. Decay scheme of ^{55}Fe.

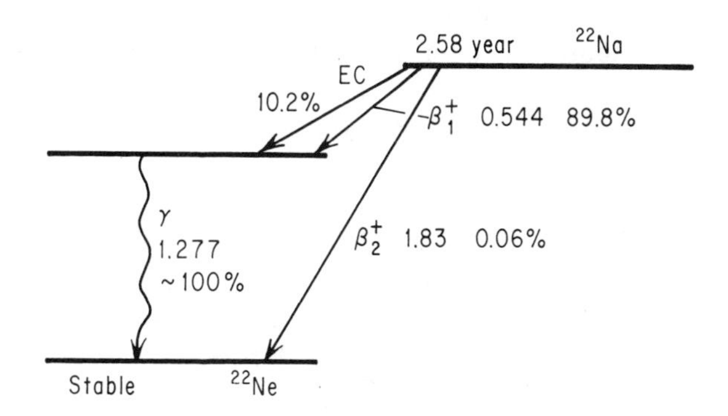

Fig. 2-8. Decay scheme of ^{22}Na.

4. Gamma Ray Decay

In γ-ray decay, the nucleus gets rid of its excess excitation energy by the emission of photons. The photons all have the same discrete energy. In γ-ray decay, the number of neutrons and protons in the nucleus does not change. Only the excitation energy of the nucleus changes, corresponding to a jump from a higher to a lower excited state. Any excited nucleus can decay by γ-ray emission, and γ-ray emission frequently follows β^--emission. When γ-ray decay takes place between two states of measurable lifetime, it is said to be an *isomeric transition* (IT). Clearly, then, isomeric transitions are only special cases of γ-ray decay.

5. Internal Conversion (IC)

In modes of decay where γ-rays are normally emitted from the nucleus, an alternative possibility exists. The excited nucleus may interact directly with an inner orbital electron, with the result that all the excitation energy is transferred to the electron. Thereupon the electron is ejected from the atom. This event is known as *internal conversion* and the emitted electrons are known as *conversion electrons*. Note that no γ-ray is emitted during internal conversion. Internal conversion is more probable for nuclides of higher atomic number and transitions involving lower γ-ray energy. In contrast to the energy-spectrum characteristics of negatron and positron emission (see Chapter 3), conversion electrons are monoenergetic.

6. Decay by Alpha Particle Emission

Among the elements of higher atomic weight only, another means of decay occurs. This is the emission of α-particles. An α-particle represents a naked helium nucleus—that is, a tightly bound unit of two neutrons and two protons. As will be seen, emission of an α-particle results in a decrease of two units of atomic number (Z) and four units of atomic mass (A). Alpha emitters are not commonly used in radioactive tracer work, but they do have some biological importance. This type of decay is typical of the natural radioactive decay series previously mentioned. Figure 2-9 shows an example of this decay category. The subscripts on the α-designations indicate the excitation energy of the product nucleus following emission of that α-particle. Thus α_{401} indicates decay to the 401-keV excited state of ^{215}Po.

7. Extranuclear Effects of Radioactive Decay

In electron capture and internal conversion decay, the orbital electron cloud surrounding the nucleus is disrupted. An electron from an inner shell has either been captured by the nucleus (electron capture) or been ejected from the atom (internal conversion), leaving a vacant space in one of the inner orbitals. Electrons from orbitals farther away from the nucleus move

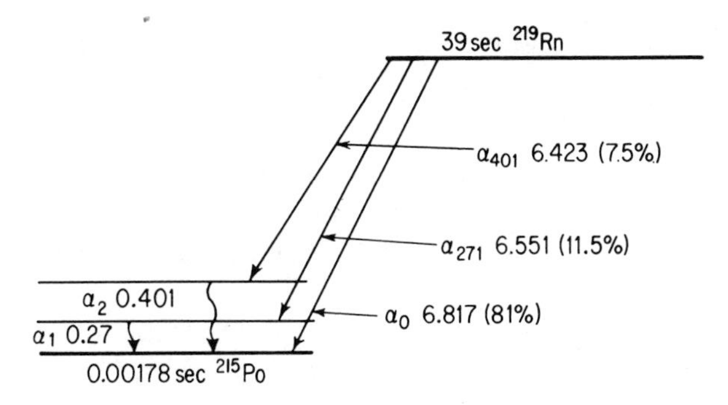

Fig. 2-9. Partial decay scheme of ^{219}Rn.

into the vacant space, creating, in turn, vacancies in the outer orbitals that are filled by electrons from the outermost orbitals. All in all, there is a total rearrangement or shifting around of the orbital electrons following these two types of nuclear decay (see Figure 2-10).

As the electrons go from an outer shell to an inner shell, their potential energy decreases and they must get rid of excess energy. They do so by either X-ray emission or Auger electron emission. In X-ray emission, the difference in *electron* binding energy between the two shells is radiated in the form of electromagnetic radiation. Thus the energy of the X ray ($E_{\text{X ray}}$) is given by

$$E_{\text{X ray}} = E_B^1 - E_B^2$$

where E_B^2 is the original electron binding energy and E_B^1 is the binding energy of the electron in its new orbital (see Figure 2-10).

Auger electron emission is in competition with X-ray emission. In Auger electron emission, the energy normally released as X-radiation is transferred

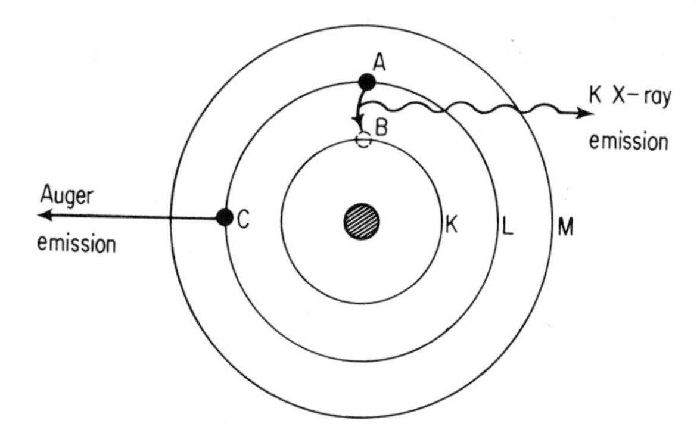

Fig. 2-10. Schematic diagram to show X-ray emission and Auger electron emission to fill vacancy caused by nuclear decay. An *L*-shell electron (*A*) is shown filling a *K*-shell vacancy (*B*). In doing so, it either emits a characteristic *K* X ray or transfers its excess energy to another *L*-shell electron (*C*), ejecting it.

to another outer electron, kicking it out of the atom. Consequently, Auger electron emission is an atomic analogy of internal conversion. The monoenergetic electron energy (E_{Auger}) is given by

$$E_{\text{Auger}} = E_{\text{X ray}} - E_B^3$$

where $E_{\text{X ray}}$ is the energy that normally would have been released as an X ray and E_B^3 is the binding energy of the electron ejected in the Auger process (see Figure 2-10).

The relative probability of having either X-ray emission or Auger electron emission is designated as the *fluorescence yield.* The fluorescence yield is defined as the fraction of electron vacancies filled with the emission of an X ray or

$$\text{Fluorescence yield} = \frac{(\text{number X-rays emitted})}{(\text{number electron vacancies})}$$

The dependence of the fluorescence yield on the atomic number, Z, of the nucleus is shown in Figure 2-11. Note that the Auger effect dominates in the light elements, whereas X-ray emission is dominant in the heavier elements.

The X-rays and Auger electrons can be detected, thereby allowing one to use radionuclides decaying by internal conversion or electron capture as radiotracers. Examples of such radionuclides are ^{55}Fe and ^{109}Cd.

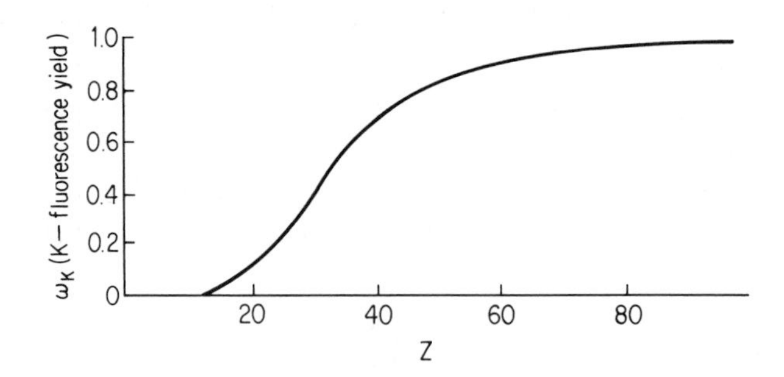

Fig. 2-11. *K*-fluorescence yield versus the atomic number.

C. RATE OF RADIOACTIVE DECAY

The number of nuclei in a radioactive sample that disintegrate during a given time interval decreases exponentially with time. This rate is essentially independent of pressure, temperature, the mass action law, or any other rate-limiting factors that commonly affect chemical and physical changes. As a result, this decay rate serves as a most useful means of identifying a given nuclide. As seen in Table 2-1, the decay rate is related to the degree of instability of the specific nuclide. Since radioactive decay represents the transformation of an unstable radioactive nuclide to a more stable nuclide, which may also be radioactive, it is an irreversible event for each nuclide.

The unstable nuclei in a radioactive sample do not all decay simultaneously. Instead the decay of a given nucleus is an entirely random event. Consequently, studies of radioactive decay events require the use of statistical methods. Thus one may observe a large number of radioactive nuclei and predict with fair assurance that, after a given length of time, a definite fraction of them will have disintegrated.

1. Radioactive Decay Kinetics

Radioactive decay is what chemists refer to as a first-order reaction; that is, the rate of radioactive decay is proportional to the number of radioactive nuclei present in a given sample. So if we double the number of radioactive nuclei in a sample, we double the number of particles emitted by the sample per unit time.* This relation may be expressed as follows:

$$\begin{matrix}\text{The rate of}\\ \text{particle emission}\end{matrix} \equiv \begin{pmatrix}\text{the rate of}\\ \text{disintegration of}\\ \text{radioactive nuclei}\end{pmatrix} \propto \begin{pmatrix}\text{number of}\\ \text{radioactive}\\ \text{nuclei present}\end{pmatrix}$$

Note that the foregoing statement is only a proportion. By introducing the decay constant, it is possible to convert this expression into an equation, as follows:

$$\begin{pmatrix}\text{The rate of}\\ \text{disintegration of}\\ \text{radioactive nuclei}\end{pmatrix} = \begin{pmatrix}\text{decay}\\ \text{constant}\end{pmatrix} \times \begin{pmatrix}\text{number of}\\ \text{radioactive}\\ \text{nuclei present}\end{pmatrix} \qquad (2\text{-}5)$$

The decay constant, λ, represents the average probability per nucleus of decay occurring. Therefore we are taking the probability of decay per nucleus, λ, and multiplying it by the number of nuclei present so as to get the rate of particle emission.

In order to convert the preceding word equations to mathematical statements using symbols, let N represent the number of radioactive nuclei present at time t. Then, using differential calculus, the preceding word equations may be written as

$$-\frac{dN}{dt} \propto N$$
$$-\frac{dN}{dt} = \lambda N \qquad (2\text{-}6)$$

Note that N is constantly reducing in magnitude as a function of time. Rearrangement of Equation (2-6) gives

$$\frac{dN}{N} = -\lambda\, dt \qquad (2\text{-}7)$$

*In order to make this statement completely correct, we should say that as we double the number of nuclei present, we double the *rate* of particle emission. This *rate* is equal to the number of particles emitted per unit time, provided that the time interval is small.

If we say that at time $t = 0$ we have N radioactive nuclei present, then integration of Equation (2-7) gives the radioactive decay law

$$\boxed{N = N_0 e^{-\lambda t}} \tag{2-8}$$

This equation gives us the number of radioactive nuclei present at time t. However, in radiotracer experiments, we want to know the count rate that we will get in our detectors as a function of time. In other words, we want to know the *activity* of our samples.

Still, it is easy to show that the counting rate in one's radiation detector is equal to the rate of disintegration of the radioactive nuclei present in a sample multiplied by a constant related to the efficiency of the radiation measuring system. Thus

$$A = c\left(-\frac{dN}{dt}\right) = c\lambda N \tag{2-9}$$

where c is the efficiency constant (see Chapter 6). Substituting into Equation (2-9), we get

$$\boxed{A = A_0 e^{-\lambda t}} \tag{2-10}$$

where A is the counting rate at some time t due to a radioactive sample that gave counting rate A_0 at time $t = 0$. Equations (2-8) and (2-10) are the basic equations governing the number of nuclei present in a radioactive sample and the number of counts observed in one's detector as a function of time. Equation (2-10) is shown graphically as Figure 2-12.

As seen in Figure 2-12, this curve flattens out and approaches zero. If the same plot is made on a semilogarithmic scale (Figure 2-13), the decay curve becomes a straight line, with a slope equal to the value of $-(\lambda/2.303)$.

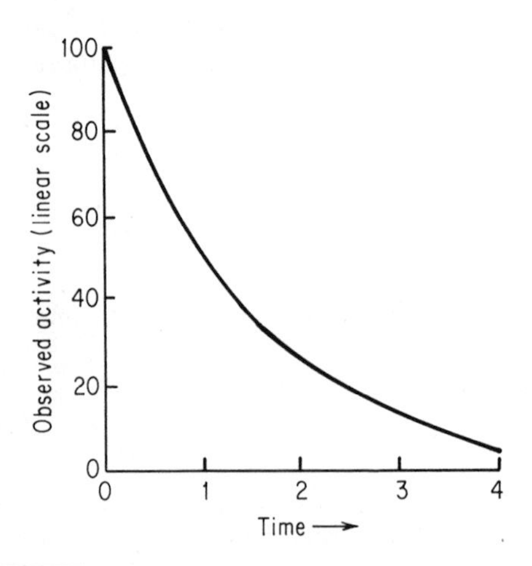

Fig. 2-12. Linear decay curve.

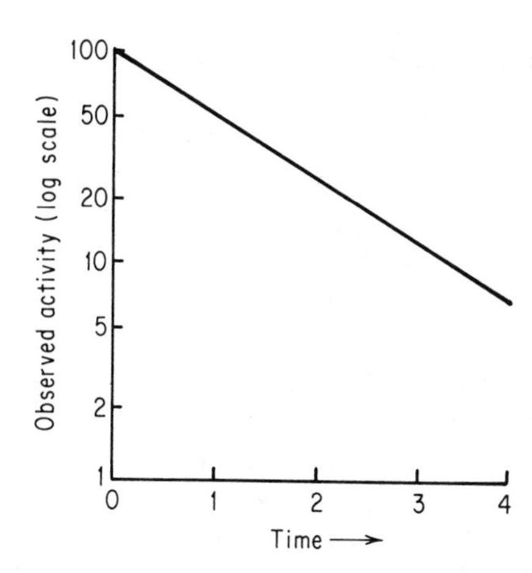

Fig. 2-13. Semilogarithmic decay curve.

Table 2-2 lists, for several commonly used radionuclides, decay constants that have been measured.

We might ask how one uses Equations (2-8) and (2-10) in practical calculations. There are several ways, all of which are equivalent but which emphasize different methods of calculation. The methods are summarized below.

a. Use of Log Tables. Rewrite Equations (2-8) and (2-10) in terms of logarithms as

$$2.303 \log_{10} \frac{N}{N_0} = -\lambda t \quad \text{or} \quad \log_{10} N = \log_{10} N_0 - 0.4343\lambda t \qquad (2\text{-}11)$$

$$2.303 \log_{10} \frac{A}{A_0} = -\lambda t \quad \text{or} \quad \log_{10} A = \log_{10} A_0 - 0.4343\lambda t \qquad (2\text{-}12)$$

Then, knowing any three of the quantities (A, A_0, λ, and t) or (N, N_0, λ, t), we can solve for the remaining quantity by the use of log tables.

As an example, assume that ^{131}I is being used in a short experiment and that one wishes to calculate how the ^{131}I activity will decrease in a day. From

TABLE 2-2

Decay Constants for Selected Radionuclides

Nuclide	Decay Constant (λ)
^{32}P	0.04854/day
^{35}S	0.00789/day
^{45}Ca	0.00420/day
^{55}Fe	0.267/yr
^{59}Fe	0.0152/day
^{60}Co	0.1317/yr
^{65}Zn	0.00283/day
^{131}I	0.0861/day

Table 2-2, the decay constant of ^{131}I is 0.0861/day. So we would say

$$2.303 \log_{10} \frac{A}{A_0} = -\lambda t = -(0.0861/\text{day})(1 \text{ day})$$

$$\log_{10}\left(\frac{A}{A_0}\right) = -\frac{0.0861}{2.303} = -0.03739 = 9.96261 - 10$$

$$\frac{A}{A_0} = 0.918$$

Thus after one day the ^{131}I activity present would be 91.8% of its original value. Note that units of λ, the decay constant, must agree with the units of time, t. For example, if t is expressed in days, λ must be given in day^{-1}.

b. Use of a Log-Log Slide Rule. Using a log-log slide rule, set the hairline on the value of λt on the D scale. Then read off the appropriate scale (LL/0 or LL/1 or LL/2 or LL/3) the value of $e^{-\lambda t}$. The choice of which scale to use is governed by the value of λt. The table given shows which scale to use.

λt	Scale
1–10	LL/3
0.1–1	LL/2
0.01–0.1	LL/1
0.001–0.01	LL/0

Using the value of $e^{-\lambda t}$, the equations can be used to calculate A or A_0 and N or N_0.

c. Use of Decay Correction Tables. If an investigator expects to make extensive use of a specific radioisotope, it is often convenient to construct a decay correction table for that isotope. Such tables (easily prepared from the known decay constant for the nuclide) allow rapid, yet accurate determination of the fraction of original activity remaining in a sample after a given interval of time. Tables 2-3 and 2-4 illustrate such compilations for ^{32}P and ^{131}I, respectively.

2. Half-life

The *half-life* ($t_{1/2}$) is another way to express the decay constant. The half-life of a radionuclide is the time required for its activity to decrease by one-half. Thus after one half-life, 50% of the initial activity remains. After two half-lives, only 25% of the initial activity remains. After three half-lives, only 12.5% is yet present and so forth. Figure 2-14 shows this relation graphically.

The half-life for a given nuclide can be derived from Equation (2-10)

TABLE 2-3

Decay Correction Table for Phosphorus-32

Hours	0	3	6	9	12	15	18	21
Days								
0	1.0000	.9940	.9879	.9820	.9760	.9701	.9642	.9584
1	.9526	.9469	.9411	.9354	.9298	.9242	.9186	.9130
2	.9075	.9020	.8965	.8911	.8857	.8804	.8750	.8697
3	.8645	.8593	.8541	.8489	.8438	.8387	.8336	.8285
4	.8235	.8185	.8136	.8087	.8038	.7989	.7941	.7893
5	.7845	.7798	.7750	.7704	.7657	.7611	.7565	.7519
6	.7473	.7428	.7383	.7339	.7294	.7250	.7206	.7163
7	.7119	.7076	.7033	.6991	.6949	.6907	.6865	.6823
8	.6782	.6741	.6700	.6660	.6619	.6579	.6539	.6500
9	.6461	.6422	.6383	.6344	.6306	.6268	.6230	.6192
10	.6155	.6117	.6080	.6043	.6007	.5971	.5934	.5899
11	.5863	.5827	.5792	.5757	.5722	.5688	.5653	.5619
12	.5585	.5551	.5518	.5484	.5451	.5418	.5385	.5353
13	.5320	.5288	.5256	.5225	.5193	.5162	.5130	.5099
14	.5068	.5038	.5007	.4977	.4947	.4917	.4887	.4858
15	.4828	.4799	.4770	.4741	.4712	.4684	.4656	.4627
16	.4599	.4572	.4544	.4517	.4489	.4462	.4435	.4408
17	.4382	.4355	.4329	.4303	.4277	.4251	.4225	.4199
18	.4174	.4149	.4124	.4099	.4074	.4049	.4025	.4000
19	.3976	.3952	.3928	.3904	.3881	.3857	.3834	.3811
20	.3788	.3765	.3742	.3719	.3697	.3675	.3652	.3630
21	.3608	.3587	.3565	.3543	.3522	.3501	.3479	.3458

TABLE 2-4

Decay Correction Table for Iodine-131

Hours	0	2	4	6	8	10	12	14	16	18	20	22
Days												
0	1.0000	.9929	.9858	.9787	.9717	.9648	.9579	.9510	.9442	.9375	.9308	.9241
1	.9175	.9109	.9044	.8980	.8915	.8852	.8788	.8726	.8663	.8601	.8540	.8479
2	.8418	.8358	.8298	.8239	.8180	.8121	.8063	.8006	.7949	.7892	.7835	.7779
3	.7724	.7668	.7614	.7559	.7505	.7451	.7398	.7345	.7293	.7241	.7189	.7137
4	.7086	.7036	.6985	.6936	.6886	.6837	.6788	.6739	.6691	.6643	.6596	.6549
5	.6502	.6455	.6409	.6363	.6318	.6273	.6228	.6183	.6139	.6095	.6052	.6008
6	.5965	.5923	.5880	.5838	.5797	.5755	.5714	.5673	.5633	.5592	.5552	.5513
7	.5473	.5434	.5395	.5357	.5318	.5280	.5243	.5205	.5168	.5131	.5094	.5058
8	.5022	.4986	.4950	.4915	.4880	.4845	.4810	.4776	.4742	.4708	.4674	.4641
9	.4607	.4575	.4542	.4509	.4477	.4445	.4413	.4382	.4350	.4319	.4288	.4258
10	.4227	.4197	.4167	.4137	.4108	.4078	.4049	.4020	.3992	.3963	.3935	.3907
11	.3879	.3851	.3823	.3796	.3769	.3742	.3715	.3689	.3662	.3636	.3610	.3584
12	.3559	.3533	.3508	.3483	.3458	.3433	.3409	.3384	.3360	.3336	.3312	.3289
13	.3265	.3242	.3219	.3196	.3173	.3150	.3128	.3105	.3083	.3061	.3039	.3017
14	.2996	.2974	.2953	.2932	.2911	.2890	.2869	.2849	.2829	.2808	.2788	.2768
15	.2749	.2729	.2709	.2690	.2671	.2652	.2633	.2614	.2595	.2577	.2558	.2540
16	.2522	.2504	.2486	.2468	.2450	.2433	.2416	.2398	.2381	.2364	.2347	.2330
17	.2314	.2297	.2281	.2265	.2248	.2232	.2216	.2200	.2185	.2169	.2154	.2138
18	.2123	.2108	.2093	.2078	.2063	.2048	.2033	.2019	.2004	.1990	.1976	.1962
19	.1948	.1934	.1920	.1906	.1893	.1879	.1866	.1852	.1839	.1826	.1813	.1800
20	.1787	.1774	.1762	.1749	.1737	.1724	.1712	.1700	.1687	.1675	.1663	.1651
21	.1640	.1628	.1616	.1605	.1593	.1582	.1571	.1559	.1548	.1537	.1526	.1515
22	.1504	.1494	.1483	.1472	.1462	.1451	.1441	.1431	.1420	.1410	.1400	.1390

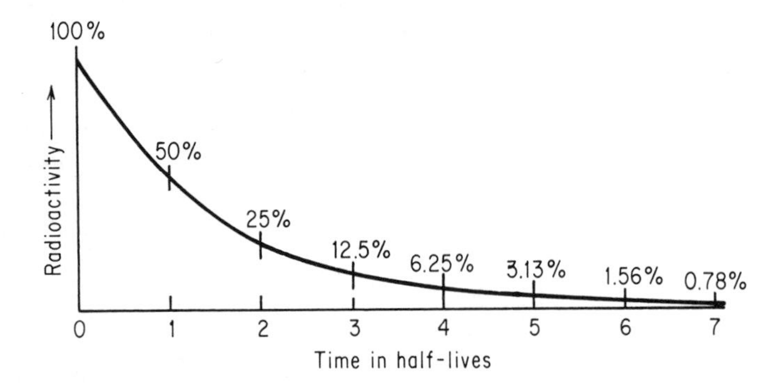

Fig. 2-14. Relation between half-life and radioactivity.

when the value of the decay constant is known. In accordance with the definition of the term half-life, when $A/A_0 = \frac{1}{2}$, then $t = t_{1/2}$. Substituting these values into Equation (2-12) gives

$$2.303 \log_{10} \tfrac{1}{2} = -\lambda t_{1/2} \quad \text{or} \quad 2.303 \log_{10} 2 = \lambda t_{1/2} \qquad (2\text{-}13)$$

Hence

$$t_{1/2} = \frac{0.693}{\lambda} \qquad (2\text{-}14)$$

Note here that the value thus calculated for $t_{1/2}$ will have the units of $1/\lambda$ or units of (time).

The half-life for different nuclides ranges from under 10^{-6} sec to 10^{10} yr. This value has been ascertained for all the commonly used radionuclides. When an unknown radioactive isotope is encountered, a determination of its half-life is normally the first step in its identification. This determination can be done by preparing a semilog plot of a series of activity observations made over a period of time. A short-lived nuclide may be observed as it decays through a complete half-life and the time interval observed directly (Figure 2-15).

It is difficult to measure the half-life of a very long-lived radionuclide. Here variation in disintegration rate may not be noticeable within a reasonable length of time. In this case, the decay constant must be calculated from the absolute decay rate according to Equation (2-6). The absolute number of atoms of the radioisotope present (N) in a given sample can be calculated according to

$$N = \frac{6.02 \times 10^{23} \text{ (Avogadro's number)}}{\text{atomic weight radioisotope}} \times \text{mass of the radioisotope} \qquad (2\text{-}15)$$

The mass of the radioisotope in the given sample can be determined once the isotopic composition of the sample is ascertained by such means as mass spectrometry. When the decay constant is known, the half-life can then be readily calculated [Equation (2-14)]. It is obvious, however, that the degree

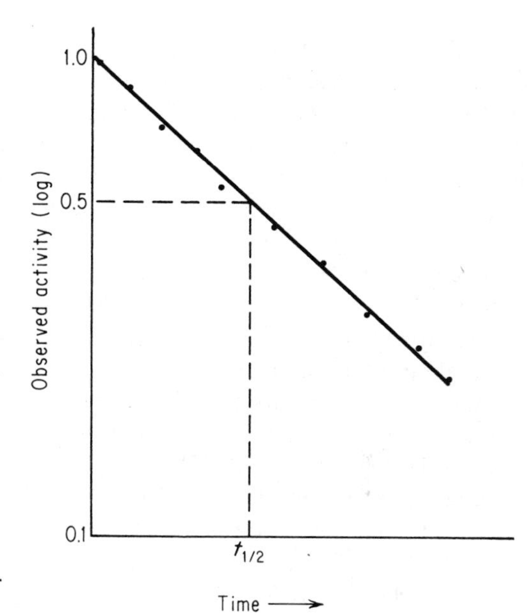

Fig. 2-15. Direct graphic determination of half-life.

of accuracy of the calculated half-life is limited by the accuracy associated with the decay constant determination.

Precise determinations of half-life values are difficult tasks, as evidenced by the variation in published data from time to time. The values indicated in Table 2-5 are the best values as given by Lederer, Hollander, and Perlman (1).

TABLE 2-5

Half-Life Values for Some Selected Radioisotopes

Nuclide	$t_{1/2}$
^{42}K	12.36 ± 0.03 hr
^{24}Na	14.96 ± 0.02 hr
^{90}Y	64.03 ± 0.05 hr
^{198}Au	2.697 ± 0.002 day
^{131}I	8.05 ± 0.02 day
^{32}P	14.28 ± 0.015 day
^{86}Rb	18.66 ± 0.02 day
^{51}Cr	27.8 ± 0.1 day
^{59}Fe	45.6 ± 0.2 day
^{35}S	87.9 ± 0.1 day
^{45}Ca	165 ± 4 days
^{65}Zn	245 ± 0.5 day
^{22}Na	2.62 ± 0.03 yr
^{60}Co	5.263 ± 0.01 yr
^{3}H	12.262 ± 0.004 yr
^{90}Sr	27.7 ± 0.3 yr
^{14}C	5730 ± 30 yr
^{36}Cl	3.08 ± 0.03 × 10^5 yr

Several important practical considerations arise from the foregoing discussion of radioactive decay. In the first place, it has already been mentioned that corrections for decay must be made when radionuclides with short half-lives are being used in tracer studies extending over a period of time.

The rate of radioactive decay is also a factor in evaluating the biological effect of a given radioisotope. If one experimentally administers a high level of radioactivity to an organism, a certain amount of radiation damage to tissues may be anticipated, depending on the rate of elimination of the nuclide from the organism (biological half-life). This biological effect tends to be minimized where the isotope used decays rapidly. Thus with all other factors equal, an organism could more readily tolerate a large dose of a short-lived radioisotope than an equivalent dose of a longer-lived nuclide of the same element. Where a choice of radioisotopes of the element in question exists, as in the case of sodium (^{22}Na or ^{24}Na), this factor may become important in designing the experiment.

The third consideration involving rate of decay concerns the problem of disposal of radionuclide wastes at the end of an experiment. This topic will be considered in greater detail in a later chapter. It will readily be seen, however, that, particularly in the case of isotopes with long half-lives, disposal can be a problem. Where the half-life is a matter of a few days or weeks, and only small tracer amounts of isotope are involved, it is possible to store the wastes in a restricted area for a time equal to about 10 half-lives. The rate of decay is such that at the end of this period the amount of activity will be reduced by a factor of 1000, and the wastes may often be disposed of through public channels (see Chapter 16).

3. Composite Decay

Where two or more radioisotopes with different half-lives are present in a sample, and one does not or cannot distinguish the radioactivity emitted by each isotope, a composite decay rate will be observed. The decay curve drawn on a semilogarithmic plot will, in this situation, not be a straight line. The decay curves of each of the isotopes present may be resolved by graphic means if their half-lives differ sufficiently and if not more than three radioactive components are present. In the graphic example shown in Figure 2-16, line C represents the observed activity. Only the activity of the longer-lived component A is observed after the shorter-lived component B has become exhausted through decay. Extrapolation of this linear portion of the curve back to zero time gives the decay curve for component A. The curve for component B is drawn by subtracting out point by point the activity values of component A from the composite curve.

If the half-lives of the two components in such samples are not sufficiently different to allow graphic resolution, the differential detection method may

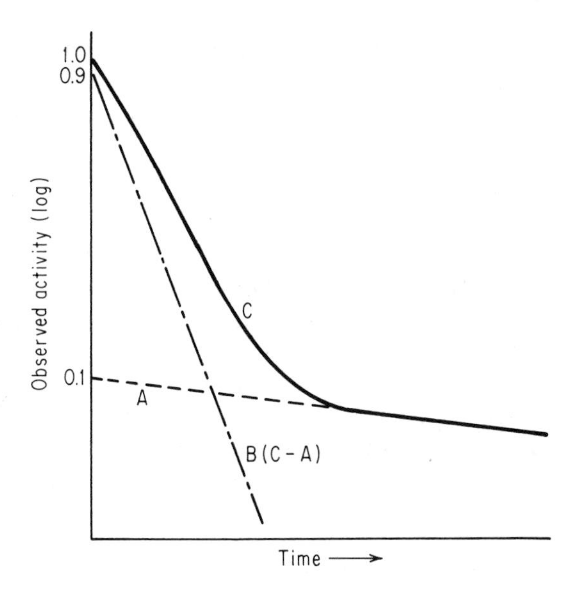

Fig. 2-16. Graphic resolution of a composite decay curve.

be applicable. If the radiation characteristics of the isotopes in the mixture are suitably distinct, it may be possible to measure the activity of one component without interference from the radiation emitted by the other component. A case in point would be where one nuclide was a pure β-emitter, while the other emitted both β- and γ-rays. In the case where the half-lives of the components are known but are not sufficiently different to allow graphical resolution of the decay curve, computer techniques that utilize least-squares fitting to resolve such a case are also available.

Other types of composite decay occur, but they are not normally encountered in applications of radiotracers. Perhaps the only such situation met with any frequency concerns "parent" radionuclides that decay into "daughter" nuclides that are also radioactive. In this case, it is the combined activity of the parent and daughter nuclides that will be observed. Two of the more common examples of parent-daughter mixtures that the biologist may encounter are ^{90}Sr (parent)—^{90}Y (daughter) and ^{137}Cs (parent)—^{137}Ba (daughter). Kirby (4) describes methods of resolving such parent-daughter mixtures mathematically.

4. Average Life

Although the half-life of a given radionuclide composed of large numbers of atoms is a defined value, the actual moment of disintegration for a particular atom can be anywhere from the very beginning of the isotope's life to infinity. The average life of a population can, however, be calculated. The average life (T_A) is naturally related to the decay constant and is, in fact, the reciprocal of the decay constant:

$$T_A = \frac{1}{\lambda} \tag{2-16}$$

or this can also be expressed as

$$T_A = 1.443t_{1/2} \tag{2-17}$$

One can understand the preceding relationship by recalling that the decay constant, λ, was defined as the average probability of decay per unit time, so that $1/\lambda$ is the average time between decays. The concept of average life is useful in the calculation of the total number of particles emitted during a defined decay period. This number is essential in determining total radiation dose delivered by a radioisotope sample, as in medical research and therapy. Table 2-6 lists the average life values for several commonly utilized nuclides.

TABLE 2-6

Average Life Values for Some Selected Radioisotopes

Nuclide	T_A (days)	Nuclide	T_A (years)
^{131}I	11.6	^{22}Na	3.78
^{59}Fe	65.8	^{3}H	17.694
^{35}S	127	^{90}Sr	40.0
^{45}Ca	238	^{14}C	8268

D. THE STANDARD UNIT OF RADIOACTIVITY—THE CURIE

In the early years of this century, radium was chosen as the standard for comparison of amounts of radioactivity. Although radium-226 is a rare element, it can be highly purified and has a long half-life (~1600 years). As a standard for radioactivity, the long half-life is a most important factor in that only very slight changes in radioactivity will be detected in standard preparations over long periods.

The *curie* (Ci) was originally defined as the number of disintegrations occurring in 1 gram of pure radium per second. This value has been measured experimentally many times. It may also be calculated by use of the decay equation. If we use the earlier half-life value of 1600 years for radium and substitute it into Equation (2-14), we obtain

$$1600 \text{ years} = \frac{0.693}{\lambda} \tag{2-18}$$

Solving for λ gives

$$\lambda = \frac{0.693}{1600 \text{ years}} = \frac{0.693}{(1600)(\pi)(10^7)} = 1.38 \times 10^{-11}/\text{sec} \qquad \text{(2-19)}$$

(Curiously enough, the number of seconds in a year is approximately $\pi \times 10^7$.) Since the atomic weight of radium is 226 and according to Avogadro's number, there are 6.02×10^{23} atoms in 1 g atom, it can readily be calculated that 1 g of radium contains approximately 2.66×10^{21} atoms. Employing the relationship expressed in Equation (2-9), the decay rate of 1 g of radium would be equal to the product of the decay constant times the number of atoms in 1 g, or

$$\frac{dN}{dt}\text{(that is, disintegrations occurring in 1 g radium/sec)}$$
$$= (1.38 \times 10^{-11})(2.66 \times 10^{21})$$
$$= 3.70 \times 10^{10} \text{ disintegrations per second (dps)}$$

Such a calculation of the value of the curie clearly depends on the experimentally determined half-life of ^{226}Ra, which is subject to subsequent refinement. (In fact, the half-life was later determined to be 1620 years.) Thus the definition of the curie, in terms of the decay rate of a specific nuclide, provided a changing standard. In order to avoid this difficulty, the term *curie* was defined arbitrarily in 1950 by the Joint Commission of the International Union of Pure and Applied Chemistry and the Union of Pure and Applied Physics as 3.700×10^{10} dps.

Since the curie is a relatively large unit, several subdivisions of it are in common use, as shown in Table 2-7.

TABLE 2-7

Commonly Used Units of Radioactivity

Units	Fraction of Ci	dps	dpm
Millicurie (mCi)	10^{-3}	3.700×10^7	2.220×10^9
Microcurie (μCi)	10^{-6}	$\times 10^4$	$\times 10^6$
Nanocurie (nCi)	10^{-9}	$\times 10^1$	$\times 10^3$
Picocurie (pCi) (also known as $\mu\mu$Ci)	10^{-12}	$\times 10^{-2}$	$\times 10^0$

Note that the curie unit refers to the number of disintegrations actually occurring in a sample rather than to the disintegrations detected by a radiation counter, which are usually only a fraction of the total disintegrations occurring. The difficulties of actually detecting the absolute number of disintegrations occurring in a sample will be pointed out in a later chapter.

It should also be emphasized that the curie is based on the rate of nuclear disintegration, not on the rate of emission of β-particles, γ-rays, or other radiation. To calculate such emission rates requires reference to the decay scheme of the nuclide. For example, ^{198}Au usually emits a β^--particle and one γ-ray for each nuclear disintegration, whereas ^{60}Co commonly emits a β^--particle and two γ-rays of different energies in cascades per nuclear disintegration. Thus for the same curie level of these two isotopes, the ^{60}Co would emit twice the number of γ-photons as the ^{198}Au. This problem is further complicated where a nuclide has alternate paths of decay. Consideration of this factor is important in connection with the calculation of disintegration rate from the counting rate as measured by a detecting instrument, and in radiation dose calculations. Nuclear data tabulations, such as Reference 1, contain a listing of what fraction of the total number of disintegrations gives rise to β^--particles, γ-rays, and other radiation. An abbreviated version of these tabulations is given below for radionuclides commonly used in tracer studies in Table 2-8.

TABLE 2-8

Fraction of Disintegrations Giving Rise to Different Types of Radiation

Radionuclide	Number of β-particles Per Disintegration	Number of γ-rays per Disintegration
^{3}H	1	0
^{14}C	1	0
^{22}Na	1	1 (1.2746 MeV)
		1.8 (0.511 MeV)
^{24}Na	1	1 (1.3685 MeV)
		1 (2.7539 MeV)
^{32}P	1	0
^{35}S	1	0
^{36}Cl	0.981 β^-, 0.019 EC	0
^{42}K	1	0.18 (1.524 MeV)
		0.002 (0.31 MeV)
^{45}Ca	1	0
^{51}Cr	1	0.09 (0.3198 MeV)
^{59}Fe	1	0.008 (0.143 MeV)
		0.028 (0.192 MeV)
		0.56 (1.095 MeV)
		0.44 (1.292 MeV)
^{60}Co	1	1 (1.73 MeV)
		1 (1.332 MeV)
^{65}Zn	0.017 (β^+)	0.034 (0.511 MeV)
		0.49 (1.115 MeV)
^{86}Rb	1	0.088 (1.078 MeV)
^{90}Sr	1	0
^{90}Y	1	0
^{131}I	1	0.026 (0.080 MeV)
		0.054 (0.284 MeV)
		0.82 (0.364 MeV)
		0.068 (0.637 MeV)
		0.016 (0.723 MeV)
^{198}Au	1	0.95 (0.412 MeV)
		0.01 (0.676 MeV)
		0.002 (1.088 MeV)

Note also that because radionuclides, in general, have differing half-lives, the number of nuclei per curie will differ from one species to another. For example, let us calculate how many nuclei are in 1 mCi of tritium ($t_{1/2} =$ 12.26 yr.). We know that

$$N = \frac{(-dN/dt)}{\lambda} = \frac{(3.7)(10^7)}{\lambda}$$

But $$\lambda = \frac{0.693}{t_{1/2}} = \frac{(0.693)}{(12.26)(\pi)(10^7)} = 1.799 \times 10^{-9} \text{ sec}$$

Thus $$N = \frac{(3.7)(10^7)}{(1.799)(10^{-9})} = (2.06)(10^{16}) \text{ nuclei}$$

[The same calculation carried out for ^{14}C ($t_{1/2} = 5730$ yr) would give 3.06×10^{11} nuclei/mCi.]

It is of further interest to calculate the mass associated with 1 mCi of tritium. We then have

$$M = \frac{(N)(\text{atomic weight})}{(\text{Avogadro's number})} = \frac{(2.06)(10^{16})(3)}{(6.02)(10^{23})} = 1.03 \times 10^{-7} \text{ g}$$

In other words, 1 mCi of tritium contains about 0.1 μg of tritium. When considering that in most tracer experiments we are dealing with 10^{-2} to 10^{-3} μCi rather than 1 mCi (i.e., a factor of 10^5 to 10^6 less material), we begin to see an important feature of radiotracers—the idea that we routinely work with 10^{-13} to 10^{-14} g of material.

E. SPECIFIC ACTIVITY

Unless a radionuclide is in a carrier-free state, it is mixed homogeneously with a certain amount of the stable nuclides of the same element. It is therefore desirable to have a simple expression to show the relative abundance of the radioisotope and the stable isotopes. This step is readily accomplished by using the concept of *specific activity*, which refers to the amount of radioactivity per given mass or other similar units of the sample. Specific activity is usually expressed in terms of the disintegration rate (dps or dpm), or counting rate (counts/min, cpm, or counts/sec, cps), or curies (or mCi, μCi) of the specific radionuclide per unit mass of the element. It is also, but less correctly, expressed as the amount of radioactivity per given volume or any other expression of quantity.

Obviously, when comparing the radioactivity of several samples, it is much more meaningful to employ the concept of specific activity. For example, suppose that one has a sample of L-alanine-^{14}C having a specific activity of 1000 dpm/millimole (mM) of compound but does not know where the ^{14}C labeling is located. Upon degradation of the compound, one manages to recover the carboxyl carbon atoms as CO_2 by means of a ninhydrin decarboxylation reaction in 50 % yield. The total radioactivity observed in the CO_2 so recovered is 500 dpm. This figure does not permit one to draw any conclusions. If, however, one expresses the results in terms of specific activity and finds that the CO_2 has a specific activity of 1000 dpm/mM, the latter figure immediately permits one to draw the conclusion that the L-alanine sample contains its ^{14}C activity exclusively in the carboxyl carbon atom.

It should be further realized that confusion may exist in comparing specific activity data. For example, in the foregoing case, the specific activity of the L-alanine-^{14}C sample can also be expressed as 333 dpm/mM of carbon atoms, inasmuch as L-alanine has three carbon atoms. For the indicated purpose, however, it is obvious that it will be desirable to examine the L-alanine-^{14}C sample in units of dpm/mM/compound.

In the design of radiotracer experiments (see Chapter 14), it is essential to calculate the specific activity of the tracer required. Use of a very low specific activity may result in counting samples with too low an activity level to be detected. In general, it is advisable to utilize a fairly high specific activity initially so as to allow for the dilution factor inherent in most radiotracer experiments. Unfortunately, for some nuclides, one finds that the higher the specific activity of the sample material, the greater the expense of producing it. Generally experimental design must include consideration of this economic factor.

BIBLIOGRAPHY

1. LEDERER, C. M., J. M. HOLLANDER, and I. PERLMAN. *Table of Isotopes*. 6th ed. New York: Wiley, 1968. Once again this book should be cited as the most easily available, up-to-date compilation of nuclear decay schemes, half-lives, radiation energies, etc.
2. EVANS, R. D. *The Atomic Nucleus*. New York: McGraw-Hill, 1955. Chapter 15 of this excellent text contains the definitive treatment of radioactive decay kinetics.
3. "Applications of Computers to Nuclear and Radiochemistry," *NAS-NS 3107*. A conference report with many fine papers on the subject of resolving multicomponent decay curves.
4. KIRBY, H. W. *Anal. Chem.* **24**, 1678 (1952).
5. HULL, M. H. *The Calculus of Physics*. New York: Benjamin, 1969. A nice little paperback treatment of the mathematical background for much of the discussion in this chapter.
6. STEVENSON, P. C. *Processing of Counting Data*. National Academy of Sciences Publication NAS-NS-3109, 1965. An excellent monograph on the mathematical methods for treating counting data with particular attention given to statistics and resolution of decay curves.

3

Characteristics of Ionizing Radiation

As noted, nuclear radiation occurs as a result of spontaneous disintegrations of atomic nuclei. These nuclear changes can give rise to several types of radiation, which have already been indicated: (a) α-particles; (b) negative β-particles, or negatrons; (c) positive β-particles, or positrons, (d) X rays resulting from electron capture; (e) γ-rays, either from isomeric transitions or, more commonly, as excess energy following particle emission; and (f) internal conversion electrons, resulting from an electromagnetic interaction between the nucleus and the orbital electrons.

The types of radiation most commonly encountered when radioisotopes are used as tracers are α-particles, β-particles (either β^+ or β^-), and γ-rays. These nuclear emissions differ radically in their physical characteristics and hence in the manner in which they interact with matter. Alpha radiation is made up of rather massive particles (helium nuclei) with a doubly positive charge that move at a relatively slow velocity (only a small fraction of the speed of light). By contrast, beta radiation consists of singly charged particles of extremely small mass, which are emitted with velocities approaching the speed of light. Gamma rays are electromagnetic radiation (photons); they are uncharged and travel at the speed of light. The characteristics of these three types of radiation will be discussed in more detail in this chapter.

A. ALPHA PARTICLES

1. Energy

In general, α-emitting isotopes are not used as tracers in most investigations largely because most α-emitting nuclides are elements of high atomic number (Z above 82) that are not normally significant in biological and chemical systems. Their presence in the environment, however, constitutes a considerable biological hazard, and they are studied primarily from this standpoint.

The α-particle has previously been described as a helium nucleus composed of two neutrons and two protons—a rather massive particle. The primary mechanism for energy loss by α-particles passing through matter is by inelastic collisions with atomic electrons. Because of their large mass, α-particles tend to follow very straight paths. Normally they are deflected only by a rare direct collision with a nucleus.

Alpha particles are emitted from radionuclides with considerable kinetic energy. The range of energy from natural α-emitting sources lies between about 4 to 8 MeV. The most significant feature of α-particle energy is probably that it is discrete; that is, all the α-particles emitted by a specific nuclide will emerge at one or a few well-defined energies. As an example, ^{238}U emits α-particles with energies of 4.20, 4.15, and 4.04 MeV. Such discrete α-particle energies serve as a means of identifying specific nuclides.

2. Half-life and Energy Relationships

It was noted quite clearly that the half-life of a specific nuclide generally seemed inversely related to the energy of the α-particle emitted in the decay of that nuclide. The current concept of the mechanism of α-particle emission provides an explanation of this relationship. According to this concept, the dense concentration of positively charged protons in the nucleus of an atom of high atomic weight produces an extremely strong potential barrier immediately around the nucleus. This high positive charge acts to repel positively charged particles approaching the nucleus. Moreover, the barrier serves to entrap the nucleons within it in an "energy well," thereby preventing their escape in the form of an α-particle. The term *barrier* is used only in a figurative sense to represent a repulsive force field.

Figure 3-1 is a graphic representation of a cross section through this barrier and the nuclear energy well. The stippled area denotes excess energy in the nucleus that may be imparted to an α-particle when the nucleus undergoes α-decay. Such an event will reduce the energy of the nucleus to a more stable level. Since this excess energy in the nucleus is much lower than the energy of the potential barrier, an α-particle, as such, cannot escape from the nucleus. According to the principles of quantum mechanics, however, a finite probability exists that the α-particle may "tunnel" through the barrier and emerge as a particle that is then actively repelled. The probability of such an occurrence is inversely related to the magnitude of the potential barrier to be penetrated, as depicted in Figure 3-1. Therefore the higher the excess energy within the "energy well," the less the potential barrier to be penetrated and, hence the greater the probability of α-particle escape—that

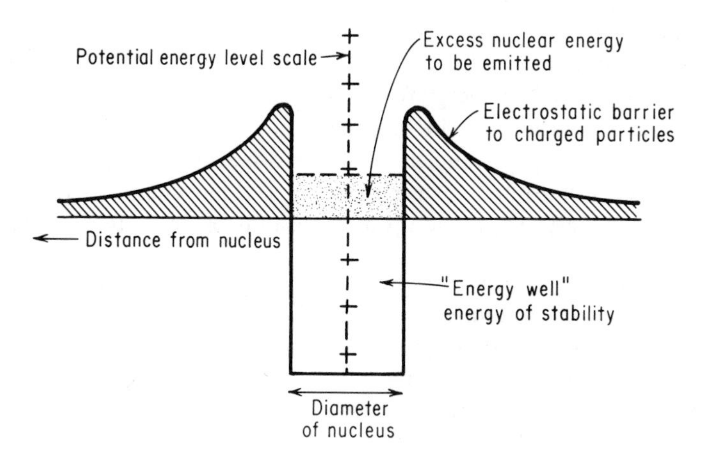

Fig. 3-1. Diagram of the nuclear potential barrier and "energy well" concept.

is, a shorter half-life. On the other hand, where the α-particle energy is relatively low, the probability of escape is small and the half-life is longer.

3. Interaction with Matter

The primary mechanism for energy loss by an α-particle transversing matter is by inelastic collisions with atomic electrons. In most cases, the electrons will leave the atom, thus producing an ion (*ionization*). Much less frequently, the electrons are merely excited to higher excited states of the atom. Such interactions with orbital electrons dissipate the kinetic energy of the α-particle.

a. Excitation. The term *excitation* is used to describe the interaction whereby orbital electrons take up energy from the passing α-particle but are not removed completely from their atoms. Afterward the excited electrons fall back into their former orbits and get rid of their excess energy. The amount of energy transferred by the excitation process is usually small.

b. Ionization. When an α-particle strips an orbital electron from an interacting atom, the loss of the negatively charged electron leaves the atom as a positively charged ion. The electron and the positive ion together are known as an *ion pair*, and the process is known as *ionization*. The electron and positive ion created by the initial α-particle-atom interaction can go on to cause further ionization themselves. These secondary processes actually account for 60 to 80 % of the total ionization caused by the α-particle. The formation of each ion pair in a gas requires, on the average, about 34 eV of the α-particle's kinetic energy. Stated another way, a 6.8-MeV α-particle will produce about 2×10^5 ion pairs in air before its energy is completely dissipated. Consequently, ionization constitutes by far the most important process in the transfer of energy from the α-particle to the interacting matter. The effect of ionization can be visualized in a cloud chamber, where each ion produced serves as a nucleus for the formation of a fog droplet in a supersaturated atmosphere. An α-track in a cloud chamber appears as a straight, dense fog track, made up of thousands of droplets per centimeter.

c. Specific Ionization. In describing the intensity of ionization, the term *specific ionization* is generally used. *Specific ionization* refers to the number of ion pairs formed per unit length (centimeter) of the α-particle's path (or that of any other ionizing particle) in air at standard pressure (or any other medium). Figure 3-2 shows the variation in specific ionization for α-particles

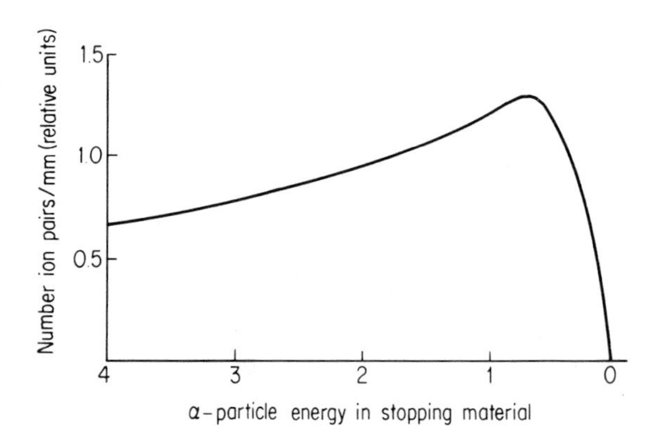

Fig. 3-2. Typical specific ionization curve for α-particles in air.

over the extent of their range in the air. Obviously, the specific ionization of a beam of α-particles increases toward the end of their range. This condition occurs because, as a result of many collisions with gas molecules, the α-particles have lost much kinetic energy and their velocity has decreased. This reduced velocity causes them to remain in the vicinity of the molecules along their path for a longer period of time and thus have a much greater probability of interacting with these molecules. After the magnitude of specific ionization reaches a peak, it declines sharply to zero. At this point, the α-particles have expended their kinetic energy, picked up two electrons, and become neutral helium-4 atoms.

4. Range

a. Determination. Because of the discrete energy of α-particle emission, α-particles from a given radioactive source will travel through a clearly defined range in matter. The α-particle range can be experimentally determined by measuring the intensity of α-radiation at increasing distances from an α-emitter. Figure 3-3 represents a plot of such data. It will be seen that the number of α-particles detected remains constant for a certain distance from the emitting source and then declines sharply to zero.

A certain amount of straggling of the α-particles produces the tail on the range curve in Figure 3-3. The result is to make absolute determination of the true range more difficult; consequently, the expression *mean range* has often been used. The mean range is the distance from the source at which the initial α-radiation intensity is reduced to one-half. The mean ranges of α-particles from common α-emitters have been accurately determined, and they are generally of the magnitude of a few centimeters in air.

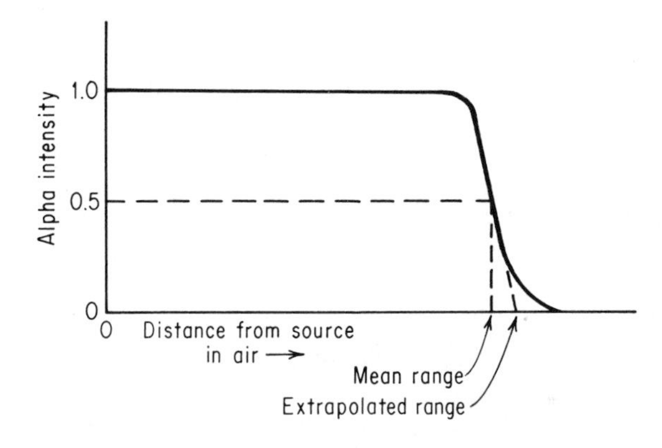

Fig. 3-3. Typical range curve for α-particles in air.

b. Range-Energy Relations. The range of α-particles is, of course, directly related to the kinetic energy with which they are emitted. Unfortunately, this is not a simple relationship. The use of Equation (3-1) allows the calculation of an approximate range for α-particles having energies from 4 to 15 MeV.

$$R = (0.005E + 0.285)E^{3/2} \tag{3-1}$$

In this equation R equals mean range in air in centimeters and E is the α-particle energy in MeV. As an example, the mean range in air of 7-MeV α-radiation could be calculated as follows:

$$R = [(0.005)(7) + 0.285](\sqrt[2]{(7.0)^3}) = (0.320)(18.52) = \underline{5.9 \text{ cm}}$$

The range-energy relationship in air for α-particles with initial kinetic energies between 0.4 and 10 MeV is shown in Figure 3-4.

c. Absorption by Other Materials. Alpha particle ranges in materials other than gases will necessarily be very much shorter owing to the greater density of liquids and solids. In fact, these ranges are so short that they are normally

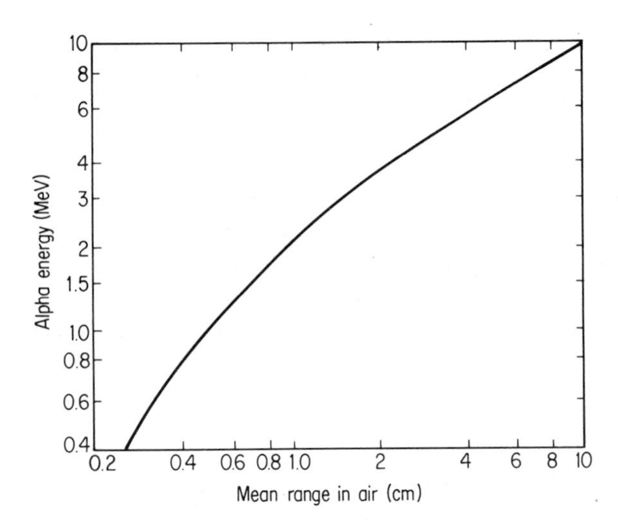

Fig. 3-4. Range-energy curve for α-particles in standard air.

stated in microns (one micron is 1/1000 millimeter). Approximate calculated ranges for α-particles of 7 MeV in several selected absorbers are listed in Table 3-1.

Although, in Table 3-1, range has been expressed as a linear distance, it is clear from the table that the density of the stopping material is a most significant factor. In addition, since the minute ranges shown are difficult to measure, another unit for range is used. Range is expressed in units of *areal density*—that is, mass per unit area. From a practical point of view, this definition of *thickness* is quite convenient, for one need only weigh a given area of material in order to determine its areal density. Clearly, the range in areal density units (such as mg/cm^2) equals the value of the range in linear units (such as cm) multiplied by the density of the stopping material (mg/cm^3). The approximate values of the range of a 7-MeV α-particle in various materials are given in units of areal density in Table 3-2. The values here should be compared with the previous list of linear ranges for the same material (Table 3-1).

Often we will want to know precisely the range of an α-particle in something other than air. Some useful semiempirical formulas exist that allow us

TABLE 3-1

Linear Range of 7-MeV α-Particles in Some Common Absorbers

Air	Water (tissue)	Aluminum	Mica	Copper	Lead
59,000μ	74μ	34μ	29μ	14μ	2μ

TABLE 3-2

Range in Areal Density Units (mg/cm²) for 7-MeV α-Particles in Some Common Absorbers

Air	Mica	Aluminum	Copper	Silver	Gold
7.1	8.2	9.2	12.5	19.1	27.1

to calculate the range of an α-particle in another material once the range of that energy α-particle in air has been calculated [using Equation (3-1) or Figure 3-4]. One such formula says that for a pure element with $10 < Z \leq 15$,

$$\frac{R_Z}{R_{\text{air}}} = 0.90 + 0.0275Z + (0.06 - 0.0086Z)\log_{10}\left(\frac{E_\alpha}{4}\right) \tag{3-2}$$

where R_Z is the range in a *pure element* of atomic number Z expressed in units of mg/cm², R_{air} is the range in air in mg/cm², and E_α is the α-particle energy in MeV. For $Z < 10$, substitute 1.00 for the term $(0.09 + 0.0275Z)$. For $Z > 15$, replace the term R_Z by $(R_Z = R_Z + 0.005Z)$. If we wish to know the range of an α-particle in some compound or elemental mixture, the relation

$$\frac{1}{R_{\text{tot}}} = \frac{P_1}{R_1} + \frac{P_2}{R_2} + \frac{P_3}{R_3} + \cdots + = \sum_i \frac{P_i}{R_i} \tag{3-3}$$

is used, where R_{tot} is the range in the mixture or compound expressed in mg/cm², P_1, P_2, P_3 are the weight fractions of each pure element in the mixture, and R_1, R_2, R_3, etc. are the ranges of the energy α-particle in pure elements 1, 2, 3, etc.

Computing the range of another heavy charged particle, such as a proton or deuteron, in a stopping material is easy. The best means of doing so is by reference to the extensive tabulations of such information in the literature. The best as of 1972 is that of Williamson, Boujot, and Picard (1). This kind of calculation may become increasingly important as the use of charged particle beams from nuclear accelerators in radiation therapy, activation analysis, and X-ray fluorescence studies increases.

5. Practical Considerations

The very short range of α-particles in matter creates certain practical problems in designing radiation-detection equipment. Because they dissipate their energy in the wall or window of the typical gas ionization detector, α-particles do not penetrate to the sensitive volume of the detector. A 7-MeV α-beam, for example, would not penetrate a mica detector tube window 20 μ, or 9 mg/cm² thick. Furthermore, the preparation of sources for α-counting is quite difficult because the α-particles are so easily absorbed in the sample.

Alpha radiation poses no great external health hazard, since the outermost horny layers of epidermis are thick enough to absorb almost all external α-radiation, even from sources deposited directly on the surface of the body. Safe storage of α-emitting isotopes is possible because thin glass or metal containers will absorb all the α-radiation being emitted from sources within them.

When α-emitters enter the body by ingestion or inhalation, the situation is quite different. Alpha particles dissipate their energy in such an exceedingly small volume of tissue that very great local damage can occur. Moreover, many natural α-emitters have long half-lives, and some (radium and plutonium) are incorporated into metabolically less active bone tissue, thus increasing the internal hazard. At least one α-emitting nuclide of considerable prominence, ^{239}Pu, is highly toxic from a purely chemical standpoint, in addition to its radiation hazard.

The maximum permissible concentrations of most α-emitters in the body is quite low. Table 3-3 shows the maximum permissible body concentrations as set by law of some typical α-emitting radionuclides.

TABLE 3-3

Maximum Permissible Concentrations of Some α-Emitting Radionuclides in the Human Body

Nuclide	Maximum Permissible Body Concentration (μCi/70-kg body weight)
^{210}Pb (RaDEF)	0.4
^{226}Ra	0.1
^{228}Th	0.02
^{232}Th	0.04
^{233}U	0.03
^{235}U	0.05
^{239}Pu	0.04
^{252}Cf	0.01

B. BETA PARTICLES

1. General Nature

a. Negatrons and Positrons. The term *beta particle* has been used for two different entities, positrons and negatrons. Both represent particles formed by nuclear changes and ejected from the nucleus with velocities approaching the speed of light. They are physically alike in every respect except charge. Positrons carry a positive charge; negatrons are negatively charged—that is, electrons.

Beta particles are only approximately 1/7300 the mass of α-particles. Because of this very small mass, they are quite easily deflected on passing near other atoms. Their track in a cloud chamber is a tortuous one indeed, wholly unlike the straight α-track. In addition, the smaller mass and higher velocity of the β-particle result in a smaller probability of interaction with the orbital electrons of the atoms it passes near. Consequently, it has a much greater penetrating power and hence a much longer range through matter. Where the range in air of α-particles may be only several centimeters, the range of β-particles from some beta emitters may be as long as several meters.

Since the *negatron* is like a normal orbital electron in all respects but origin, it will persist after dissipating all its kinetic energy. As a rule, it ultimately becomes attached to a positive ion as an orbital electron. A *positron*, on the other hand, has only a transient existence. After expending all its kinetic energy, it interacts with an electron and is "annihilated." The mass of both particles is converted to energy in the form of two 0.51-MeV γ-rays, which are emitted at angles of 180° to each other ("annihilation radiation"). Hereafter, following common practice, the term *beta particle* will be used for the negatron.

b. Conversion Electrons. A similar, yet distinct type of high-velocity particle is the *internal conversion electron.* Energy from a disintegrating nucleus that would otherwise be emitted as a γ-ray may be quantitatively transferred to an inner orbital electron of the same atom. The orbital electron thus energized is immediately ejected from the atom at high velocity. Note that no γ-ray is emitted in internal conversion. Since γ-rays are emitted with a discrete energy, internal conversion electrons are ejected with discrete energies from a given nuclide—in contrast to β-particle emission. Thus the term *beta particle* is restricted to electrons originating in the nucleus.

2. Energy of Beta Decay

a. Spectral Distribution of Energy. Perhaps the most striking character of β-radiation is the particle energy. It has been seen that a given nuclide may eject α-particles of one or a few discrete energies. By contrast, in β-decay, the particles are emitted over a continuous range of kinetic energy up to a maximum value (E_{max}) characteristic of the nuclide in question. Values for beta E_{max} range from 0.019 MeV for ^{3}H up to 4.81 MeV for ^{38}Cl.

The *magnetic spectrometer* has been used experimentally to determine the β-energy distribution for many nuclides. In this instrument, a collimated beam of β-particles is deflected through 180° by a strong magnetic field. The more energetic particles in the beam swing in a wider arc; the less energetic travel over a shorter arc. As a result, the beam particles of various

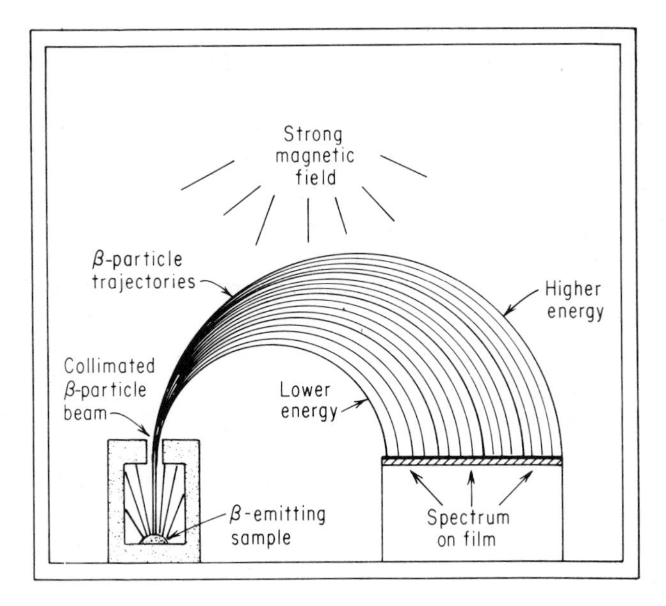

Fig. 3-5. Simplified diagram of a magnetic spectrometer.

energies are spread out to form a spectrum, which is detected by exposure of a strip of photographic film (Figure 3-5). Alternatively, a β-sensitive detector may be placed in fixed position, instead of the film, and the magnetic field varied so that successive energy levels of the β-spectrum are detected. The energy of the various β-particles can be calculated from the known strength of the magnetic field and the radius of the arc followed by the particles.

Figure 3-6 shows some typical energy distributions for the particles emitted in β^--and β^+-decay. Note that the fraction of β-particles emitted near *the maximum energy* (E_{max}) is very small. A much larger fraction is emitted with *the average energy* ($\bar{E}$). In negatron decay, the average energy is approximately $\frac{1}{3}E_{max}$, whereas in positron decay, the average energy is approximately $\frac{2}{3}E_{max}$. The β^+-energy spectrum [Figure 3-6(b)] is shifted to higher energies due to the electrostatic repulsion between the β^+-particles and the positively charged nucleus. The range of β-particles quite obviously depends on the value of E_{max}, but in calculating the actual radiation dose from a β-emitter, the $\bar{E}$ value is more suitable.

The energy spectra for several β-emitters commonly used as radiotracers are given in Figure 3-7. Note that the general shape of the curves varies from nuclide to nuclide. The low-energy (soft) emitters have particularly noteworthy spectra. A dashed line at the low-energy portion of each spectrum indicates that this portion is theoretically calculated, for experi-

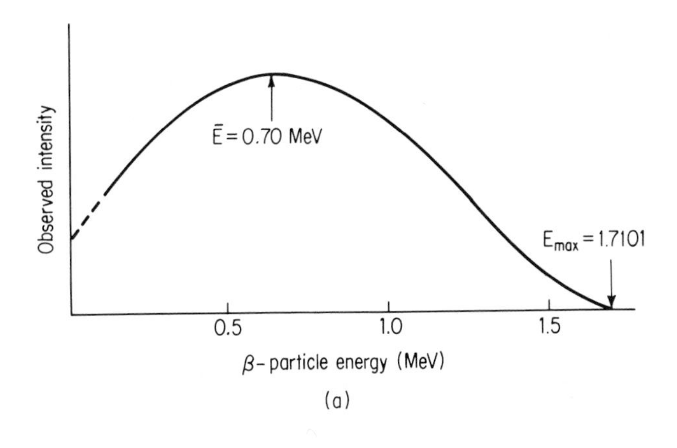

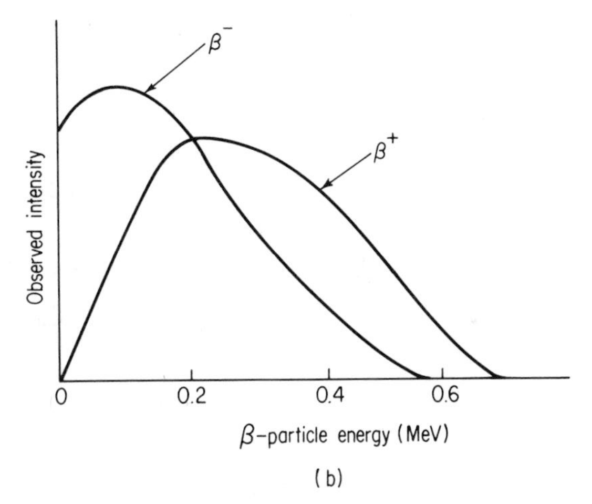

Fig. 3-6. (a) Simple β-particle energy distribution curve for ^{32}P showing E_{max} and $\bar{E}$. Note that $\bar{E}$ does not always correspond to the most probable energy (highest point on the curve).
(b) A schematic view of the emitted particles energy distribution in β^--and β^+-decay showing differences between two cases.

mental determination is not readily feasible. Marshall (8) presents methods for calculating the shape of β-spectra.

b. The Neutrino and Beta Decay. Since discrete energy levels were known to exist within the nucleus (see previous discussion of α-emission energy), the emission of β-particles with a continuous energy spectrum from the same nuclide was somewhat of an enigma. In 1931, to explain this situation, Pauli suggested that each β-decay actually occurred with a total energy

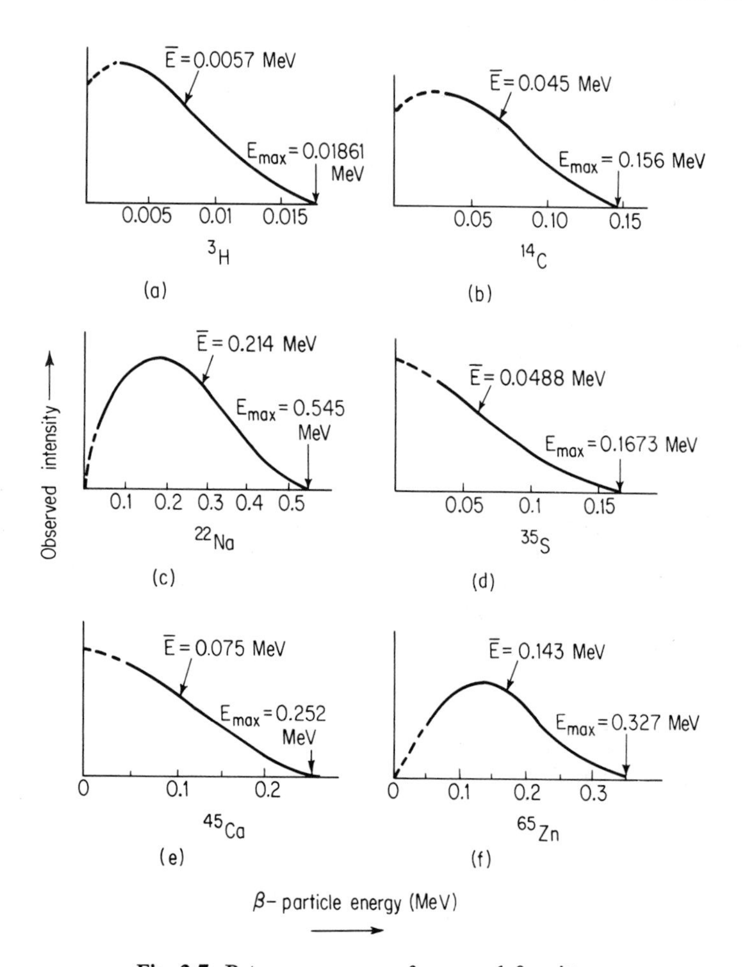

Fig. 3-7. Beta energy curves for several β-emitters.

involvement equivalent to the E_{max} of the β-particle in question. He postulated that a hitherto undiscovered particle, the *neutrino*, shares this total energy with the β-particle in varying proportions from disintegration to disintegration. For example, a β-particle emitted from the radionuclide ^{32}P ($E_{max} = 1.71$ MeV) with an energy of 1 MeV would be accompanied by a neutrino equivalent to 0.71-MeV energy, making a total energy emission of 1.71 MeV.

Pauli indicated that the neutrino had no charge and negligible mass; as a result, its interaction with matter would be virtually nil and its detection a most difficult task. The elusive neutrino long defied detection, but its existence was finally demonstrated by means of an elaborate experimental technique in 1956. The current concept of β-decay can therefore be expressed as follows:

$$\text{Neutron} \longrightarrow \text{proton} + \overbrace{\text{negatron} + \text{neutrino}^*}^{\text{energy} = E_{\max}}$$

$$\text{Proton} \longrightarrow \text{neutron} + \overbrace{\text{positron} + \text{neutrino}^*}^{\text{energy} = E_{\max}}$$

3. Interaction with Matter

a. Modes of Interaction. As with α-particles, β-particles lose energy largely by ionization and excitation of the atoms with which they interact. However other minor mechanisms for energy loss are also possible. First, there is the emission of *bremsstrahlung* (German for "braking radiation"). In the bremsstrahlung emission process, as the negatively charged, high-energy electron from β^--decay passes by the positively charged nucleus, it is attracted and accelerated by the nuclear coulomb force field. In the course of this acceleration, the electron radiates excess energy in the form of electromagnetic radiation (the bremsstrahlung). The process is shown schematically in Figure 3-8.

The energy spectrum of the bremsstrahlung radiation is continuous and runs from 0 up to the original β^--energy. Figure 3-9 shows a typical bremsstrahlung energy spectrum.

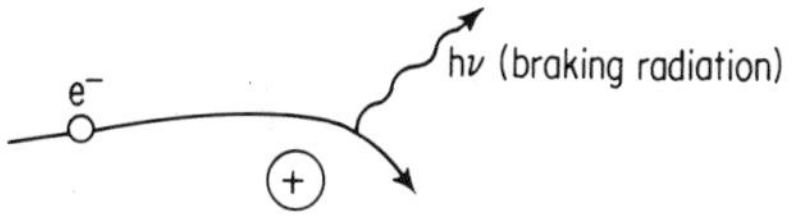

Fig. 3-8. Schematic view of bremsstrahlung emission process.

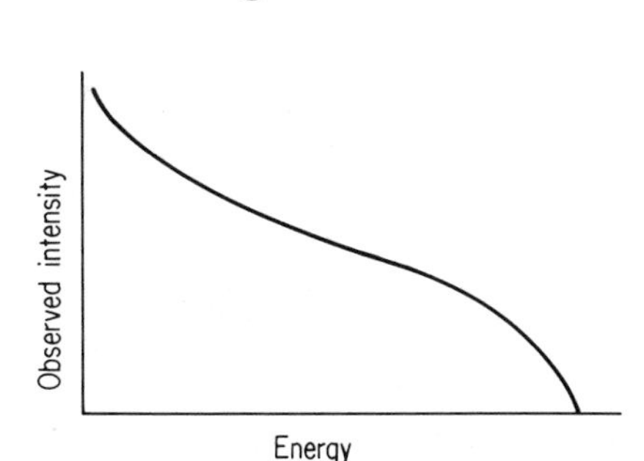

Fig. 3-9. The bremsstrahlung energy spectrum.

*The most correct statement of the β^- and β^+ decay processes is that

$$n \longrightarrow p + \beta^- + \bar{\nu}_e \qquad (\beta^- \text{ decay})$$
$$p \longrightarrow n + \beta^+ + \nu_e \qquad (\beta^+ \text{ decay})$$

where $\bar{\nu}_e$ and ν_e stand for the antielectron neutrino and the electron neutrino, respectively.

The importance of bremsstrahlung as a mode of energy loss for electrons is revealed in the following equation:

$$\frac{\text{(amount of energy loss by bremsstrahlung)}}{\text{(amount of energy loss by ionization)}} \approx \frac{E_e \cdot Z}{800} \tag{3-4}$$

where E_e is the electron energy in MeV and Z is the atomic number of the stopping material. Thus even for high Z materials ($Z \approx 80$), the amount of energy loss by bremsstrahlung is small ($\sim 10\ \%$ for $E_e = 1$ MeV).

Bremsstrahlung is of no particular value in detecting β^--radiation. Instead it constitutes an important nuisance in that whenever modest energy β^--particles are present, some highly penetrating bremsstrahlung that perturbs absorption curves (see Figure 3-12) will occur.

It is clear from Equation (3-4) that the use of low Z materials, such as lucite, near the β^--source will reduce the amount of bremsstrahlung.

A fourth mechanism for electron energy loss in matter also occurs—the emission of Cerenkov radiation. When a beam of fast-charged particles with velocity v near the speed of light c enters another medium with index of refraction n, the particle velocity will exceed the speed of light in the new medium (which is c/n). Then the electron radiates its "excess" energy as a blue-white light called *Cerenkov radiation*. Thus the emission of Cerenkov radiation is similar to the "sonic boom" phenomena in acoustics. Furthermore, the light is localized in a cone (of half-angle θ) around the direction of motion of the incident electron such that

$$\cos \theta = \frac{c}{nv}$$

The process is shown schematically in Figure 3-10.

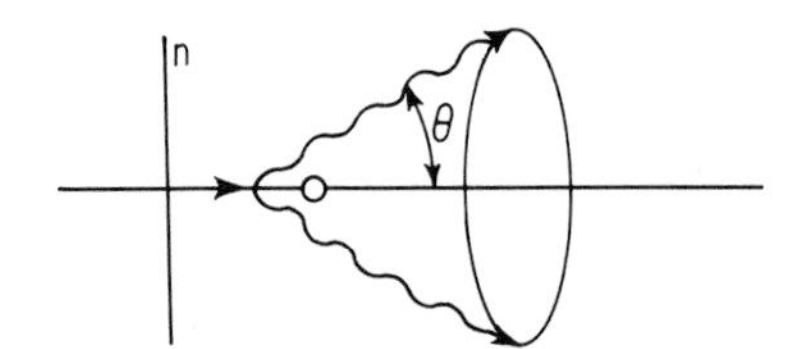

Fig. 3-10. Schematic view of emission of Cerenkov radiation.

Cerenkov radiation is a favorite subject for science fiction writers and movie set designers in that it is the source of the intense blue-white glow surrounding the core of a "swimming pool"-type nuclear reactor. Here the Compton electrons produced in the reactor core propagate into the water, where $(v \approx c) > (c/n)_{\text{water}}$.

b. Specific Ionization. As previously mentioned, the track of a β^--particle as seen in a cloud chamber is tortuous, poorly defined, and much longer than that of an α-particle. In considering an initially unidirectional beam of β^--radiation, it is evident that many of the β^--particles will be deflected and scattered out of the beam. This process will lead to an apparent decrease in intensity of the radiation that is not strictly due to absorption. The situation is further complicated by the continuous energy spectrum associated with the particles. Consequently, the specific ionization pattern for a beam of β^--particles is not directly comparable to that of a beam of monoenergetic α-particles.

Figure 3-11 shows that, with low-energy β^--particles, specific ionization decreases sharply as β^--energy increases. The maximum specific ionization,

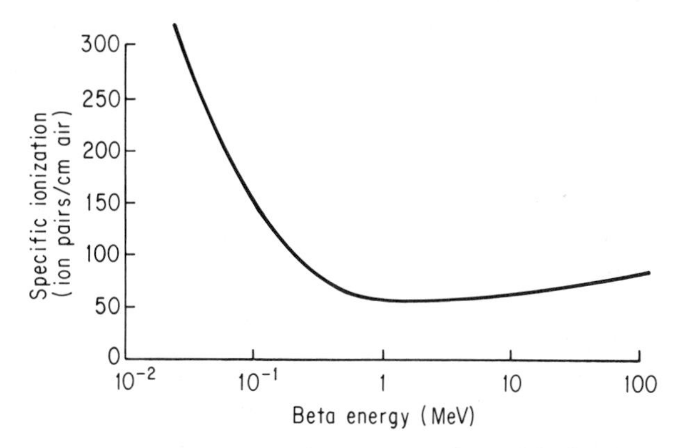

Fig. 3-11. Specific ionization curve for β^--particles in air.

7700 ion pairs cm, occurs with β^--particles having energies of 146 eV. Consequently, the preponderance of ionization, and hence energy loss, from a given β^--particle occurs toward the end of the β^--track when its energy content drops below a few thousand electron volts. With β^--particles having higher energy, particularly those having energies above about 2 MeV, relativistic considerations lead to a gradually increasing specific ionization with increasing particle energy.

By comparing Figure 3-2 and 3-11, we see that specific ionization for α-particles is many times greater than that for β^--particles having the same kinetic energy. This point is understandable, for not only do α-particles carry twice as much charge as β^--particles but, in particular, specific ionization is inversely related to particle velocity as well. Thus, for a given kinetic energy, the velocity of a β^--particle is much higher than that of an α-particle.

4. Range

The range of α-radiation is rather clear-cut, but β^--particle range both in air and in metal absorbers is quite ill defined. Since linear β^--particle ranges in air are extremely difficult to determine experimentally, range is best expressed in areal density (in mg/cm^2) of an absorber. Aluminum is the most commonly used absorber for this purpose. It should be emphasized that since the absorption process is a function of the density of the absorbing material, the range expressed in mg/cm² for β^--radiation in other absorbers of similar Z values (such as air and mica) varies only slightly from that in aluminum. Figure 3-12 portrays a typical β^--absorption curve obtained in range determination. Note the marked contrast to the α-absorption curve (Figure 3-3), where beam intensity remains constant nearly to the end of the range and then drops rapidly to zero.

Several significant items should be noted in Figure 3-12. First, much of the semilog plot is linear; that is, the absorption process is pseudoexponential in nature, because of the combined effects of a continuous spectrum of energies and particle scattering. It will be seen later that γ-absorption is a truly exponential process. Second, the curve flattens out to constant activity chiefly because of the production of the very penetrating bremsstrahlung. Third, although it would seem that extrapolation of the linear portion of the curve would give the true β^--particle range, unfortunately, such is not the case.

One method of identifying β^--emitting nuclides is a measurement of the range of the β^--particles in matter. The range can be related to the maximum β^--particle energy (E_{max}), which is, in general, a distinctive characteristic of a specific nuclide. One method of accomplishing this goal is called *Feather analysis*. In Feather analysis, the aluminum absorption curve of an unknown β^--emitter is compared with that of a β^--emitter of known range. A detailed, simple presentation of the technique is described by Chase and Rabinowitz (7). To repeat the point made earlier, *one important feature of β^--ranges is that values of the β^--range expressed in units of mg/cm² are roughly independent of the nature of the stopping material.* This means that if a β^--particle has a range in Al of 6 mg/cm², its range in air will be ~6 mg/cm².

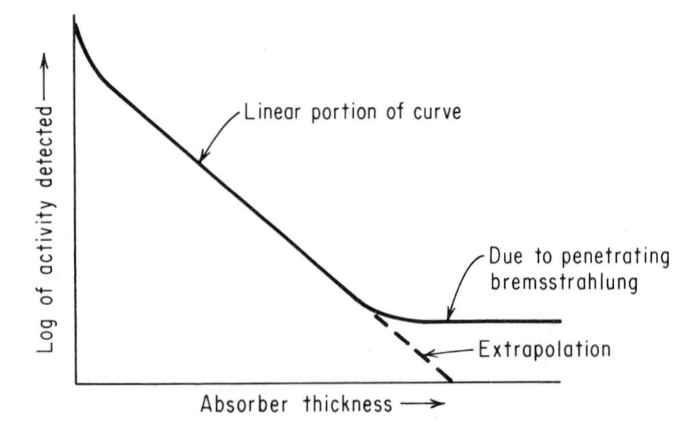

Fig. 3-12. Plot of absorption of β^--particles in matter.

Two commonly used ways of relating the β^--range to the maximum β^--energy, E_{max}, exist. The first involves the use of graphs, such as that shown in Figure 3-13. The second method involves the use of semiempirical relationships, such as the Glendenin equations. These equations state

$$R = 407E_{max}^{1.38} \qquad \text{for} \quad 0.15 \leq E_{max} \leq 0.8 \text{ MeV} \tag{3-5}$$

$$R = 542E_{max} - 133 \qquad \text{for} \quad E_{max} > 0.8 \text{ MeV} \tag{3-6}$$

where R is the range in mg/cm² of Al and E_{max} is expressed in MeV.

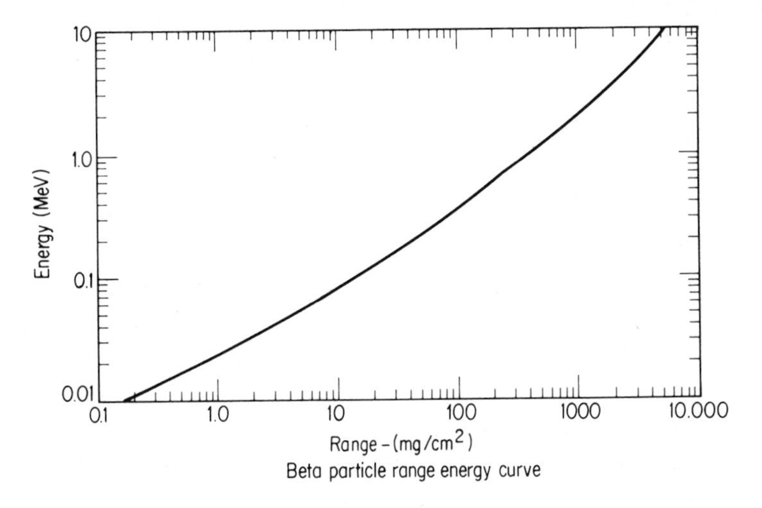

Fig. 3-13. Range-energy relationships for β^--particles interacting with Al.

5. Practical Considerations

a. Detection Problems. In the application of radioisotopes as tracers, normally the identity of the nuclide being used is known. Thus it is not necessary to determine experimentally the R_{max} and E_{max} of the nuclide in question. Instead the investigator focuses on the limitations on detection imposed by the characteristic absorption process of β^--particles in matter. Since the detection of β^--radiation sometimes necessitates the particles entering an enclosed detector through a "window," it is important to know and respect the sizable fraction of the incident β^--radiation that will be absorbed by such windows. Transmission curves are available to supply this information for β^--particles of various energies and windows of different thicknesses. Curves for β^--transmission values of three mica window thicknesses are shown in Figure 3-14. Window thickness is stated in areal density units (mg/cm²). The effect of increasing window thickness on β^--particle transmission is striking.

Obviously, for a low-energy β^--emitter, such as ^3H ($E_{max} = 0.019$ MeV),

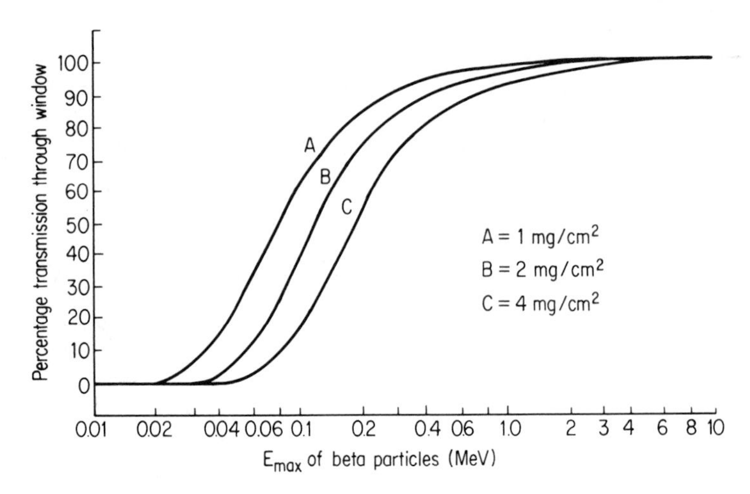

Fig. 3-14. β^--transmission curves for three mica window thicknesses.

all the β^--radiation would be absorbed in even the thinnest window shown in Figure 3-14. Unfortunately, in addition to ^3H, several other radioisotopes of biological importance, such as ^{14}C (E_{max} 0.156 MeV), ^{35}S (E_{max} 0.167 MeV), and ^{45}Ca (E_{max} 0.252 MeV), are also "soft" β^--emitters. Much, if not all, of the radiation from these isotopes would be absorbed before actually entering the sensitive volume of a gas-filled ionization-type detector. It is often necessary to employ special detecting techniques for these soft emitters. Such techniques, which will be further described in later chapters, customarily involve using an ultrathin window detector, placing the radioactive sample inside the detector (windowless type of detector), or mixing the sample in intimate contact with the detecting medium (liquid scintillation method).

The short range of soft β^--radiation is used to advantage in autoradiography, where a sensitive film emulsion is employed to record the track of the radioactive particles (see Chapter 10). Such a high resolution is possible by this method that individual β^--particle tracks may be traced.

b. Biological Hazards. The biological hazards attending the use of β^--emitters, as for α-emitters, differ considerably, depending on whether one considers external or internal hazards. The radiation hazard from an external β^--source in a laboratory is normally not significant, provided that one is not directly handling the source. Heavy glass or metal containers will absorb most, if not all, of the β^--radiation from enclosed radionuclides. Mere distance from exposed sources can be a good safeguard because of their limited range in air. Accidental body-surface contamination will generally lead to irradiation of only superficial tissues. One should, however, be quite cautious in handling β^--emitters, particularly those of high energy, to ensure that the radiation dosage to the hands and other parts of the body is within legally permissible limits (see Chapter 16). It is important in this regard to consider carefully the very penetrating bremsstrahlung associated with intense β^--emitters.

Ingested β^--emitters pose a greater hazard. Many of the frequently used β^--emitting nuclides are isotopes of elements commonly found in living tissue, such as C, H, S, and P. As a result, these radionuclides may be readily incorporated into the constituents of the body. If this situation leads to local deposition or concentration, such as the incorporation of tritium in the DNA of chromosomes, or calcium-45 in the bone, radiation damage may be quite extensive.

C. GAMMA RAYS

1. Nature and Source

a. Electromagnetic Nature. In contrast to α- and β^--radiation, γ-rays are a form of electromagnetic radiation—that is, photons. Photons are electrically neutral particles with zero mass when the particle is at rest, and, consequently, they can penetrate matter readily with little interaction. This characteristic allows γ-rays to have an effective range in matter that is much greater than that of α- or β^--particles of comparable energy. For example, whereas the range in air of α-particles is 2 to 8 cm and the range of β^--particles in air is 0 to 10 m, the distance traveled by typical γ-radiation is from cm to 100 m. (Actually, the spread in distance traveled in matter for γ-rays is so large that the concept of a definite range is not valid.)

The spectrum of electromagnetic radiation shown in Figure 3-15 indicates the relation of γ-radiation to other forms of electromagnetic radiation (light, radio, etc.). It should be noted that light, X rays, γ-rays, and similiar items are all forms of electromagnetic radiation that differ only in frequency and wavelength. The distinction between X rays and γ-rays is not made on the basis of wavelength, however, but on point of origin of the radiation. Gamma rays are electromagnetic radiation resulting from transformations in the nucleus, whereas X rays are electromagnetic radiation resulting, in general, from transformations involving the orbital electrons of the atom.

A fundamental question that may be asked at this point is just what is electromagnetic radiation. Certainly we recognize some of its forms, like light, radio waves, and X rays. But can we understand just how this radiation comes about? The explanation goes something like this. An electric charge creates a "condition" in the space around it such that when we place another charge in this surrounding space, it will feel a force acting on it. That "potential" for producing a force is called an *electromagnetic field*

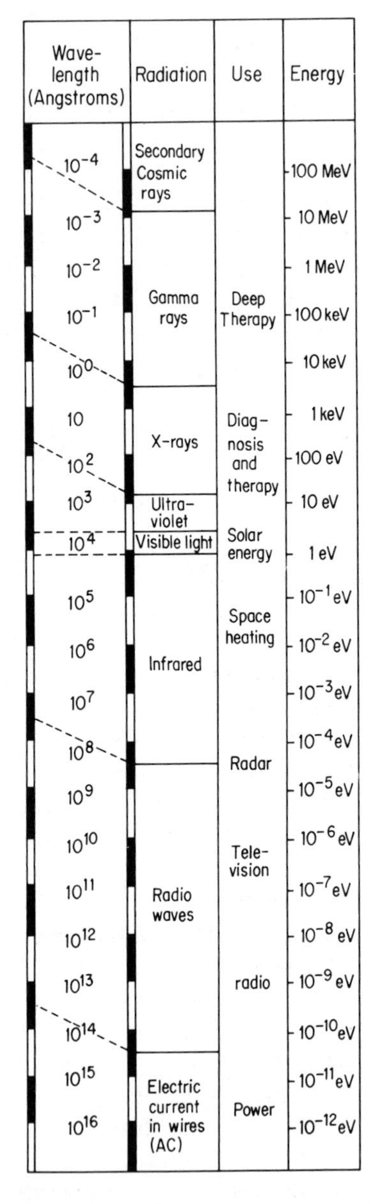

Fig. 3-15. The electromagnetic radiation spectrum.

and it surrounds any charged object. Now if we shake that charged object, it will set up waves in the electromagnetic field, called *electromagnetic radiation.* How fast we shake the generating charge will determine the frequency of the waves (how many waves pass a given point per second) and the wavelength (the spatial distance between crests of different waves). The situation is quite analogous to moving a stick rapidly in a pond. All electromagnetic radiation consists of waves set up by the rapid motion of charged objects.

We must hasten to add that, in some instances, these waves behave in a most unwavelike manner; that is, at high frequencies, they behave like particles. Quantum mechanics offers us a way out of this apparent dilemma by pointing out that no real distinction between the wave and particle descriptions of physical phenomena exists. They are simply two different but completely equivalent ways of describing some natural phenomena.

As an example, we are all familiar with the wave aspects of light, such as rainbows and diffraction patterns. However, Planck and Einstein pointed out that light could be thought of as containing little bundles of energy, called quanta or *photons*. Some phenomena, such as the photoelectric effect (see p. 60), are best discussed in terms of this "particle picture" of light.*

Whether light is "really" a wave or a particle, is a question for theology. Science is only charged with the responsibility of providing the best, simple rational explanation of phenomena. And so light is sometimes best described as a wave, whereas at other times it is best described as a particle. It is as simple as that.

b. Source of gamma emission. In Chapter 2 it was pointed out that γ-rays are most frequently emitted immediately following α- or β^--particle emission from a nucleus. In the case of certain nuclides, ^{131}I for example, the excess energy of the excited state is carried off by a series or "cascade" of γ-rays. Such γ-rays are normally emitted within 10^{-12} sec following the particle emission. It therefore appears that the γ-radiation represents the readjustment of energy content in the radionuclide from excited states to more stable states. With some nuclides, the consequent γ-emission is delayed considerably (up to several hours); in that case, the radionuclide is presumably maintained at an "excited state" over a prolonged period. Such delayed γ-ray transitions are known as *isomeric transitions.*

Spectral analysis of γ-rays reveals that they are emitted with discrete energies. As is the case in α-emission, this phenomenon presumably reflects transitions between the discrete energy levels existing within the nuclear energy well (see Figure 3-1). Gamma rays from most radioactive nuclides have energies in the range of 10 keV to 3 MeV. A very few range up to 7 MeV. Because of alternate pathways of decay, many isotopes emit γ-rays of several different energies (see Figure 2-5).

c. X Rays. X rays from radioactive nuclides are commonly emitted as a result of electron capture and internal conversion, and their energies are

*In a similar vein, we normally think of an electron as a particle. In many circumstances, it is quite convenient to consider the electron as a wave of wavelength

$$\lambda = \frac{h}{mv} \tag{3-7}$$

where λ is the de Broglie wavelength of the electron, h is Planck's constant, and m, v are the electron mass and velocity, respectively.

those characteristic of the energy differences between the inner orbital electron levels. The energy range is from a few electron volts to about 120 keV, quite in contrast to the gamma energy range. Two common examples are the 8-keV X rays from ^{65}Zn and the 5-keV X rays from ^{51}Cr.

2. Interaction of Gamma Radiation with Matter

Gamma rays interact with matter in several different ways. Of these ways, some of which are listed below, only the last three are important mechanisms for γ-ray energy loss in matter.

a. Nuclear Transformation. Very high energy γ-rays may directly interact with a nucleus, causing excitation of the nucleons. The results may be the ejection of a particle, usually a neutron, and the transmutation of the atom to another nuclide (a (γ, n) reaction). An exceptional situation exists in the case of ^{3}H and ^{9}Be. These nuclides have photodisintegration thresholds of 2.23 and 1.67 MeV, respectively, with respect to γ-ray energy. Thus a convenient neutron source can be prepared by the use of ^{9}Be and ^{124}Sb, a γ-emitter having a half-life of 60.9 days. The neutron yield realized from this source can be as high as 3.2×10^{6} neutrons/sec/Ci.

b. Bragg Scattering (diffraction). Low-energy γ-rays may be scattered by a crystal lattice with no loss of energy. The diffraction of X rays has been used effectively in the study of molecular structure; the phenomenon is, however, of little importance in radiotracer methodology.

c. Photoelectric Effect. In the photoelectric effect, the γ-ray interacts with an orbital electron, transferring all its energy to the electron and disappearing in the process (see Figure 3-16). The electron is ejected from the atom with a kinetic energy, E_e, given by

$$E_e = E_\gamma - E_{\mathrm{BE}} \tag{3-8}$$

where E_γ is the incoming γ-ray energy and E_{BE} is the binding energy of the ejected electron. Such a photoelectron will interact with other atoms in its path, leading to further ionization. The photoelectric effect is most likely to occur when the γ-ray energy matches the electron binding energy, since, in the photoelectric effect, the electron must absorb *all* the γ-ray's energy. Furthermore, the most tightly bound electrons will be most likely to absorb

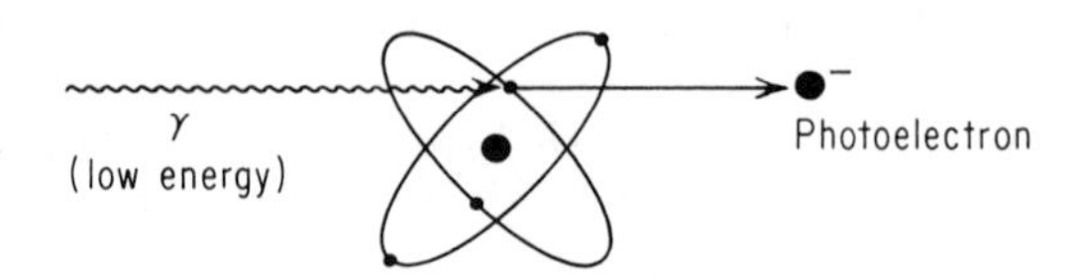

Fig. 3-16. Photoelectric effect.

the γ-ray energy; that is, the photoelectric-effect probability will be greatest for K-shell electrons. Mathematically, we say

$$\text{Overall photoelectric-effect probability} \propto \frac{Z^5}{E_\gamma^{7/2}}$$

where Z is the atomic number of the stopping material and E_γ is the γ-ray energy (see Chapter 7 for further use of this equation). Thus the photoelectric effect is most important at low γ-ray energies ($0 \leq E_\gamma \leq 0.5$ MeV) and is relatively more important in heavy elements. The photoelectric effect is accompanied by X-ray emission and/or Auger electron emission. These radiations are associated with necessary rearrangements in the atomic electrons due to the ejection of one electron.

d. Compton Effect. Gamma rays of medium energy (0.5 to 1.5 MeV) may undergo elastic collisions with loosely bound orbital electrons (see Figure 3-18). In such cases, only a portion of the γ-ray energy is transferred to the electron, which is ejected. The γ-ray photon itself is deflected in a new direction with a reduced energy. These recoil electrons may carry away from such an encounter any amount of energy up to a defined maximum. The spectrum of recoil electrons is shown in Figure 3-17. The maximum energy transfer

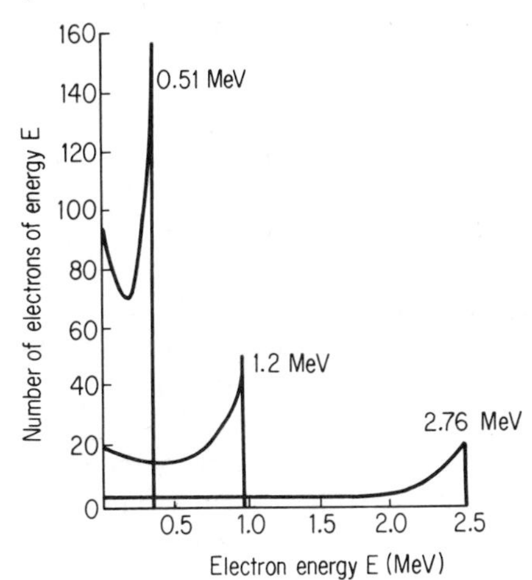

Fig. 3-17. Energy distribution of Compton electrons produced by primary photons whose energies are 0.51, 1.2, and 2.76 MeV.

to the electron occurs when the photon undergoes 180° backscattering, and then the electron energy E_e is given by

$$E_e = \frac{E_\gamma}{1 + 0.511/2E_\gamma} \tag{3-9}$$

where E_γ is the incident γ-ray energy in MeV. Thus Compton recoil electrons appear with a wide energy spread, although they are derived from a monoenergetic beam of incident γ-radiation. Considerable ionization can naturally be realized as these electrons dissipate their energy on interaction with matter. Moreover, the attenuated γ-ray may undergo several more such collisions before finally losing all its energy. The Compton effect is a favored mode of interaction for γ-rays of medium energy interacting with absorbers of medium-to-low atomic number. As in the photoelectric effect, if an inner orbital electron is ejected, X-ray and Auger electron emission will result.

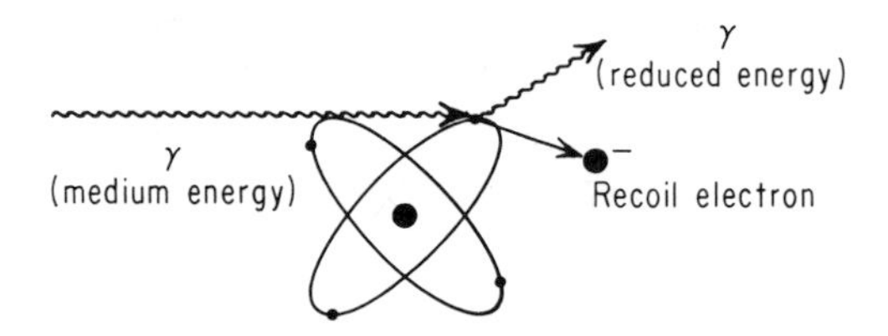

Fig. 3-18. Compton effect.

e. Pair Production. A unique phenomenon may occur when a γ-ray interacts directly with the nuclear force field. In such an event, the photon may cease to exist and may have all its energy converted into two particles, a positron and an electron, which are ejected from the site with varying energy (see Figure 3.19). In order for this process to take place, the incident γ-ray must have an

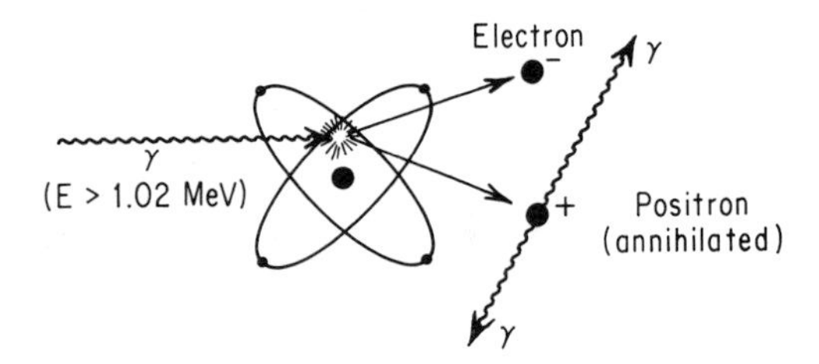

Fig. 3-19. Pair production.

energy equal to, or greater than, 1.02 MeV, which is the energy equivalent to the rest mass of one electron and one positron. This occurrence illustrates the interconvertibility of energy and mass. The γ-ray energy in excess of 1.02 MeV goes into the e^- and e^+ kinetic energies. After slowing down to thermal energies, the positron combines with a nearby electron and undergoes annihilation, with consequent production of two 0.511-MeV photons, which travel away in opposite directions. Both the ejected positron and the ejected electron may cause ionization of atoms near their path. Pair production increases with the square of atomic number of the absorber and linearly with increasing energy of the incident γ-ray photon above the absolute minimum level of 1.02 MeV. Above 4 MeV, the probability of pair production is proportional to log E_γ.

In summary, the last three methods of interaction (photoelectric effect, Compton effect, pair production) are the principal ways in which γ-ray energy is dissipated. The ionization effects of the secondary electrons produced by γ-absorption provide ready detection of γ-radiation. The relative importance

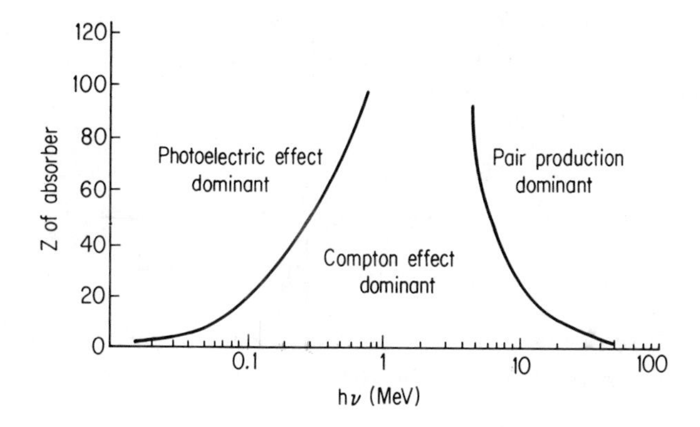

Fig. 3-20. Relative importance of the three major types of γ-ray interaction. From *The Atomic Nucleus* by R. D. Evans. Copyright (c) 1955 by McGraw-Hill, Inc. Used with permission of McGraw-Hill Book Company.

of these methods of interaction for different absorbers and γ-ray energies is shown in Figure 3-20.

3. Absorption Relations

a. Linear Absorption Coefficient. Since the absorption of γ-radiation is exponential in nature, γ-rays have no clear-cut range. This situation is in contrast to α- and β^--radiation. In order to examine certain quantitative aspects of γ-ray absorption, the ideal absorption situation is pictured (Figure 3-21), in which a collimated beam (all parallel rays) of γ-radiation is incident on a thin slab of absorber material.

As the incident γ-ray beam of intensity I_0 passes through the absorber with thickness x, some of the γ-rays are absorbed. Consider that each photon is independent of all other photons and that the *average* probability that one photon will be taken out of the beam per unit path length is a constant, μ_l, *the linear absorption coefficient.* Then the rate of decrease of intensity of a

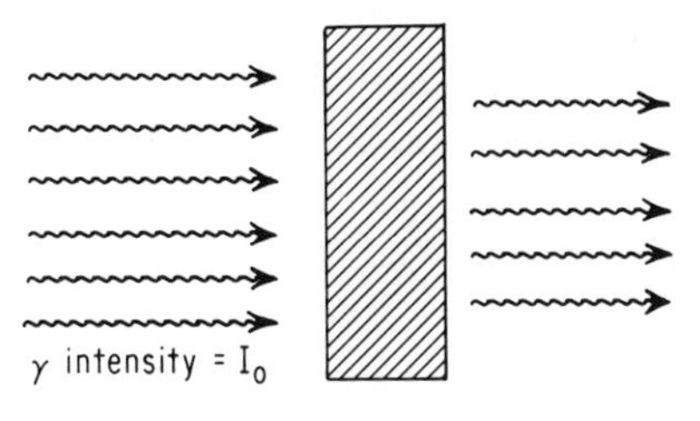

Fig. 3-21. Gamma-ray absorption.

beam of photons, $-dI/dx$, will be equal to (the *average* probability per photon per unit path length of being absorbed) × (the number of photons in the beam). In other words,

$$\frac{-dI}{dx} = \mu_l I \tag{3-10}$$

Integration of this equation gives

$$\ln\left(\frac{I}{I_0}\right) = -\mu_l x \tag{3-11}$$

where I_0 represents the incident γ-ray intensity and I represents the intensity that will emerge from an absorber of thickness x. Rearranging Equation (3-11) gives*

$$I = I_0 e^{-\mu_l x} \tag{3-12}$$

The γ-ray intensity decreases exponentially with absorber thickness, as shown in Figure 3-22. Note that in Equations (3-11) and (3-12), the product $\mu_l x$

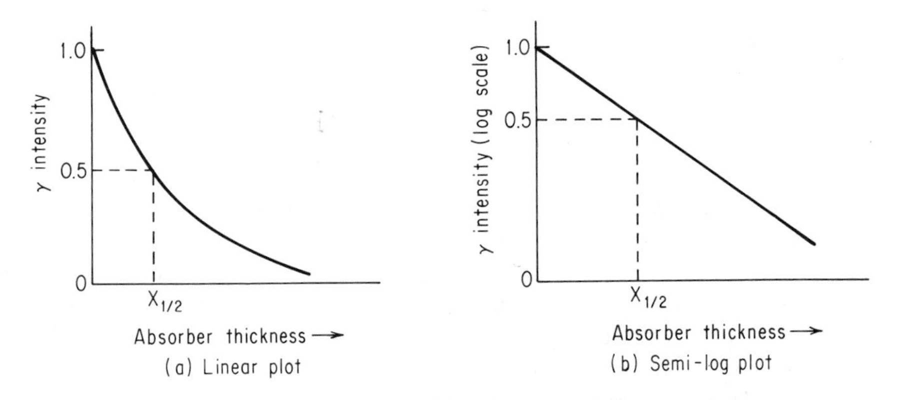

Fig. 3-22. Exponential absorption of γ-radiation.

must be dimensionless. If x is expressed in cm, μ_l must be expressed in cm^{-1}; if x is expressed in feet, μ_l must be expressed in $feet^{-1}$. The linear absorption coefficient is commonly expressed in cm^{-1}. The numerical value of μ_l depends on the γ-ray energy and the stopping material. Some typical values are shown in Table 3-4.

As an example of the use of Equation (3-11), consider the following problem: What fraction of a 1.5-MeV γ-ray beam will pass through a 5-cm-

*The reader will note the obvious similarity between this discussion and that concerning radioactive decay. Here μ_l is the average probability per unit x that a photon will be removed from the beam, while λ is the average probability per unit time that a nucleus will decay.

TABLE 3-4

Linear Absorption Coefficients (cm^{-1}) for Selected Absorbers

Incident Gamma-Ray Energy (MeV)	H_2O	Al	Fe	Pb
1.0	0.071	0.168	0.44	0.79
1.5	0.057	0.136	0.40	0.590
2.0	0.050	0.117	0.33	0.504

thick lead brick? From Table 3-4 we note that μ_l for 1.5-MeV γ-rays in lead is 0.590 cm^{-1}. Thus

$$\ln\left(\frac{I}{I_0}\right) = -\mu_l x$$

$$\log_{10}\left(\frac{I}{I_0}\right) = -0.434\mu_l x = (-0.434)(0.590\ \text{cm}^{-1})(5\ \text{cm})$$

$$\log_{10}\left(\frac{I}{I_0}\right) = -1.280 = 8.7197 - 10$$

$$\left(\frac{I}{I_0}\right) = 0.052$$

In other words, $I \simeq 5\%\, I_0$.

b. Mass Absorption Coefficient. As can be seen in Table 3-4, the linear absorption coefficient varies considerably for different absorber materials. Since the absorption of γ-rays is primarily a function of the mass of the absorber, by taking the density of the absorbing material into account, we can arrive at more comparable values of the absorption coefficients for the different absorber materials. Accordingly, we define the *mass absorption coefficient* (μ_m) as the linear absorption coefficient, μ_l, divided by the density of the absorber, ρ, or

$$\mu_m = \frac{\mu_l}{\rho} \quad \text{or} \quad \mu_l = \rho\mu_\mu \tag{3-13}$$

Thus our basic absorption equation becomes

$$I = I_0 e^{-\mu_l x} = I_0 e^{-\rho\mu_m x} = I_0 e^{-\mu_m d} \tag{3-14}$$

where d is the absorber thickness expressed in units of $\rho \cdot x$ (i.e., g/cm^2). Table 3-5 shows that the mass absorption coefficients are nearly the same for γ-rays of the energies listed in different absorbers. The mass absorption coefficients are commonly expressed in units of cm^2/g, or occasionally as cm^2/mg.

TABLE 3-5

Mass Absorption Coefficients in cm^2/g

Incident Gamma-Ray Energy (MeV)	H_2O	Al	Fe	Pb
1.0	0.071	0.062	0.062	0.070
1.5	0.057	0.050	0.056	0.052
2.0	0.050	0.043	0.046	0.046

Example. What fraction of γ-radiation from a 1-MeV beam would pass through 2 cm of lead under ideal conditions? The μ_m value for 1-MeV γ-radiation in lead is 0.07 cm^2/g, and the density (ρ) of lead is 11.3 g/cm^3. Let I_0 be unity.

$$\log_{10} \frac{1.00}{I} = 0.434 \times 0.07 \text{ cm}^2/\text{g} \times 11.3 \text{ g/cm}^3 \times 2.0 \text{ cm}$$

$$\log_{10} 1.00 - \log_{10} \text{I} = 0.687$$

$$\log_{10} I = -0.687 = 9.313 - 10$$

$$I = 0.206, \text{ or } \sim 21\% \text{ of the original } \gamma\text{-ray intensity}$$

c. Half-Thickness. Since Equation (3-14) is identical in form to Equation (2-10) describing radioactive decay, we can use the concept of half-thickness to evaluate the extent of absorption just as we used the concept of half-life to evaluate the extent of radioactive decay. The *half-thickness* ($x_{1/2}$), or "half-value layer (HVL)," is defined as the thickness of absorber material that will reduce the incident radiation intensity by a factor of 2 (see Figure 3-24). Half-thickness is a useful expression in calculating the shielding necessary to reduce γ-ray intensity to desired levels.

Half-thickness can be defined as either linear half-thickness ($x_{1/2}$) or mass half-thickness ($d_{1/2}$). For the reasons outlined concerning the mass absorption coefficient, the mass half-thickness is the more useful unit. The mass half-thickness can be derived from Equation (3-14). If the γ-radiation intensity is to be reduced by a factor of 2, then, according to Equation (3-14),

$$\ln \frac{I_0}{I_0/2} = \mu_m d_{1/2} \tag{3-15a}$$

$$\ln 2 = \mu_m d_{1/2}, \quad \text{or} \quad 2.303 \log_{10} 2 = \mu_m d_{1/2} \tag{3-15b}$$

$$d_{1/2} = \frac{0.693}{\mu_m} \tag{3-15c}$$

This result shows that the value of the mass half-thickness can be directly calculated from the mass absorption coefficient.

In γ-ray shielding problems, the use of the mass half-thickness allows rapid calculation of the approximate shielding necessary to reduce personnel exposure to γ-radiation. If one half-thickness will reduce the γ-ray intensity to one-half its original value, then two half-thicknesses will reduce it to one-fourth, and so on. The number of half-thicknesses (N) of absorber required to reduce the γ-ray intensity by the factor of X would be equivalent to a total absorber thickness of D. The value of N can be readily derived from Equation (3-15b). First, by substitution,

$$\ln X = \mu_m D \tag{3-16}$$

Introducing the value of μ_m from Equation (3-15c) gives

$$\ln X = \frac{0.693}{d_{1/2}} D \tag{3-17}$$

but since $N = D/d_{1/2}$, this result can be reduced to

$$\ln X = 0.693N \tag{3-18}$$

Converting to common logarithms and solving for N yield

$$N = 3.32 \log_{10} X \tag{3-19}$$

This equation therefore provides ready means for the calculation of shielding requirements. Thus when it is desired to reduce a given γ-ray intensity to one-eighth of its magnitude,

$$N = 3.32 \log_{10} 8 = 3.0 \text{ half-thicknesses required}$$

d. Dependence on Gamma Energy and Absorber Density. The exponential nature of γ-ray absorption is evident from Figure 3-22(b).

It is quite remarkable that γ-ray absorption should be exponential, since it is actually the result of several distinctly different processes. Of these interaction processes, only three (photoelectric effect, Compton effect, and pair production) are normally significant at the γ-ray energies associated with most radionuclides used in radiotracer applications. Each of these processes is highly dependent on the γ-ray energy involved and the nature of the absorber used. These dependences are clearly seen in Figure 3-23, which indicates the plot of μ_l values against γ-ray energy for the three significant γ-ray interaction processes for two different absorbers, Al and Pb.

Note that for Pb the photoelectric effect is most pronounced at lower energies, the Compton effect is the predominant interaction type at intermediate energies, and pair production becomes an increasingly important factor above the minimum energy of 1.02 MeV. Note that in Al, Compton scattering is the dominant mode of interaction at all energies.

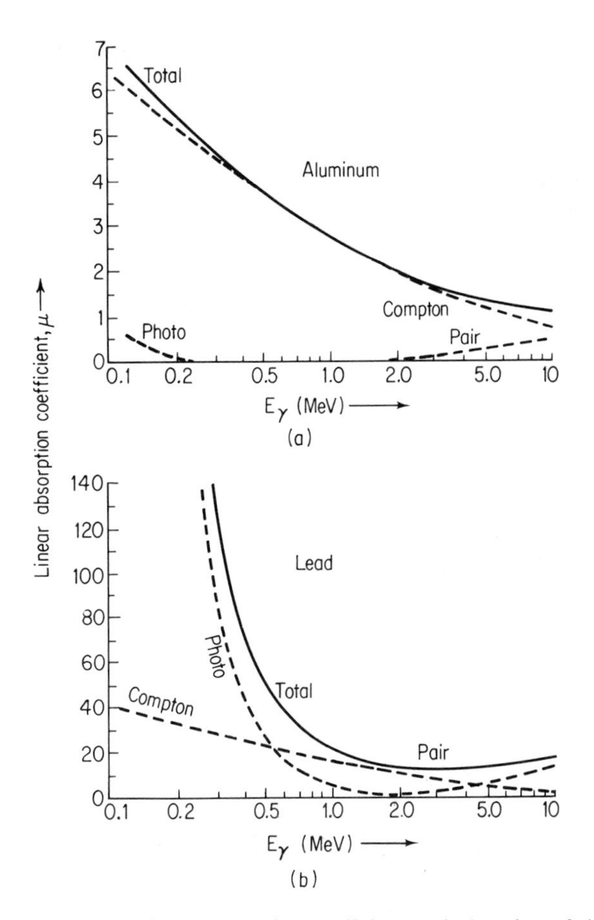

Fig. 3-23. Total linear absorption coefficients μ in function of the γ-ray energy E_γ for γ-rays passing through (a) aluminum and (b) lead. Used by permission of North-Holland Publishing Co.

The dependence of γ-ray absorption on the density of the absorber can be see in a plot of mass half-thickness against incident γ-ray energy in various absorbers (Figure 3-24). Note that for water, aluminum, and iron, the $d_{1/2}$ increases steadily with γ-energy. For lead, however, $d_{1/2}$ increases up to 3 MeV and then begins to decrease, indicating that γ-ray absorption in lead is least efficient at about 3 MeV. This peculiarity is explained by remembering that pair production, which becomes increasingly more operative as a means of γ-absorption with increasing energies, is far more pronounced in absorbers of high atomic weight. Up to 3 MeV, the photoelectric and Compton effects are declining in effectiveness as absorption mechanisms. At this point, however, the effectiveness of pair production offsets the decline of the other two effects and reverses the trend of the curve.

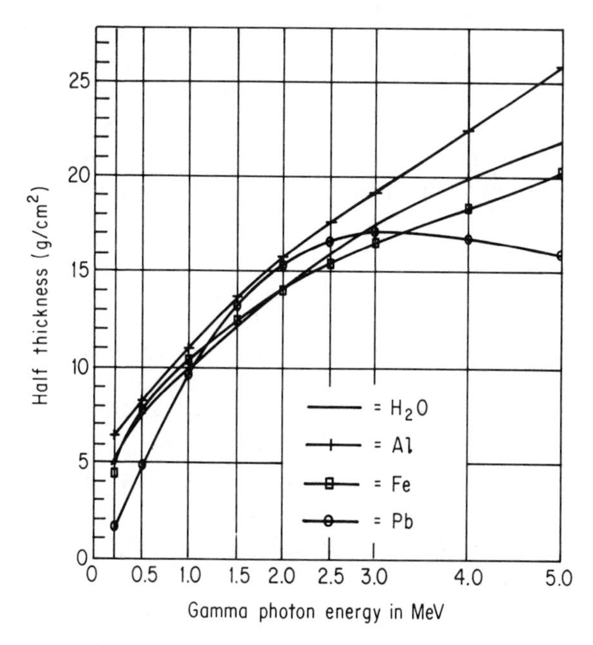

Fig. 3-24. Relation between $d_{1/2}$ and γ-ray energy in various absorbers.

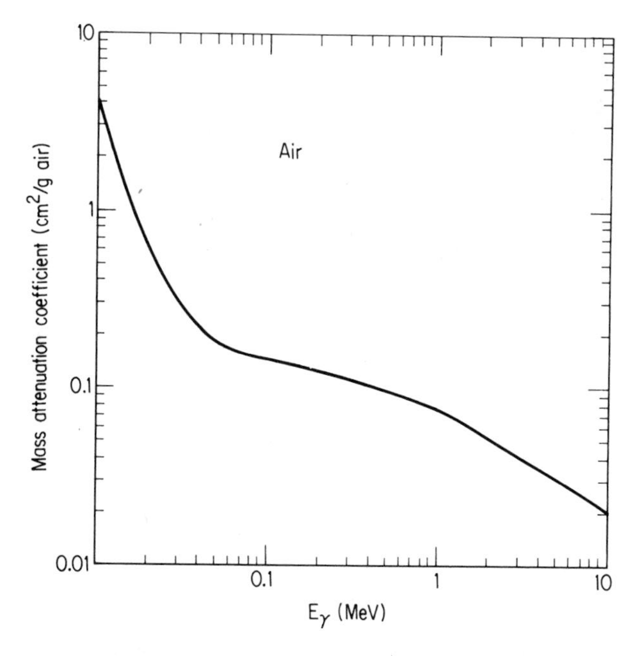

Fig. 3-25. Mass attenuation coefficients for photons in air at STP of volume composition 78.04% N_2, 21.02% O_2, 0.94% Ar, calculated and plotted by Evans from tables of atomic cross sections compiled by White. From *The Atomic Nucleus* by R. D. Evans. Copyright (c) 1955 by McGraw-Hill, Inc. Used with permission of McGraw-Hill Book Company.

Detailed plots of mass absorption coefficients versus γ-ray energy for air, water, aluminum, and lead are shown in Figures 3-25, 3-26, 3-27, and 3-28.

For a more extensive discussion of γ-ray absorption processes, see Fano's two excellent articles (3, 4). Green (5) has prepared a useful nomogram for such determinations, which is shown in Figure 3-29.

In order to use the nomogram, one simply draws a straight line between points on two of the scales and reads off the third scale, the desired quantity. For example, to see how much Pb is necessary to reduce the intensity of 1.33-MeV γ-rays by a factor of 10, one looks at Figure 3-28 and determines that $\mu_m = 0.056\ \text{cm}^2/\text{g}$. Since the density of lead is 11.3 g/cm³, $\mu_l = (0.056)(11.3) = 0.63\ \text{cm}^{-1}$. Locating the points 0.63 on the μ scale and 10% on the intensity scale, we draw a straight line between them and extend it to the thickness scale, reading off a thickness value of 3.7 g/cm² of Pb.

The general features of γ-ray absorption also apply to X-ray absorption. The major difference is that X rays are not emitted from radionuclides with sufficient energy to cause pair production. Because the X-ray energies are generally low, diffraction is a more significant phenomenon. This latter

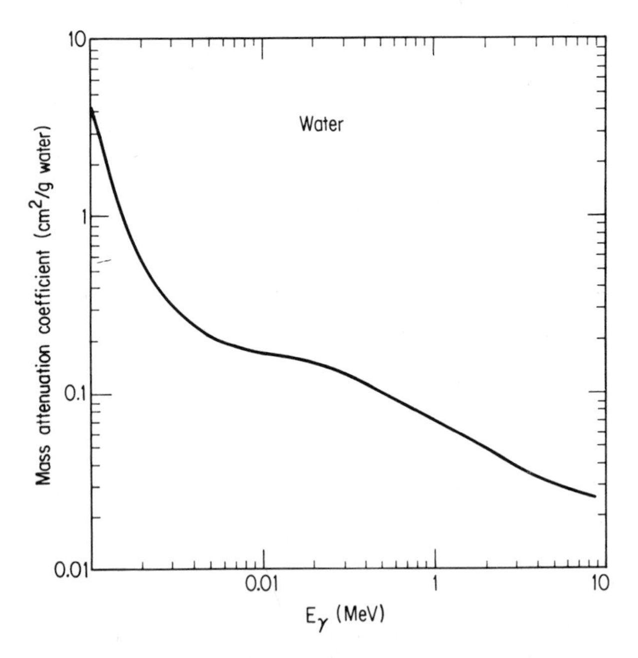

Fig. 3-26. Mass attenuation coefficients for photons in water or biological tissue, calculated and plotted by Evans from tables of atomic cross sections compiled by White. From *The Atomic Nucleus* by R. D. Evans. Copyright (c) 1955 by McGraw-Hill, Inc. Used with permission of McGraw-Hill Book Company.

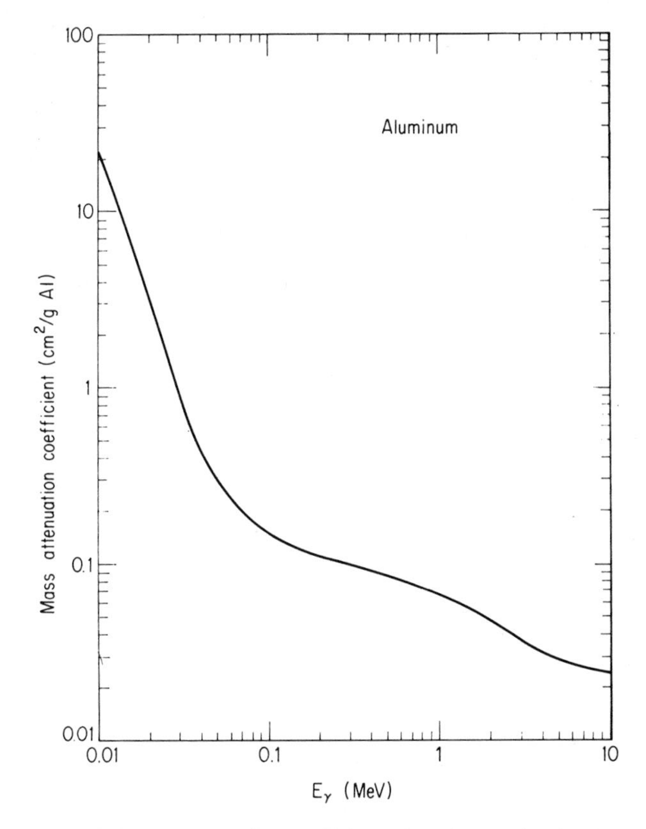

Fig. 3-27. Mass attenuation coefficients for photons in aluminum (density $\rho = 2.70\ g/cm^{-3}$), calculated and plotted by Evans from tables of atomic cross sections compiled by White. From *The Atomic Nucleus* by R. D. Evans. Copyright (c) 1955 by McGraw-Hill, Inc. Used with permission of McGraw-Hill Book Company.

characteristic is utilized in the study of crystalline structure. The diffraction pattern produced when a beam of soft X rays strikes a crystal face permits a very exact determination of the spacing of atoms within the crystal.

4. Practical Considerations

The presence of even moderate quantities of γ-emitting nuclides in a laboratory poses a problem because of the highly penetrating nature of γ-radiation. Generally the radioactive material must be stored in a shielded container of lead or other dense material. If the radiation level is sufficiently high, manipulation of the γ-emitting material may require shielding and/or remote control apparatus of a simple type. The external hazard from γ-emitting radionuclides is in contrast to that from α- and β^--emitters. The

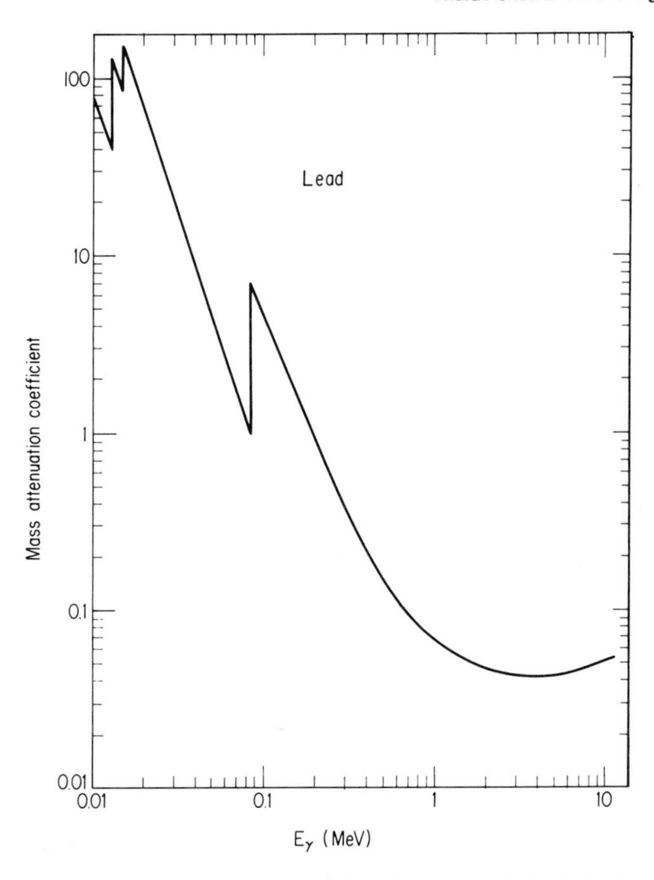

Fig. 3-28. Mass attenuation coefficients for photons in lead (density $\rho = 11.35$ g/cm^{-3}), of isotopic composition 1.5% $^{204}_{82}$Pb, 23.6% $^{206}_{82}$Pb, 22.6% $^{207}_{82}$Pb, 52.3% $^{208}_{82}$Pb, calculated and plotted by Evans from tables of atomic cross sections compiled by White. The characteristic K and L edges associated with the photoelectric effect can be observed at low energies. From *The Atomic Nucleus* by R. D. Evans. Copyright (c) 1955 by McGraw-Hill, Inc. Used with permission of McGraw-Hill Book Company.

entire volume of body tissue can be irradiated from an external γ-source. When taken internally, γ-emitters can produce essentially whole-body radiation effects, regardless of the localization of the material. The presence of γ-emitting nuclides in the laboratory also tends to increase the normal background radiation level, which prolongs the necessary counting time for a desired accuracy in radioactivity assays (see Chapter 12).

In detecting γ-rays, interaction is such that the gas-filled ionization chambers used to detect α- and β^--radiation are most inefficient. It is desirable to use detectors of great density for maximum absorption. Sodium

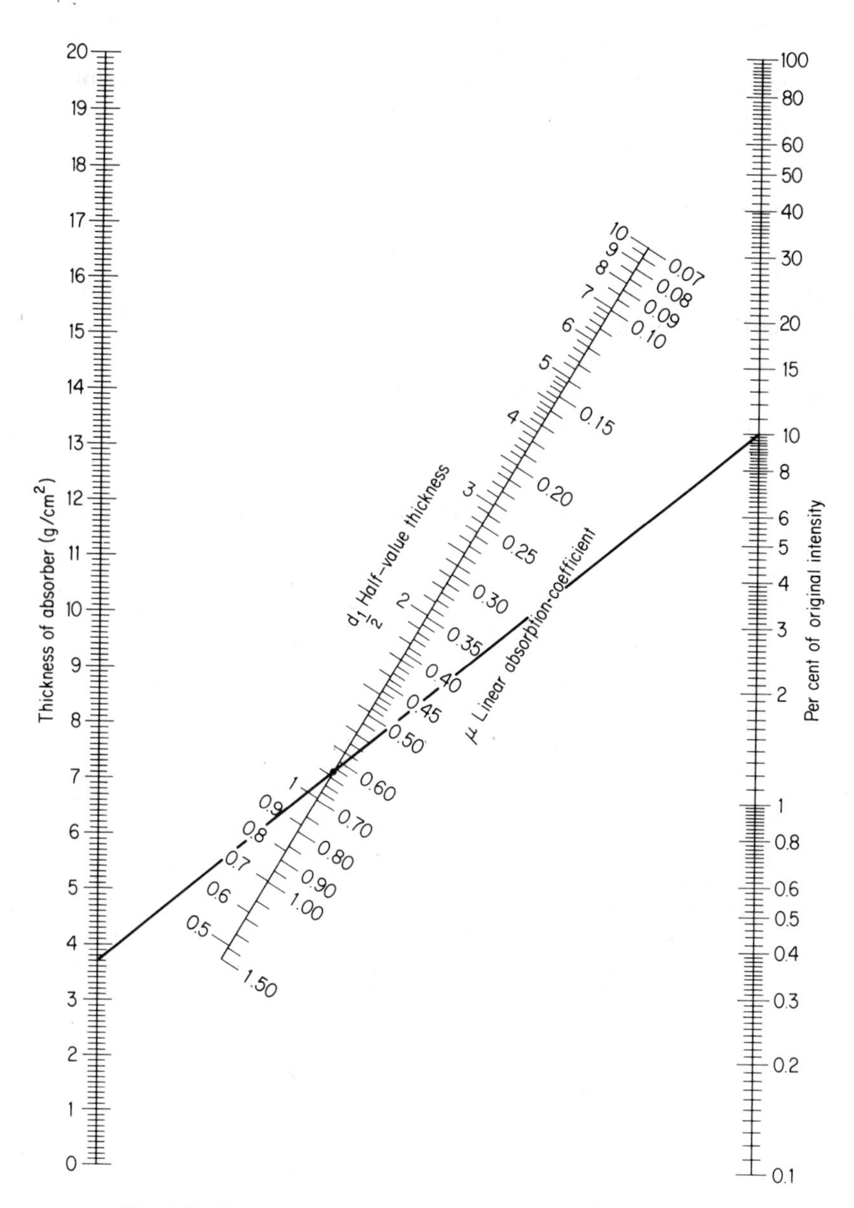

Fig. 3-29. Nomogram for absorption of monoenergetic X rays and γ-rays by lead. From M. H. Green. "Data Sheet No. 34," *Nucleonics* **17**, 77(October 1959).

iodide crystals are most commonly employed for this purpose. The excitation energy from absorbed γ-ray quanta in such a crystal is emitted as minute flashes of visible light (photons). These scintillations may be converted to an electron flow, which can be amplified and measured (see Chapter 6). More recently, semiconductor detectors are being employed (see Chapter 8).

D. SUMMARY

It will be recalled that α-, β- and γ-radiation all lose energy in passing through matter by means of ionization. The range of α-particles is well defined, that of β-particles not as well defined, and γ-radiation has no clearly defined range but is continuously and exponentially absorbed. Alpha particles have very short ranges in matter, β-particles have ranges of intermediate length, and γ-rays are very penetrating.

BIBLIOGRAPHY

1. Williamson, C. F., J. P. Boujot, and J. Picard. "Tables of range and stopping power of chemical elements," *Saclay Report CEA-R-3042*. A definitive compilation of data on the stopping of heavy charged particles in matter.
2. Marmier, P., and E. Sheldon. *Physics of Nuclei and Particles*. Vol. I. New York: Academic, 1970. One of the best, more advanced treatments of the interaction of radiation with matter.
3. Fano, U. "Gamma-ray attenuation." Part I. Basic processes. *Nucleonics* **11**(8), 8–12 (1953a).
4. Fano, U. "Gamma-ray attenuation." Part II. Analysis of penetration. *Nucleonics* **11**(9), 55–61 (1953b). Two excellent, simple articles on the interaction of gamma radiation with matter.
5. Green, Marvin H. "Absorption of mono-energetic X- and γ-rays," *Nucleonics* **17**(10), 77 (1959). Source of Figure 3-29.
6. Katz, L., and A. S. Penfold. "Range-energy relations for electrons and the determination of beta-ray end-point energies by absorption." *Rev. Mod. Phys.* **24**(1), 28–44 (1952). A definitive work on the ranges of monoenergetic electrons and β^--particles.
7. Chase, G. D., and J. L. Rabinowitz. *Principles of Radioisotope Methodology*. 3rd ed. Minneapolis: Burgess, 1967. An excellent discussion of Feather analysis on pp. 208–214.
8. Marshall, John H. "How to figure shapes of beta-ray spectra," *Nucleonics* **13**(8), 28(1955).

4

Nuclear Instrumentation

A. INTRODUCTION

Most radiation detectors produce an electrical signal, which must be processed in order to give meaningful information about the radiation being detected. This signal processing usually consists of some combination of three basic operations:

1. *Amplification*—making the signal bigger in magnitude (i.e., changing an 0.1-volt pulse into a 5-volt pulse).
2. *Shaping*—changing the "shape" of the electrical signal. A detector's electrical signal consists of an electrical charge that increases and then decreases with time. Examples of detector signals are shown in Figure 4-1. Pulse shaping, whose effect is shown in Figure 4-1(b), changes the time behavior of the signal.
3. *Analysis*—sorting of the pulses by their height or area, and so forth.

This amplification, shaping, and analysis of the detector's signal is done by electronic circuitry.

In order to perform radiotracer experiments effectively, it is highly desirable to understand the general purpose and operation of each electronic component in the counting system connected to the detector. Most counting-system failure and malfunction occur in the electronic components, not in the radiation detector. The experimenter who does not generally understand how his electronic apparatus operates is at a disadvantage in discerning

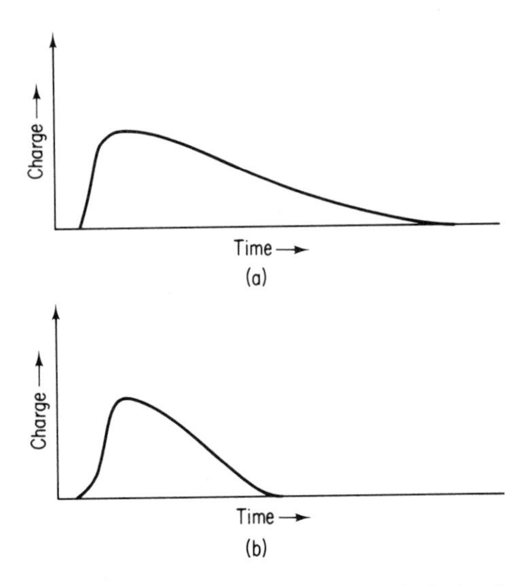

Fig. 4-1. (a) Typical unshaped detector electrical signals and (b) the same signals after pulse shaping.

whether a significant equipment malfunction has occurred. Furthermore, he will be handicapped in efforts to repair the equipment. Even if a skilled electronics technician is available to aid in equipment repair, he may not understand the problem by examining the experimental data, unless you, the experimenter, can relate the equipment malfunction to him in electronic terms.

All the electronic equipment that extracts information from the detector signals is classified into certain general types by the function it performs. Consider a typical set of electronic apparatus as shown in Figure 4-2. Electrical signals from the *detector* pass to a *preamplifier*, where a preliminary shaping and amplification take place. From the preamplifier, the signals go to an amplifier, where final shaping and amplification are done. On leaving the amplifier, the signals may go to a *pulse height selector*, which electronically selects pulses of a certain size and records the number of such pulses with a

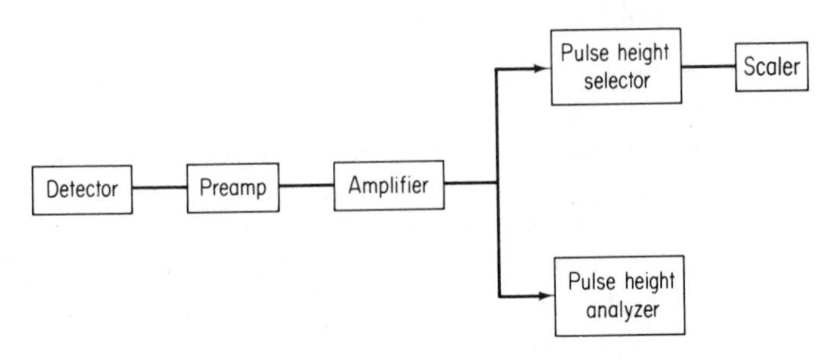

Fig. 4-2. A schematic view of the functions of nuclear electronics.

scaler. Alternatively, the signals may pass to a *pulse height analyzer*, where they are sorted in groups by height and the number of pulses in each group is recorded.

Many of these functions may be performed by components contained in a single box. For pedagogical purposes, and because there is a growing trend toward the use of linked sets of single-function boxes, we shall consider each functional unit separately. Current instrumentation utilizes transistorized electronics rather than vacuum-tube electronics, because of the improved reliability, small size, and low-heat dissipation associated with the former. In the 1970s the use of integrated circuit devices has made extreme miniaturization possible. Let us consider the various functional electronic components used in nuclear-radiation counting systems.

B. THE PREAMPLIFIER

The interaction of radiation (whether α, β, or γ) with the radiation detector creates ionization (as described in Chapter 3). The detector is basically a device that senses the ionization created in the detector and converts it into a pulse of electrical charge, q. (The charge q changes as a function of time and thus is most properly denoted as $q[t]$.) The magnitude of the charge, $q(t)$, is in the range 10^{-10} to 10^{-15} coulomb/pulse. The preamplifier produces a voltage signal $V(t)$ by passing the electrical charge from the detector $q(t)$ to a capacitor C. Elementary electrostatistics tell us

$$V(t) = \frac{q(t)}{C} \qquad (4\text{-}1)$$

where $V(t)$ is the voltage in volts, $q(t)$ the charge in coulombs, and C the capacitance in farads. Typical detector signals are shown in Figure 4-3, along with typical preamplifier output signals. The preamplifier signal characteristics for various detectors are also tabulated in Table 4-1. Note that the magnitude and shape of the pulses are strong functions of detector type. We will show presently how these factors will influence possible uses of the detectors.

The preamplifier is usually located very close to the detector and is connected to it by a short, electrically shielded cable. The reason for this arrangement is to minimize the distortion of the signal entering the preamplifier by electrical "noise" in the cable. Electrical "noise" is broadly defined as any signal in the equipment not due to nuclear radiation and its interaction with the detector. Some common sources of noise include vibration of circuit components, random thermal agitation of electrons in the circuits, and unwanted discharge of electrons by various circuit components. These sources of noise introduce small signals into the circuit that can add to and distort the main detector signals or that can be analyzed themselves. The preamplifier changes the shape of the detector signal to allow the circuits in the amplifier

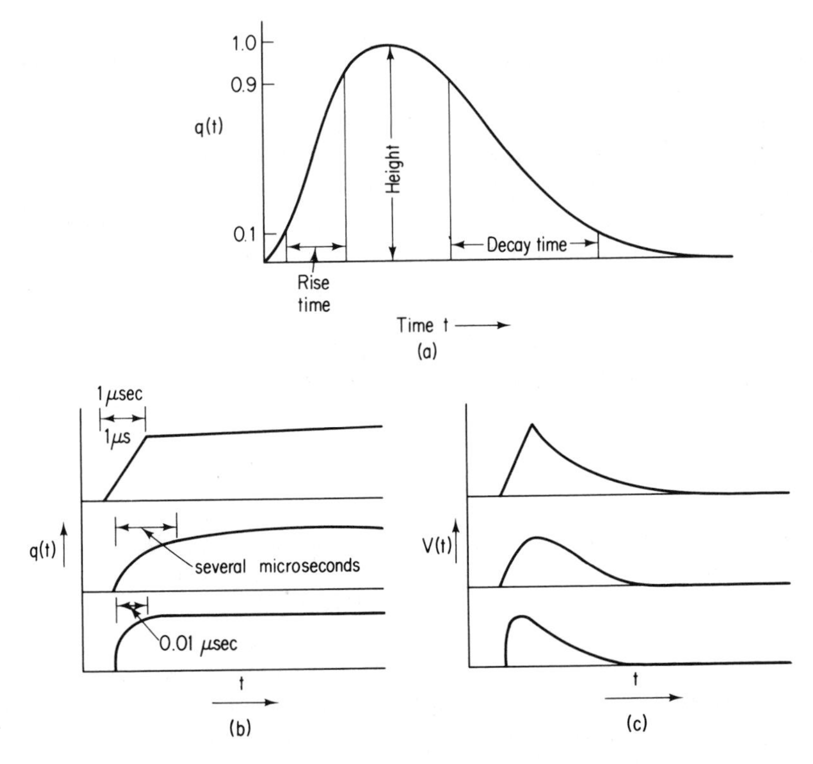

Fig. 4-3. (a) Schematic diagram of a typical pulse showing the definition of height, rise time, and decay time. (b) and (c) Typical pulse outputs from various detectors before and after preamplification. Top: parallel-plate ionization chamber; middle: proportional counter; bottom: scintillation counter.

TABLE 4-1

Typical Preamplifier Output Signal Characteristics

Type of Detector	Magnitude (volts)[a]	Rise Time[b]
Geiger-Müller	0–10	Slow
Proportional	0–0.1	Slow
Semiconductor	0–0.025	Fast
Solid scintillation	0–2	Fast

[a]For a capacitance of 20 pF.
[b]See Figure 4-3(a) for a graphical description of this term.

to operate properly (they would not operate properly with the raw detector signal) and amplifies the detector signal to make the *signal-to-cable noise ratio* as high as possible when entering the amplifier.

C. THE AMPLIFIER

The amplifier acts on the signal from the preamplifier to further change its shape and size. The purpose of this additional amplification and shaping is twofold. First, further amplification improves the signal-to-cable noise ratio. Second, further shaping acts to prevent pulse pileup. Since pulses from the radiation detector occur randomly, one pulse from the detector may begin before the preceding detector pulse has terminated. This situation is shown in Figure 4-4 and is called pulse overlap or *pileup*. By shortening the time duration of each amplifier pulse ("clipping" the pulse), pulse overlap can be minimized [see Figure 4-4(b)].

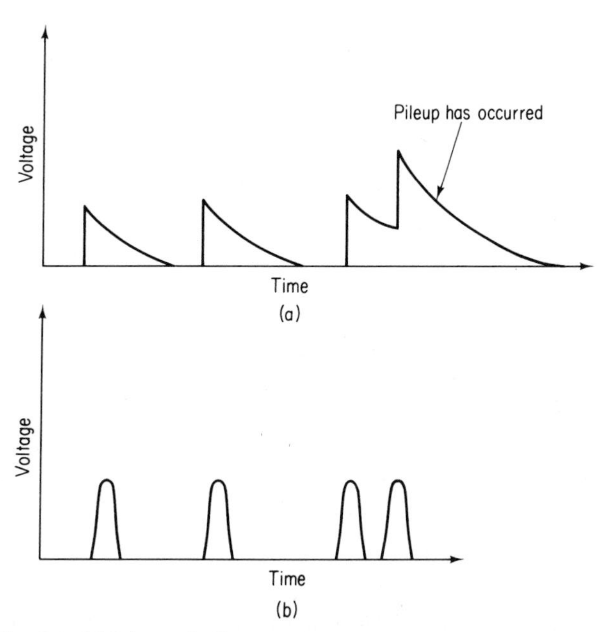

Fig. 4-4. (a) Schematic diagram of typical preamplifier pulses showing the occurrence of pileup. (b) Schematic diagram of amplifier output pulses derived from preamplifier pulses shown in (a). Note how clipping has eliminated the pulse pileup of (a).

The amplification achieved in the amplifier is specified by stating the *gain* of the amplifier, which is formally defined as the ratio of the output pulse height to the input pulse height. The gain of an amplifier is usually the same for all pulses entering the amplifier, regardless of size. When the output pulse height is linearly proportional to the input pulse height, we say that the amplifier is *linear*. In a few special cases, a logarithmic amplifier—that is, an amplifier whose output pulse height is proportional to the logarithm of the input pulse height—is used. In most cases, the gain of the amplifier is selected by the experimenter.

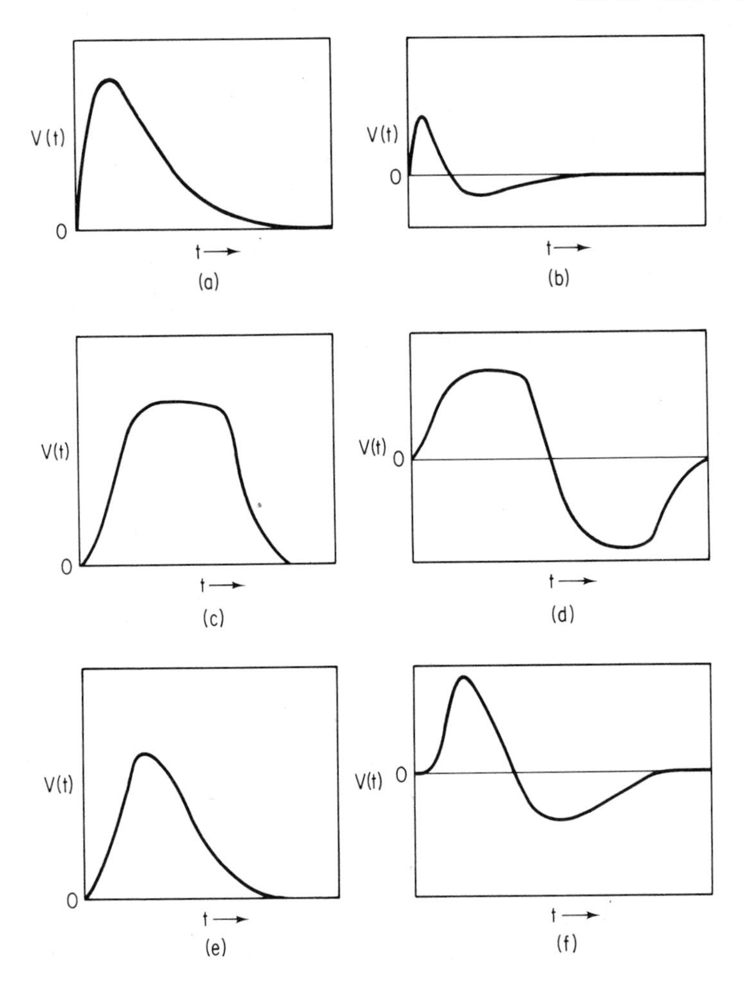

Fig. 4-5. Output pulses from amplifiers using (a) unipolar *RC* shaping, (b) bipolar *RC* shaping, (c) unipolar delay-line shaping, (d) bipolar delay-line shaping, (e) unipolar semigaussian shaping, and (f) bipolar semigaussian shaping. After O'Kelley (2).

The type of pulse shaping employed in the pulse amplifier may also be selected by the experimenter. Typical types of amplifier output pulse shapes are shown in Figure 4-5. The basic types are *RC*, delay line, and semigaussian, which are available in unipolar or bipolar format.* Which shape does one choose? In any counting system operating at count rates of greater than 100 counts per second (cps), when choosing the amplifier pulse shape, there

*The names *RC* (resistance-capacitance) and delay line refer to types of circuit elements used to create these pulse shapes, while the term gaussian refers to the pulse shape achieved. Unipolar means that the pulse has either a positive or negative part, while the term bipolar indicates the presence of both a positive and a negative portion to the pulse (see Figure 4-5).

is always a conflict between increasing the signal-to-noise ratio and preventing pulse pileup. With caution, however, a few general rules can be used to make pulse shape selection easier. These rules are as follows:

1. For scintillation counters, especially those that must operate at high count rates, bipolar delay-line shaping is preferred.
2. For semiconductor detectors, semigaussian pulse shaping is best because one achieves ~18% better energy resolution with semigaussian shaping than with *RC* pulse shaping (see Chapter 7 for a discussion of resolution). If a cheaper amplifier is desired, *RC* shaping can be employed.
3. In general, bipolar shaping gives less pulse overlap but poorer signal-to-noise ratios than unipolar shaping. Conversely, unipolar shaping gives the best signal-to-noise ratio.

D. PULSE HEIGHT SELECTORS

The basic pulse-height selector device is the *discriminator*. It produces a standard-sized pulse every time a signal whose height is greater than some preset height enters the unit. The standard output pulse can be counted with a scaler or used to operate other electronic circuitry. A schematic view of the action of the discriminator is shown in Figure 4-6. Figure 4-6(b) shows a typical set

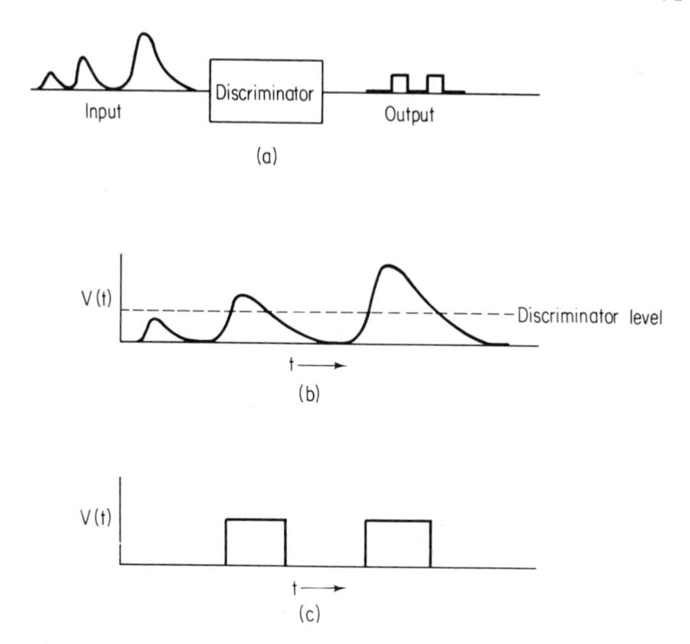

Fig. 4-6. A schematic view of the discriminator. (a) Block diagram, (b) a typical set of input pulses, and (c) discriminator output pulses corresponding to input pulses as in (b).

of amplifier output pulses of varying magnitudes as they enter the discriminator circuit. Each time one of the input pulses exceeds a certain preset voltage, called *the discriminator level*, the discriminator unit will produce a standard output pulse. Note that the discriminator destroys all information about the relative heights of the second and third pulse in Figure 4-6(b). The value of the discriminator level is normally set by the experimenter, using a control on the front of the instrument labeled "DISC" or "BASELINE."

E. SCALERS

Scalers are devices for recording the number of pulses arriving during a fixed time interval. Scaler operation may be controlled by simple manual switches or by means of electronic timers that turn on the scaler for a preset time period. Scalers can be classified into two types—the binary scaler or the decade scaler.

1. Binary Scalers

In a binary scaler, the counting of pulses is done by using binary arithmetic—that is, a number system with base 2. Figure 4-7(a) shows a typical binary

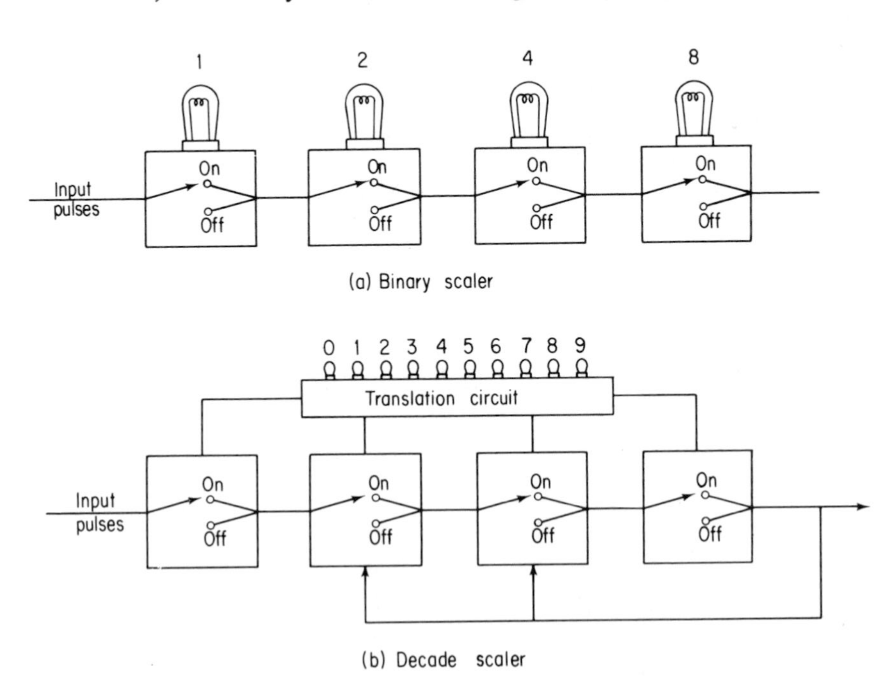

Fig. 4-7. Schematic diagrams of (a) a binary scaler and (b) a decade scaler. See text for explanation.

counting system. Each circuit has two stable states—"OFF" and "ON"—corresponding to the binary numbers 0 and 1, respectively. Consider a group of four pulses striking the scaler shown in Figure 4-7(a). Pulse No. 1 turns the switch in the 1 circuit to the ON position. Pulse No. 2 turns the switch in the 1 circuit to the OFF position and generates a "carry" pulse that turns the switch in the 2 circuit to the ON position. Pulse No. 3 just turns the switch in the 1 circuit to the ON position (thus allowing the 1 and 2 circuits to be "on" simultaneously). Pulse No. 4 turns "off" the switch in the 1 circuit, thereby generating a "carry" pulse that turns "off" the 2 circuit, which, in turn, generates a "carry" pulse to turn "on" the 4 circuit. If we attach light bulbs to each circuit (1, 2, etc.), then, by adding together the numerical values of the lamps that are lit, we can count the number of pulses that have arrived. A string of n circuits allows one to count to 2^n. Clearly, such scalers are awkward to read. Binary scalers are found mainly in older instruments.

2. Decade Scalers

If very large numbers must be counted, the binary system is tedious. Generally it is easier for the experimenter if the number of counts is expressed directly in decimal form, as done by the decade scaler. A schematic diagram of a decade scaler is shown in Figure 4-7(b). It is basically a four-unit binary scaler with some modifications. Pulses Nos. 1 to 7 arrive at the scaler count in the same manner as in a binary scaler. Pulse No. 8 turns "on" the 8 circuit and trips a short circuit that also turns "on" the 2 and 4 circuits. (Thus 8 "looks" like 14.) Pulse No. 9 is counted in the regular manner, turning "on" the 1 circuit. Pulse No. 10 turns "off" all circuits and generates a "carry" for the next decade. A little translation circuit senses the various circuit conditions and causes the appropriate decimal lamps corresponding to the numbers 0 through 9 to be lit. Several decades can be linked together to form a full decimal number. A picture of a typical, commercially available decade scaler is shown in Figure 4-8.

Most scalers have discriminators built into the scaler module. The purpose of the scaler discriminator is to prevent electrical noise pulses from being counted. The most important characteristic of a scaler is its maximum counting rate. Typical values of this quantity for good electronic scalers range from 10^5 to 2×10^7 cps. As a rule, the maximum counting rate of any scaler used should be at least 10^1 to 10^2 times the average sample counting rate encountered. Occasionally one finds scalers with mechanical registers (as shown in Figure 4-8). These registers are adequate for the higher decades (10^4, 10^5, 10^6, etc.) but definitely should not be used for the least significant decades (10^0, 10^1, 10^2, 10^3).* This rule is derived from the well-known pro-

*In a binary scaler, this rule means that mechanical registers should not be used for any register less than 2^{12}.

Fig. 4-8. Front-panel view of a typical, commercially available scaler unit. Photo courtesy of ORTEC, Inc.

pensity of mechanical scalers to jam or skip a digit when operated at high speeds.

F. COUNT RATE METERS

A count rate meter is a device that displays counting rate directly, rather than the number of counts in some fixed time, as a scaler does. It has the advantage of recording the count rate continuously and can be used in conjunction with chart recorders and similar devices so as to provide graphical displays of changes in count rate. On the other hand, the highest accuracy in count-rate measurements is obtained with scaler-timer combinations.

Count rate meters are arranged so that each pulse adds a charge to a capacitor that is leaking charge through a resistor R. If the count rate is steady, an equilibrium situation will eventually be reached, where the rate of charge leakage equals the rate of charge gain from the incoming pulses. The voltage across the capacitor, V, will have some steady value, given as

$$V = nqR \tag{4-2}$$

where n is the number of pulses per unit time and q is the charge per pulse. If a linear meter is used to measure V, a linear relationship will exist between the meter voltage and the count rate. When a wide range of counting rates is to be measured, logarithmic count rate meters are used. A typical commercial device is shown as Figure 4-9.

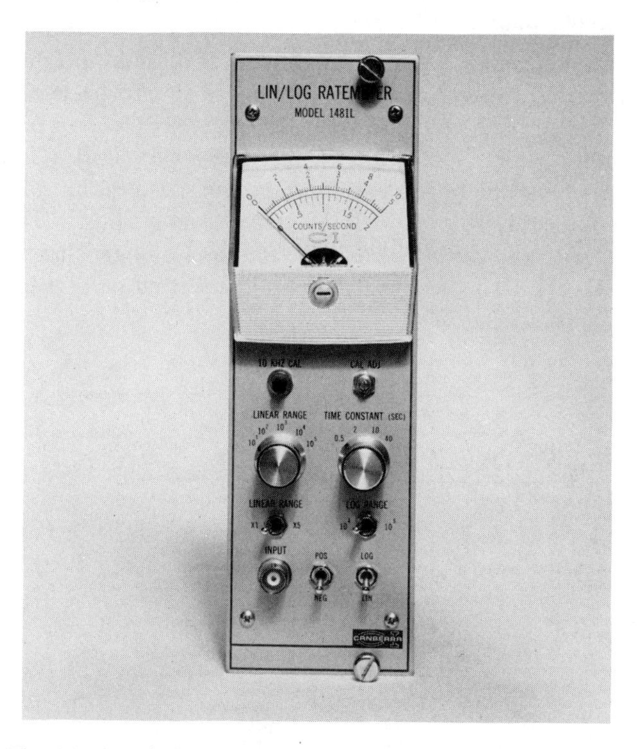

Fig. 4-9. A typical, commercially available rate meter that provides either a linear or a logarithmic presentation of the count rate. Photo courtesy of Canberra Industries.

One parameter that must be selected in using a rate meter is the value of the time constant. The time constant controls the time interval over which the charge is allowed to build up on the capacitor. If the time constant picked is too short, not enough charge will be collected on the capacitor, thus causing the meter readings to be erratic and fluctuating. If the time constant picked is too long, the meter will respond too slowly to changes in the count rate. A happy medium is struck when the value of the meter time constant is ~ 5 times the input pulse width.

G. PULSE HEIGHT ANALYZERS

The discriminator only indicates how many pulses greater than a given height occurred. Often we wish to know how many pulses of a given height or range of pulse heights occurred. Pulse height analyzers give this information. There are two principal types of pulse height analyzers—the single channel and the multichannel.

1. The Single-Channel Analyzer

The single-channel analyzer is the simplest of all pulse height analyzers. A single-channel analyzer gives out one standard-shaped pulse each time an input signal is *above* one voltage level (called the *lower level* or *baseline*) and is *below* another preset voltage level (called the *upper level*). (Occasionally, instead of specifying the upper-level voltage, one specifies the voltage difference between the lower-level and the upper-level voltages. This voltage difference is called the *window width.*) The action of a single-channel analyzer is shown in Figure 4-10. Note that because the second pulse was below the

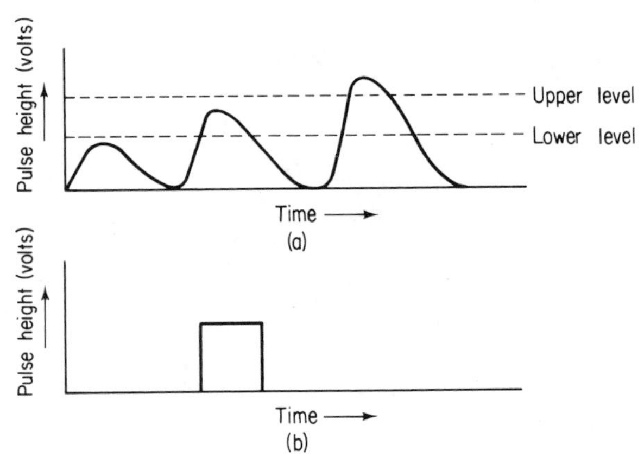

Fig. 4-10. Schematic view of single-channel analyzer action. See text for explanation.

upper level and above the lower level, it produced an output pulse. Since the first pulse was below the lower level and the third pulse was above the upper level, they did not produce an output pulse. Note that the primary difference between a discriminator and a single-channel analyzer is that a discriminator produces an output pulse whenever the input pulse is greater than some given voltage, whereas the single-channel analyzer requires not only that a lower level be exceeded but also that an upper voltage level *not* be exceeded.

In many experiments, the pulses from the radiation detector will form a pulse height spectrum whose features must be discerned. In principle, it is possible to trace out the shape of the spectrum by sequential measurements with single-channel analyzers. However, the use of a single-channel analyzer to measure spectra is tedious, time consuming, and, in some cases, such as rapidly decaying radionuclides, not possible. Thus one is led to the use of a multichannel pulse height analyzer.

2. Multichannel Pulse Height Analyzers

The multichannel pulse height analyzer is a small digital computer used to measure rapidly the spectrum of pulse heights emerging from a nuclear pulse amplifier. It differs from other small computers used in laboratories for data acquisition and analysis in that the machine has been taught, and can only be taught, one set of instructions on how to acquire and analyze data. The multichannel pulse height analyzer costs anywhere from \$2000 to \$50,000 and is a very complex instrument. The basic operation of this machine can be understood by considering a schematic diagram of its principal components, as shown in Figure 4-11, taken from O'Kelley's excellent monograph (2).

The heart of the analyzer is the *analog-to-digital converter* (ADC), which converts the incoming analog amplifier signal to a group of standard-shaped

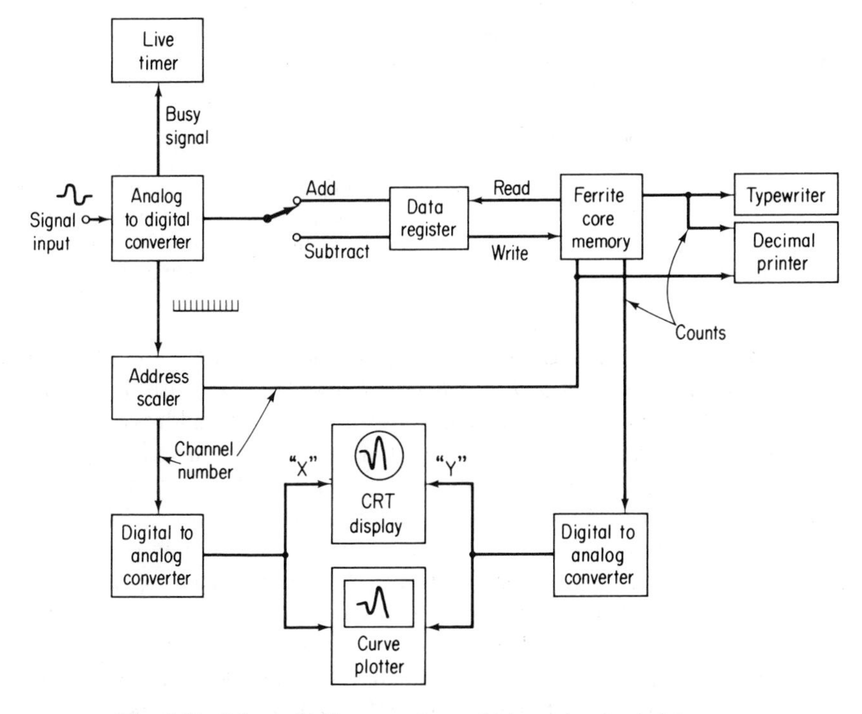

Fig. 4-11. Schematic diagram of a multichannel pulse height analyzer showing relationships between principal subassemblies and accessory equipment. From O'Kelley (2).

pulses. If the incoming pulse is 4 volts high, the ADC might produce 400 standard pulses; or if the incoming pulse was 3 volts high, 300 pulses would be produced. In this way, analog information (the signal height) is converted to digital information (the number of pulses). Thus the unit is called an analog-to-digital converter.

The most commonly used ADC design is that of Wilkinson (10). Figure 4-12 shows how the Wilkinson-type ADC works. First, the input signal is given a flat-top, using a pulse-stretching network. When the input signal reaches its peak amplitude, an oscillator is turned on, thereby producing a train of standard pulses, and a linearly rising voltage is also turned on. When the voltage of the input signal and the linearly rising signal become equal, the oscillator is turned off. Consequently, the bigger the initial signal voltage, the longer it will take the linearly rising voltage to equal the input signal amplitude and the more pulses the oscillator will produce.

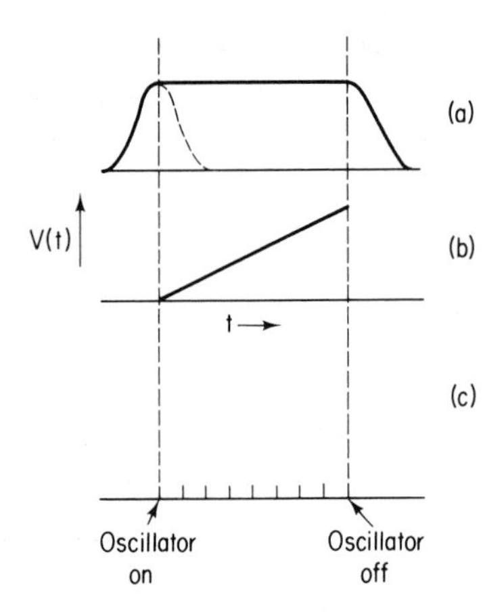

Fig. 4-12. Schematic view of the action of a Wilkinson ADC showing (a) input amplifier pulse after stretching, (b) linearly rising voltage ramp, and (c) oscillator pulses.

The pulses from the ADC are counted by the *address scaler*, and the resulting number is said to be the *address* (*channel number*) of that pulse. The analyzer circuitry then interrogates the address scaler, notes the address, and goes to the analyzer's ferrite core memory unit. In particular, this circuitry goes to the memory location corresponding to the address, reads out the number of counts stored at that location into a scaler called the *data register*, increments the number of counts by one, and restores the new number back into the memory. This memory cycle takes approximately 10 to 30 μsec.

Once pulses have been analyzed in this manner, the analyzer memory will contain the pulse height distribution in the form of the number of counts per memory location corresponding to the number of pulses of a given height. For example, using the data from the ADC example given earlier, where a 4-volt pulse gave 400 pulses out of the ADC, we might have found 1000 counts at location 400 in the memory and 10,000 counts at location 300.

Thus we would know that out of the 11,000 pulses entering the analyzer, 10,000 were 3 volts high and 1000 were 4 volts high. Since, in most experiments, the pulse heights are proportional to the energy deposits in the detector, such a pulse height spectrum corresponds to a spectrum of radiation energies.

In typical multichannel analyzers, the number of memory locations, or *channels*, as they are called, ranges from 200 to 8000, with capacities of 10^5 to 10^6 counts per channel. Because the analyzer is actually a small computer, some analyzers have the ability to subtract, add, multiply, and divide. Therefore, one spectrum can be stored in one portion of the memory, multiplied by some normalizing factor, and then subtracted from another spectrum stored in another portion of the memory. The most common application of this feature is the subtraction of "backgrounds" due to other radionuclides or incidental radiation.

While the analyzer is busy converting and storing a pulse, it is rendered insensitive to new input pulses. Thus it becomes necessary to correct for this analyzer "dead time" when making accurate measurements of the number of counts in a given time. Most analyzers have a feature that does so automatically and that records the time that the analyzer is "live" (i.e., able to accept pulses) during the measurement.

The contents of the analyzer memory may be read out in a variety of ways. A printed digital readout of the number of counts in each channel of the analyzer memory may be obtained by using a typewriter or printer. Information for input to another computer may be obtained in a punched paper-tape format or on magnetic tape. All analyzers can graphically display a plot of the number of counts in a given memory location versus the memory location on a cathode-ray tube. Some analyzers also have curve plotters attached to them so as to make a permanent copy of this graphical information.

H. OUTPUT DEVICES

After being shaped, amplified, and analyzed, the signals from the detectors are recorded by scalers, count rate meters, or multichannel analyzers. Each device furnishes a visual display of the recorded data in the form of a meter reading, a set of lighted lamps, or a cathode-ray-tube display of the stored pulse height spectrum. In most experiments, however, a permanent copy of the data, such as a sheet of paper with numbers or a graph, is wanted. Output devices (i.e., devices that provide permanent copy output) are generally of two kinds—computer oriented or noncomputer oriented.

The noncomputer-oriented output devices include typewriters, teletype units, and high-speed printers (speed of $\geq$20 lines/sec) that provide printed lists in digital form of the data stored in a scaler or multichannel analyzer. Plotters will draw graphs of multichannel analyzer spectra. The computer-oriented devices provide the data in a format for entry into a digital computer and include paper and tape punch systems, magnetic tape systems, and so on. The advantages of the computer-oriented output for handling large quantities of data are manyfold.

I. COINCIDENCE CIRCUITRY

Coincidence circuitry permits the time relationship between two signals to be examined. This factor is useful in studying nuclear events in which two or more simultaneous radiations are given off (such as β^+-annihiliation) or in which two or more sequential radiations occur (such as γ-ray-decay cascade). Coincidence circuits are also used in liquid scintillation counters to reduce the background due to electrical noise originating in the photomultipliers used (see Chapter 9).

A coincidence circuit indicates whether two signals sent into it occurred within some preset time interval, Δt, known as *the resolving time*. The operation of the coincidence circuit is shown schematically in Figure 4-13.

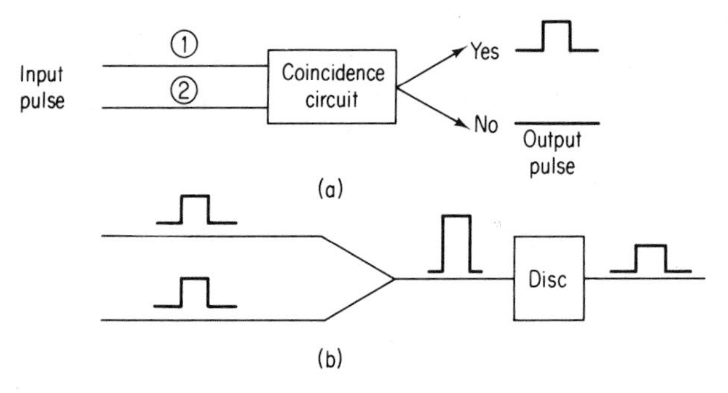

Fig. 4-13. (a) Schematic block diagram of a coincidence circuit and (b) simplified detailed schematic diagram of how a coincidence circuit works. See text for explanation.

Two pulses enter the circuit, as shown in Figure 4-13(a). If both pulses arrive at the circuit within the time interval Δt, the circuit produces a standard output pulse (the YES condition). If the signals did not arrive with time interval Δt, there is no output from the circuit (the NO condition). The details of how the YES–NO decision is made are shown in Figure 4-13(b). Imagine two 1-volt-high pulses arriving at the circuit. If they are in time coincidence, they will add together to give a 2-volt-high pulse. This 2-volt-high pulse goes to a discriminator set at 1.5 volts, trips it, and a standard output pulse is produced. If the original two 1-volt signals were not in time coincidence, they would not add together. When they reached the discriminator set at 1.5 volts, no output pulse would be produced. (Actual coincidence circuits are more complicated than this explanation would indicate, but the principle of their operation is similar.) Coincidence systems are usually classified by the value of the resolving time used. A "slow" coincidence system, such as that used in liquid scintillation counters, has a resolving time of greater than 0.2 μsec; systems with resolving times of less than 0.01 μsec are termed "fast." The intermediate range is termed a "medium-fast" coincidence system.

1. Circuitry Used

In general, several other pieces of electronic equipment are combined with a coincidence circuit to form a coincidence counting system. Figure 4-14 shows a typical block diagram of what a simple coincidence counting system looks like. Assume that both detectors are struck by simultaneous radiations. The signals from each detector are shaped and amplified by their respective preamplifiers and amplifiers. Each single-channel analyzer can be used to set an energy requirement for that portion of the circuit. The signals emerging from the single-channel analyzers are sent to the coincidence circuit. As a result, a coincidence count will mean not only that the two signals occurred within the resolving time Δt but also that they each met energy requirements set by their respective single-channel analyzers.

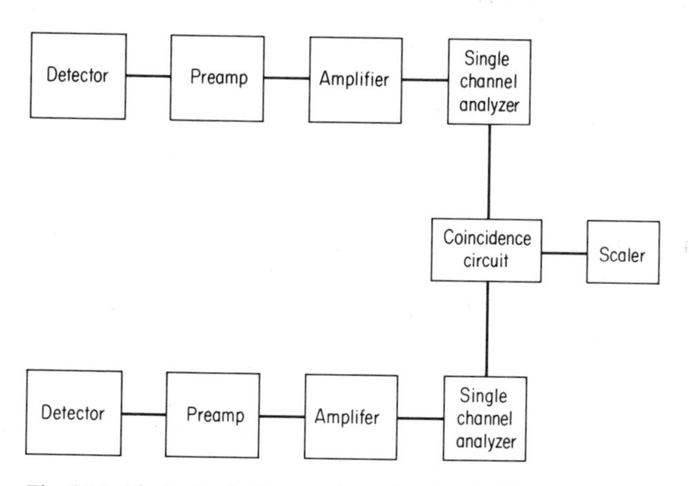

Fig. 4-14. Block circuit diagram for a simple coincidence counting system.

2. Techniques and Problems

a. Chance or Accidental Coincidences. In any coincidence circuit in which noncoincident pulses are present in each half of the circuit, a certain probability exists that by pure chance two random, uncorrelated pulses may enter the coincidence circuit within its resolving time Δt. It can be straightforwardly shown that the rate of such "chance" or "random" coincidences, R_c, is given by the formula

$$R_c = R_1 R_2 \, \Delta t \tag{4-3}$$

where R_1 and R_2 are the count rates (expressed in counts per second, cps) in each half of the circuit and Δt is the resolving time (in seconds). Thus if one has a resolving time Δt equal to 1 μsec and count rates of 10^4 cps in each half of the circuit, the chance coincidence rate will be

$$R_c = (10^4)(10^4)(10^{-6}) = 100 \text{ cps}$$

So, in many cases, the chance coincidence counting rate is appreciable and must be taken into account as a background count in a normal measurement would. Obviously, the smaller Δt is, the smaller the chance coincidence rate is. Hence the importance of using low values of Δt in coincidence counting experiments becomes apparent.

The total coincidence counting rate in any experiment will be the sum of two terms—the "true" coincidence rate due to time-correlated radiations striking the two detectors and the chance coincidence rate due to this accidental overlap in time of the signals. The chance rate can be calculated from a measurement of R_1, R_2, and Δt by the use of Equation (4-3). Thus the true coincidence rate, R_t, is given by

$$R_t = R_{\text{tot}} - R_c \tag{4-4}$$

where R_{tot} is the total coincidence rate. An important characteristic of a coincidence measurement is R_t/R_c, the "true-to-chance" ratio. If the true-to-chance ratio is too low (~ 1), the statistical uncertainty in the true ratio may preclude a meaningful determination of this quantity.

b. Adjusting the Delays in a Coincidence Circuit. When dealing with a slow coincidence circuit, there is little for the experimenter to do except connect up the circuitry and begin taking data. But for medium-fast or fast coincidence circuits, such is not the case. In such systems, it is very unusual to connect the two circuit inputs and find that the cable lengths and other features, were so equal in both halves of the circuit that a perfect coincidence was set up and the apparatus began working immediately. In general, one must adjust the signal transit times in each half of the circuit to be equal. This operation is called "taking a cable curve," after the original method of physically putting large bundles of cables into the circuit to equalize the signal transit times. The procedure to be used in taking a cable curve is as follows:

1. Apply real coincident signals to each half (or "leg," as it is called) of the coincidence circuit.

2. Measure the coincidence counting rate versus the time delay electronically added in one leg of the circuit. The measurement should be similar to the one in Figure 4-15.
3. The optimum electronic delay is the one corresponding to the midpoint of the flat-topped cable curve.

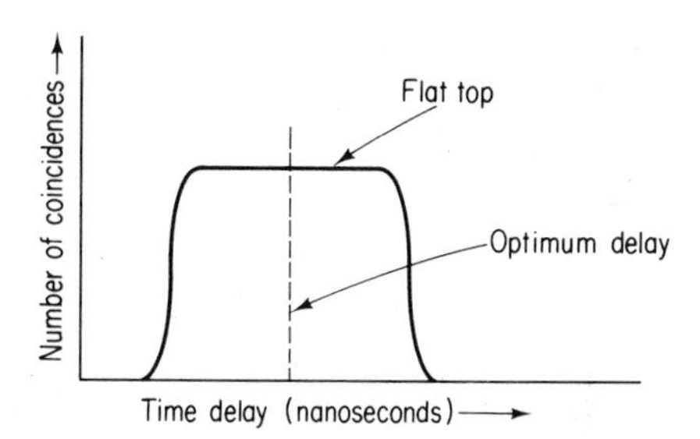

Fig. 4-15. Coincidence circuit cable curve.

c. Sensing the Time of Arrival of a Pulse. Let us explore how the coincidence circuit senses the time of arrival of the pulses. Two common methods for this sensing are used: (a) *leading edge timing* and (b) *crossover timing.* In leading edge timing, a discriminator is set at a very low level, as shown in Figure 4-16(a). When the pulse crosses the discriminator level for the first time, it triggers a circuit that produces a standard pulse marking the time of arrival. This method is an accurate one for sensing pulse arrival times when the range of pulse heights is small. Difficulty occurs, however, when pulses of varying amplitude arrive at the discriminator, as shown in Figure 4-16(b). The two pulses shown in Figure 4-16(b) actually start at the same time, but because of their different pulse amplitudes, they cross the discriminator level at different times. The result leads to an apparent difference in their time of arrival as sensed by the discriminator circuit. This phony time difference

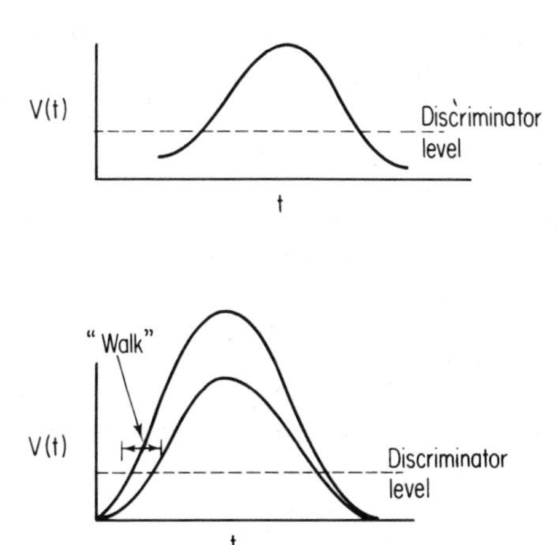

Fig. 4-16. Schematic views of how to sense the arrival of a pulse.

is referred to as the "pulse height walk," or "jitter," of a circuit and should be minimized. (Good coincidence circuits have pulse height walks of <10 nsec for a 10:1 ratio of pulse heights.)

One way that low-pulse height walk is achieved is through the use of crossover timing. In crossover timing, one uses bipolar-amplifier pulse shaping and senses electronically the pulse crossover point (see Figure 4-17). The pulse crossover point is roughly independent of pulse height, and hence walk is minimized. Unfortunately, the use of bipolar pulse shaping causes a deterioration in pulse resolution.

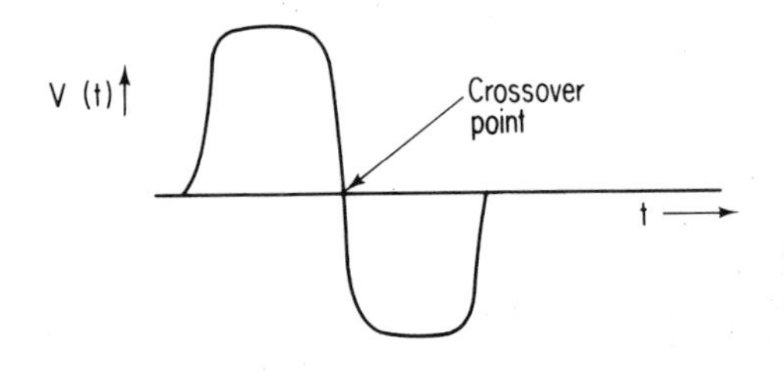

Fig. 4-17. Schematic view of bipolar pulse showing crossover point.

BIBLIOGRAPHY

1. PRICE, W. J. *Nuclear Radiation Detection.* 2nd ed. New York: McGraw-Hill, 1964, Chapter 11. An excellent, somewhat advanced summary of the basic principles of nuclear electronics.
2. O'KELLEY, G. D. *Detection and Measurement of Nuclear Radiation.* National Academy of Sciences Publication NAS-NS-3105, 1962. An excellent, simple discussion of nuclear electronics, on which much of the material in this chapter was based.
3. CHIANG, HAI HUNG. *Basic Nuclear Electronics.* New York: Wiley, 1969. An excellent, very technical, advanced discussion that is somewhat dated.
4. KOWALSKI, E. *Nuclear Electronics.* New York: Springer-Verlag, 1969. A complete, very advanced monograph on radiation detectors and auxiliary circuits, with emphasis on digital circuitry and complex data-acquisition systems.
5. MEILING, W., and F. STARY. *Nanosecond Pulse Techniques.* New York: Gordon and Breach, 1968. A moderately interesting, technical discussion of nuclear pulse electronics.
6. *Semiconductor Nuclear Particle Detectors and Circuits.* National Academy of Sciences Publication 1593, 1969. A conference proceeding that contains many interesting papers on nuclear electronics, including those aspects related to high-resolution semiconductor detector spectroscopy.
7. CHASE, R. L. *Nuclear Pulse Spectrometry.* New York: McGraw-Hill, 1961. A classic, advanced treatment of nuclear electronics that is now somewhat out of date.
8. FAIRSTEIN, E. and J. HALIN. "Nuclear pulse amplifiers," *Nucleonics* **23**(7), 56(July 1965); (9), 81(September 1965); (11), 50(November 1965); **24**(1), 54(January 1966); (3), 68(March 1966). An important series of articles on proper techniques of pulse shaping.
9. OVERMAN, R. T. *Laboratory Manual A.* Oak Ridge, Tennessee. ORTEC Inc., 1968. A very good, very simple discussion of pulse shaping, circuitry, and so on. Well worth reading despite the ORTEC propaganda sandwiched in the discussions.

10. Wilkinson, D. H. *Proc. Cambridge Phil. Soc.* **46**, 508(1950). The article in which the basic principle of the ADC was described.

11. Farley, F. J. M. *Elements of Pulse Circuits.* London: Methuen, 1962. A good, advanced general discussion of circuit theory.

12. Phillips, L. F. *Electronics for Experimenters in Chemistry, Physics, and Biology.* New York: Wiley, 1966. A good, introductory discussion of circuit theory.

13. Delaney, C. F. G. *Electronics for the Physicist.* Baltimore: Penguin, 1969. A slightly more advanced, up-to-date discussion of circuit theory.

5

Measurement of Radioactivity: General Considerations and the Methods Based on Gas Ionization

A. TYPES OF RADIOACTIVITY MEASUREMENTS

In measuring radioactivity, several different types of measurement can be made. A partial list of the factors one might measure is given below:

1. *The type of radiation.* Is the radiation α, β, or γ?
2. *The energy of the radiation.*
3. *The time of arrival at the detector.* Did two different radiations arrive at the same time in one or two detectors? For example, if trying to detect the presence of a nuclide that decays by β^+-emission, one may look for the two 0.511-MeV photons resulting from the β^+-annihiliation in matter. The time of arrival of each photon at a γ-ray detector must be measured in order to verify that both photons came from the same event.
4. *The number of particles per unit time emitted by the source.* This measurement can be made in two different ways—that is, on an *absolute* basis, where every disintegration occurring in the sample is recorded, or on a *relative* basis, where only a fraction of the true disintegration rate of a sample is measured. Relative counting can be used to compare the radioactivity of a series of samples by utilizing identical counting conditions for all samples. The results of absolute counting are expressed as disintegrations per minute (dpm), while the results of relative counting are usually stated as counts per minute (cpm).
5. *The amount of energy deposited by a given radiation in some system per*

unit time. This measurement is called the radiation dose. The radiation dose depends heavily on the energy and type of the incident radiation, as well as the nature of the absorbing material; hence precise determinations are much more difficult to make. Radiation-dose values are commonly given in roentgen or rad units (see Chapter 16).

The measurement of the radiation dose is important chiefly to radiation biologists and health physicists. (Further discussion of this topic will be found in Chapter 16.) Radiotracer work is generally concerned with measurement of property (4), using properties (1), (2), and sometimes (3) to identify the radiation.

B. GENERAL DETECTOR CHARACTERISTICS

Although the various types of radiation detectors differ in many respects, several common criteria are used to evaluate the performance of any detector type. The criteria used for this purpose are as follows:

1. *The sensitivity of the detector*. What types of radiation will the detector detect? For example, solid scintillation detectors normally are not used to detect α-particles because the α-particles cannot penetrate the detector covering.
2. *The energy resolution of the detector*. Will the detector measure the energy of the radiation striking it, and if so, how precisely does it do this? If two γ-rays of energies 1.10 MeV and 1.15 MeV strike the detector, will it be able to distinguish between them?
3. *The time resolution of the detector or its pulse-resolving time*. How high a counting rate will be measured by the detector without error? How accurately and precisely can one measure the time of arrival of a particle at the detector?
4. *The detector efficiency*. If 100 γ-rays strike a detector, exactly how many will be detected?

Each detector discussed here and in the following chapters will be evaluated using these basic criteria.

C. BASIC PRINCIPLES OF RADIATION DETECTION

In a previous chapter we saw that α-, β-, and γ-radiation can interact with matter in a variety of ways. Of these types of interaction, the ones most commonly utilized in detecting radiation are those that ionize gas atoms, cause orbital electron excitation in solids or liquids, or induce specific chemical reactions in sensitive emulsions. The detectors themselves will be described in detail after a brief survey of the detector types involved.

1. Gas Ionization

Several detector types take advantage of the ionizing effect of radiation on gases. The ion pairs so produced can be separately collected. When a potential gradient is applied between the two electrodes in a gas-filled ion chamber, the positively charged molecules move to the cathode and the negative ions (electrons) move swiftly to the anode, thereby creating a measurable pulse. Such pulses can be readily measured by the associated devices as individual events or integrated current.

2. Scintillation

a. In a Solid Fluor. A portion of the energy of ionizing radiation can be transferred to fluor molecules (i.e., compounds that can give rise to fluorescence) in a crystalline solid. The absorbed energy causes excitation of orbital electrons in the fluor. De-excitation gives rise to the emission of the absorbed energy as electromagnetic radiation in the visible or near ultraviolet region (*scintillations*). It is possible to observe these weak scintillations visually under certain circumstances, but visual observation normally is not a feasible detection method. Instead a photomultiplier tube in close proximity to the solid fluor is employed. In the photomultiplier, the photons are converted to photoelectrons, which are greatly amplified by secondary electron emission through a series of electrodes (dynodes) to a sizable electrical pulse. Thus the original excitation energy is transformed into a measurable pulse. (This means of detection is further described in Chapter 6.)

b. In a Liquid Fluor. This detection mechanism is quite similar in principle to the preceding one. Here, however, the radioactive sample and the fluor are the solute in a liquid medium, usually a nonpolar solvent. The energy of nuclear radiation first excites the solvent molecules. This excitation energy eventually appears as photons emitted from the fluor following an intermediate transfer stage. The photons are detected by means of a photomultiplier arrangement. (Further discussion of the detection method will be found in Chapter 9.)

3. Semiconductor Detectors

In the semiconductor radiation detector, incident radiation interacts with the detector material, a semiconductor such as Si or Ge, to create hole-electron pairs. These hole-electron pairs are collected by charged electrodes with the electrons migrating to the positive electrode and the holes to the negative electrode, thereby creating an electrical pulse. Such pulses contain information on the type, energy, time of arrival, and number of particles arriving per unit time. The important features of semiconductor detectors are their superior energy resolution and compact size.

4. Nuclear Emulsions and Autoradiography

The process involved here is a chemical one. Ionizing radiation from a sample interacts with the silver halide grains in a photographic emulsion to bring about a chemical reaction. Subsequent development of the film produces an image and so permits a semiquantitative estimation of the radiation coming from the sample. When the sample is placed in close contact with the emulsion, so that the resulting image reflects the distribution of radioactivity in a given sample, the process is called *autoradiography*. Historically, this is the way in which radioactivity was first detected (by Becquerel in 1896). Unlike detectors of the gas ionization or scintillation types, which require relatively complex electronic components, nuclear emulsions can be used without specialized apparatus and are adequate for crude detection. (Chapter 10 includes a more detailed discussion of autoradiography and nuclear emulsions.)

D. GAS IONIZATION

As an energetic charged particle passes through a gas, its electrostatic field will dislodge orbital electrons from atoms sufficiently close to its path. In each case, the negatively charged electron dislodged and the more massive positive ion comprising the remainder of the atom form an *ion pair*. The minimum energy (in electron volts) required for such ion pair formation in a given gas is called the *ionization potential*. This value differs markedly for different gases and is dependent on the type and energy of the charged particle. A more meaningful value is the average energy lost by the particle in producing one ion pair (see Table 5-1), which is nearly independent of particle energy and type.

TABLE 5-1

Average Energy Lost by 5-MeV α-Particles in Producing One Ion Pair in Some Common Gases

Gas	Average Energy Lost (eV)
Argon	26.2
Methane	29.2
Oxygen	32.3
Ethyl alcohol	32.6
Air	35.2
Nitrogen	36.4
Hydrogen	36.6
Neon	36.8
Helium	44.4

In a defined volume of gas, the amount of ionization that will occur as a result of the passage of an α-particle, a β-particle, or a γ-photon of the same energy differs strikingly. The α-particle will create intense ionization (about 10^4 to 10^5 ion pairs/cm), whereas the β-particle will produce a rather diffuse ionization (about 10^2 to 10^3 ion pairs/cm), and the γ-ray will give rise to little ionization (about 1 to 10 ion pairs/cm) and that only by secondary mechanisms. Accordingly, it can be seen that α- and β-particles produce sufficiently intense specific ionization for detection, but γ-rays are poorly detected at best by this means.

1. Without Gas Amplification

a. Pulse-Type Ionization Chambers. The essential parts of a gas ionization chamber are the two electrodes insulated from one another, thus defining a gas-filled space between them. A parallel-plate ionization chamber operating in the pulse mode is shown in Figure 5-1. Note that one electrode has been connected to the negative terminal of the voltage source V_p, making it the cathode, while the other electrode acts as the anode.

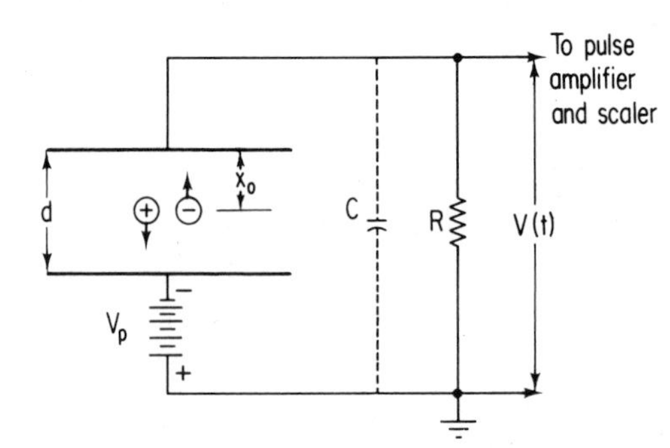

Fig. 5-1. Schematic representation of a parallel-plate ionization chamber in which one ion pair has just been formed. V_p is the voltage source, R denotes resistor, and C denotes capacitor. From O'Kelley (5).

If a 3.5-MeV α-particle traverses the chamber, intense ionization will occur along its short path. Since about 35 eV are expended, on the average, in forming an ion pair in air, the 3.5-MeV α-particle could form approximately 1×10^5 such ion pairs before dissipating all its energy. Because of the potential impressed on the chamber electrodes, these ions migrate rapidly to the respective electrodes. The less-massive electrons move very quickly to the anode and produce a rapid buildup of charge there. (Because the positive ions move about 1000 times slower than the electrons, their effect can be neglected for the moment.) The time for collection of the electron charge is on the order of 0.1 to 1 μsec. The magnitude of this charge due to the electrons can be calculated as follows:

$$\begin{aligned} \text{One } e^- \text{ charge} &= 1.6 \times 10^{-19} \text{ coulomb} \\ 1 \times 10^5 e^- \times 1.6 \times 10^{-19} \text{ coulomb}/e^- &= 1.6 \times 10^{-14} \text{ coulomb} \end{aligned} \tag{5-1}$$

The collected charge flows through the external circuit as a surge, or pulse.

If a 20-pF capacitor is being used, the potential of the pulse, $V(t)$, is found as follows:

$$\frac{\text{Charge in coulombs}}{\text{Capacitor size in farads}} = \text{pulse size in volts} = V(t)$$
$$\frac{1.6 \times 10^{-14} \text{ coulomb}}{20 \times 10^{-12} \text{ farad}} = 8 \times 10^{-4} \text{ volt, or } 0.0008 \text{ volt} \qquad (5\text{-}2)$$

The precise measurement of such minute pulses constitutes a basic difficulty in using the simple ionization chamber for detecting ionizing radiation.

If, instead, a β-emitting source is brought near the ionization chamber, a much smaller degree of ionization will occur along β-tracks in the chamber. It is not possible to calculate the actual number of ion pairs that will be formed for an individual β^--particle and the consequent charge collected, since β^--particles are emitted over a continuous energy spectrum. The mean kinetic energy associated with β^--particles is generally much lower than for α-particles, and the ionization power of the particles (singly charged electrons) is inherently low. As a result, the pulse size produced by an individual β^--particle will be smaller by a hundred- or a thousandfold than that produced by an alpha particle. Gamma rays produce only secondary ionization and have such very low specific ionization that the pulse produced in an ion chamber, before their escape, is very small indeed.

skip The discussion up to now has not been completely accurate in that the effect of the positive ions on charge collection has been totally neglected. In practice, the positive ions are troublesome. Although they move very slowly to the cathode, as they move, they induce a charge on the negative electrode. If no correction is made for this induced charge, the magnitude of the output pulse will depend on the position of the particle track in the chamber volume. A simple method for eliminating this positive ion induction is the addition of a grid to the ionization chamber, as shown in Figure 5-2.

The sample is placed on the cathode or high-voltage electrode. The grid is charged positive with respect to the cathode, but it is less positive than the anode or collecting electrode. The grid acts to shield the collecting electrode from the effects of the positive ions and also accelerates electrons toward the anode.

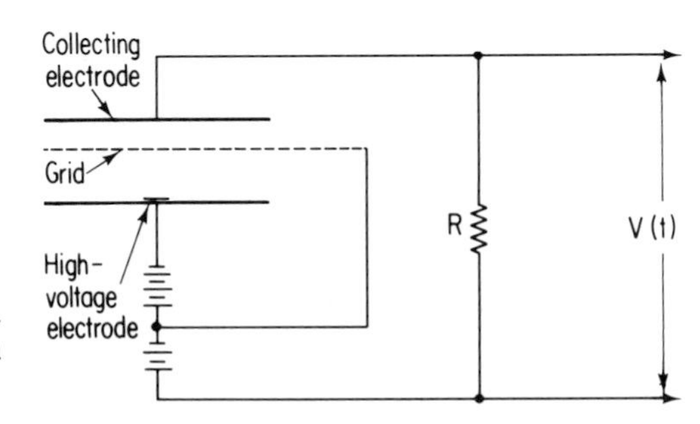

Fig. 5-2. Schematic diagram of a gridded ionization chamber. From O'Kelley (5).

Gridded pulse-type ionization chambers are used in some laboratories for routine analysis of α-emitting radionuclides, particularly in applications where very low disintegration rates are encountered. (Typical counting efficiencies are $\sim 50\%$.) The energy resolution of these counters is poor, however, and they are difficult to operate in a stable manner. Thus, for most purposes, semiconductor detectors are preferred for α-particle assay.

b. Current-Type Ionization Chambers. In many applications, instead of recording pulses from each particle that strikes an ionization chamber, the charge from several events is integrated or added together. The total current from the chamber is then measured as a function of time. A parallel-plate ionization chamber arranged to operate as a current chamber is shown in Figure 5-3.

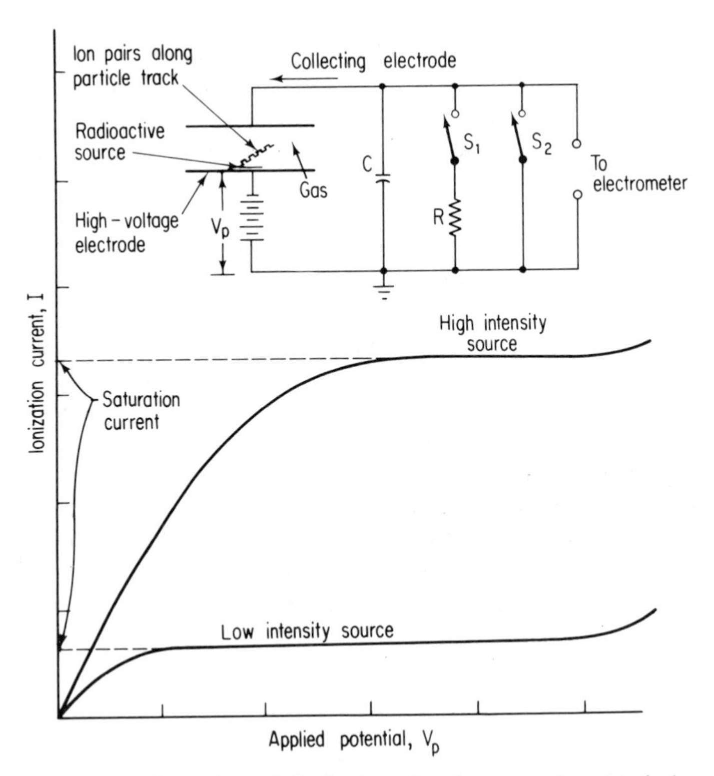

Fig. 5-3. Illustration of ionization chamber operation. Typical current-voltage curves are shown for different source intensities. The insert shows how a parallel-plate chamber is arranged for current measurement by the "IR-drop" method; for measurements by the "rate-of-drift" technique, both switches S_1 and S_2 must be opened. From O'Kelley (5).

So far it has merely been stated that a potential is applied across the electrodes of the ionization chamber. We now examine the relationship between the potential in the ionization chamber and the amount of charge collected on the anode. If no potential is applied, a negligible charge will be collected, although ionization does occur in the chamber. Recombination of the ions will take place rapidly in the wake of the ionizing particle. At only a few volts potential, some ions will be collected, but most will still recombine before reaching the electrodes. Continued increase in potential gradient will result in an increasingly larger fraction of the ions being collected, until eventually all ions are collected and almost no recombination occurs. From this point, an increase of chamber potential over perhaps several hundred volts gives essentially no increase in charge collected. Under these conditions a *saturation current* is realized at the collecting electrode. Ionization chambers are usually operated toward the middle of the saturation current region, so that any fluctuation in the supplied potential will not affect the ion current. (Figure 5-3 shows this relationship.) One advantage of the simple ionization chamber, then, is that a highly stable chamber potential supply is not required.

The numerical value of the saturation current is obtained simply by multiplying the electronic charge in coulombs ($e = 1.60 \times 10^{-19}$ coulomb) by the number of ion pairs created per second, n. n can be calculated from

$$n = \frac{N\bar{E}}{W} \tag{5-3}$$

where N is the number of particles entering the chamber per second, $\bar{E}$ is their average energy, and W is the average energy required to create an ion pair in the chamber gas. Thus if two sources emit the same energy radiation, the measured saturation currents will be proportional to the source strengths. This is the basis for the use of this type of instrument for radiotracer assay. Typical currents lie in the range of 10^{-8} to 10^{-14} amp. Measurement of these tiny currents is commonly done with either the electroscope or the electrometer.

c. Lauritsen Electroscopes. The Lauritsen electroscope is merely a highly refined version of the classic gold-leaf electroscope. In this instrument, the total ionization is integrated over a defined period of time. The details of this type of electroscope are seen in Figure 5-4. Within the ionization chamber, an insulated L-shaped wire supports a delicate gold-plated quartz fiber. A microscope tube is so mounted that the end of the quartz fiber is in focus against a transparent scale. The quartz fiber and supporting wire assembly are charged by an external battery through a charging terminal. When charged, the end of the flexible quartz fiber is repelled by the supporting wire. As a source of ionizing radiation is brought near and ionization occurs within

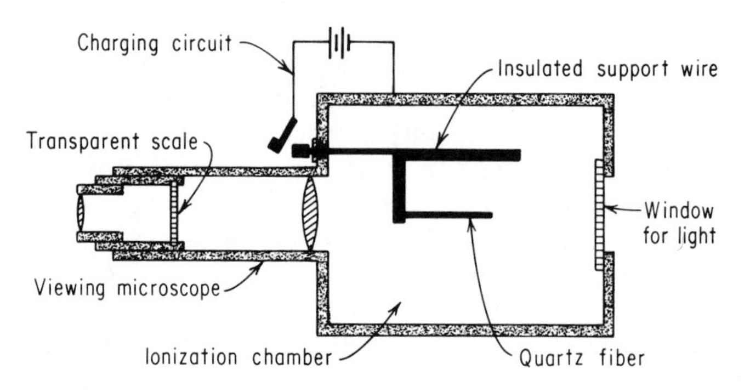

Fig. 5-4. Diagram of a Lauritsen electroscope.

the chamber, the ions are collected on the quartz fiber assembly. This step reduces its charge, and the quartz fiber moves back toward the supporting wire. This motion can be observed against the scale. The rate of the movement is directly proportional to the amount of ionization occurring in the ionization chamber during the observation period.

Electroscopes are not in common use for radiotracer assay because of the length of time required for each determination. Such instruments, however, can measure respectable levels of activity without being overwhelmed (in contrast to the common Geiger-Müller counter) and can be constructed in a readily portable form. Since the electroscope measures total ionization occurring per unit time, it is useful for determinations of radiation dose. The familiar pocket dosimeter, worn by personnel in radiation hazard areas, is essentially a miniaturized version of the electroscope just described.

d. Ion Chamber-Electrometers. The second major type of integrating instrument, the electrometer, amplifies and measures the current from the ionization chamber. Two methods are generally used for current measurement. In the first one, the *IR drop or high-resistance leak* method, the voltage drop V across a large resistor R is measured (see Figure 5-3 for a typical setup). Then the current I is given simply by Ohm's law

$$I = \frac{V}{R} \tag{5-4}$$

If R has the value of 10^{10} ohms and V has the value of 1 volt, then $I = 10^{-10}$ amp. Unfortunately, this technique is not good for measuring currents less than 10^{-12} amp, since the detection of such currents requires the use of precision resistances of greater than 10^{12} ohms, a situation fraught with problems.

For currents less than 10^{-12} amps, which are commonly encountered in radiotracer experiments, the *rate-of-drift* or *rate-of-charge* method is used.

Using the chamber setup in Figure 5-3, switch S_1 is opened, removing resistance R from the circuit. The collecting electrode is then grounded by closing switch S_2, thus making the voltage across capacitor C equal to zero. At the start of the measurement, S_2 is opened; then the voltage after a time t is given by

$$V = \frac{It}{C} \tag{5-5}$$

The time dependence of the voltage is shown in Figure 5-5. Hence the voltage V at any time t is proportional to the current that flowed since the start of the measurement. Because C is of the order of 10^{-11} farads, and because this method allows the electrometer to use all the charge collected in the time t, it is the most sensitive method of ion-chamber-current measurement.

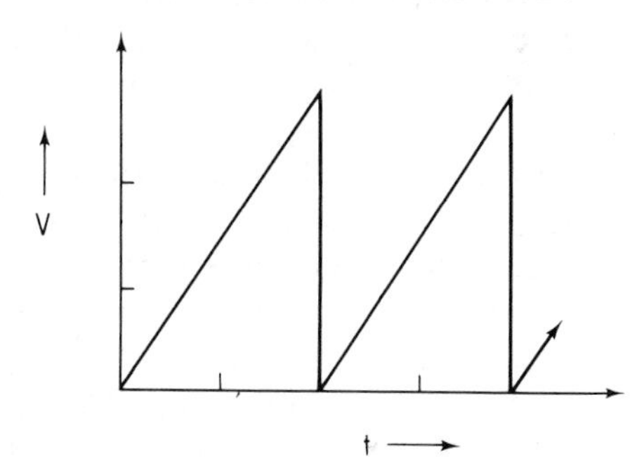

Fig. 5-5. Typical recording of rate-of-charge method.

Both methods described above have a common disadvantage—the problem of amplifying small dc currents. An ingenious instrument that can be used with either of the techniques, the *vibrating-reed* electrometer, circumvents this problem. Here one plate of a capacitor is mechanically driven, causing it to vibrate at a frequency of several hundred hertz. The capacitance becomes an oscillating function of time. Charge from the ionization flows to this continuously varying capacitor, and, consequently, an ac signal is produced at the capacitor output. This ac signal may be easily amplified by stable ac amplifiers and subsequently rectified to operate a recorder. Alternatively, the analog signal can be converted to digital form and recorded on a scaler.

A vibrating-reed electrometer system has several noteworthy features. It allows highly precise measurements to be made over a wider range of sample activities ($5 \times 10^{-5} \ldots 10^3$ μCi) than any other type of detector. This system is sufficiently versatile to accommodate solid, liquid, or gaseous samples. A common practice is to fill the ion chamber with gaseous samples. In practice, these gaseous samples have been almost entirely limited to ^{14}C- or ^{3}H- labeled compounds, since biological samples containing them can usually be readily transformed into a gaseous state. The counting efficiency for stationary gas samples of such low-energy β-emitting isotopes approaches 100% with a 1-liter ion chamber. It is also possible to measure the radioactivity associated with flowing gas samples. Note that the detection efficiency declines sharply with smaller sizes of ion chambers. Here the counting efficiency is heavily dependent on both flow rate and ion chamber volume and is generally somewhat lower than for stationary samples. With liquid or solid samples, it is necessary to cover the samples with an open-end ion chamber, so that the ionization of the chamber atmosphere by the liquid or solid radioactive samples can be measured.

A common use of the electrometer system is to measure $^{14}CO_2$ in the respiratory flow of air from an organism metabolizing a ^{14}C-labeled compound. Brownell and Lockhart (6) have reviewed methods of $^{14}CO_2$ measurement by means of the electrometer; Springell (7) has described ion current measurements of aqueous solutions containing ^{14}C. Tolbert and Siri (8) give the most thorough and practical coverage of electrometer assay of radioactive gases.

In summary, simple ionization-chamber instruments are quite stable and normally require only minimal associated electronic circuitry. Their relatively low-voltage requirements make them well suited for use as battery-operated, portable radiation detectors. As such, they are more frequently used for radiation health purposes than for accurate radiotracer analyses. In general, they are employed not as differentiating pulse detectors but to measure the integrated charge collected, or total amount of ionization occurring.

2. With Gas Amplification

It has been noted previously that the basic problem with the simple ionization chamber is the exceedingly small pulse produced in it by an individual ionizing particle. A high degree of external amplification is required in order to measure these individual pulses. Such amplification is difficult because of the noises and instabilities associated with electronic circuitry. A simple type of amplification within the detector itself would be more desirable, so that the output pulse would already be of measurable size. Fortunately, such amplification is possible.

a. The Nature of Gas Amplification. In a simple ionization chamber operating at the potential of the saturation current, essentially all the ions produced by an incident ionizing particle are collected. This situation is due to the effect of the potential gradient between the chamber electrodes, which accelerates the ions and allows them to be collected at the electrodes before substantial ion recombination can occur. If the potential in the chamber is increased beyond the amount required to produce saturation current, the negative ions (electrons) will be accelerated toward the anode at a rather high velocity. The result will be the ionization of gas atoms in the chamber, similar to the ionization caused by the primary β-particles. The secondary ions so formed are accelerated by the prevailing potential gradient, thereby producing still more ionization. Thus, from a few primary ion pairs, a geometrical increase results in a veritable torrent of ions moving toward the chamber electrodes. The process described is known as *gas amplification*; the flood of ions produced is termed the *Townsend avalanche*, in honor of the discoverer of this phenomenon. As a result of gas amplification, a very large number of electrons are collected at the anode within a microsecond or less from the entrance of a single β^--particle into the chamber. A strong pulse is thereby formed and fed into the external circuit, which can be directly measured after only a low magnitude of amplification.

As the potential gradient between the electrodes of the ionization chamber is further increased, the number of electrons, mostly secondary, reaching the anode rises sharply for a given original ionization event from either an α- or a β^--particle. Eventually a potential will be reached at which the chamber undergoes continuous discharge and is no longer usable as a detector. Figure 5-6 shows that there are three distinct potential regions between the saturation current and continuous discharge—namely, the proportional region, the

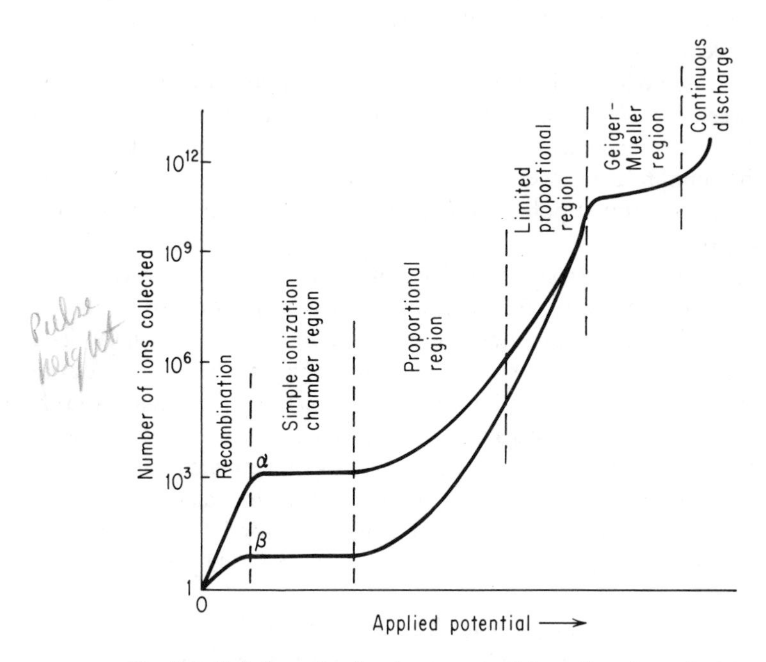

Fig. 5-6. Relation of pulse size to potential gradient in an ionization chamber.

limited proportional region, and the Geiger-Müller region. One should not deduce from the graph that a given ionization chamber can be readily operated in any of these regions by a mere change of potential gradient. Normally chambers are designed to be operated in a particular region under specific conditions of internal pressure and potential. Moreover, the potential gradient needed to define each region depends on many factors, including the filling gas, chamber size, and so on.

b. The Proportional Region. This region is so named because of the proportionality observed between the charge collected and the extent of the initial ionization. In detectors operating in the *proportional region*, the number of ions forming an output pulse is very much greater than, yet proportional to, the number formed by the initial ionization. Gas amplification factors (A values) on the order of 10^3 to 10^4 are generally obtained. The amplification factor is primarily dependent on the composition of the chamber filling gas and the potential gradient. At a given potential, the A value is the same for all ionizing events. Consequently, if an α-particle traversing the ionization chamber causes 10^5 primary ion pairs, with an amplification factor of 10^3, a charge equivalent to 10^8 electrons would be collected at the anode. An incident β^--particle, on the other hand, producing only 10^3 ion pairs, would, after amplification by the factor of 10^3, result in a collected charge equivalent to only 10^6 electrons.

As in the case of simple ionization chambers, then, it is possible to differentiate between α- and β^--particles in the proportional region on the basis of pulse size. This is one advantage of operating a detector in the region. Because the amplification factor in the proportional region is so heavily dependent on the applied potential, it is necessary to employ highly stable high-voltage supplies.

The avalanche of electrons in proportional detectors is collected only on a portion of the anode wire. Furthermore, only a small fraction of the gas volume of the ionization chamber is involved in the formation of ions. These factors result in a very short *dead time*—that is, the interval during which ion pairs from a previous ionization event are being collected and the chamber is rendered unresponsive to a new ionizing particle. Ionization chambers operating in the proportional region are thus inactivated for only 1 to 2 μsec following each ionization event. Dead times as low as 0.2 to 0.5 μsec can be achieved, but if a proportional counter is used for spectroscopy purposes, the average time between pulses should be $\sim$100 μsec or greater.

c. The Limited Proportional Region. The next potential region is similar to the proportional region in operating characteristics with one important exception: the extent of gas amplification possible has an upper limit set by the size of the detector chamber, the form of the collecting electrode, and the number of gas atoms present. As this limit is approached, ionization extends essentially throughout the entire chamber, involving nearly all the available gas atoms, and electrons are collected along the entire length of anode wire. At potentials above this limit, insufficient "ion space" exists in the chamber to accommodate all the ion pairs that, theoretically, could be formed. At first, the result is to limit the amplification possible only for highly ionizing radiation, such as α-particles.

Assume, for example, that the total number of ion pairs that a given chamber can accommodate is limited to 10^{10} ion pairs and that an A value of 10^7 is obtained at the potential being used. An incident α-particle that created 10^5 primary ion pairs could be amplified by a factor of only 10^5, to a maximum of 10^{10} (instead of 10^{12}) ion pairs. Under the same conditions a low E_{max} β^--particle initiating only 10^2 primary ion pairs would be amplified to a collected charge of 10^9 electrons. Consequently, in these circumstances, the proportionality between primary ionization and collected charge is limited, and the distinction between α- and β^--particles is minimized. This situation can be noted by the convergence of the curves for α- and β^--radiation in Figure 5-6. For this reason, the limited proportional region is generally not used for detection purposes.

d. The Geiger-Müller Region. At still higher potential gradients, the A value may reach 10^8. Now even a weak β-particle or γ-ray is able to initiate sufficient ion pair formation to fill completely the available "ion space" in the chamber. Consequently, the size of the charge collected on the anode no longer depends on the number of primary ions produced, and thus, it is no longer possible to distinguish between the various types of radiation. This potential level is called the *Geiger-Müller region*, after the German physicists who first investigated it. Ionization chambers operated in this potential region are commonly called Geiger-Müller (G-M) detectors. Since maximal gas amplification is realized in this region, the size of the output pulse from the detector will remain the same over a considerable voltage range until continuous discharge occurs. This fact makes it possible to use a less-expensive high-voltage supply than that required for proportional detectors.

Dead time: Use of a very high amplification factor in the Geiger region is not without problems. One is the longer dead time of the chamber. Following the passage of an ionizing particle through a detector, an electron avalanche occurs along the entire anode wire, thereby resulting in a cylindrical sheath of positive ions around the anode. The number of such positive ions per pulse will be one or more orders of magnitude greater than in chambers operated in the proportional region. In order to be discharged, this mass of positive ions must migrate to the cathode wall. Being much more massive than the electrons, these ions move at a considerably slower velocity in the electrical field. During this migration, the chamber is unresponsive to any new ionizing particles passing through it. Thus the dead time of a detector operated in the Geiger region is from 100 to 300 μsec or more. The importance of this characteristic in radioactivity determinations is considered in Chapter 13.

Quenching: Another important problem is the perpetuated chamber ionization resulting from complications associated with the discharge of positive ions at the cathode wall of a detector. As a positive ion approaches the cathode, it attracts an electron from the wall and becomes a neutral atom. The newly acquired electron, however, tends to reside in an orbit having a higher energy than that of the electron originally ejected during ionization. The newly acquired electron will soon fall to the ground state orbital in the atom, and the surplus energy so resulting is released in the form of electromagnetic radiation. The magnitude of the energy involvement is generally in the region of ultraviolet or soft X radiation. Such radiations are basically similar to γ-radiation and are capable of creating ionization events via the intermediary formation of secondary electrons. Thus a new avalanche of ion pairs will be formed, and self-perpetuating ionization will prevail in the chamber, thereby rendering the detector unresponsive to incoming ionizing particles.

As a result, we need a means of terminating, or "*quenching*," the perpetual ionization in the detector. Earlier methods for this purpose were in the nature of an *external quench*, which relied on the momentary reduction of the potential gradient across the detecting chamber. This step was accomplished either by a suitable external resistance or by an external electronic circuit that lowered the potential across the electrodes for a sufficiently long period, thus allowing all positive ion discharge and secondary photoelectron production to cease. The entire process was rather slow, however, and accentuated the dead time of this type of detector.

Subsequently, certain gaseous compounds were introduced into the commonly used inert gas atmosphere of G-M detectors to serve as quenching agents. This process is generally know as *internal quenching*, or self-quenching. Two different types of quenching gases are in use today: polyatomic organic compounds and halogen gases.

Choosing ethyl alcohol vapor (one of the most common organic compounds used) as an example, we can follow the chain of events in quenching action. As they move toward the cathode wall, positive ions of the inert filling gas (usually argon, helium, or neon) collide with alcohol molecules.

Since the quenching gas has a lower ionization potential than the inert gas—argon, for example—the net result is that the positive argon ions are discharged into neutral argon atoms, whereas the alcohol molecules are ionized. The positive alcohol ions then migrate to the cathode wall, pick up electrons, and become neutral ethyl alcohol molecules. But instead of releasing their surplus excitation energy as electromagnetic radiation, thus leading to a perpetuated pulse in the tube, the alcohol molecules irreversibly dissociate chemically into degradation products. This is an effective quenching mechanism, but it has the obvious disadvantage that, since a quenching gas molecule is destroyed after each discharge event, the life of the detector is finite. The useful life of such an organic-quenched tube is about 10^8 to 10^{10} pulses.

In halogen-quenched tubes, on the other hand, the dissociated molecules of the quenching gas readily recombine. In this way, quenching occurs without irreversible loss of the quenching gas. Theoretically, at least, the useful life of such a G-M detector is infinite. Bromine or chlorine is usually employed for this purpose, with $\sim 0.1\%$ halogen mixed in neon. Unlike the uniform quenching action in organic-filled detectors, the halogen-quenched G-M tubes may show considerable variation in quenching efficiency for various portions of the chamber volume. This fact limits their use to applications that do not require a high degree of precision, such as radiation monitoring.

3. Proportional Detectors

a. Construction. Detectors made to operate in the proportional region have taken many forms. The commercially produced detectors used in radiotracer assay are generally either the cylindrical end-window type [much like the G-M detector shown in Figure 5-11(a)] or the hemispherical windowed or windowless variety [Figure 5-7(b)]. A variety of specialized proportional detectors has been developed, of which the Bernstein-Ballentine internal gas counter [Figure 5-7(a)] is a good example.

A distinctive characteristic of the proportional detector is the very fine wire used for the anode. It is essential in order to produce a steep potential gradient immediately around the anode for maximal gas amplification. By contrast, the collecting electrode of the simple ionization chamber often consists of a thick rod. The cylindrical form of a proportional detector [Figure 5-11(a)] utilizes a straight-wire anode, whereas in the hemispherical detector [Figure 5-7(b)], the anode wire is shaped into a loop for optimal performance.

Since one advantage of operating a detector in the proportional region is the ability to distinguish α-particles either alone or in the presence of β^--

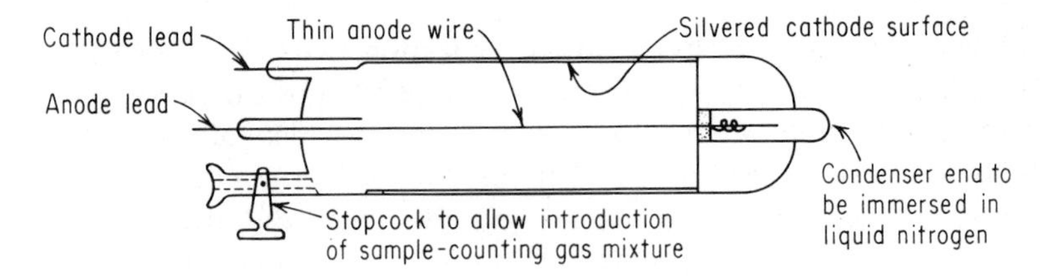

(a) Internal gas counter (Bernstein-Ballentine type)

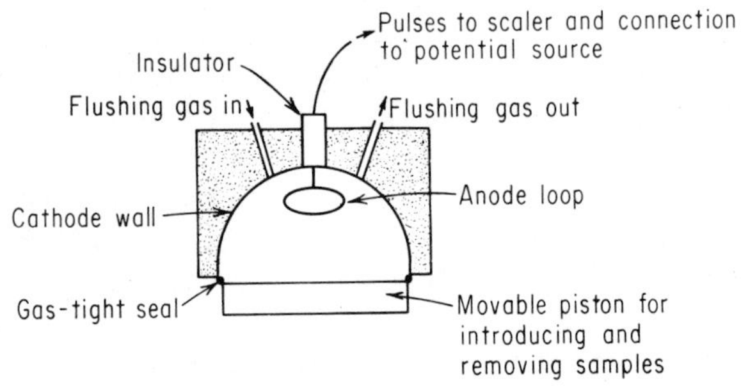

(b) Windowless hemispherical detector
(also used as G-M detector)

Fig. 5-7. Common forms of proportional detectors.

radiation (see Figure 5-6), it is necessary that weakly penetrating α-particles be permitted to reach the sensitive volume of the detector. With the cylindrical variety of detector, a very thin window of split mica or Mylar plastic covers one end of the tube. It can be so thin (down to 150 μg/cm^2) that the absorption of α-particles by the window is not too extensive. An even more efficient arrangement is found with the hemispherical detector, where the radioactive sample can be introduced directly into the detector chamber. In the hemispherical detector, one detects $\sim 50\%$ of all the particles emitted by the source. Such windowless detectors are widely used for α- and weak β^--particle counting.

With either ultrathin end-window or windowless detectors, a certain amount of air leaks into the counting chamber. Both the oxygen and the nitrogen of the air reduce the detection efficiency. Detectors of this variety, therefore, must be purged with an appropriate counting gas before counting is started and must be continually flushed at a lower flow rate during the counting operation. Consequently, such chambers are often called *gas flow detectors*. The operating potential of the chamber is determined, to a large extent, by the gases used for the foregoing purpose. Argon, methane, a 90% argon–10% methane mixture known as P-10 gas, or a 4% isobutane–96% helium mixture known as Q-gas are some commonly used counting gases.

A unique feature of the proportional detector is that a radioactive gaseous sample can actually be incorporated in the detector filling-gas mixture. A detector operated in this manner is known as an *internal gas counter*. The detector designed by Bernstein and Ballentine (9), shown in Figure 5-7(a), utilizes a $^{14}CO_2$-methane counting gas mixture. The gas mixture is introduced into the detector though the stopcock from a vacuum-line manifold. In other internal gas counters, carbon dioxide alone is used as the filling gas.

The electronic instrumentation necessary for the operation of the proportional counter is shown in Figure 5-8. Pulses from the detector pass through a preamplifier and amplifier, where they are shaped and amplified. Emerging from the amplifier, the pulses go to a discriminator. The discriminator is set so as not to trip on noise pulses but to trip on radiation pulses of any size. The number of discriminator pulses produced is recorded by the scaler.

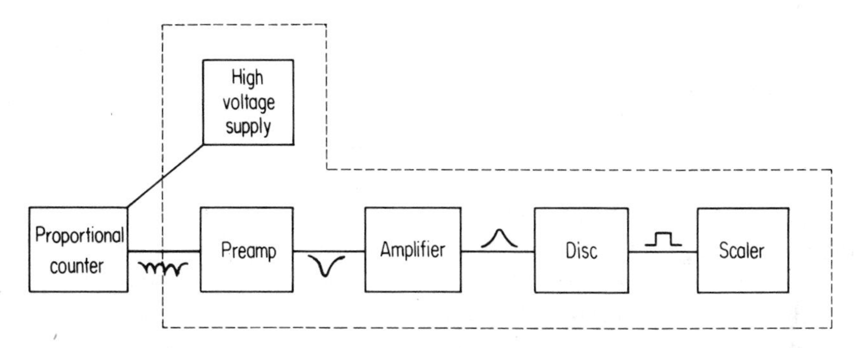

Fig. 5-8. Schematic block diagram of the components of a proportional counter. Typical pulse shapes at various points in the system are shown. The components enclosed within the dotted lines frequently are housed in a single "black box."

b. Operating Characteristics. When the count rate of a sample emitting both α- and β^--particles is determined over the voltage range of a proportional detector and the data are plotted, the results are as seen in Figure 5-9. This curve should be carefully distinguished from the plot illustrated in Figure 5-6, which shows the variation of charge collected for a *single* ionizing particle. The *characteristic curve* for a proportional detector exhibits two plateaus. The plateau at the lower potentials represents α-radiation alone because, at this potential range, only the α-particles, with their much greater specific ionization, produce pulses large enough to trip the discriminator. Not only may the α-particles thus be counted apart from accompanying β^--radiation at this potential, but the background radiation counting rate

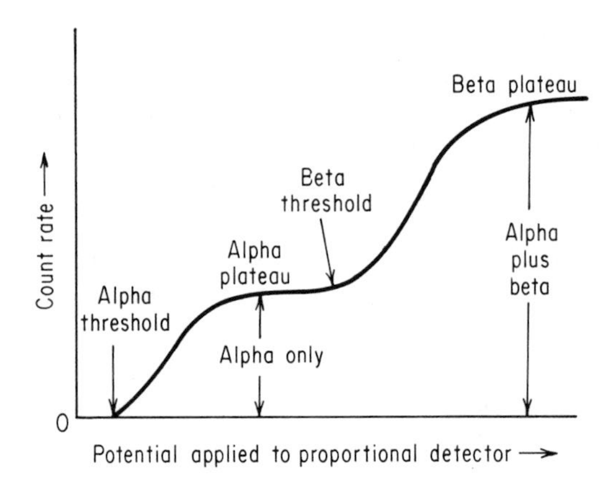

Fig. 5-9. Characteristic curve for a proportional detector.

(primarily cosmic rays and γ-rays) is extremely low—on the order of a few counts per hour—as well.

As the potential gradient is increased, the amplification factor becomes greater. Eventually the primary ions produced by the most energetic β^--particles are amplified sufficiently to produce pulses large enough to be recorded. This point represents the *beta threshold.* Further increase in potential gradient allows even the pulses from the weaker beta particles to be registered. The *beta plateau* has now been reached. The count rate here actually represents alpha plus beta radiation. A good proportional counter has a beta plateau slope of less than 0.2% per 100 volts. The efficiency of the proportional detector for gamma radiation is so low that it is seldom used for gamma counting.

Often, in discussions of the proportional counter, one forgets to mention that the proportional counter is an excellent spectrometer (i.e., energy-measuring instrument) for low-energy radiation. Proportional counters are superior spectrometers when compared to scintillation detectors for the measurement of X rays and electrons with energies in the range of 1 to 100 keV. Managan and Crouthamal have described the construction of a simple, inexpensive high-resolution proportional spectrometer (10). Figure 5-10 shows a typical X-ray spectrum measured with this detector.

4. Geiger-Müller Detectors

a. Construction. This detector is the most widely used of all types of radiation detectors. In general construction, it does not differ significantly from the proportional detectors previously discussed. A G-M tube conventionally

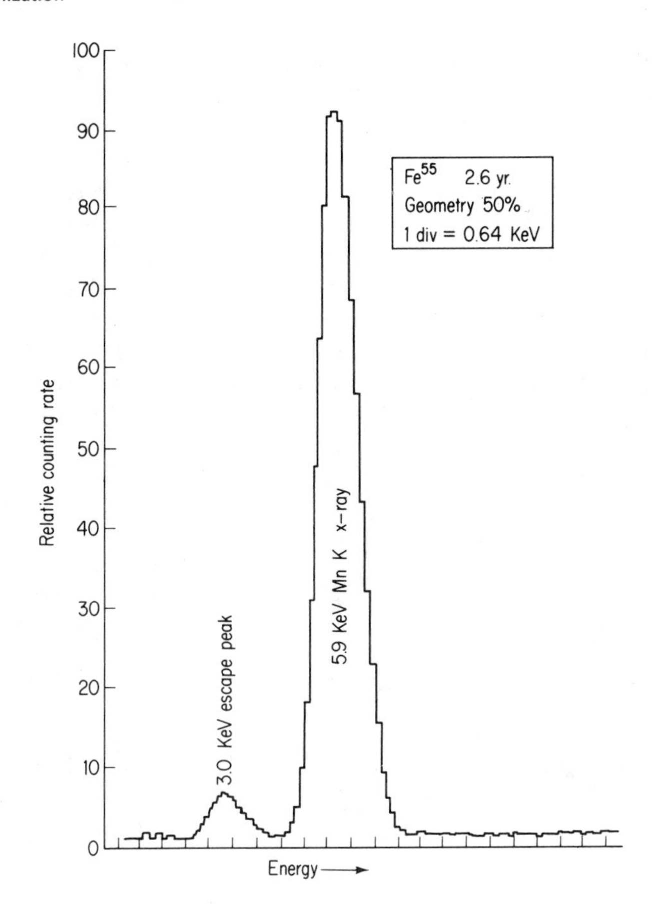

Fig. 5-10. Spectrum of X-rays occurring in the decay of ^{55}Fe. Measurement is made using detector constructed by Managan and Crouthamel described in text. Reprinted with permission from F. Adams and R. Dams, *Applied Gamma Ray Spectrometry*. 2nd ed., Oxford: Pergamon, 1970.

takes the forms of a cylindrical cathode about 1 to 1.5 in. in diameter by 4 to 8 in. in length. A fine central wire is suspended from insulation at one or both ends of the tube so as to act as the collecting anode. The cylindrical cathode may be of metal or glass with an applied metallic film. Sensitivity to gamma rays may be increased by using a bismuth coating on the inside of the cylinder as a cathode.

The G-M detector tube is filled with a gas mixture, usually at a reduced pressure. A readily ionizable, inert gas (argon, helium, neon) makes up the bulk of the gas filling; only a small amount of organic or halogen gas is introduced as the quenching agent. Windowless G-M flow detectors, mostly hemispherical in shape, are similar to the corresponding proportional detectors in construction. They also must be flushed continuously with a special gas mixture made up of about 99% helium and 1% butane, or isobutane.

A broad range of different G-M tube varieties is commercially available (see Figure 5-11). The end-window type (a) is the most widely used for radiotracer assay. A variety of end-window materials are used. By employing a window of split mica 1 to 2 mg/cm^2 in thickness, it is possible to obtain the transmission of approximately two-thirds to three-fourths of even the weak β^--particles from ^{14}C into the sensitive volume of the detector (see Figure 3-11). Tubes with thicker windows would be utilized primarily for assay of more energetic β^--emitters. Ultrathin windows (less than 150 μg/cm^2) of

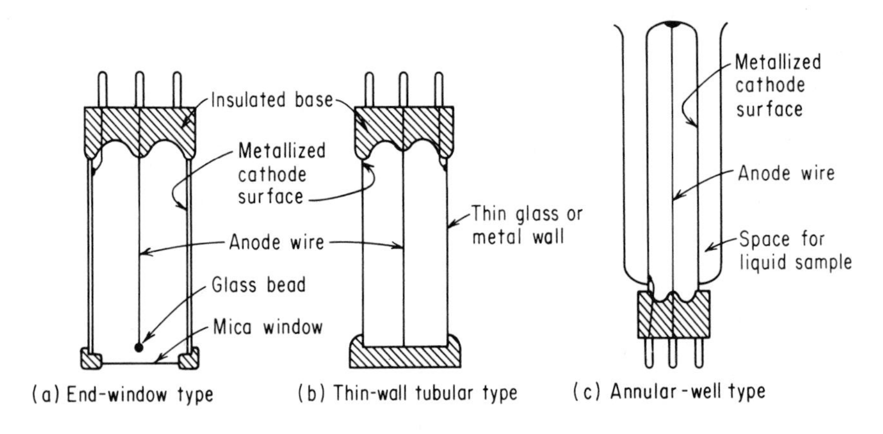

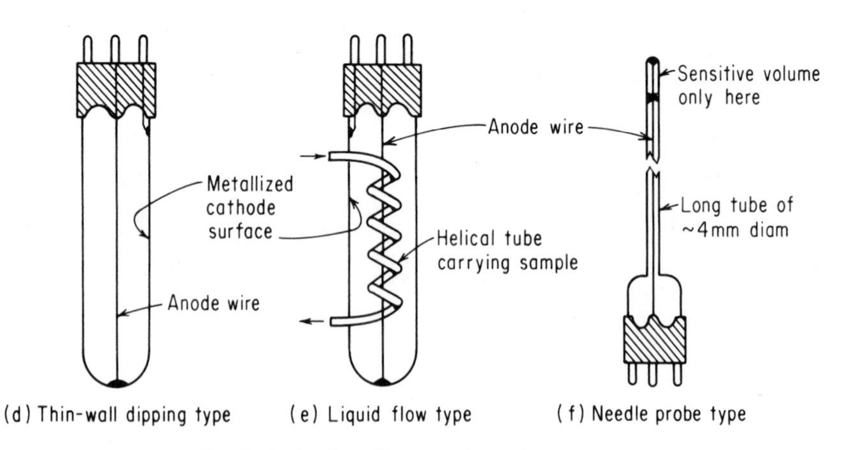

Fig. 5-11. Various forms of G-M detectors.

Mylar plastic will transmit even α-particles, but such tubes must be operated as flow detectors in order to counteract the inward diffusion of air. With the windowless flow chamber [see Figure 5-7(b)], the maximum detection efficiency is obtained, but this detector type allows the possibility of contamination of the interior of the chamber by radioactive dust or absorbed vapors.

G-M tubes designed primarily for monitoring (b) are usually constructed with thin glass (about 30 mg/cm^2) or metal cylindrical walls. This fact restricts their use to detection of more energetic β^--particles or γ-rays. Such tubes are commonly mounted in probes equipped with adjustable metal "beta shields," which, optionally, allow γ-rays to be detected to the exclusion of β^--radiation.

Tube types (c), (d), and (e) were developed for use with radioactive solutions. In the case of type (c), the solution is poured into the annular space around the tube; type (d) is actually dipped into a radioactive solution in a separate container. The design of type (e) was developed for use with a flow of radioactive liquid. Their necessarily thick, glass walls render these tubes rather inefficient for beta detection. Because they are gas filled, gamma rays are not readily detected by them either. In addition, there is the constant problem of radioactive contamination adhering to the outside of the tube. Hence these three tube types are no longer in common use.

The needle-probe detector (f) is an example of a highly specialized G-M tube; it illustrates the versatility of design possible with this type of detector. Such tubes have a minute sensitive volume located at the tip of the long, thin probe. Thus they are useful in precisely locating radioactive implants or concentrations of radioisotopes in patients.

Lead shielding around the detector proper is often used to reduce the background-radiation count rate. In some circumstances, even this reduced background radiation level cannot be tolerated because of the very low specific activity of the counting sample. Such is particularly the case in ^{14}C age-dating measurements. Here specially designed *low background counters* may prove necessary. Typically, they consist of an "umbrella" of guard G-M detectors surrounding the sample detector. Anticoincidence circuitry is employed so that external background radiation, mainly high-energy particles, passing through both the guard detectors and the sample detector is not registered. Only pulses originating in the sample detector alone are registered. Thus background count rates of only a few counts per hour are obtainable.

The G-M tube must be connected to some accessory components, such as a ratemeter or scaler, so that its output pulses may be registered. An external high-voltage supply is also required.

b. Operating Characteristics. A plot of the count rate from a fixed radioactive source with increasing potential applied to the G-M tube is shown in Figure 5-12. Do not confuse it with the plot illustrated in Figure 5-6, which

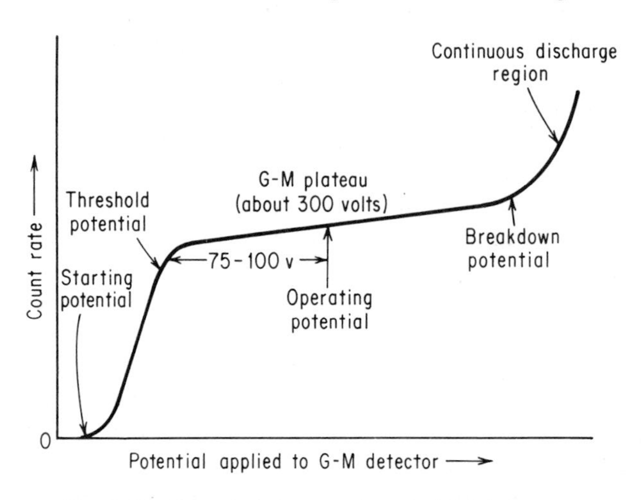

Fig. 5-12. Characteristic curve for a G-M detector.

shows the size of the pulse collected for a *single* ionizing particle at different detector potentials. The characteristic curve first shows a sharp rise in counting rate (*starting potential*), where only the most energetic beta particles cause the tube to go into discharge. At the *threshold potential*, this rise levels off onto a plateau that is perhaps 300 volts long with a slope of ~1 to 2% per 100 volts for an organically quenched tube. With a halogen-quenched tube, the plateau length will be ~100 to 200 volts with a slope of ~3 to 4% per 100 volts. The extent and slope of this plateau are a measure of the quality of a G-M tube. When the detector is operated within the plateau range, every ionization event occurring in the tube causes a measurable pulse. Eventually the curve displays a second rise with a higher potential gradient. Here the potential across the detector electrodes is so high that spontaneous discharges can occur in the tube without immediate ionizing radiation. The tube should never be operated in this region; otherwise it may be irreparably damaged. The exact potentials in volts on the abscissa of the plot are not stated, for they will vary widely with different tube types and filling-gas combinations. In general, the starting potential of commercially available G-M detectors ranges from 700 volts for halogen-quenched tubes to 1200 volts for organic-quenched tubes.

G-M tubes are nearly 100% efficient in detecting α- and β^--particles that reach their sensitive volume but only about 1% efficient in detecting γ-rays.

The chief difficulty in the detection of α- and soft (low-energy) β^--radiation involves the extent of absorption occurring in the tube window or wall. Once inside, the high specific ionization of α- and β^--radiation ensures the initiation of a discharge. Unfortunately, the great penetrating power of γ-rays, which enables them to pass readily through the walls of a G-M tube, allows nearly all these photons to traverse the filling gas without causing ionization and thus escape detection.

5. Summary of Gas Ionization Detectors

In review, we compare the advantages and disadvantages of detectors operated in the simple ion chamber, proportional, and Geiger regions. We present specific examples of detectors most commonly employed for radiotracer assay in these three regions.

The *ion chamber* used in conjunction with a vibrating-reed electrometer is best employed for precise measurement of low-energy β^--particles. It is highly stable and requires only a low chamber potential. A disadvantage of this instrument is that, in order to achieve maximum counting efficiency, the sample must be introduced in a gaseous form into the ion chamber. Under these conditions it is also possible to determine rather large total activities (as much as 1 mCi in a sample). Since no gas amplification occurs in the ion chamber here, the weak ionization current produced requires a very high gain amplification.

Proportional detectors are commonly operated as thin-window or windowless flow counters. In this form, their high α- and β^--detection efficiency are best used to advantage. In additon, α- and β^--radiation can be readily distinguished by means of proportional detectors. Perhaps their chief advantage is that the very short resolving time inherent in these detectors allows determination of very high count rates without significant coincidence loss. Their principal disadvantage lies in the higher gain amplification required, which necessitates more costly electronic accessory components than for G-M detectors.

The most widely used G-M tube for radiotracer assay is the thin end-window variety. With this type, beta particles are readily detected, with, of course, decreased efficiency for low-energy β^--radiation. The end-window tube is also generally simpler to operate compared to windowless flow detectors. Because of higher gas amplification factors in the G-M chamber, its output pulses need only moderate amplification, thus resulting in less costly accessory electronic components. Perhaps the greatest disadvantage of the G-M detector is its long resolving time, which seriously limits the activity of samples that can be assayed by it. Some of these considerations are summarized in Table 5-2.

TABLE 5-2

Summary and Comparison of Gas Ionization Detectors

	Ionization Chamber	Proportional Counter	Geiger Counter
Gas amplification factor	1	10^3–10^4	10^8
Approximate dead time (μsec)	1	1–2	100–300
Radiations detected efficiently	α, β	α, β, X	β
Energy resolution	α ~0.4%	X-rays, 5% at 10 keV β^-, 1.8% at 100 keV	None
Detection efficiency	Can approach 100% for gas-filled chamber	Best is 50% for α and β in windowless counter	γ 0.01 β can approach proportional counter
Cost	Expensive	Expensive, with necessary amplifiers, etc.	Cheap (< $100 for tube)

BIBLIOGRAPHY

1. PRICE, W. J. *Nuclear Radiation Detection.* 2nd ed. New York: McGraw-Hill, 1964, Chapters 2, 4, 5, 6. An excellent, advanced, detailed discussion of ionization chambers, proportional counters, and Geiger-Müller counters. The best place to start to look for specific information on these detectors.
2. WILKINSON, D. H. *Ionization Chambers and Counters.* Cambridge: Cambridge University Press, 1950. The classic discussion of the principles of operation of the ionization chamber, proportional counter, and Geiger-Müller counter.
3. FLÜGGE, S., and E. CREUTZ (Eds.). *Encyclopedia of Physics.* Vol. XLV, Nuclear Instrumentation II. Berlin: Springer-Verlag, 1958. An old but nonetheless authorative collection of articles by various experts on ionization chambers, proportional counters, and Geiger counters, among other things.
4. FAIRSTEIN, E. "Electrometers and amplifiers." In A. H. Snell (Ed.), *Nuclear Instrumentation and Methods.* New York: Wiley, 1961. A helpful review of the theory and design of electrometers and the properties of insulators suitable for ionization chambers.
5. O'KELLEY, G. D. *Detection and Measurement of Nuclear Radiation.* National Academy of Sciences Document NAS-NS-3105, April 1962. An excellent, *simple*, yet thorough discussion of ionization chambers, proportional counters, and Geiger counters.
6. BROWNELL, GORDON L., and HELEN S., LOCKHART. "CO_2 ion chamber techniques for radiocarbon measurement," *Nucleonics* **10** (2), 26(1952).
7. SPRINGELL, P. H. "Ionization current measurement of aqueous solutions containing ^{14}C," *Intern. J. Appl. Rad. Iso.* **9**, 88(1960).
8. TOLBERT, B. M. and W. E. SIRI. "Radioactivity." In A. Weissberger (Ed.), *Physical Methods of Organic Chemistry.* 3rd ed. New York: Interscience, 1960, p. 3335.
9. BERNSTEIN, W., and R. BALLENTINE. "Gas phase counting of low energy beta emitters," *Rev. Sci. Instr.* **21**, 158(1950).
10. ADAMS, F. and R. DAMS. *Applied Gamma Ray Spectrometry.* 2nd ed. Oxford: Pergamon, 1970, p. 111. An excellent, straightforward discussion of the use of proportional counters as spectrometers for low-energy radiation.

6

Gamma Ray Counting Using Solid Scintillators

A. BASIC FACETS OF THE SCINTILLATION PHENOMENON

Historically, the scintillation phenomenon was in common use early in this century as a means of detecting α-particles. Because the method involved the tedious visual observation and recording of faint scintillations, it fell into disuse with the advent of gas ionization detectors. The development during World War II of suitable photomultiplier tubes, which permitted electronic detection and recording of individual scintillation events, brought a resurgence of interest in the scintillation method. Since then a rapid expansion of scintillation detector types and applications has occurred.

In the radiation detectors described in Chapter 5, the detection process involved the collection of ion pairs produced as a result of the interaction of radiation with gas in an enclosed chamber. Scintillation detection, on the contrary, is based on the interaction of radiation with substances known as fluors (solid or liquid), or scintillators. (The term phosphor has often been used for compounds of this type, but the implied relation between phosphor and phosphorescence* may create a misleading impression.) Excitation of the fluor molecules leads to the subsequent emission of a flash of light (scintillation). An adjacent photomultiplier tube converts such photons into

*In fluorescence, emission of light from an excited state takes place promptly after the initial transition to the excited state. In phosphorescence, the molecule becomes "trapped" in an excited state, and emission of light takes place some time after the initial excitation.

an electronic pulse whose magnitude is proportional to the energy lost by the incident radiation in the excitation of the fluor. Thus scintillation detection is proportional in nature and, when used with a pulse height analyzer, allows determination of some features of the energy spectrum of the incident radiation. Comprehensive discussions of the scintillation mechanism and its application to radiation detectors are found in the work of Curran (1) and Birks (2).

It is difficult to arrive at a consistent classification of scintillation detectors. The fluor substances may be grouped into two broad categories: inorganic (ionic) and organic fluors. Inorganic fluors are normally used in the form of single large crystals. Organic fluors, by contrast, may be employed as single crystals, as a compact mass of fine crystals, as a copolymer in a solid plastic, or dissolved in an organic solvent (liquid fluor). The scintillation process in organic and inorganic fluors differs somewhat. In organic fluors, scintillation is primarily a molecular phenomenon and is marked by a transfer of excitation energy from solvent molecule to solvent molecule to organic fluor. Scintillation in inorganic fluors, on the other hand, results primarily from the electronic excitation in the solid state and the presence of specific activating "impurities" in the crystalline lattice. Furthermore, photon decay times in crystals are of the order of microseconds, whereas in plastic fluors and liquid scintillators they are of the order of millimicroseconds. Thus crystal fluors are relatively slow.

Scintillation detectors may also be categorized according to the relation of the sample and the fluor. In the case of *external-sample* scintillation detectors, radiation from an external source interacts with a fluor (usually a single crystal or plastic block), which is coupled to a photomultiplier. If the radioactive sample and the fluor are in intimate contact (dissolved in a common solvent, or one suspended in a solution of the other) and placed adjacent to one or more photomultipliers, an *internal-sample* scintillation detector results. This arrangement is most commonly referred to as a *liquid scintillation* detector. In this book scintillation detectors are somewhat arbitrarily divided into external-sample (employing solid fluors) and internal-sample (usually involving liquid fluors) detectors. The first type is described in the remainder of this chapter and the next; Chapter 9 deals with internal-sample detectors.

B. SOLID SCINTILLATION DETECTORS

1. Mechanism of Solid Scintillation Detection

Scintillation detection involves a series of energy conversions from the initial interaction of the radiation with the fluor until an output electronic pulse leaves the photomultiplier. These steps will be discussed in their normal sequence. Figure 6-1 will aid the reader in visualizing the various portions of a typical scintillation detector in relation to these energy-conversion steps. Detailed information concerning the various detector components themselves is presented later in the chapter.

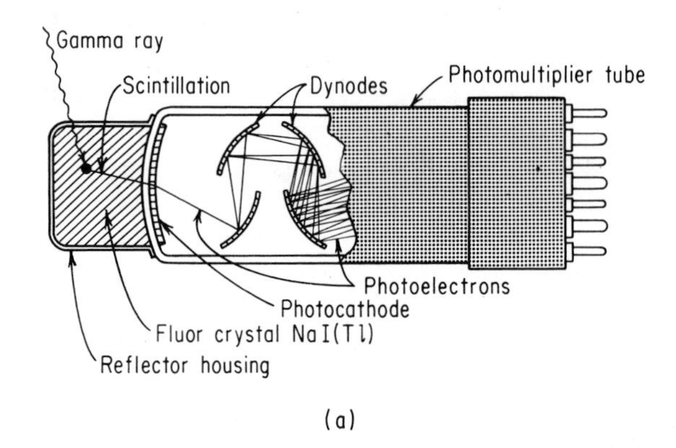

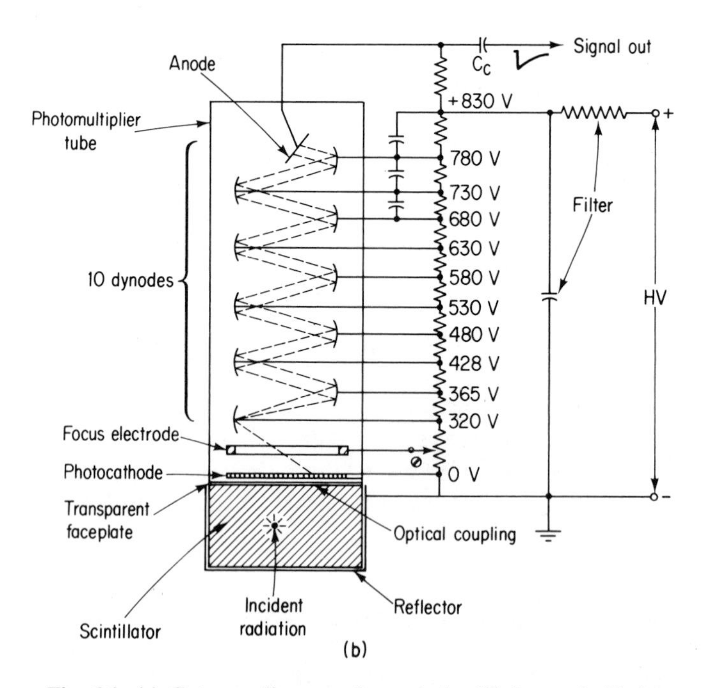

Fig. 6-1. (a) Cutaway diagram of a typical solid-fluor scintillation detector. (b) Diagram of a scintillation detector illustrating schematically the way in which light from the scintillator is transmitted to a photomultiplier tube. A typical wiring diagram is shown for the 10-stage photomultiplier operated with a positive high-voltage supply. From O'Kelley (11).

a. Energy Transfer in the Fluor Crystal. As radiation traverses a fluor crystal, it may dissipate some of its energy (see Chapter 3). The mechanism of energy loss will depend on the type of radiation. Recall that γ-rays commonly interact with matter by means of three distinct processes: pair production, the Compton effect, and the photoelectric process. In each case, electrons are ejected, and these electrons, in turn, can produce excitation or ionization of adjacent portions of the fluor crystal. The direct interaction of β^--particles or electrons with the fluor produces either ionization or excitation. Following the radiation-crystal interaction, the excitation of the fluor and its subsequent de-excitation depend on the type of scintillator.

In the scintillation process in organic crystals, molecules of the organic solid are excited from their ground state to their *electronic* excited states (see Figure 6-2). The decay of these excited states occurs in a time of the order of 10^{-8} sec (*fluorescence*) and is done by the emission of light quanta. Some of the initial energy absorbed by the molecule is dissipated as vibrational lattice energy prior to or after the decay by photon emission. As a result, the crystal will transmit its own fluorescent radiation without absorption.

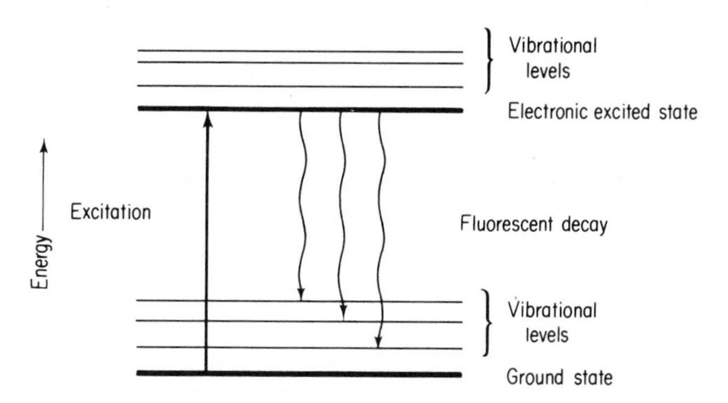

Fig. 6-2. A schematic view of the scintillation mechanism in organic crystals. The solid upward-point arrow shows excitation of the fluor to an electronic excited state, while wavy arrows represent decay of the excited electronic level to various vibrational levels near the ground state. Note that since the outgoing energy is less than the excitation energy, the radiation will pass through the crystal without attenuation.

The scintillation process in inorganic crystals differs from that observed for organic scintillators. In order to understand how inorganic scintillators work, one must consider the band theory of solids as applied to typical inorganic scintillators, such as the alkali halides. In the band theory of solids, the outer or "valence" electrons have energies that lie in a band called *the valence band* (see Figure 6-3). Somewhere above the valence band in energy lies a band of electron energy levels called the *conduction band.* The region of energies between the valence and conduction bands is called *the forbidden gap*

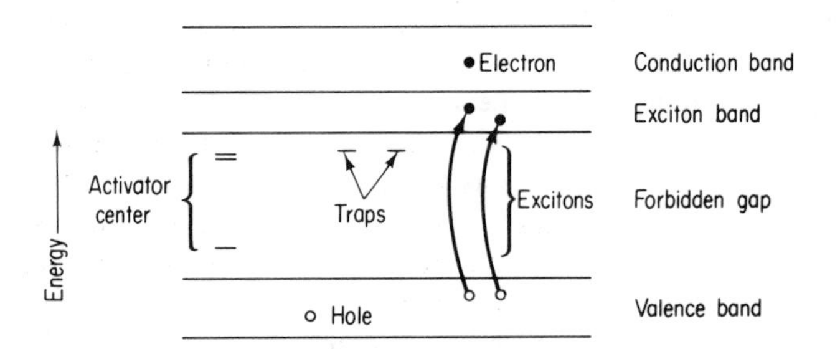

Fig. 6-3. The electronic energy band system in an ionic crystal insulator.

and represents values of the electron energies that are not allowed in a pure crystal. In the crystal's ground state, the valence band is completely filled with electrons, while the conduction band is empty.

When an energetic electron passes through the crystal, it may raise valence electrons from the valence band to the conduction band. This process is called *ionization.* The electron vacancy in the valence band resulting from this ionization is called a "*hole*" in the valence band. The electron in the conduction band and the hole in the valence band can migrate independently throught the crystal.

Alternatively, another process called *excitation* can occur whereby a valence band electron is excited to an energy level lower than the conduction band. The electron remains bound to the hole in the valence band. This neutral electron-hole pair is called an *exciton* and it can move through the crystal. Associated with the exciton is a band of energy levels called the exciton band (see Figure 6-3).

The presence of lattice defects and/or intentionally placed impurities in the alkali halide crystal will cause the formation of local energy levels in the forbidden gap, called traps or *activator centers.* Figure 6-3 shows the energy levels of an alkali halide crystal, including the activator centers and traps.

Excitons, holes, and electrons produced by the interaction of radiation with the crystal wander through the crystal until they are trapped at an activator center or trap. Migration of an exciton in a crystal may be thought of as a 6 to 8 eV excited iodide ion, I^{-*}, transferring its energy to an adjacent stable I^-, which, in turn, becomes excited. Thus energy may be transferred from I^- to I^- in the crystal lattice to final capture by either an activator center or crystal impurity. By exciton capture or hole-electron capture, the activator centers are raised from their ground state G to an excited state E. The de-excitation of this activator center by emission of light occurs in a time of the order of 0.3 μsec. Hence the energy deposited by the radiation in the scintillator is emitted as light by the activator center. The amount of light emitted by the entire crystal is directly proportional to the amount of energy deposited in the crystal by the incident radiation.

A summary of the scintillation mechanisms in the two scintillator types is shown in Figure 6-4.

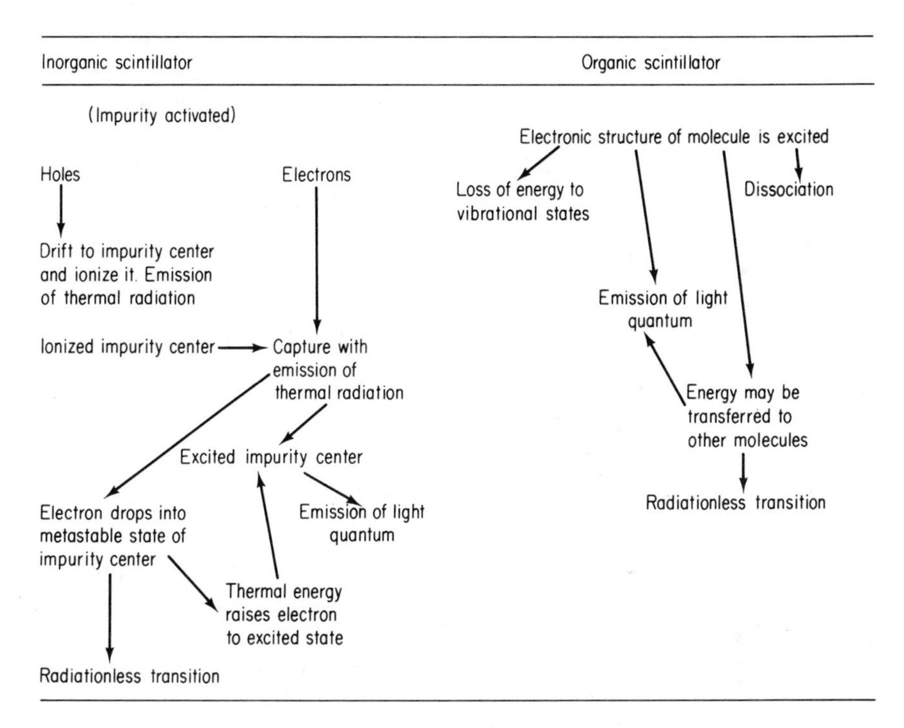

Fig. 6-4. The series of processes leading to the emission of light when a charged particle traverses a scintillator material. From J. Sharpe, *Nuclear Radiation Detectors*. Methuen (Courtesy of the Publishers).

b. Energy Transfer in the Photomultiplier. The photons of visible light emitted by the activator centers, such as Tl^+ ions in a NaI(Tl) crystal, pass through the transparent fluor substance and out through a clear window to impinge on an adjacent photocathode. The typical photocathode is composed of a thin, photosensitive layer (commonly a cesium-antimony alloy) on the inner surface of the end of the photomultiplier tube. Here impinging photons, particularly those having wavelengths between 3000 and 6000 Å, are absorbed, with a consequent emission of photoelectrons. The number of photoelectrons ejected is less than, but directly related to, the number of incident photons. Such a burst of photoelectrons resulting from a single γ-ray interaction is still far too weak to be registered directly.

Amplification occurs by means of a series of electrodes, called *dynodes*, spaced along the length of the photomultiplier tube (see Figure 6-1). Each dynode is maintained at a higher potential (usually about 50 volts higher) than the preceding one. The photoelectrons emitted from the photocathode are focused by the focusing electrode to hit the first dynode. Striking the dynode surface, they bring about the secondary emission of a larger number of electrons. This new burst of electrons is attracted by the potential gradient to the second dynode in the series, where a still larger number of electrons is dislodged. This electron-multiplying process continues at each dynode until at last the collecting anode is struck by 10^5 to 10^6 electrons for each original photoelectron ejected from the photocathode. Thus the magnitude of the output pulse from the photomultiplier is directly related to the quantity of energy dissipated by the incident γ-ray photon in the fluor.

c. Proportionality of Energy Conversion. In order to present a quantitative example of the energy conversions involved in scintillation detection, we trace the results of the interaction of a single 1.17-MeV γ-ray from ^{60}Co with a thallium-activated sodium iodide crystal [NaI(Tl)]:

1. If 20% of the energy of the γ-ray results in exciton production in a fluor crystal, and if it is assumed that 7 eV are needed to produce an exciton, then approximately 33,000 excitons could result from this γ-ray photon.
2. Assuming that only 10% of the excitation events results in the production of photons of visible light that are seen by the adjacent photocathode, this would mean that about 3300 photons would reach the photocathode.
3. This number of photons striking a photocathode with a conversion efficiency of 10% would eject approximately 330 photoelectrons.
4. The successive dynodes of a photomultiplier operated at an overall gain of 10^6 could then amplify this quantity of photoelectrons so that $\sim 3.3 \times 10^8$ electrons would be collected at the photomultiplier anode, or a charge of $\sim 5 \times 10^{-11}$ coulomb.
5. This charge could then be transformed by a preamplifier circuit with a capacitance of 30 pF into an output pulse of 1.8 volts. A pulse of this size would be capable of directly triggering a scaler.

Cobalt-60 emits two γ-rays per disintegration. The other γ-ray has an energy of 1.33 MeV. Following the preceding calculations, this 1.33-MeV γ-ray would result in an output pulse of about 2.05 volts from the detector. It should be noted that such energy calculations are based on several variables and are only crude approximations of true values. These calculations are important because they indicate that the size of the photomultiplier output pulse is directly proportional to the amount of energy dissipated in the fluor by the incident gamma photon. This relationship makes possible the determination of gamma ray energy when a pulse height analyzer is used with the detector, of which more will be said in Chapter 7.

2. Components of Solid Scintillation Detectors

a. Detector Housing. The entire scintillation detector is housed in a light-tight, thin metal cylinder. The cylinder gives mechanical protection to all components and also prevents stray light from reaching the photocathode while it is in operation. Also, since inorganic crystals, such as NaI, are hygroscopic, it prevents moisture from attacking the crystal. The inside of the detector can is coated with a light-reflecting material, such as Al_2O_3 or MgO. Protection against the effect of exterior magnetic fluxes on the photomultiplier is offered by a Mu-metal shield (a magnetic-field shield). Where directional sensitivity is desired, a lead collimator may be fitted over the scintillation crystal. Incident radiation must pass through the light-tight enclosure around the fluor in order to interact and be detected. In the case of γ-rays, the thin aluminum "can" (about 0.08 cm thick) presents no absorption problem. For crystal fluors used to detect α- or β^-- particles, however, ultrathin entrace windows of aluminum foil or aluminized Mylar film must be employed. When low-background counting is to be done, the detector can is made of high-purity copper rather than aluminum, since most aluminum contains trace amounts of radium.

b. Fluor Crystals. Scintillation crystals are commonly, but erroneously, referred to as phosphors. These crystals are used for radiation detection because they show short-lived fluorescence following energy absorption and, technically, are thus fluors. If they were true phosphors, producing long-lasting phosphorescence, they would be quite unsuitable for this purpose.

As noted previously, the solid fluor substances may be divided into two broad categories: inorganic and organic scintillators. The organic scintillators are particularly useful for counting β-particles and electrons due to their low atomic number. This low Z ensures that there is little chance for β^--particles to scatter out of the crystal, leaving only a small fraction of their energy in the crystal. Because of their low Z, organic scintillators have low probability of interacting with γ-rays, thus allowing one to count β^--particles in the presence of a high γ-ray background.

Some properties of the most popular solid scintillators are shown in Table 6-1. Note that the decay time of the scintillator (the time for the intensity of light output from one scintillation to decay to 0.37 of its original value—that is, the average lifetime of the excited scintillator molecule) is very short for the organic scintillators. Thus the organic scintillators are frequently used when precise measurements of time are needed.

TABLE 6-1

Solid Scintillators

Material	Relative Light Output	Decay Time (ns)	Wavelength of Maximum Emission (Å)
NaI(Tl)	210	250	4100
Anthracene	100	23–38	4450
Trans-stilbene	46	3–8	3850
p-terphenyl	30	4•5	4000
p-quaterphenyl	94	4•2	4350

The fluorescent property of the inorganic crystals is lost when the crystals are dissolved, but the organic fluors may either be used in crystalline form or dissolved. When dissolved in various organic solvents, they form *liquid fluors*, which will be further discussed in Chapter 9. Organic fluors, such as anthracene or *p*-terphenyl, have also been used to impregnate polystyrene or polyvinyltoluene. In this form, they are called *plastic fluors* and have the advantage of being formed or machined into a wide variety of shapes and sizes (3).

Anthracene is the most commonly used solid organic scintillator. A typical β^--particle and conversion electron spectrum measured by an anthracene crystal is shown in Figure 6-5. Energy resolutions on the order of 10 to 14% are possible.

The other major classification of scintillators is the inorganic scintillators. These are crystals of inorganic salts, particularly the alkali halides, which intentionally contain small amounts of impurities that act as activators for the fluorescence process. The high density and high Z of the inorganic scintillators greatly enhance the probability of gamma-ray interaction with the crystal and make this type of scintillator useful in the detection of γ-radiation.

The most commonly used inorganic scintillator is sodium iodide, containing approximately 0.1% thallium iodide impurity, which acts as an activator center (denoted as NaI[Tl]). The light output of this scintillator is the largest of all scintillators, being about twice that of anthracene. The emitted light has an average wavelength of ~4100 Å and a decay time of 0.25 μsec. In order to reduce unwanted background counts from ^{40}K in the crystal, the NaI crystal should contain less than 1 ppm KI.

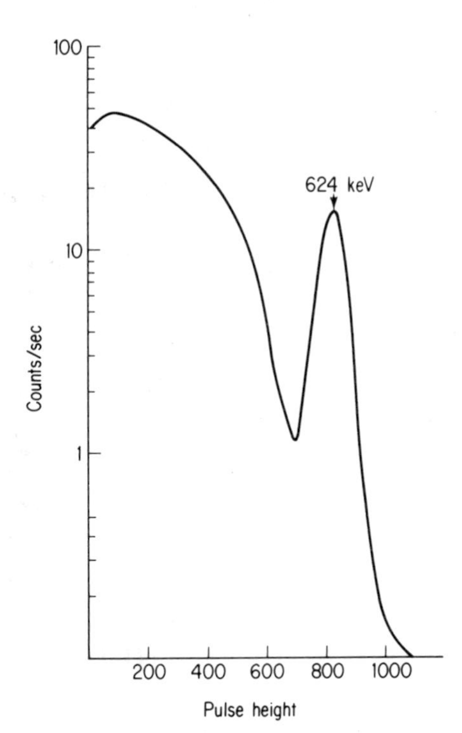

Fig. 6-5. Spectrum of a Cs^{137} source, measured on a flat anthracene crystal. From O'Kelley (11).

A large variety of shapes and sizes are available for the solid scintillators. A typical shape for the crystal is that of a right circular cylinder. Sizes up to 75 cm in diameter and 75 cm high are available for NaI(Tl) crystals, although the 7.6-cm diameter and 7.6-cm-thick crystals (the so-called 3 × 3-in. crystal) is most widely used for routine assay. For additional information on the response characteristics and variety of crystal shapes and sizes for NaI (Tl) fluors, see the booklet by the Harshaw Chemical Company (4), one of the leading commercial producers.

When thin radioactive samples are placed against a crystal fluor for assay, a maximum of 50% of the radiation is emitted in the direction of the detector (2π geometry). In an effort to increase detection efficiency, fluors with a "well" drilled into the crystal are commonly available. Assay samples may then be placed essentially within the crystal, so that a much greater proportion of the radiation emitted may be detected. Figure 6-6 shows such a well crystal. This type of *well crystal* is particularly suited for direct assay of liquid samples, homogenized tissue, or ashed specimens. NaI(Tl) crystals with well sizes capable of accommodating fluid samples up to 25 ml are generally available. Plastic fluors can be fabricated with very much larger well sizes.

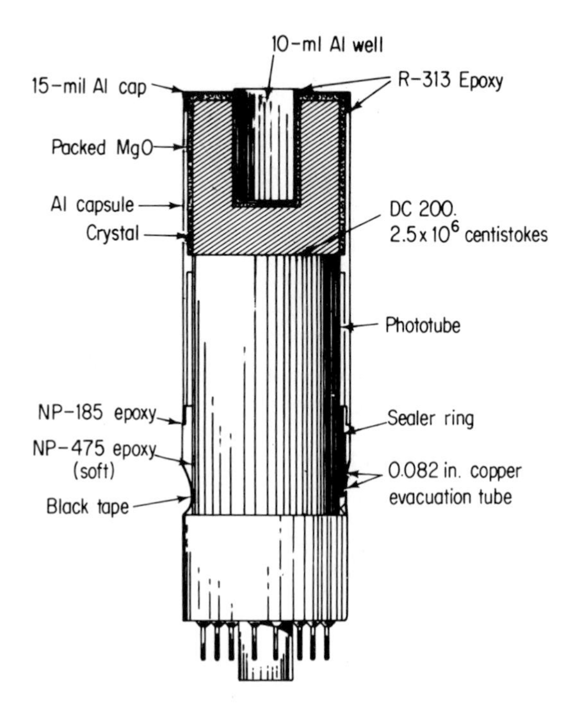

Fig. 6-6. An integral mounting arrangement for a 2 × 2-in. NaI(Tl) "well-type" crystal on a 2-in. photomultiplier tube. From V. A. McKay, Oak Ridge National Laboratory.

c. Photocoupling. In order to direct the maximum number of photons from the scintillation crystal to the photocathode, certain optical features must be included in the detector. The side of the crystal housing facing the photocathode is formed by an optical window of clear glass or quartz. All other surfaces of the crystal are covered with a light-reflecting layer, usually Al_2O_3 or aluminum foil. Good optical contact between the crystal and the photomultiplier tube is obtained by means of a transparent viscous medium, such as Dow-Corning "200" silicone fluid. In applications where it is desirable to have the scintillation crystal in probe form so as to reach less-accessible positions, a "light pipe" of lucite or quartz is used to connect it with the photomultiplier tube. For counting low levels of radioactivity, an optical window of quartz, pure SiO_2, should be obtained in order to reduce the 1.46-MeV γ-ray contribution of ^{40}K to the background of the detector. Glass generally contains substantial amounts of potassium and therefore, ^{40}K.

d. Photomultiplier Tubes. A photomultiplier tube consists of a photosensitive cathode, a series of dynodes maintained at increasingly positive potentials to the cathode, and an electron-collecting anode, all sealed in a glass envelope. Both the photocathode and the dynodes have sensitive surfaces that are capable of emitting electrons when struck by incident photons or electrons. It is this property of secondary emission that allows electron multiplication to occur through the dynode series, as previously noted. [Photomultiplier tubes for scintillation detection are presently available from several manufacturers, such as Radio Corporation of America (RCA), DuMont, CBS Labs, Amperex, and EMI Electronics, Ltd. (EMI).]

A variety of *photocathode* materials has been used in the past, but antimony-cesium or silver-magnesium alloys are still the most widely employed both because of their relatively high light sensitivity (40 to 60 μa/lumen) and quantum efficiency (10 to 20%)—that is, the ratio of photoelectrons ejected to incident photons—and because their spectral sensitivity matches the emission spectra of the most commonly used fluors. Figure 6-7 graphically

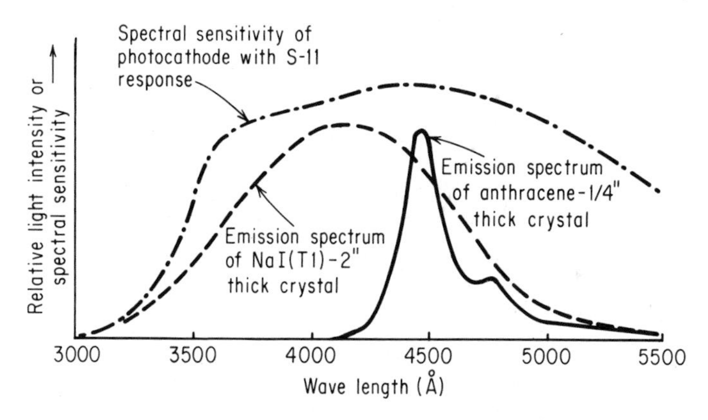

Fig. 6-7. Comparison of emission spectra of NaI(Tl) and anthracene with the spectral sensitivity of an Sb-Cs alloy photocathode with S-11 response.

illustrates this latter relationship in the case of NaI(Tl) and anthracene fluors. Recently the development of bialkali (Sb-K_2-Cs) or trialkali (Sb-K-Na-Cs) photocathodes has lead to improved quantum efficiencies, and these photocathode materials are finding increased use. Although photocathodes over 10 cm in diameter are available, the larger sizes usually lack uniformity of sensitivity over their surface. When large-sized crystals or fluor solutions are employed, a number of small photomultipliers can be arranged to detect the scintillations.

Two principal types of construction are used for the *dynode series* in photomultipliers. Tubes of American manufacture normally contain curved dynodes arranged so that an electrostatic field focuses the electron burst from one dynode to another. Many of the EMI tubes use flat dynodes arranged like a series of venetian blinds. In either case, the total amplification depends on the number of dynode stages and the potential applied between successive stages. From 6 to 14 dynodes may be included and the potential applied in approximately equal steps of 50 to 150 volts. Thus an overall potential gradient of up to nearly 3000 volts may be required, depending on the specific photomultiplier used. The stability of the high-voltage supply is essential in view of the direct relation between total amplification and dynode potential. A change of 0.01 volt, or less, in high-voltage output per 1-volt change in the line voltage is desirable. Although gains of 2 to 3×10^6 are most common, with higher voltages and larger numbers of dynodes, overall gains of 10^7 to 10^8 are possible.

C. INTEGRAL COUNTING

Solid scintillation detectors may be used either as *counters*—that is, merely to record the occurrence of an event wherein radiation struck the detector—or as *spectrometers*, where the energy deposit of the radiation in the scintillator is recorded. The uses of the solid scintillation counter as a spectrometer are so rich and varied that they will be covered in a separate chapter, Chapter 7. In the remainder of this chapter we shall discuss the use of a scintillation detector as a counter—that is, the recording of counts without regard to energy (*integral counting*). A word of caution should be added, however. Doing integral counting without understanding spectroscopy is similar to driving a car without knowing how it works. Most of the time you are all right, but if anything unusual develops, you may be in severe trouble.

Integral counting, although destroying any information about the energy deposit of the radiation in the detector, has certain advantages. The first is a large improvement in detector efficiency over the technique of γ-ray spectroscopy (see Figure 6-8). This improvement in detector efficiency is due to the fact that in integral counting all the pulses coming from the detector above a low-energy discriminator level are counted, whereas in spectroscopy only those pulses in the photopeak are used for analysis (the photopeak constitutes a small fraction of the total spectrum, as pointed out in Chapter 7).

The experimental setup used for integral counting is shown in Figure 6-9. Note that pulses from the detector are amplified, shaped, and passed on to a discriminator. The discriminator level is set to exclude electrical noise pulses but to trigger on all detector pulses above the discriminator level, regardless of size. The discriminator output pulses are counted by a scaler.

The determination of the optimal operating conditions for integral counting with solid scintillation detectors involves two parameters: the potential

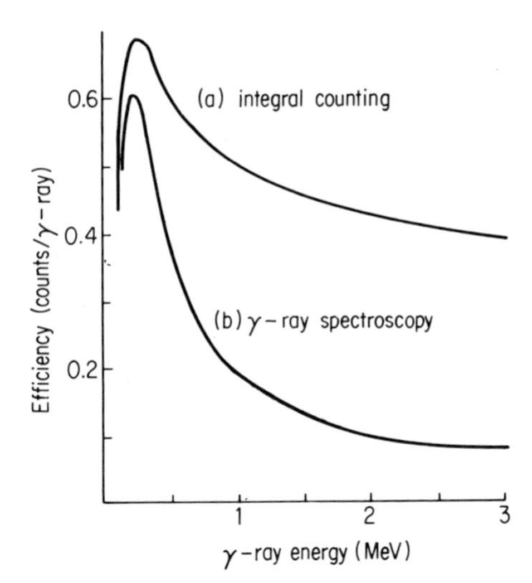

Fig. 6-8. Counting efficiencies for a 3 × 3-in. NaI(Tl) well-type scintillation detector: (a) integral counting with baseline set at 0.05 MeV; (b) γ-ray spectroscopy using photopeak counting. From Kruger (8).

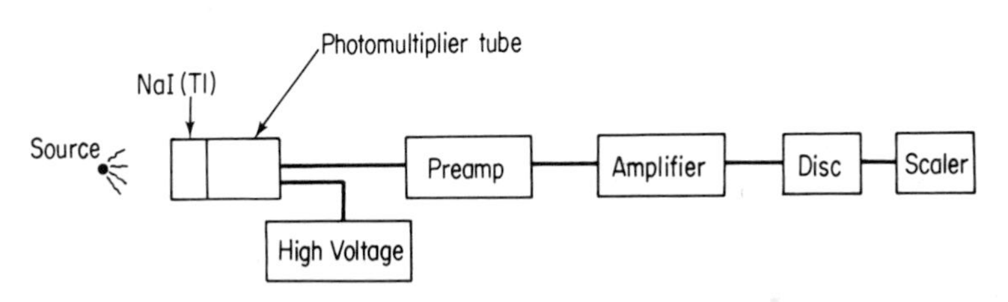

Fig. 6-9. Schematic diagram of an integral counting setup.

applied to the photomultiplier and the gain of the amplifier. In contrast, such determination for G-M counters involves only the potential applied across the G-M detector electrodes.

1. Effect of Photomultiplier Potential

At a fixed gain setting, a series of activity determinations made at successively higher photomultiplier potentials will yield data that may be plotted as for the solid-line curve in Figure 6-10. This curve shows the rapid rise in counting rate at lower potentials, the nearly flat *plateau* region of essentially constant counting rate with increasing potential, and, finally, the second region of rising count rate that superficially appears to resemble the characteristic G-M plateau curve (Figure 5-12). Actually, the two curves are completely unrelated.

The counting rate from total background radiation and that due to thermal noise in the photomultiplier alone are also seen in Figure 6-10. Obviously, the rapidly rising thermal noise level at the higher voltages is primarily

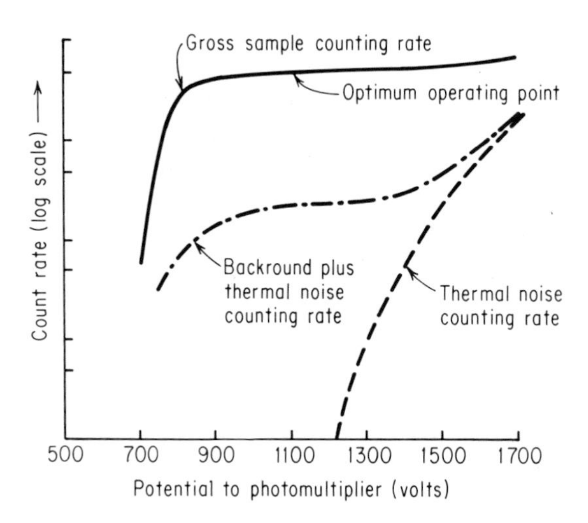

Fig. 6-10. Effect of photomultiplier potential on counting rate for a typical NaI(Tl) scintillation counter.

responsible for the increase in total counting rate of the "background" at the end of the plateau. Such being the case, it is best to operate the scintillation detector at the lowest possible potential on the plateau. The optimal operating potential is chosen on statistical grounds as that potential where the ratio of the square of the sample count rate to the total background count rate (both from radiation and thermal noise) is at maximum ($S^2/B = \text{max}$).

2. Effect of Amplifier Gain

The effect of different amplifier-gain settings on the scintillation-counter plateau curve is seen in Figure 6-11. It will be found, in general, that increased

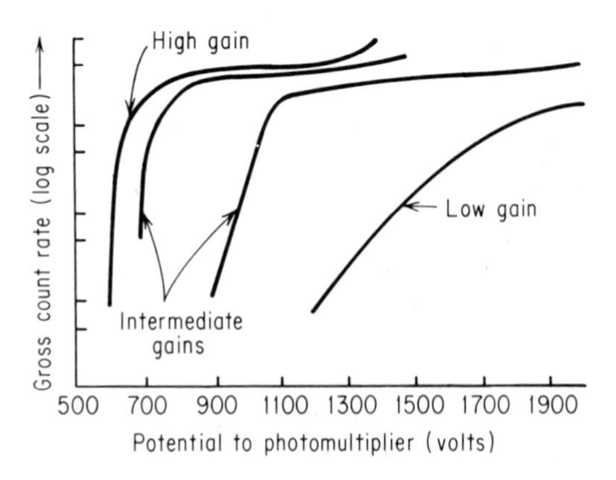

Fig. 6-11. Effect of amplifier gain on counting rate for a NaI(Tl) scintillation counter.

amplifier gain results in shortened plateau length, increased counting rate, and decreased plateau slope. Since the effects on counting rate of photomultiplier potential and amplifier gain are somewhat similar, but thermal noise is directly related to photomomultiplier potential, it is advisable to operate a scintillation counter at the lowest possible potential consistent with proper phototube operation and to increase the amplifier gain as required in order to compensate.

3. Detection Efficiency

The efficiency of an integral counting system, expressed as number of counts per source γ-ray emitted, is a function of several variables. Equation (6-1) presents the relationship between the observed detector counting rate C (counts per second) and the source disintegration rate D (distintegrations per second).

$$C = D\epsilon_T f_g \Omega \tag{6-1}$$

where ϵ_T is the total probability that a γ-ray striking the detector will give rise to a measurable pulse (*the total efficiency*), Ω is the solid angle subtended by the detector with respect to the source, and f_g is the fraction of disintegration giving rise to a γ-ray.*

For a point source and a large source-detector distance, the solid angle Ω is simply that fraction of the total area of a sphere (centered at the source with radius r equal to the source-detector distance) that is covered by the detector area. Thus

$$\Omega \approx \frac{\text{detector area}}{4\pi r^2} \tag{6-2}$$

where r is the source-detector distance. More general solid-angle formulas for disc sources and small source-detector distances are given in Reference 5. The effect of different-sized disc sources and different source-detector distances on the detector efficiency is shown in Figure 6-12.

Plots of the product $\epsilon_T\Omega$ for various detectors and source-detector distances are available in the literature (9). Figure 6-13 shows some typical data for the commonly used 3 × 3-in. NaI(Tl) detector. Note how the $\epsilon_T\Omega$ product decreases with increasing γ-ray energy and increasing source-detector distance. These factors must be considered when planning a radiotracer experiment (see Chapter 14). Note also that when the sample is placed at a distance of 10 cm or greater from the crystal, small changes in sample volume do not significantly affect the detection efficiency. Note, however, that when the

*This formula is accurate to within a few percent for most γ-emitting radionuclides. For a more accurate treatment, corrections for the emission of more than one γ-ray by the source should be made and are described in References 6 and 8.

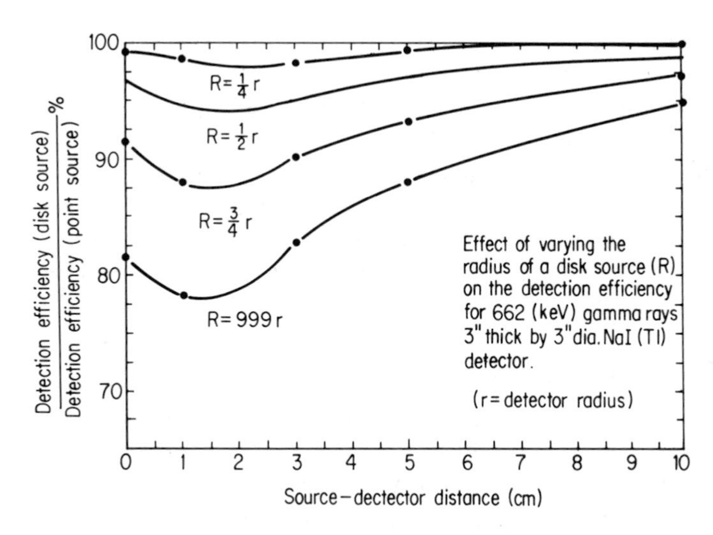

Fig. 6-12. Detection efficiency for a 3 × 3-in. detector (for 662 keV γ-rays) as a function of source radius and distance from the detector.

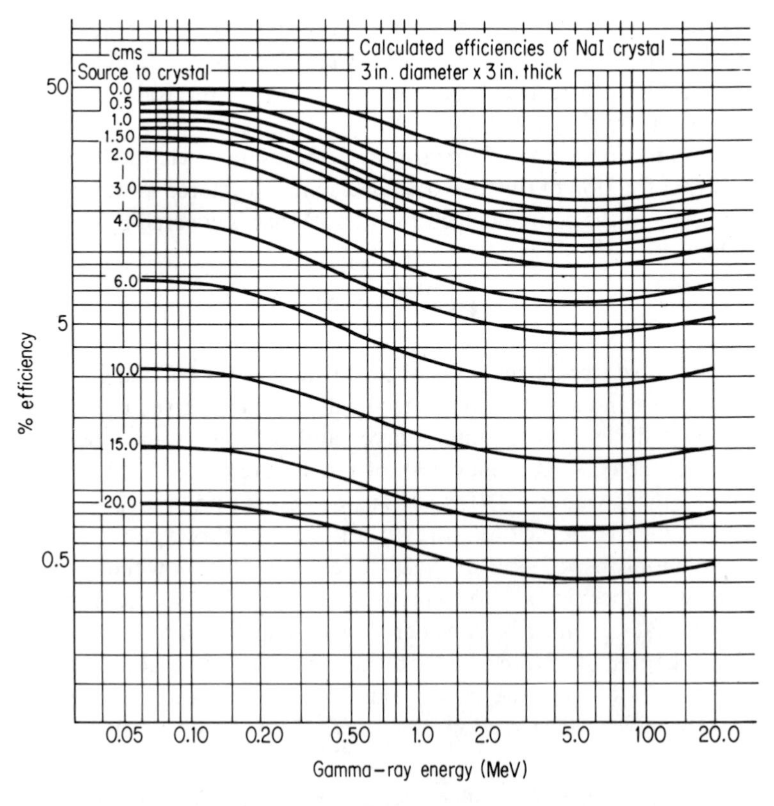

Fig. 6-13. From "Calculated efficiencies of NaI crystals" by E. A. Wolicki, R. Jastrow, and F. Brooks. NRL Report 4833..

sample is placed directly on top of the crystal, minute changes in sample size cause large changes in detection efficiency. This is a strong reason for preferring the use of well crystals for routine assay.

Well crystals do have the advantage that, in contrast to the solid crystal, the effect of a change in sample volume on detection efficiency will be small, provided that the sample volume is not too large compared to the well volume. This effect is shown in Figure 6-14 for various-sized well crystals.

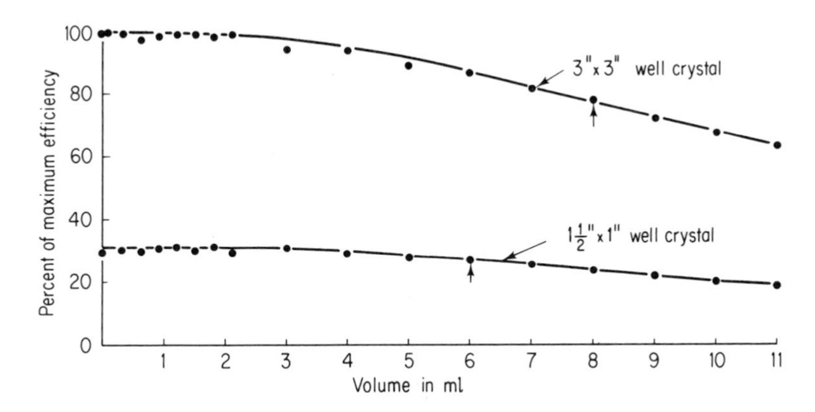

Fig. 6-14. Comparison of relative counting efficiencies of two different sized well crystals with respect to sample volume. Data were taken using ^{137}Cs tracer. The arrows indicate the point at which the sample volume exceeded the well volume. (R. Jones, Oregon State University, private communication.)

Note that if the volume of the sample becomes large enough, the efficiency will decrease significantly. When counting a series of sample of varying volumes in a well crystal, curves such as those show in Figure 6-14 can and should be used to correct for changes in efficiency with sample volume. No correction is necessary, however, for samples with constant volume. Well crystals have the additional advantage of providing a reproducible geometry for sample placement, and they offer impressive improvements in the value of the $\epsilon_T \Omega$ product for low-energy γ-rays ($E_\gamma \leq 0.5$ MeV). A sample counted in the well of a 3 × 3-in. [NaI(Tl)] crystal will have over twice the counting rate of an identical sample placed on top of a 3 × 3-in. solid crystal for counting.

The branching ratio, f_g, represents the fraction of disintegrations giving rise to the γ-ray of interest. Values of f_g may be obtained from nuclear data tabulations (10). Values of f_g for common γ-emitting radionuclides are tabulated in Table 6-2. This f_g factor is important in planning how much radiotracer to use in a given experiment (see Chapter 14).

TABLE 6-2

Branching Ratios for Common Radionuclides

Radionuclide	E_γ (MeV)	f_g
^{22}Na	0.511	1.80
	1.275	0.90
^{24}Na	1.369	1.00
	2.754	1.00
^{42}K	0.310	0.002
	1.524	0.18
^{51}Cr	0.3198	0.09
^{59}Fe	0.143	0.008
	0.192	0.028
	1.095	0.56
	1.292	0.44
^{60}Co	1.173	1.00
	1.332	1.00
^{95}Zr	0.724	0.49
	0.756	0.49
^{131}I	0.080	0.026
	0.284	0.054
	0.364	0.82
	0.637	0.068
	0.723	0.016
^{137}Cs	0.662	0.85
^{198}Au	0.412	0.95
	0.676	0.01
	1.088	0.002

Let us see if we can use these data to solve a practical problem in integral γ-ray counting. Suppose that we want to have a sample counting rate of 1000 cpm from an ^{51}Cr-containing sample that we position 10 cm above a 3×3-in. solid NaI(Tl) crystal. How much ^{51}Cr must be present in the sample?

Equation (6-1) states that

$$C = D\epsilon_T \Omega f_g$$

or, in other words,

$$D = \frac{C}{\epsilon_T \Omega f_g}$$

From Figure 6-13 we see that for the 0.320-MeV gamma ray of ^{51}Cr,

$$\epsilon_T \Omega \approx 0.025$$

Table 6-2 tells us that

$$f_g \approx 0.09$$

Thus

$$D = \frac{1000}{(0.025)(0.09)} = = 4.4 \times 10^5 \text{ dpm} = 0.20\ \mu\text{Ci}$$

One should be cautious about the use of integral counting. Generally integral counting should only be used to compare the activities of a series of samples containing the same γ-emitting radionuclide. Furthermore, one must be cognizant of the trouble associated with noise pulses. If the discriminator level shifts or the electrical noise of the system momentarily increases, the integral count will be increased by the counting of spurious, nonreproducible electrical noise pulses. On the other hand, integral counting is easily automated. Many commercial systems exist that automatically place the sample near the detector, count for a fixed length of time, or fixed number of counts, or either, print out the integral number of counts, and proceed to the next sample (similar automated systems are available for spectroscopy).

BIBLIOGRAPHY

1. CURRAN, S. C., *Luminescence and the Scintillation Counter.* New York: Academic, 1953. An old classic.
2. BIRKS, J. B., *Scintillation Counters.* New York: McGraw-Hill, 1953. Another old classic on scintillation counting.
3. BUCK, W. L., and R. K. SWANK. "Preparation and performance of efficient plastic scintillators," *Nucleonics* **11**(11), 48(1953).
4. Harshaw Chemical Company. *Harshaw Scintillation Phosphors.* A nice little book about NaI detectors.
5. HEATH, R. L., *Scintillation Spectrometry.* USAEC Report IDO–16880, 1964. The bible of scintillation spectroscopy. A Ge(Li) version should be available soon.
6. ADAMS, F., and R. DAMS. *Applied Gamma Ray Spectrometry.* 2nd ed., Oxford: Pergamon, 1970. An impressive monograph containing spectra and much information about NaI and Ge(Li) spectroscopy.
7. FILBY, R. H., *et al. Gamma Ray Energy Tables for Neutron Activation Analysis.* Washington State University Report WSUNRC-97(2), 1970. A table of gamma ray energies, and nuclides commonly encountered in neutron-activated materials.
8. KRUGER, P. H., *Principles of Activation Analysis.* New York: Wiley, 1971. A highly recommended, thorough discussion of activation analysis with very good material on scintillation spectroscopy.
9. WOLICKI, E. A., R. JASTROW, and F. BROOKS. "Calculated efficiences of NaI crystals." *NRL Report* **4833**.
10. LEDERER, C. M., J. M. HOLLANDER, and I. PERLMAN. *Table of Isotopes.* 6th ed. New York: Wiley, 1968.
11. O'KELLEY, G. D. *Detection and Measurement of Nuclear Radiation.* National Academy of Sciences Publication NAS-NS-3105, 1962.

7

Gamma Ray Spectrometry Using Solid Scintillation Detectors

In Chapter 6 we discussed how the solid scintillation detector could be used to count the number of particles or γ-rays striking the radiation detector (integral counting). In Chapter 7 we will discuss how the solid scintillation detector can be used to measure the *energy deposit* by the radiation in the detector, as well as to count the radiation. This additional information, the energy deposit, is extremely valuable in environmental and physical science applications, besides having a modest importance in biological applications. The radiation energy deposit can be used to identify the nuclide that emitted the radiation and the amount of that nuclide present. The measuring of the radiation energy deposit of γ-rays in detectors is called *γ-ray spectrometry*.

A. FUNDAMENTALS

Consider the interaction of γ-rays with a NaI(Tl) crystal. The height of the electrical pulse from the scintillation detector will be linearly proportional to the energy deposited in the crystal initially via the photoelectric, Compton, or pair production phenomena. Thus the *energies* of the γ-rays coming from a radioactive source, as well as their *intensities*, can be measured. The existence of many possible physical mechanisms for the interaction of γ-radiation with matter can lead, however, to a complex pattern in the pulse height distribution that is difficult to interpret. The purpose of this discussion is to show how the pulse height distribution emerging from the scintillation detector may be used to discern the energies and intensities of the γ-radiation striking the crystal (i.e., *scintillation spectrometry*).

In beginning this discussion, let us consider the various processes that can occur when a γ-ray interacts with a NaI(Tl) crystal and their relative importance. Figure 7-1 shows the relative importance of the interaction processes as a function of the incident γ-ray energy.

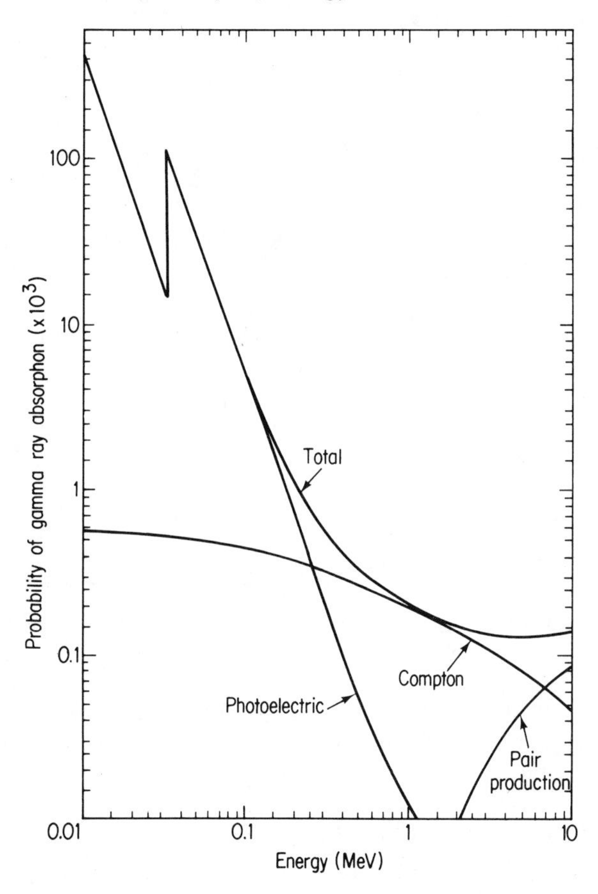

Fig. 7-1. Complete cross sections of sodium iodide showing total absorption and fractional components due to Compton absorption, photoelectric absorption, and pair production. All curve ratios have been corrected for coherent scattering. Data from NBS Circular No. 583.

The practical significance of Figure 7-1 and its meaning for the interpretation of observed pulse height spectra can be shown by considering the situation depicted in Figure 7-2. Figure 7-2, which is taken from the excellent monograph of O'Kelley (8), shows schematically two NaI(Tl) crystals of

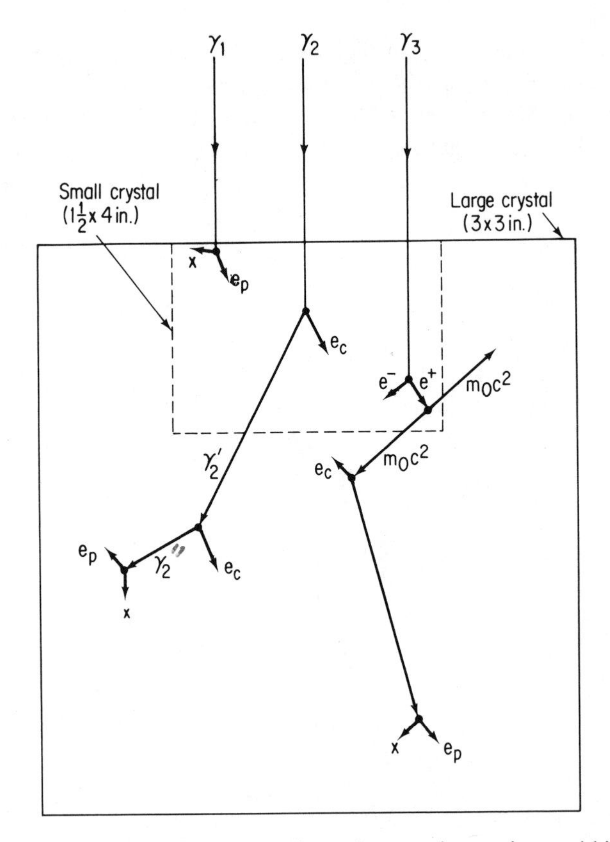

Fig. 7-2. Schematic representation of γ-ray interactions within NaI(Tl) crystals of two sizes. From O'Kelley (8).

different sizes and their interactions with a low-energy γ-ray (γ_1), a medium-energy γ-ray (γ_2), and a high-energy γ-ray (γ_3). Let us consider each of these γ-rays individually.

1. Low-Energy Gamma-Rays (γ_1)

Here $E_\gamma \leq 0.3$ MeV. By reference to Figure 7-1, we can see that the dominant mode of interaction of these γ-rays with NaI is the photoelectric effect. Remember that, in the photoelectric effect, nearly all the γ-ray energy is given to the photoelectron, e_p, which deposits all its energy in the NaI(Tl) crystal. The iodine atom* that released the photoelectron will be left with a vacancy in its K shell, which, when filled, will lead to release of an iodine K-X-ray of energy 28 keV (see Figure 7-2). The energy of this X-ray will

*Because the photoelectric-effect probability increases as Z^5, where Z is the atomic number of the stopping material, the iodine atoms ($Z = 53$) are more likely to interact with the γ-rays than the sodium atoms ($Z = 11$).

usually be completely dissipated in the crystal, thus adding to the energy deposited by the photoelectron in the crystal. The resulting output pulse corresponds to the full incident γ-ray energy ($E_{\gamma 1}$). Occasionally the iodine K-X-ray will escape the crystal without being detected, thereby leading to a total energy deposit in the crystal of ($E_{\gamma 1}$ — 28 keV) and a secondary peak in the pulse height spectrum. This situation is depicted in Figure 7-3, where

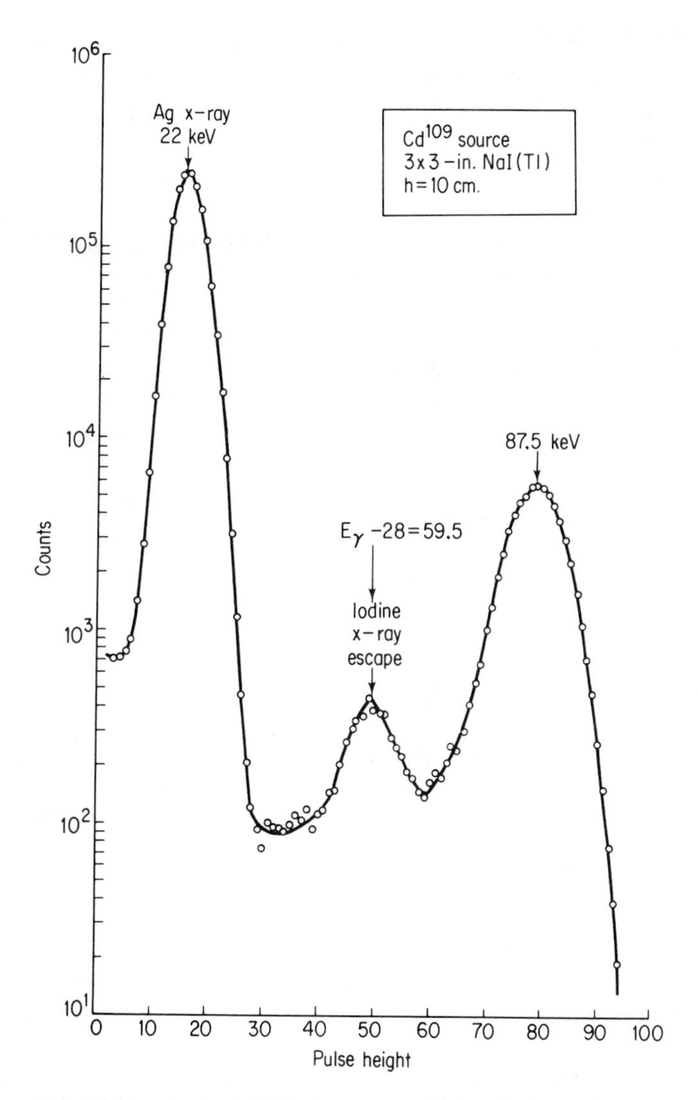

Fig. 7-3. Spectrum of 87.5-keV γ-rays and 22-keV X rays from a Cd^{109} source, illustrating the phenomenon of X-ray escape following detection of 87.5-keV γ-rays. From Heath (1).

the γ-ray spectrum for ^{109}Cd is given.* Note the peak corresponding to the escape of the iodine K-X-rays, as well as the full energy peak corresponding to $E_{\gamma 1} = 87.5$ keV. However, if $E_\gamma > 200$ keV, the escape peak may not be visible in the spectrum, since the width of the full energy peak may be so broad as to "cover" the region of the spectrum where the escape peak would appear.

2. Intermediate-Energy Gamma-Rays (γ_2)

The γ-ray energy is in the region from 0.3 to 2 MeV. Reference to Figure 7-1 shows that for γ-rays in this energy range, Compton scattering is the dominant mode of interaction with the crystal. Referring to Figure 7-2, we see that, because of its higher energy, γ_2 penetrates deeper than γ_1 into the crystal before interacting.

In the initial Compton scattering, γ_2 produces a Compton-scattered electron, e_c, which will probably be stopped in the crystal. The Compton-scattered photon, γ'_2, may escape the crystal without further interaction, depending on the crystal size, or it may undergo subsequent scatterings in the crystal until it loses the last portion of its energy via the photoelectric effect. If the latter happens, the energy deposited by all the scattering effects will be summed by the electronics attached to the crystal and counted as one full-energy pulse. Thus the theoretical pulse height distribution from a medium-energy photon should contain two components—a full-energy deposit peak and a distribution of pulse heights corresponding to the Compton electron energy distribution. This situation is shown in Figure 7-4.

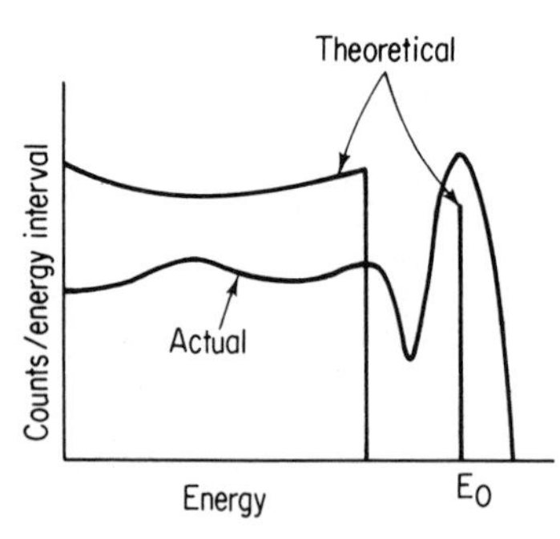

Fig. 7-4. Theoretical electron energy distribution (single events) for Compton and photoelectric interaction in a NaI detector compared with an experimental pulse height distribution obtained on a 3 × 3-in. NaI detector (0.50 MeV).

Also appearing in Figure 7-4 is the actual pulse height distribution observed with a 3 × 3-in. NaI detector, which shows the smearing effect of the detector resolution and multiple scattering events. Note that there is a

*The Ag X ray in the spectrum in Figure 7-3 is a by-product of the internal conversion process, which competes with γ-ray emission in the de-excitation of the 87.5-keV level of ^{109m}Ag.

maximum possible energy that the Compton electron can acquire in a single encounter. It shows up as a sharp upper edge on the Compton electron energy distribution (*the Compton edge*). Recall also, from Equation (3-9), that the position of this edge is given by the formula

$$E_e = \frac{E_\gamma}{1 + (0.511/2E_\gamma)} \tag{7.1}$$

where E_e is the Compton edge in MeV and E_γ is the incident γ-ray energy in MeV. Typical γ-ray spectra observed with a medium-energy γ-ray of ^{137}Cs ($E_\gamma = 0.662$ MeV) interacting with various-sized NaI(Tl) crystals are shown in Figure 7-5. Note that the smaller the crystal, the smaller the fraction of events leading to counts in the full energy peak as compared to the Compton distribution. Since the full energy peak is used to measure the γ-ray energy

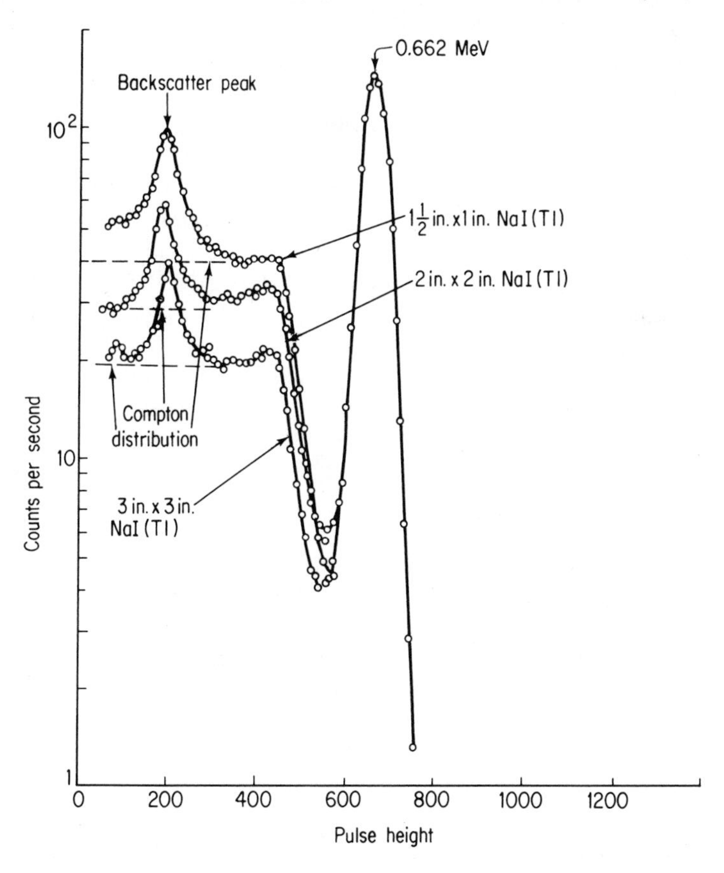

Fig. 7-5. Spectra obtained by measuring a Cs^{137} source with NaI(Tl) spectrometers of three crystal sizes. From Heath (1).

and intensity, one tries to maximize the photopeak/Compton ratio in any counting situation by using a detector of sufficiently large size.

3. High-Energy Gamma-Rays (γ_3)

Here $E_\gamma \geq 2$ MeV. Figure 7-1 tells us that pair production will become a significant mode of interaction for this energy γ-ray. Note that (Figure 7-2) γ_3 will, because of its high energy, penetrate deeply into the crystal before interacting with it. A 2.5-to 5-MeV γ-ray may even pass through a small crystal without interaction. The primary pair-production event will give rise to a positron e^+ and an electron e^-, both of which will stop in the crystal. After the positron has expended its kinetic energy, it will be annihilated, thereby giving rise to two 0.511-MeV γ-rays (m_0c^2). Several things can happen to these two γ-rays that are created within the crystal volume: (a) both γ-rays will be stopped in the detector, thus causing a full-energy peak; (b) both γ-rays will escape the crystal without further interaction, giving rise to a peak corresponding to $E_\gamma - 2(0.511$ MeV) (the *double escape peak*); (c) one γ-ray will escape the crystal and one will be captured, giving rise to *a single escape peak* with energy deposition in the detector equal to $E_\gamma - 0.511$ MeV. An example of this type of behavior is given in Figure 7-6, where γ-ray spectra of ^{24}Na are shown. ^{24}Na emits two γ-rays with energies 1.38 and 2.76 MeV. It is the 2.76-MeV γ-ray that gives rise to the single and double escape peaks. The escape peaks may be more prominent relative to the full-energy peak for the smaller crystal and are used in γ-ray spectroscopy for identification of high-energy γ-ray emitters.

Extensive compilations of the response of NaI(Tl) detectors to γ-rays from various radionuclides are available. The most useful of these spectra catalogs are those of Heath (1) and Adams and Dams (2). Some typical γ-ray spectra of frequently encountered radionuclides taken with a 4 × 4-in. NaI(Tl) detector are shown in Figure 7-7.

B. SPECIAL EFFECTS

In addition to the aforementioned effects, a series of special effects, due primarily to the physical surroundings of the scintillation detector, will contribute features to experimentally measured γ-ray spectra. A typical detector installation is shown in Figure 7-8. The kinds of special effects that occur are as follows:

1. *Backscattering*. In order to reduce the cosmic-ray background, NaI(Tl) detectors are usually encased in Pb shields of 10 cm thickness. Photons

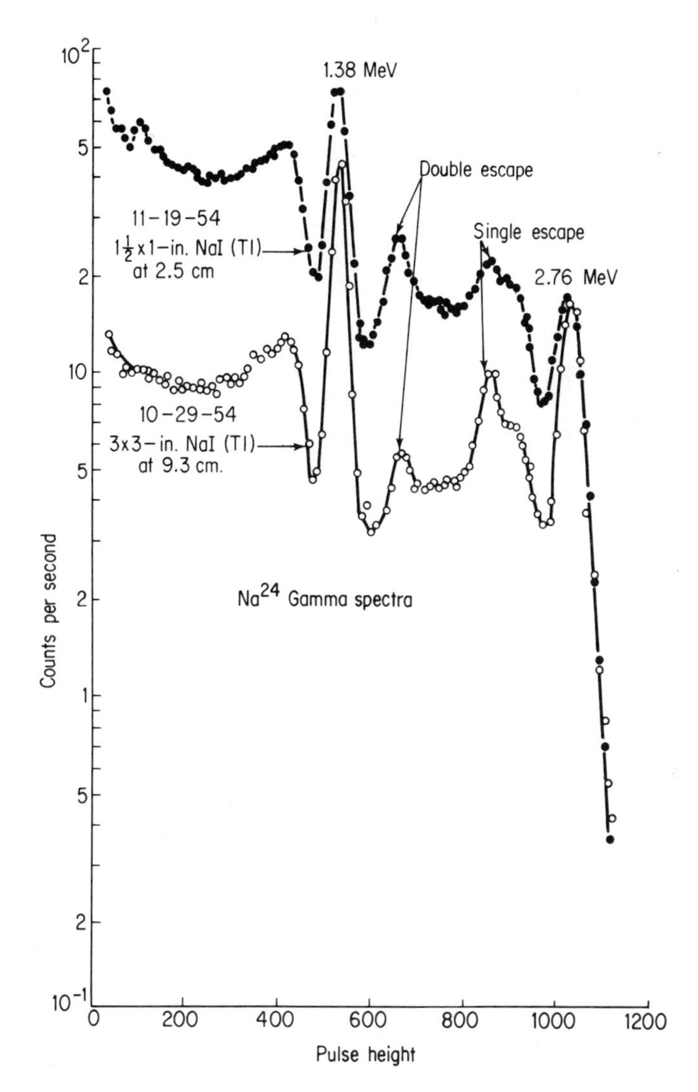

Fig. 7-6. γ-ray spectra of ^{24}Na, using $1\frac{1}{2} \times$ 1-in. and $3 \times$ 3-in. NaI(Tl) spectrometers. From Heath (1).

emitted from the source may strike the Pb shield walls and bounce back into the crystal. When they do so, they enter the crystal with an energy of $\sim$200 keV regardless of the incident photon energy. This is simply a property of large-angle Compton scattering. These backscattered photons cause the appearance of a spurious peak in the γ-ray spectrum at about

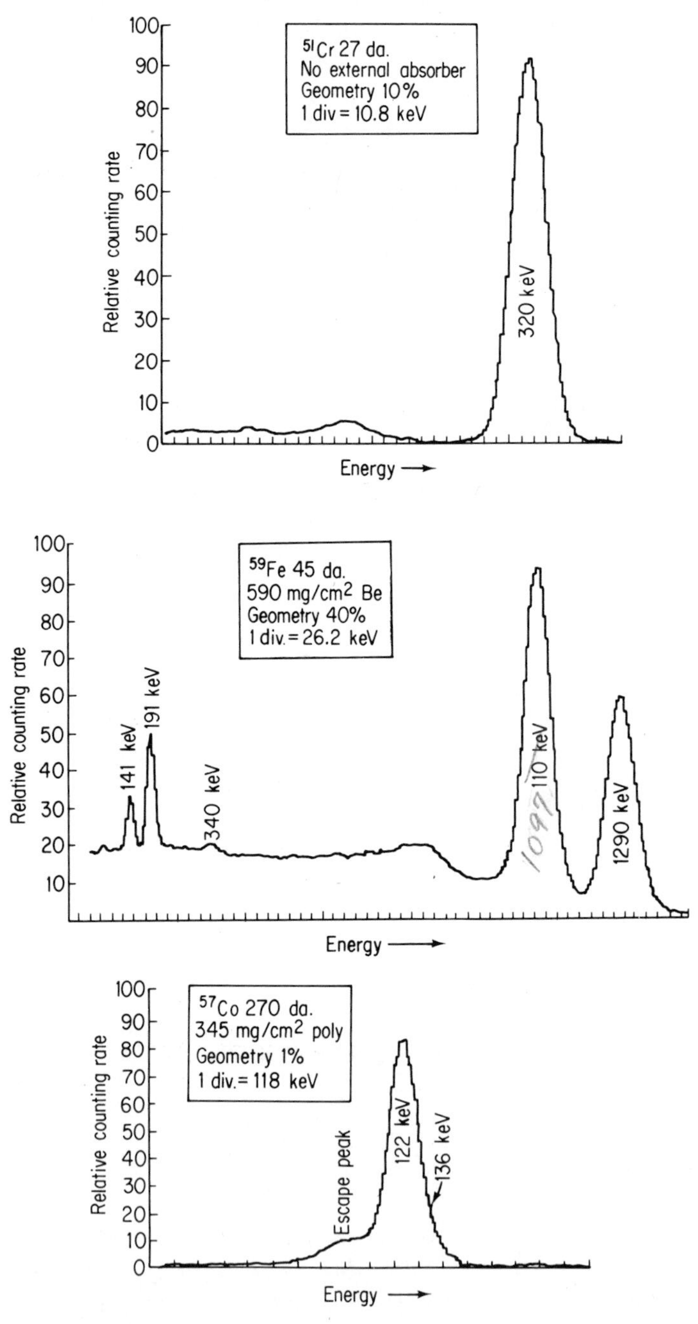

Fig. 7-7. Examples of some typical γ-ray spectra as measured with a NaI(Tl) scintillation detector. (Reprinted with permission from F. Adams and R. Dams, *Applied Gamma Ray Spectrometry*. 2nd ed. Oxford: Pergamon, 1970.

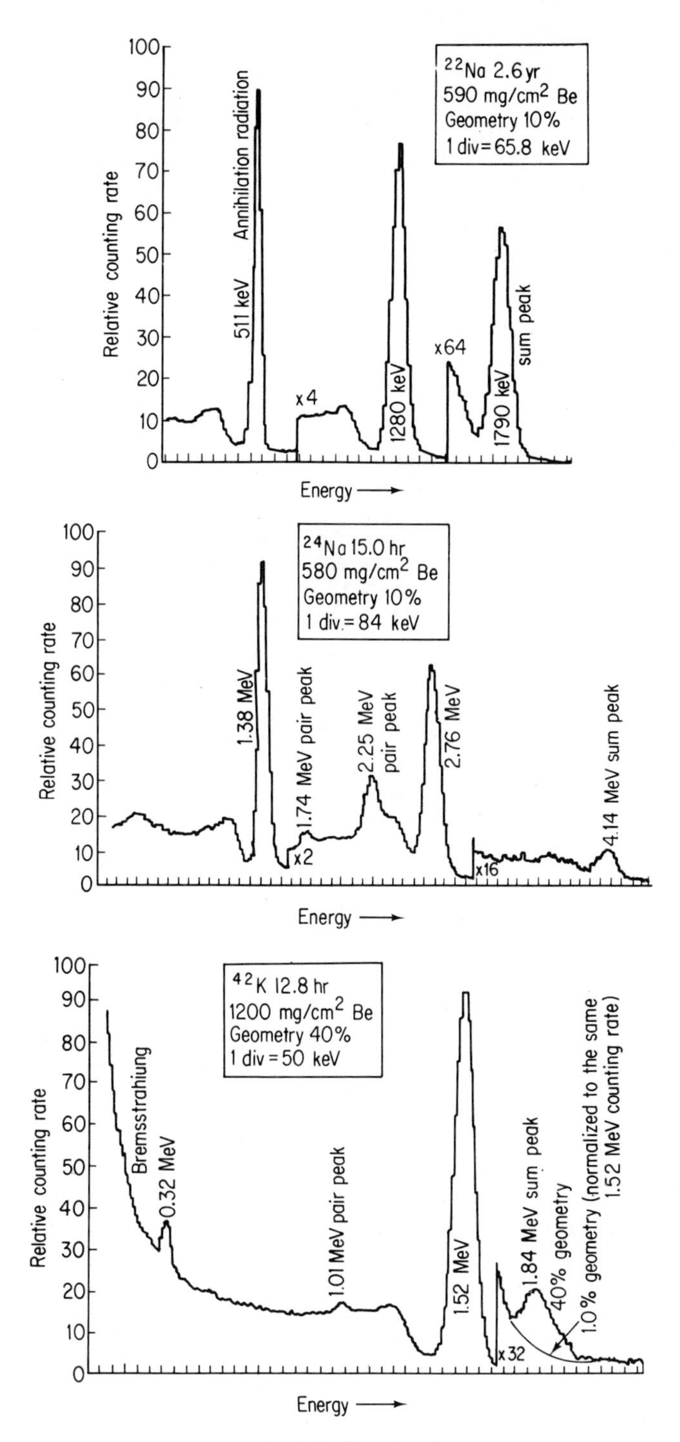

Fig. 7-7. (Continued)

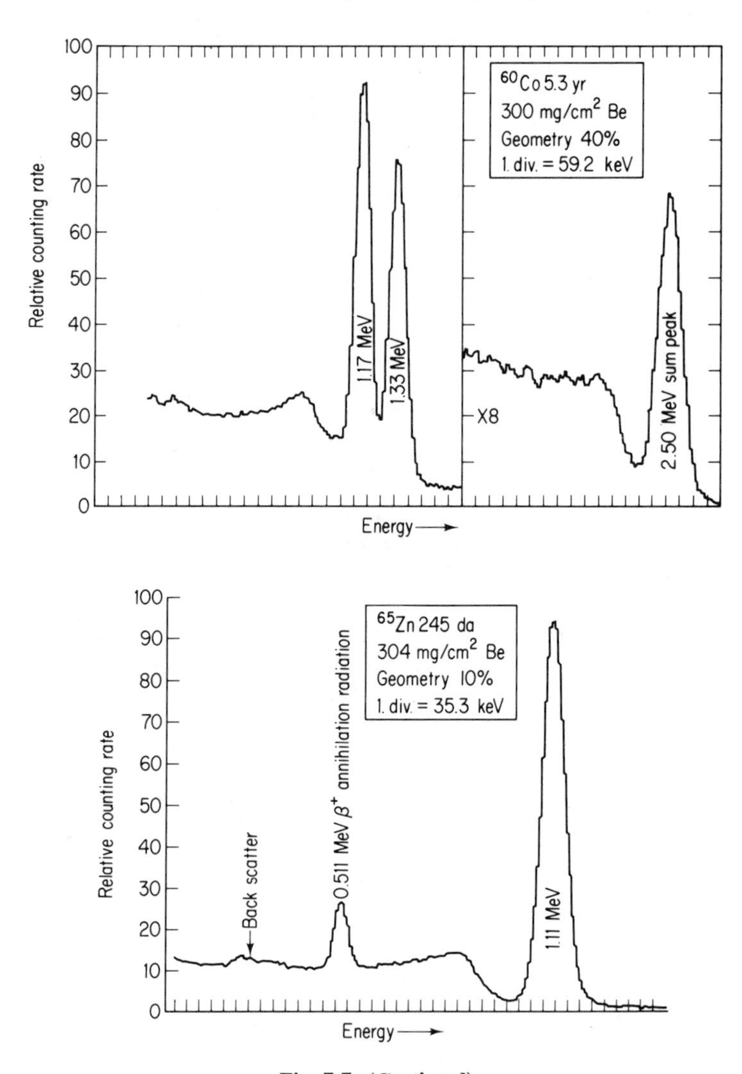

Fig. 7-7. (Continued)

200 keV. (See Figure 7-5 as an example.) About the only thing one can do to lessen this problem is to make the inside dimensions of the shield very large compared to the detector size, thus decreasing the probability of a scattered photon striking the detector. This effect is shown in Figure 7-9.

2. *Pb-X-rays.* When photons from the radioactive source strike the Pb shield, they will cause the emission of fluorescent X-rays of Pb from the walls. These Pb X-rays may be detected by the NaI(Tl) detector, causing a

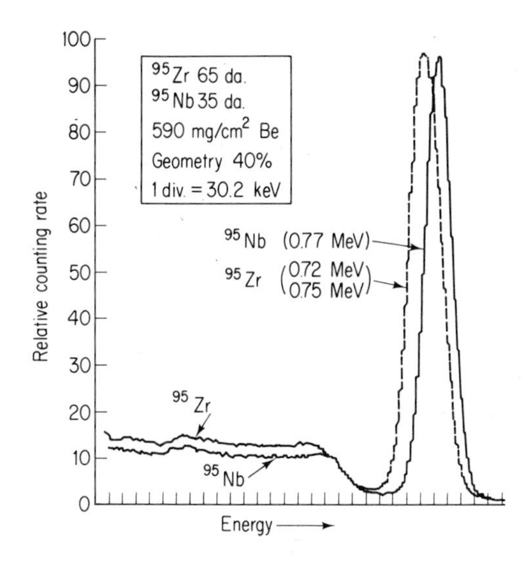

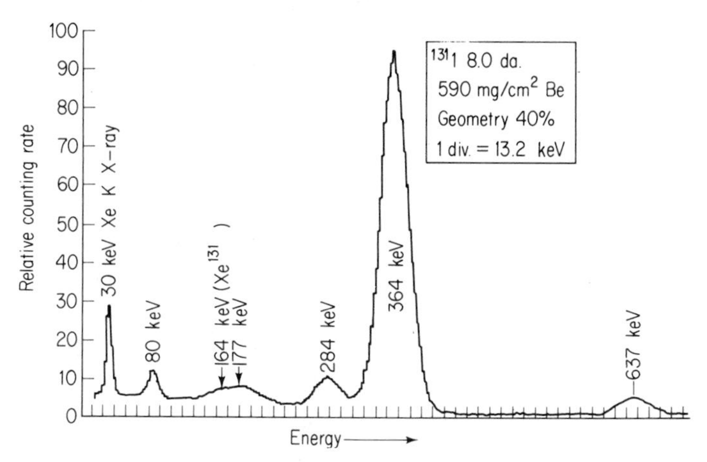

Fig. 7-7. (Continued)

spurious peak at 72 keV in the γ-ray spectrum. (Note that this peak is in addition to the 200-keV backscatter peak.) Such effects can be lessened by increasing the size of the inside dimensions of the lead shield or by using a "graded" shield. A graded shield is a Pb shield whose inside walls are covered with a layer of Cd and then a layer of Cu. The Cd absorbs the Pb X rays, while the Cu acts to absorb any Cd X-rays created. (See Figure 7-8 for details.)

3. *Peaks at 0.511 MeV*. There are several possible causes for a peak at or

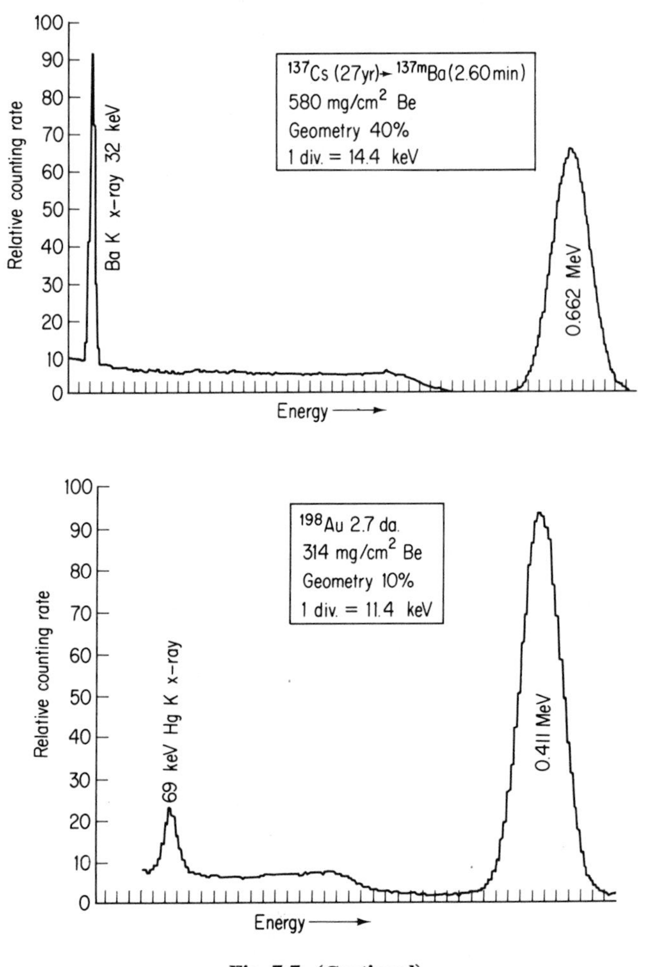

Fig. 7-7. (Continued)

near 0.511 MeV in a γ-ray spectrum. Among them are (a) detection of a real 0.5-MeV γ-ray emitted by a source nuclide, (b) interaction of a very high energy γ-ray with the shield walls, giving rise to pair production and the subsequent 0.511-MeV annihilation radiation, and (c) emission of a positron by the source, followed by subsequent annihilation radiation.

4. *Bremsstrahlung*. Any β^--emitting source can cause bremsstrahlung because of interactions in the source and the materials surrounding the detector. As noted in Chapter 3, the energy spectrum of bremsstrahlung radiation is a continuous one, extending from 0 MeV up to the electron energy. Such a "smear spectrum" can obliterate many features of the normal γ-ray spectrum. In order to attenuate the effect of bremsstrahlung,

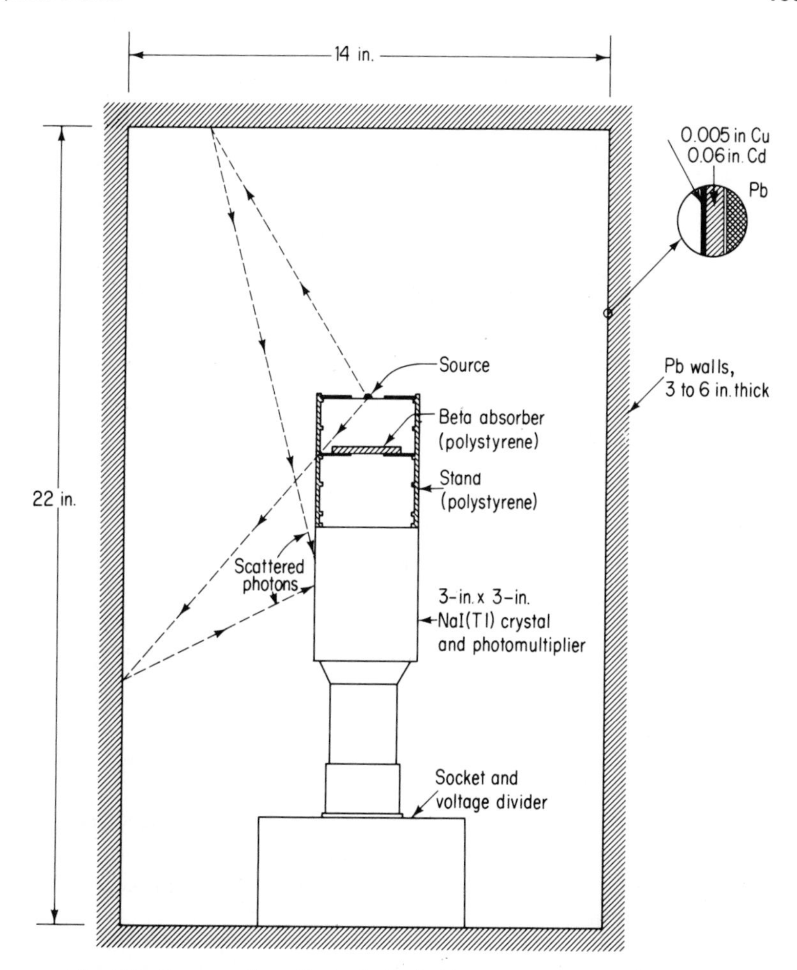

Fig. 7-8. Cross section of a typical scintillation spectrometer installation showing the 3 × 3-in NaI(Tl) detector assembly, the lead shielding with "graded" liner, and the use of a low-mass support for the source and beta absorber. The origin of scattered photons is illustrated. From O'Kelley (8).

a beta absorber of low Z material is placed near the source, and thus the effective probability of bremsstrahlung reaching the crystal is decreased. Decreasing the average atomic number of the sample matrix for the β^--emitting radionuclide by dissolution and dilution of the matrix with low Z solvent also effectively reduces the bremsstrahlung component of the γ-ray spectrum.

5. *Summation Effects.* When a radioactive source emits two γ-rays, there is a nonvanishing probability that both will strike the crystal simultaneously

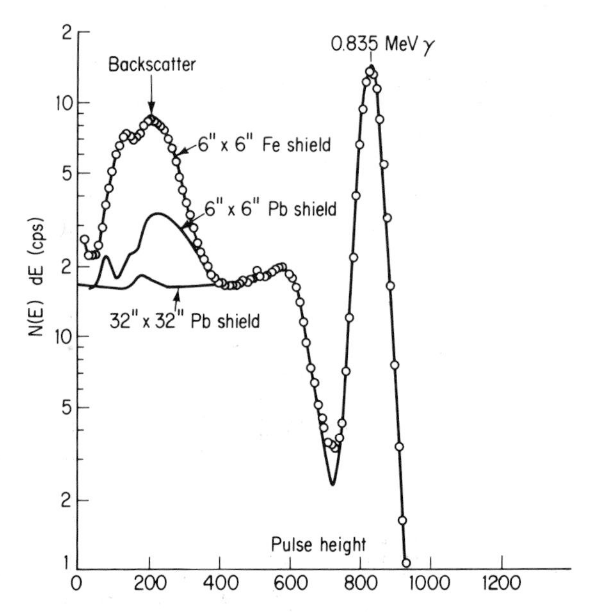

Fig. 7-9. The effect of detector-shield configuration on scattered component of pulse height spectrum obtained with γ-rays of 0.835-MeV energy. From R. L. Heath, *Scintillation Spectrometry*. Gamma-Ray Spectrum Catalog, AEC Report No. IDO-16880-1, 2nd ed., 1964.

and cause energy deposits in excess of the energy of either single γ-ray. Such counts are "sum counts," and the peak in the spectrum arising from the deposit of the full energy of both γ-rays in the crystal is called *the sum peak*. An example is shown in Figure 7-7.

C. COMPONENTS OF A SOLID SCINTILLATION SPECTROMETER

The physical arrangements of detector, sample, and electronic apparatus associated with a solid scintillation spectrometer are similar to those associated with a scintillation counter (as described in Chapter 6). A typical arrangement of the detector and shield is shown in Figure 7-8. To reduce background radiation, a shield of 5 to 15 cm of Pb is used. The Pb is usually lined with a graded layer of 0.1-cm Cd and 0.01-cm Cu, as described in Section 7B-2. For low-level analysis, the lead should be specially purified to give an activity of ~1 cpm/g. Any supporting material for the shield should be made of iron or steel prepared before 1940. (Steel manufactured after World War II contains significant amounts of the radionuclides ^{60}Co, ^{106}Ru, ^{232}Th, etc.) The inside volume of the shield should be as large as possible so as to reduce backscattering and Pb X rays.

In ultra-low-level work, such as that associated with monitoring for environmental radionuclides, the air entering the room should be filtered and purified. Counting areas for such work should not be situated near nuclear reactors, accelerators, and similar features, because of the possible occurrence of airborne radioactivity. May and Steinberger (5) have measured typical contributions of various sources to the background of a Pb-shielded 7 × 3.5-in. NaI(Tl) crystal, and these sources are shown in Table 7-1. (An unshielded crystal might have a background of ~20,000 cpm under similar circumstances.)

TABLE 7-1

Typical Contributions to NaI(Tl) Crystal Background (in counts per minute)

Source	cpm
Cosmic rays	175
Lead shield	110
Phototube	100
^{40}K in crystal	30
Rn in air	100
Residual, unaccounted for	140
Total	655

The electronic circuitry associated with a γ-ray scintillation spectrometer is shown in Figure 7-10. The high-voltage supply must be stable against fluctuations in temperature, line voltage, and load. A fluctuating high-voltage supply will worsen energy resolution and, in extreme cases, completely prevent spectral measurements. An acceptable standard for power supply drift is a drift less than 1 part in 10^5 per hour. The basic features of the preamplifier and amplifier were discussed in Chapter 4. One need only further suggest

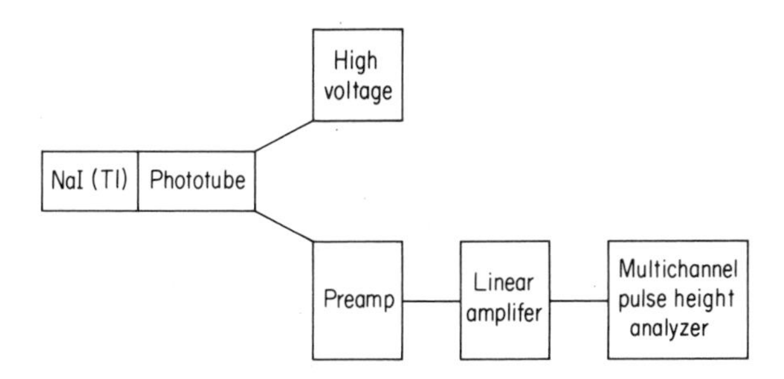

Fig. 7-10. Schematic block diagram of electronic apparatus used in γ-ray scintillation spectroscopy.

that the amplifier pulse shaping be chosen to be bipolar so as to reduce pileup effects (see Chapter 4) and that the count rates in the detector be kept to less than 10^4 cps. (At this rate, pileup distortion is $<5\%$). The multichannel pulse height analyzer should have 400 to 512 channels with a storage capacity of 10^5 to 10^7 counts per channel for scintillation spectroscopy.

D. ENERGY RESOLUTION

One of the most important aspects of scintillation counting is the width of the peak resulting from the deposit of the full γ-ray energy in the scintillator. In a source emitting many γ-rays, the width of each full-energy peak may determine whether the presence of any γ-ray can be detected. A measure of the peak width is the quantity called the *resolution* of the detector, R, which is defined as the full width at half maximum (expressed in energy units) of a given γ-ray peak, $\Delta E_{1/2}$, divided by the peak energy, E, times 100%. Thus

$$R = \frac{\Delta E_{1/2}}{E} \times 100\% \qquad (7\text{-}2)$$

Figure 7-11 shows this calculation graphically. The resolution can be calculated for each γ-ray energy and will generally differ for each one. The

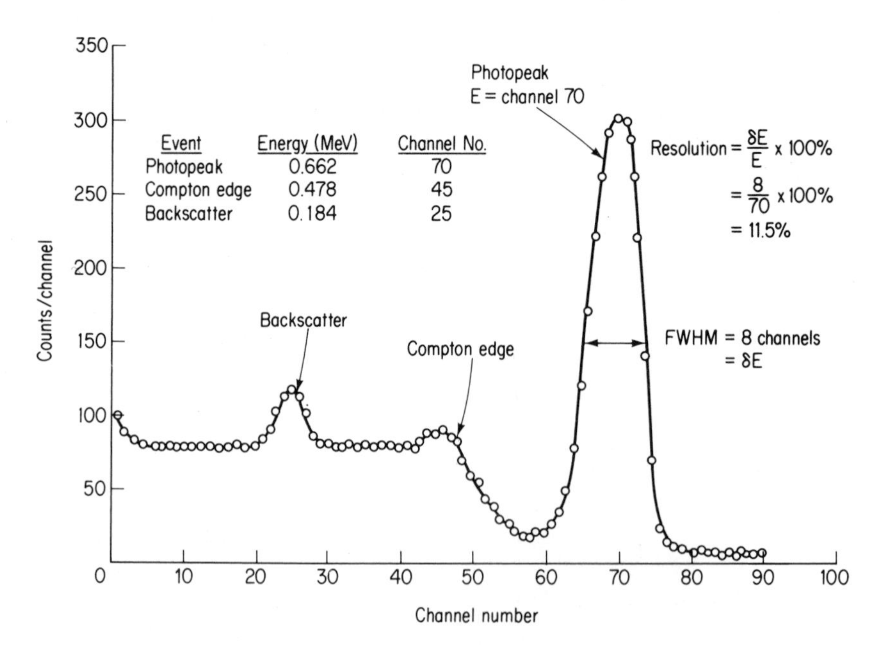

Fig. 7-11. Typical calculation of pulse height resolution. From *Experiments in Nuclear Science*, ORTEC Publication AN34, July, 1971.

resolution of a scintillation detector depends on several factors. Among them are

1. The number of photons per scintillation event.
2. The number of photons that strike the photocathode.
3. The number of photoelectrons released per photon hitting the photocathode.
4. The number of photoelectrons that strike the first dynode.
5. The multiplication factor of the photomultiplier tube.

In general, the greater each one of the preceding quantities is, the better (i.e., smaller) the resolution. Factors (1) and (2) refer to properties of the scintillator, while factors (3), (4), and (5) refer to the phototube. Mathematically, we say that

$$R^2 = \alpha + \frac{\beta}{E_\gamma} \tag{7-3}$$

where R is the resolution, α and β are constants referring to the scintillator and phototube, respectively, and E_γ is the γ-ray energy. Note that R varies inversely as $\sqrt{E_\gamma}$, and thus the higher E_γ is, the smaller R is.

Typical values of NaI(Tl) detector resolutions are 7 to 9% for $E_\gamma = 0.662$ MeV. Figure 7-12 illustrates a typical plot of energy resolution versus E_γ for a 3 × 3-in. NaI(Tl) crystal. The figure shows how difficult it is to measure the amounts of two γ-emitting radionuclides in a sample unless their γ-ray energies differ by at least the energy equivalent of the resolution for that energy γ-ray. Figure 7-12 is therefore valuable in planning whether

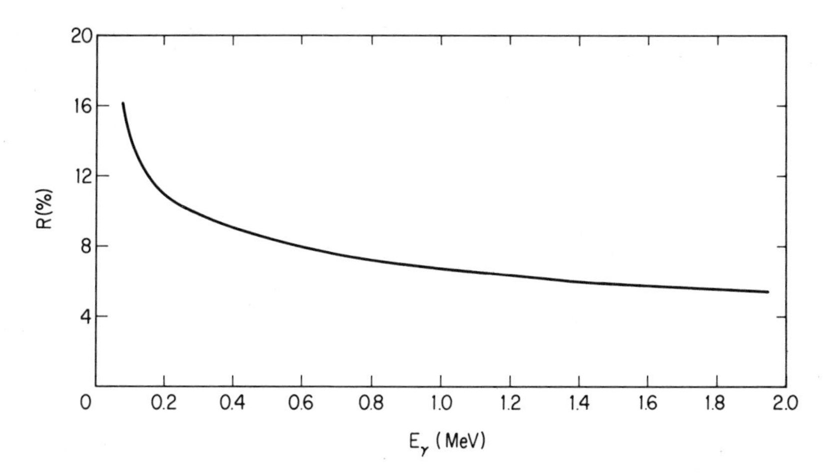

Fig. 7-12. A plot of energy resolution vs. γ-ray energy for a 3 × 3-in. solid NaI(Tl) crystal.

or not a given experiment involving detection of mixtures of γ-emitting radionuclides is possible.

E. IDENTIFICATION OF RADIONUCLIDES

In Sections A and B, we presented a qualitative discussion of the nature of γ-ray spectra as measured with NaI(Tl) scintillation detectors. We discussed the origins of the various peaks and other features in the spectra. In this section we want to discuss how to determine the *identity* of γ-emitting radionuclides from their pulse height spectra. (Section F deals with measuring the *amounts* of nuclides present.) Although we shall use NaI(Tl) detectors in our examples, the methods and techniques mentioned, with few exceptions, are equally applicable to γ-ray spectra measured with Ge(Li) detectors (as covered in Chapter 8).

Three primary characteristics of γ-ray decay of a radionuclide are used to identify its presence. These "fingerprints" of a nuclide are

1. The energy of the γ-rays emitted.
2. The half-life of the radionuclide.
3. The ratio of γ-ray intensities when a nuclide emits more than one γ-ray.

How does one go about measuring these "fingerprints" and using them as an identifier of radionuclides?

Consider a typical γ-ray spectrum of an unknown mixture of radionuclides, as shown in Figure 7-13. The first step in determining what radionuclides are present is to establish an energy calibration for the pulse height analyzer—that is, a relationship between the γ-ray energy deposit in the detector and the analyzer channel number. This is usually done by measuring, under the same conditions as the unknown sample, the γ-ray spectra of several radionuclides of known γ-ray energy. Table 7-2 shows some typical radionuclides used for spectrometer energy calibration.

TABLE 7-2

Radionuclides Used in Calibrating Scintillation Spectrometers

Nuclide	$E\gamma$(keV)	$t_{1/2}$
^{109}Cd	88.0	453 days
^{203}Hg	279.12 $\pm$ 0.05	47 days
(Annihilation radiation)	511.006 $\pm$ 0.02	—
^{137}Cs	661.59 $\pm$ 0.07	30.0 years
^{54}Mn	834.84 $\pm$ 0.07	303 days
^{60}Co	1173.13 $\pm$ 0.04	5.26 years
^{60}Co	1332.39 $\pm$ 0.05	5.26 years

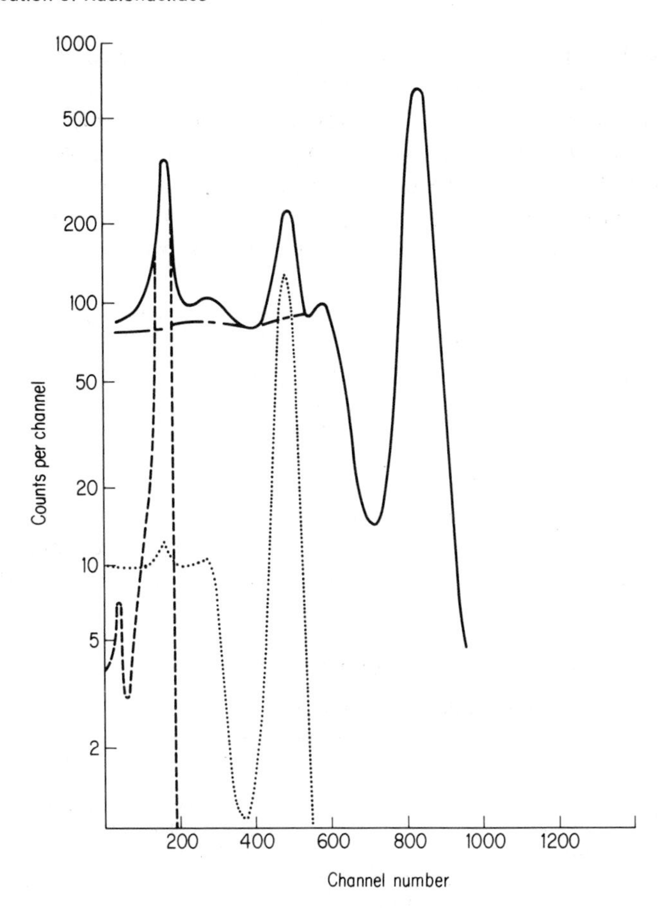

Fig. 7-13. A three-component γ-ray spectrum. The individual γ-ray spectra are shown as dashed lines, while the composite spectrum is shown as a solid line. From Heath (1).

One can associate with each photopeak in the pulse height spectrum a channel number N (corresponding to the maximum point in the peak) and a γ-ray energy, E_γ, using the linear relationship*

$$E_\gamma = aN + b \tag{7-4}$$

where the constants a and b are determined by measuring the spectra of known energy γ-ray emitters. For the spectrum shown in Figure 7-13, the energy calibration equation was determined to be

$$E_\gamma \text{ (MeV)} = 0.001N + 0.0005$$

*Usually it is sufficient to assume a linear relationship between pulse height and energy deposit in the detector for a NaI(Tl) detector. For accurate measurements of γ-ray energies, one must take into account instrumental nonlinearities and the well-known nonlinear response of NaI(Tl) crystals for $E_\gamma < 250$ keV (see Reference 2 for details).

and thus the three peaks at channels 159, 478, and 835 correspond to the deposit of 0.160, 0.478, and 0.835 MeV in the detector.

Spectra of the mixture can be taken as a function of time, and the peak areas (see Section F) can be used to determine the half-lives of the nuclides present (see Chapter 2). The peak areas can also be used to determine the relative intensities of each γ-ray present. Let us assume that such measurements have been made and that the results are as shown in the table.

E_γ(keV)	$t_{1/2}$	Relative Intensity
155	3.4 days	9
478	54 days	12
835	290 days	100

The question remains as to what radionuclides are present. Extensive compilations exist (2, 3, 7) of the γ-ray energies, intensities, and half-lives associated with all known γ-ray emitters. Table 7-3 is an abbreviated version of such compilations.

TABLE 7-3

Energies and Intensities of Common Gamma Ray Emitters

E_γ (keV)	Relative Intensities (%)	$t_{1/2}$	Energies of Other Gamma Rays (keV)	Relative Intensities of Other Gamma Rays (%)	Nuclide
35	100	60.2 days	—	—	^{125}I
37	41	2.4×10^4 yr	120, 38	22, 19	^{239}Pu
60	40	2.35 days	278, 228	31, 28	^{239}Np
106.4	50	458 yr	26	3	^{241}Am
121.9	15	270 days	14	2	^{57}Co
127.4	98	2.895 hr	137	1	^{134m}Cs
160.0	100	3.43 days	—	—	^{47}Sc
187	100	1620 yr	—	—	^{226}Ra
320.1	100	27.8 days	—	—	^{51}Cr
364	80	8.05 days	637, 284	9, 5	^{131}I
411.8	100	2.7 days	—	—	^{198}Au
442.7	14	24.99 min	—	—	^{128}I
477	100	53.6 days	—	—	^{7}Be
617.0	100	17.6 min	—	—	^{80}Br
834.8	100	303 days	—	—	^{54}Mn
846.7	100	2.576 hr	1811, 2112	30, 15.3	^{56}Mn
1099.3	57	45.1 days	1291, 192	43, 3	^{59}Fe
1173.2	100	5.26 yr	1333	100	^{60}Co
1274.6	100	2.62 yr	—	—	^{22}Na
1293.8	100	110 min	—	—	^{41}Ar
1345.8	100	12.8 hr	—	—	^{64}Cu
1368.4	100	14.96 hr	2754	100	^{24}Na
1778.7	100	2.31 min	—	—	^{28}Al
2167.6	47	37.3 min	1643	31	^{38}Cl
3084.4	89	8.8 min	4072	10	^{49}Ca

A quick comparison of the γ-ray energies, intensities, and half-lives associated with the nuclides of Figure 7-13 and given in Table 7-3 shows the mixture to contain ^{47}Sc, ^{7}Be, and ^{54}Mn. In making identifications of unknown radionuclides, one should be careful to ensure that not only do the measured γ-ray energies agree with those of known γ-ray emitters, (within $\pm 0.5\,\%$) but that the observed half-lives and intensities also agree.

F. QUANTITATIVE ANALYSIS OF GAMMA RAY SPECTRA

Let us consider how to determine the *quantity* of γ-emitting radionuclides present in a sample by using γ-ray spectroscopy. Consider a typical γ-ray pulse height spectrum, as shown in Figure 7-14. Once the identity of the radionuclide responsible for a given spectral peak has been determined, the peak area (the shaded area in Figure 7-14) can be used as a quantitative measure of how many nuclei are present.

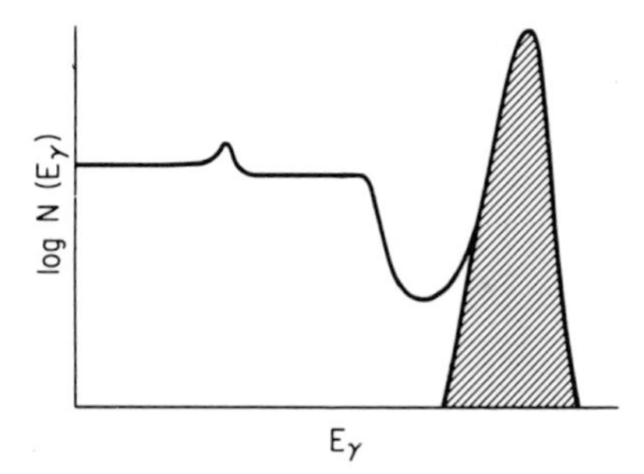

Fig. 7-14. A schematic diagram of a typical γ-ray pulse height spectrum as measured with a NaI(Tl) detector.

The actual determination of the number of nuclei present can be done on a *relative* or *absolute* basis. In activation analysis and most radiotracer work, activity determinations are made on a relative basis. A known amount of the radionuclide in question and the sample whose content is to be determined are counted under exactly the same conditions for the same length of time. Then one can say that the activity of the unknown sample, D_u, is given as

$$D_u = D_k \frac{A_u}{A_k} \tag{7-5}$$

where D_k is the activity of the known sample and A_u and A_k are the γ-ray photopeak areas for the unknown and known samples, respectively. When doing relative counting, one must be careful that both known and unknown samples have the same size and shape, are counted in identical geometries, and are similar in activity (to minimize any count rate effects, etc).

In nuclear chemical and environmental applications, activity determinations are usually made on an absolute basis. In absolute counting, the problem is to calculate the disintegration rate of the radionuclide in the source,

D, (dps), by knowing just the photopeak area, A_p, (cps), in the pulse height spectrum. These two quantities are related by the following equation:

$$A_p = D\epsilon_T f_p \Omega f_g \tag{7-6}$$

where ϵ_T is the total probability that any γ-ray of a given energy striking the detector will produce a measurable pulse (*the total efficiency*), f_p is the fraction of those events in the total spectrum that correspond to photopeak events (*the photofraction*), Ω is the solid angle subtended by the detector with respect to the source, and f_g is the fraction of disintegrations giving rise to γ-ray.

The factors ϵ_T, Ω, and f_g have been discussed in Chapter 6, and plots of $\epsilon_T\Omega$ for various detector-source distances and γ-ray energies are shown in Figure 6-13. Table 6-2 contains a list of f_g factors for common radionuclides. The photofraction, f_p, is simply the ratio of the number of counts in the photopeak to the total number of counts in the spectrum (*the peak-to-total ratio*). A plot of the photofraction versus the γ-ray energy is shown in Figure 7-15

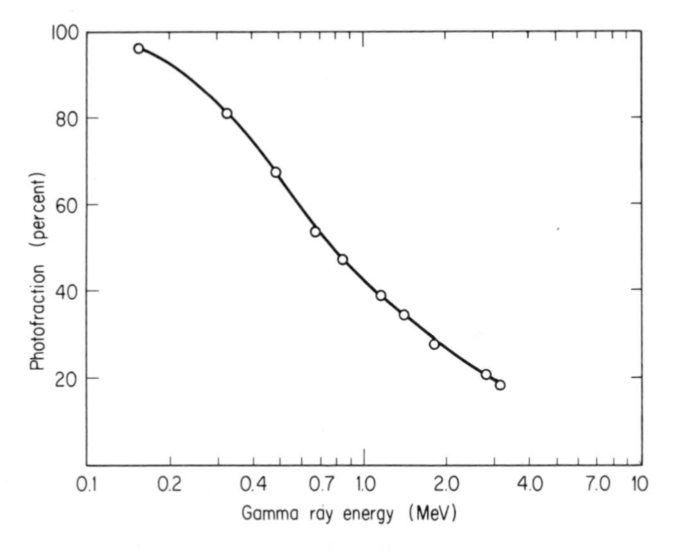

Fig. 7-15. Experimental photofractions for a thallium-activated sodium iodide crystal 3 in. dia. × 3 in. long. Point source on axis 10 cm from crystal face. From IDO-16408, R. L. Heath, *Scintillation Spectrometry*, Gamma Ray Spectrum Catalog. Phillips Petroleum Co., Atomic Energy Div., Idaho Falls, Idaho.

and demonstrates how sharply this quantity decreases with increasing γ-ray energy. Since the photopeak is used for most quantitative work with γ-ray emitters, it is important to have a detector with a large photofraction. The larger the detector, the greater the photofraction is (as shown in Figure 7-16). Note, however, that the gain in efficiency is not large for low-energy γ-rays.

Occasionally, rather than reporting the amount of a given radionuclide that is present, we want to be able to state whether the nuclide is present or

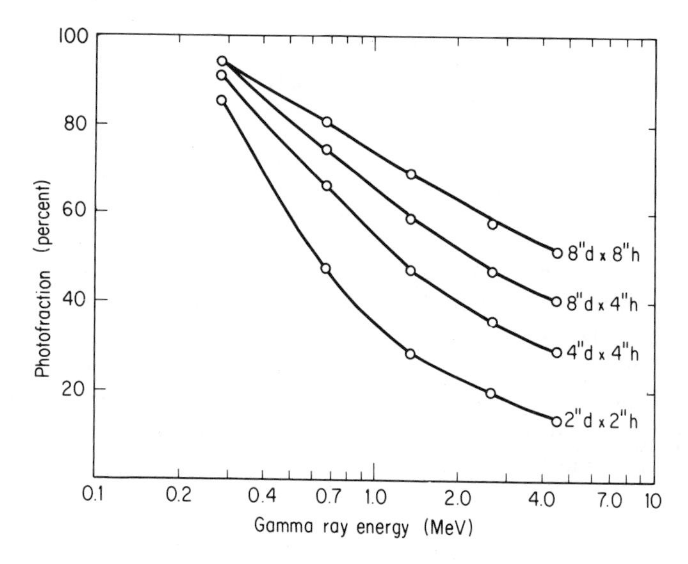

Fig. 7-16. W. F. Miller, John Reynolds, and W. J. Snow. "Calculated efficiencies and photofractions for various size Thallium Activated Sodium Iodide Crystals," Review of Scientific Instruments **28**, 717: (1957).

not. This situation is encountered frequently in environmental monitoring and raises the question of determining the *detection sensitivity* of one's spectrometer. The question of detection sensitivity is a complex one and has been treated in detail in Currie's excellent article (9). Currie points out that three limits are involved in the detection sensitivity:

1. L_C, the *a posteriori* limit. What level of *observed* activity is necessary before we can say that a nuclide has been *detected*?
2. L_D, the *a priori* limit. What level of activity would be expected, *a priori*, to allow detection of a radionuclide?
3. L_Q, the limit of quantitative determination. What level of activity would be necessary to make a precise quantitative determination of the abundance of the nuclide?

Table 7-4, taken from Currie's article, shows what level of activity is necessary for each of the limits, L_C, L_D, and L_Q. The activity levels are expressed in terms of the standard deviation of the "background" under the photopeak of interest in the γ-ray spectrum, σ_B.* (See Chapter 12 for a discussion of standard deviations.)

*If the background under the photopeak of interest was 100 cpm, $\sigma_B = \sqrt{100} =$ 10 cpm. According to Table 7-4, one would need to detect a net photopeak area of 23 cpm in order to say that a nuclide was present and would need to detect at least 141 cpm to measure the amount of nuclide present. One would need a count rate of 47 cpm to ensure, before making the measurement, that the nuclide in question could be detected.

TABLE 7-4

Activity Levels Defining L_C, L_D, L_Q

L_C	L_D	L_Q
$2.33\sigma_B$	$4.65\sigma_B$	$14.1\sigma_B$

As can be seen from Table 7-4, the detection limits depend on the detector efficiency and the detector background. The lower the counter background, the lower the levels of radioactivity that can be detected. Thus integral counting leads to a higher detection limit than γ-ray spectrometry because the background counting rate in the region around the photopeak is much lower than the total background of an integral counting system. For similar reasons, spectrometers with the best energy resolution are preferred for low-level work.

G. SINGLE-CHANNEL COUNTING

Single-channel counting is a variant of γ-ray counting that falls between integral counting and spectroscopy. In a single-channel measurement, the number of pulses from the detector that fall in one preset range of pulse heights (corresponding to a range of energies) are counted. The experimental apparatus for doing so is shown in Figure 7-17. Pulses from the detector

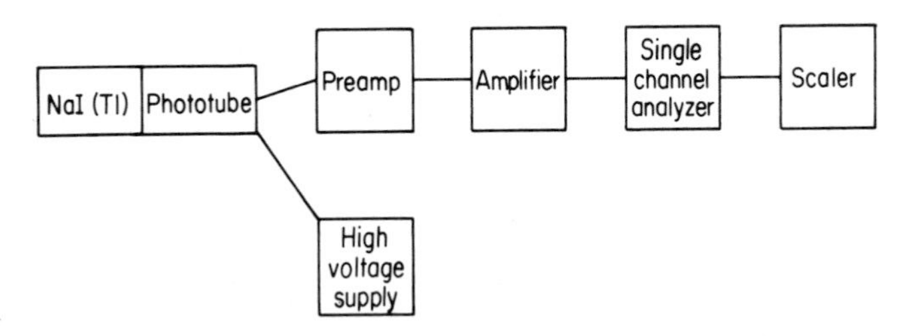

Fig. 7-17. Scnematic diagram of a single-channel counting setup.

are amplified, shaped, and passed on to a single-channel analyzer. If the pulses are within the single-channel-analyzer window, the single-channel analyzer is triggered and gives a pulse that is counted by the scaler. Since the single-channel analyzer can be used to "straddle" a photopeak in a γ-ray spectrum, it can be used specifically to count γ-rays from a given nuclide. Problems with fluctuation in electrical noise that plague integral counting are largely absent from single-channel counting. However, since one is looking at a very small portion of the spectrum through the preset single-channel-analyzer window, small changes in the photomultiplier-potential, amplifier-gain, single-channel-analyzer window can cause significant changes in counting rate due to spectral shifts. Of particular importance in this regard is the shift in photomultiplier gain as a function of counting rate. Shifts of up to 10% in the position of a ^{137}Cs photopeak have been reported for shifts in counting rate from 100 to 100,000 cps. Furthermore, one is advised never to try to analyze sets of multicomponent γ-ray spectra using single-channel counting, for shifts on the level of Compton counts from one nuclide can affect the photopeak count of another nuclide.

Since single-channel analyzers are one to two orders of magnitude cheaper than multichannel analyzers, single-channel counting is favored by experimenters with limited funds. And because of the simplicity of the circuitry involved in single-channel counting, it is frequently used in tracer experiments involving the use of a single γ-emitting radionuclide.

BIBLIOGRAPHY

1. HEATH, R. L. *Scintillation Spectrometry.* USAEC Report IDO-16880, 1964. The bible of scintillation spectroscopy.
2. ADAMS, F., and R. DAMS. *Applied Gamma Ray Spectrometry.* 2nd ed. Oxford: Pergamon, 1970. The most up-to-date, comprehensive monograph on scintillation spectroscopy.
3. FILBY, R. H., *et al. Gamma Ray Energy Tables for Neutron Activation Analysis.* Washington State University Report WSUNRC–97(2), 1970. Tables of gamma ray energies and half-lives for nuclides produced by neutron irradiation.
4. KRUGER, P. H. *Principles of Activation Analysis.* New York: Wiley, 1971. A highly recommended, thorough discussion of activation analysis with a good discussion of scintillation spectroscopy.
5. SHAFROTH, S. M. *Scintillation Spectroscopy of Gamma Radiation.* Vol. I. London: Gordon and Breach, 1966. A collection of articles by experts dealing with various aspects of scintillation spectroscopy.
6. WOLICKI, E. A., R. JASTRAW, and F. BROOKS. "Calculated efficiencies of NaI crystals," *NRL Report 4833.*
7. LEDERER, C. M., J. M. HOLLANDER, and I. PERLMAN. *Table of Isotopes.* 6th ed. New York: Wiley, 1968.
8. O'KELLEY, G. D. *Detection and Measurement of Nuclear Radiation.* National Academy of Sciences Report NAS-NS-3105.
9. CURRIE, L. A. "Limits for qualitative detection and quantitative determination," *Anal. Chem.* **40**, 587(1968).

8

Semiconductor Radiation Detectors

Most modern research in nuclear chemistry and physics utilizes semiconductor radiation detectors. Although tracer research has not made extensive use of this type of radiation detector to date, it seems inevitable, because of the superior characteristics of these detectors, that their use will increase in the coming years. The basic operating mechanism of the semiconductor radiation detector is similar to that of the gas ionization detectors discussed in Chapter 5. In both types of detector, ionizing radiation interacts with the detector material (a gas in one case and a solid in the other), and the positive- and negative-charged species that result are collected to form an electrical signal. The differences between the detectors involve the nature of the positive-and negative-charged species (an ion and electron for the gas ionization detector, an electron and a "hole" in the other case) and the method of charge collection employed.

A. BASIC NATURE OF SEMICONDUCTORS

In order to understand how semiconductor radiation detectors operate, it is necessary to review a little of the basic chemistry of semiconductors. Consider a typical Group IV element, such as Si or Ge. It will crystallize in the diamond lattice structure, as shown in Figure 8-1. Each silicon atom is bound by four electron-pair bonds to adjacent silicon atoms. The electrons are not free to migrate through the crystal, and therefore pure silicon is a

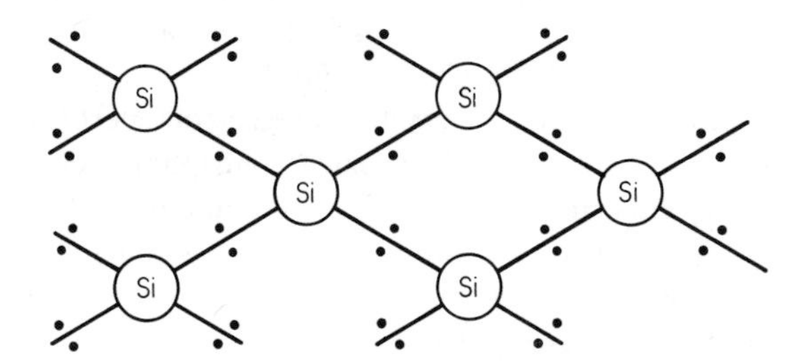

Fig. 8-1. Schematic view of the crystal lattice of Si. The dots represent electron pair bonds between the Si atoms.

poor conductor of electricity. Speaking in terms of the electron energies, we depict silicon as shown in Figure 8-2. The energy levels of the valence electrons of the entire solid are so close together as to form a "band" of energies, known as *the valence band.* There is a region of energy just above the top of the valence band in which there are no allowed energy levels in *pure* silicon.

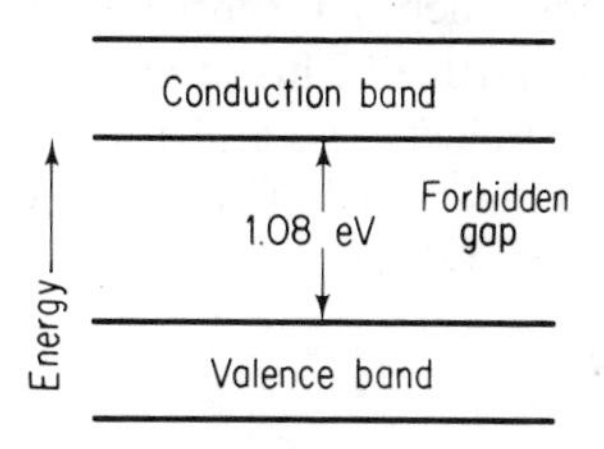

Fig. 8-2. Schematic diagram of the energy levels of crystalline silicon according to the band theory of solids. The energy of the forbidden gap, $\sim$1.08 eV, is shown in the diagram.

This energy region is known as *the forbidden gap* and has a magnitude of $\sim$1.08 eV for silicon. Just above the forbidden gap is *the conduction band,* that band of electron energies corresponding to free electron migration through the crystal—that is, conduction of electricity.

Suppose that we replace a silicon atom in the silicon lattice by a Group V atom, such as phosphorus. Then we have the situation depicted in Figure 8-3. Phosphorus has five valence electrons, so after forming bonds to the four

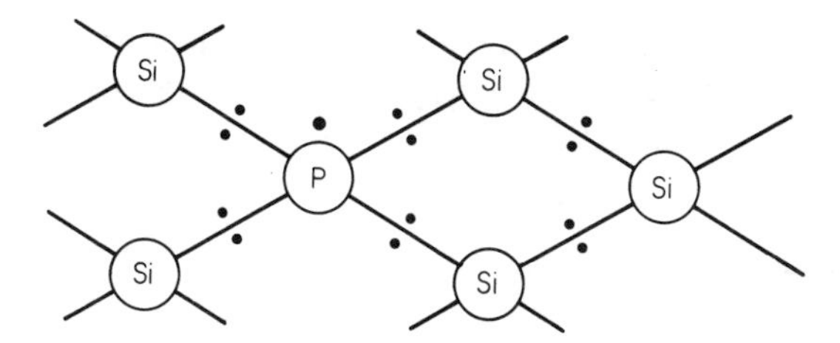

Fig. 8-3. Schematic diagram of a typical *n*-type impurity in a silicon crystal lattice.

adjacent silicon atoms, one electron is "left over." This "leftover" electron will be very loosely bound to the phosphorus atom and will easily be torn loose to conduct electricity through the crystal.

In terms of our diagrams of the crystalline-electron-energy levels, we have the situation shown in Figure 8-4. The "extra" phosphorus electron occupies a "donor level" very close to the conduction band and is easily promoted into this conduction band. Silicon containing Group V impurities, such as phosphorus, is called *n-type silicon* because the species that carries charge through the crystal (*charge carrier*) is *n*egative.

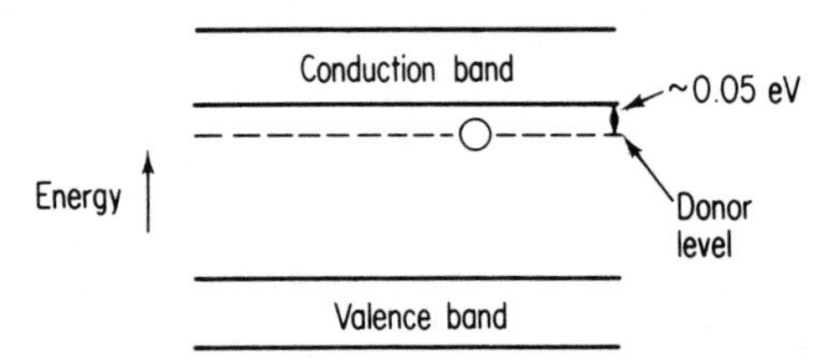

Fig. 8-4. Schematic diagram of the energy levels of crystalline Si with a donor impurity.

What happens when a Group III element like boron replaces a silicon atom in a silicon lattice? This situation is shown in Figure 8-5. Boron has only three valence electrons, and thus after forming bonds to three silicon atoms, it has no electron left to pair up with the valence electron on the fourth silicon atom. We are said to have an electron *hole* in the silicon lattice.

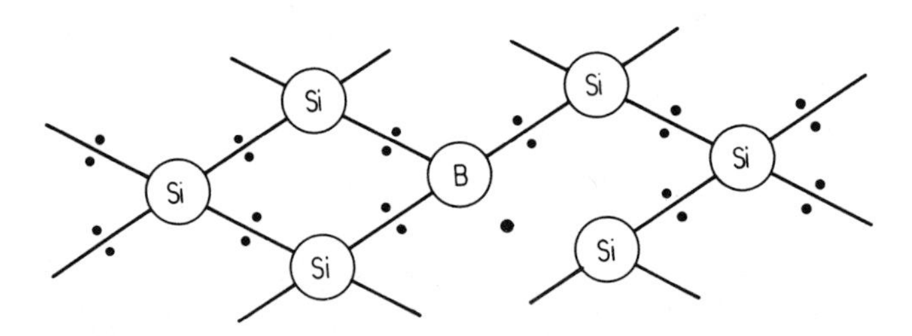

Fig. 8-5. Schematic diagram of a silicon crystal lattice with a *p*-type impurity in it.

In terms of our energy level diagrams, we have the situation illustrated in Figure 8-6. The hole occupies an energy level very close to the valence band (*the acceptor level*) and can easily be "promoted" into the valence band.

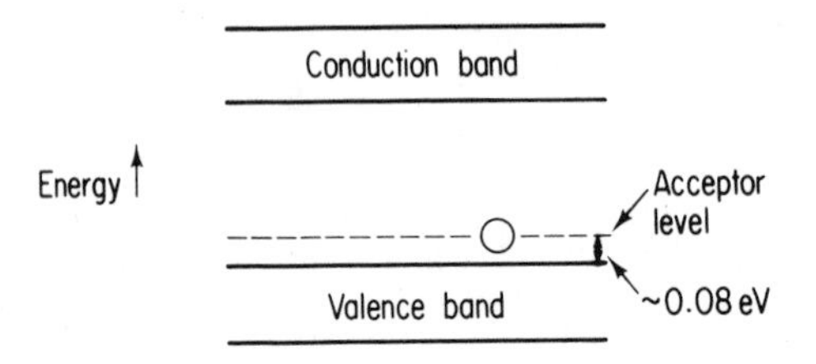

Fig. 8-6. Schematic diagram of energy levels of Si crystal lattice showing acceptor lever filled by *p*-type impurity.

("Promotion" of a hole into the valence band simply means that an electron in the valence band and a hole in an acceptor level switch levels, so that a hole is created in the valence band.)

We must realize, unphysical as it may sound, that a hole in the valence band, can conduct electricity as well as an electron in the conduction band. How does this work? Consider Figure 8-7. Imagine that electron 1 moves to fill hole 0. This step creates a hole at position 1. Electron 2 moves to fill this hole, leaving a hole at position 2. Electron 3 fills the hole at position 2, leaving a hole at position 3, and so forth. Thus as the hole moves to the right in Figure 8-7, negative charge is moving toward the left. Since electricity is the movement of charge, the motion of the hole corresponds to the flow of electricity. Silicon containing Group III impurities is said to be *p-type silicon* because of the *p*ositive charge carriers (the holes).

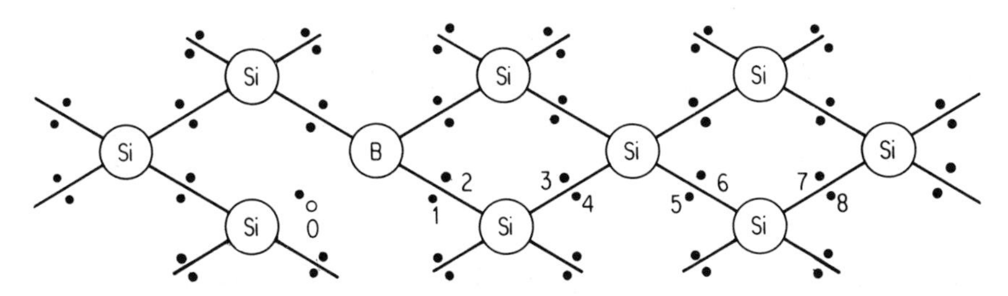

Fig. 8-7. Schematic diagram of silicon crystal lattice with a *p*-type impurity boron, at one lattice point. The hole is labeled 0, while the electrons are denoted by the numbers 1, 2, . . . , 8.

B. BASIC OPERATING PRINCIPLES OF SEMICONDUCTOR RADIATION DETECTORS

A silicon-semiconductor-radiation detector of a layer of *p*-type silicon in contact with a layer of *n*-type Si is shown in Figure 8-8. What happens when this *p-n junction* is created? The electrons from the *n*-type silicon will migrate across the junction and fill the holes in the *p*-type silicon to create an area around the *p-n* junction in which there is no excess of holes or electrons. (We say that a "*depletion region*" has been formed around the junction.) Imagine that we apply a positive voltage to the *n*-type material and a negative voltage to the *p*-type material (the junction is said to be *reverse-biased*). The electrons will be "pulled farther away" from the junction by the positive voltage on the *n*-type material, thus creating a much thicker depletion region around the *p-n* junction.* The exact thickness of the depletion region, *d*, is

*Note that if we apply a positive voltage to the *p* side and a negative voltage to the *n* side, we force a large electron-hole migration across the junction; that is, the junction conducts electricity well. Thus by changing the sign of the voltage applied to each side of the *p-n* junction, we can cause it to conduct or not to conduct. This is the basic action of a *diode*.

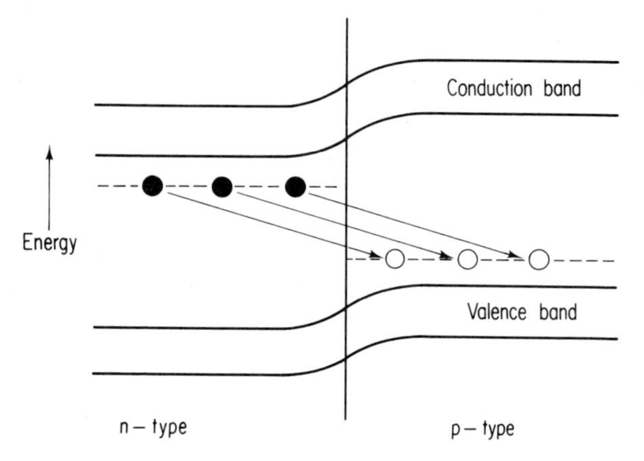

Fig. 8-8. Schematic diagram of the foundation of a *p-n* junction. The difference in energy between the conduction and valence bands in the *p*-type and *n*-type material is real.

given by

$$d \propto \sqrt{\rho V} \tag{8-1}$$

where ρ is the resistivity of the silicon and V is the magnitude of the applied reverse-bias voltage. Note that the depletion of the depletion region can be varied at will by changing the voltage applied to the detector.

The depletion region acts as the sensitive volume of the detector. The passage of ionizing radiation through this region will create holes in the valence band and electrons in the conduction band. The electrons will migrate to the positive charge on the n side, while the holes will migrate to the negative voltage on the p side, thereby creating an electrical pulse at the output of the device.

The *average* energy necessary to create a hole-electron pair in silicon is ~3.6 eV. [This average energy is about three times the forbidden gap energy (~1.1 eV) because most electrons are promoted from deep in the valence band to high in the conduction band.] The energy required to create a hole-electron pair is independent of particle charge and mass, thus causing semiconductor detector response to be independent of particle type, such as α and β^-. If we remember (Chapter 5) that the average energy to create an ion-electron pair in a gas ionization device was ~30 eV, then we see that, for the same energy deposit in the detector, we get $\sim 30/3.6 = 8.3$ times more charged pairs. If we note (as shown in Chapter 12) that the energy resolution of a detector, $\Delta E/E$, is proportional to $N^{-1/2}$, where N is the number of charge pairs formed, we can see that the energy resolution of a semiconductor is approximately $\sqrt{8.3} = 2.9$ times better than the energy resolution of a gas ionization detector. (Furthermore, if we remember that the average (γ-ray energy deposit required to liberate one photoelectron at the cathode of a photomultiplier tube is ~ 1000 eV, then we say that the resolution of a semiconductor detector is $\sqrt{1000/3.6} \approx 17$ times better than that of a scintillation detector.)

For some semiconductor detectors, germanium is used instead of silicon for the detector material. The reasons for this substitution are as follows: (a) The average energy needed to create a hole-electron pair in germanium is 2.9 eV rather than the 3.6 eV necessary for Si. Thus the energy resolution for germanium should be $\sqrt{3.6/2.9} = 1.1$ times better than silicon. (b) The atomic number of germanium (32) is much higher than that of silicon (14), leading to increased probability of γ-ray interaction with the detector material. Consequently, germanium is preferred over silicon for γ-ray detection. The forbidden gap is so small, however, for germanium (0.66 eV) that room-temperature thermal excitation leads to the formation of hole-electron pairs in the solid. Therefore germanium detectors must be operated at liquid nitrogen temperature (77°K) to prevent this phenomenon from occurring.

C. BASIC TYPES OF SEMICONDUCTOR RADIATION DETECTORS

There are three basic types of semiconductor radiation detectors: the diffused *p-n* junction detector, the surface barrier detector, and the lithium-drifted detector. In all three types of detectors, a high positive potential is created on the *n*-side of the detector, with electrons rapidly migrating to this region and being collected there. The primary differences between the different types of detectors concern the physical techniques used to form the *p-n* junction and the depleted region.

Consider the diffused *p-n* junction detector of Figure 8-9. In constructing

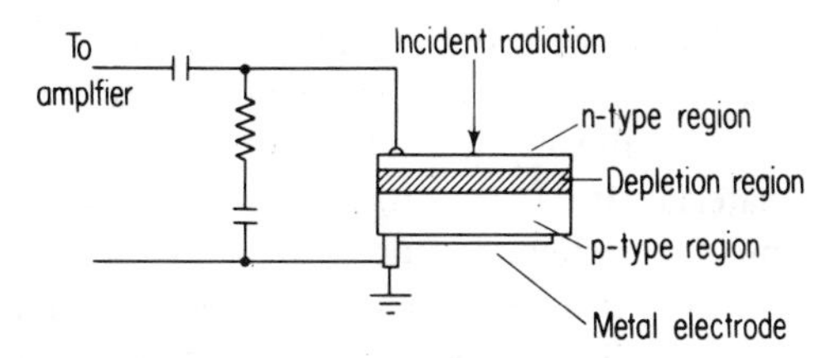

Fig. 8-9. Schematic diagram of a diffused *p-n* junction detector showing how a *p-n* junction is created in a slab of *p*-type silicon by diffusing Group V impurity, phosphorus, into the crystal lattice.

this type of detector, a Group V impurity, phosphorus, is thermally diffused partway into a slab of *p*-type silicon. This step creates an *n*-type region in the silicon, where phosphorus is the dominant impurity, and a *p-n* junction at the interface between the *p*-type silicon and the silicon containing the phosphorus impurity.

Diffused *p-n* junction detectors are quite rugged and their operating characteristics not too sensitive to ambient conditions. Such detectors have been boiled in oil, dried, and still continue to operate normally.

Surface barrier detectors are similar in basic construction to the diffused *p-n* junction detectors except that a *p*-type layer is produced in a slab of *n*-type silicon (see Figure 8-10). The *p*-type layer is produced by chemically etching the surface of the detector and allowing spontaneous oxidation to take place at the surface. The "surface states" created by this spontaneous oxidation induce a high density of holes, thus forming a *p* layer. By means of evaporation, a thin gold film is deposited on the surface in order to make electrical contact with it.

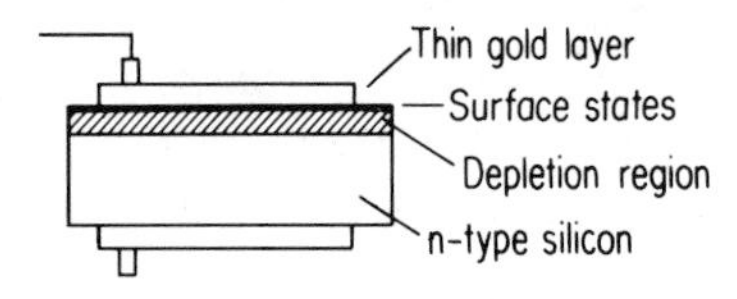

Fig. 8-10. Schematic diagram of a surface barrier detector showing depletion region at interface on *n*-type bulk material and thin *p*-type layer produced on the detector surface by oxidation.

Surface barrier detectors generally exhibit better energy resolution than a diffused *p-n* junction device of similar size. They are more sensitive to the ambient atmosphere, however, and must be stored and used in areas that are free from chemical fumes or water vapor.

Surface barrier and diffused *p-n* junction detectors are the best detectors available for low-energy, heavy charged particles, such as α-particles. Typical detector energy resolutions are of the order of $\sim$10 to 20 keV with 100% detection efficiency for all particles striking the detector. The cost of these detectors is low.

Practical limitations in the construction of surface barrier and diffused *p-n* junction detectors restrict the depletion depths to less than $\sim$2 mm. Since such depletion depths will stop only $\sim$1.1-MeV β^--particles, 18-MeV protons, and so on, other methods of detector construction must be used for detectors for energetic-charged particles, electrons, and γ-rays. The lithium-ion-drift technique is used to produce detectors with large depletion depths.

The basic processes that take place in making lithium-drifted detectors are shown in Figure 8-11. A *p-n* junction is formed in a piece of *p*-type material by diffusing lithium (as donor or "*n*" impurity) into the surface of the material. Because of its small size and high mobility, lithium rapidly diffuses into the semiconductor lattice and takes up interstitial positions. [See Figure 8-11(a).] A reverse bias is applied to this junction and the temperature is elevated. The lithium ions move toward the negative potential on the *p*-type material. This ion drift continues until the lithium (donor) concentration exactly balances the acceptor concentration in the *p*-type material. In terms of chemical equations, we have

$$\begin{array}{ccc} \text{Li} & & \text{A (Acceptor)} \\ \updownarrow & & \updownarrow \\ (\text{Li}^+ + e^-) & + & (\text{A}^- + e^+) \longrightarrow \text{Li}^+\text{A}^- \end{array} \tag{8-2}$$

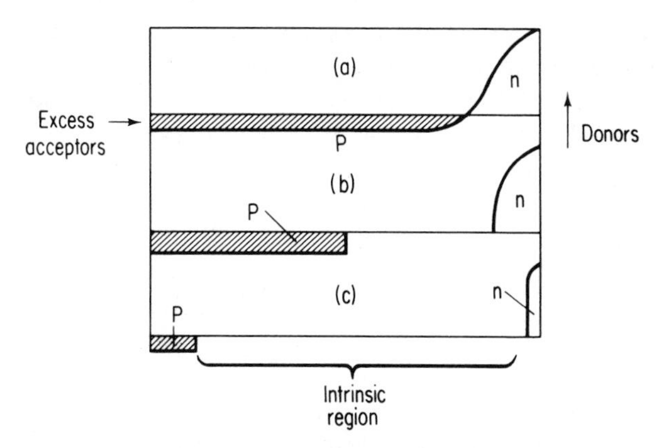

Fig. 8-11. Schematic diagram of the lithium drift technique. (a) Density of donors and acceptors after initial formation of *p-n* junction. (b) and (c) Density of donors and acceptors after diffusion of Li into crystal under reverse bias at elevated temperatures. (c) is a later time than (b). Note that an "intrinsic region" has been formed in which the donor and acceptor concentrations exactly cancel one another. From Camp (6).

Because of the electrical fields present in the crystal, the drifting process is self-governing; that is, no Li excess or deficiency can arise over the drifted region.

The result of the drifting process is the creation of an "intrinsic" or "compensated" region of considerable thickness where the number of donors exactly equals the number of acceptors. This intrinsic region acts as the detector depletion region and because of its large thickness can be used to detect penetrating radiations, such as β^-, X rays, γ rays, or energetic-charged particles. Thus one has "chemically" as well as "electrically" created a region within the detector that is "pure"—that is, no excess of one type of charge carrier.

Two types of lithium-drifted detector are used: the lithium-drifted silicon detector [Si(Li) or "silly" detector] and the lithium-drifted germanium detector (Ge(Li) or "jelly" detector). Si(Li) detectors are the detectors of choice for detecting β^- particles, X rays, and energetic-charged particles. Si(Li) detectors are favored over Ge(Li) detectors for β^- detection because of their low γ-sensitivity and their lower (by $\sim\frac{1}{3}$ to $\frac{1}{2}$) backscattering. The energy resolution of Si(Li) detectors for electrons is $\sim$1 to 2 keV for electron energies up to 1000 keV. The detection efficiency of Si(Li) detectors for β^- particles ranges from one-half that of a gas counter for a low-energy beta emitter like ^{14}C to greater than that of a gas counter for an energetic β^- emitter such as ^{32}P. The background of these detectors is exceptionally low, because of their small size for a given stopping power, and they do not require any peripheral gas supply, and so on. The energy resolutions achieved in X-ray detection with Si(Li) detectors are impressive. Measured energy resolutions of 180 eV for the 5.9-keV Mn K_α X-ray are routinely obtained with Si(Li) detectors, whereas the best energy resolution available with a scintillation counter is ~1000 eV. This superb X-ray energy resolution has opened up new vistas in radioanalytical chemistry (see Chapter 18).

For γ-ray detection, the detector of choice is the Ge(Li) detector. The reason is the higher γ-ray-absorption cross section for Ge as compared to Si. For example, in Chapter 3 we noted that the photoelectric process is proportional to Z^5. Germanium ($Z = 32$) is therefore ~62 times more efficient than silicon for photoelectric absorption.

The most spectacular feature of the use of Ge(Li) detectors for measuring γ-radiation is their superior energy resolution. Energy resolutions of ~2 to 3 keV for the 1332-keV γ-ray of ^{60}Co are routinely obtained and should

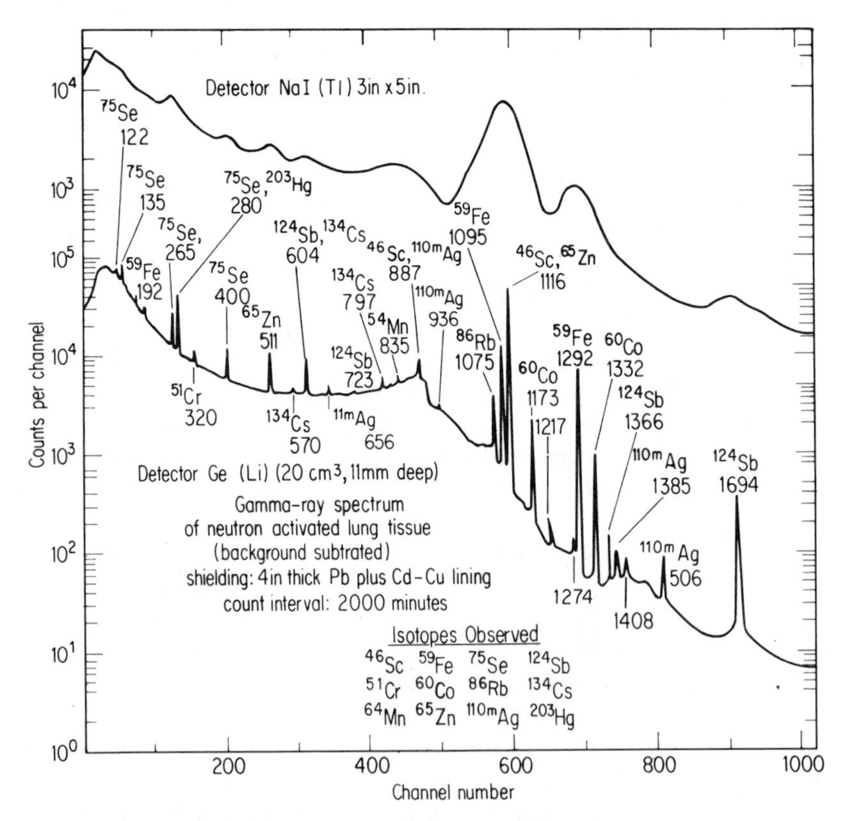

Fig. 8-12. The γ-ray spectrum of neutron-activated lung tissue as measured with a 3 in. thick by 5 in. diameter NaI(Tl) detector (top curve) and a 20 cm^3 Ge(Li) detector (bottom curve). From Cooper (7).

be compared to the typical energy resolution of 90 to 100 keV for a 3 × 3-in. NaI(Tl) crystal. What this number means in terms of actual γ-ray spectra is shown in Figures 8-12, 8-13, and 8-14.

Here typical γ-ray spectra as measured with Ge(Li) detectors are given. Note that in Figures 8-12 and 8-13 the same spectrum, as measured with a 3 × 5-in NaI(Tl) detector, is also displayed. Note that in these complex γ-ray spectra, we can discern little about the identity of the radionuclides present from the NaI(Tl) spectrum but can identify a large number of nuclides using the Ge(Li) spectrum.

The superior energy resolution of the Ge(Li) detector is obtained at the expense of detection efficiency. The efficiencies of Ge(Li) detectors are one to two orders of magnitude less than the standard 3 × 3-in. NaI(Tl) crystal. This situation is shown in Figure 8-15, where the efficiencies of several Ge(Li) detectors are plotted versus γ-ray energy, along with the efficiency of a 3 ×

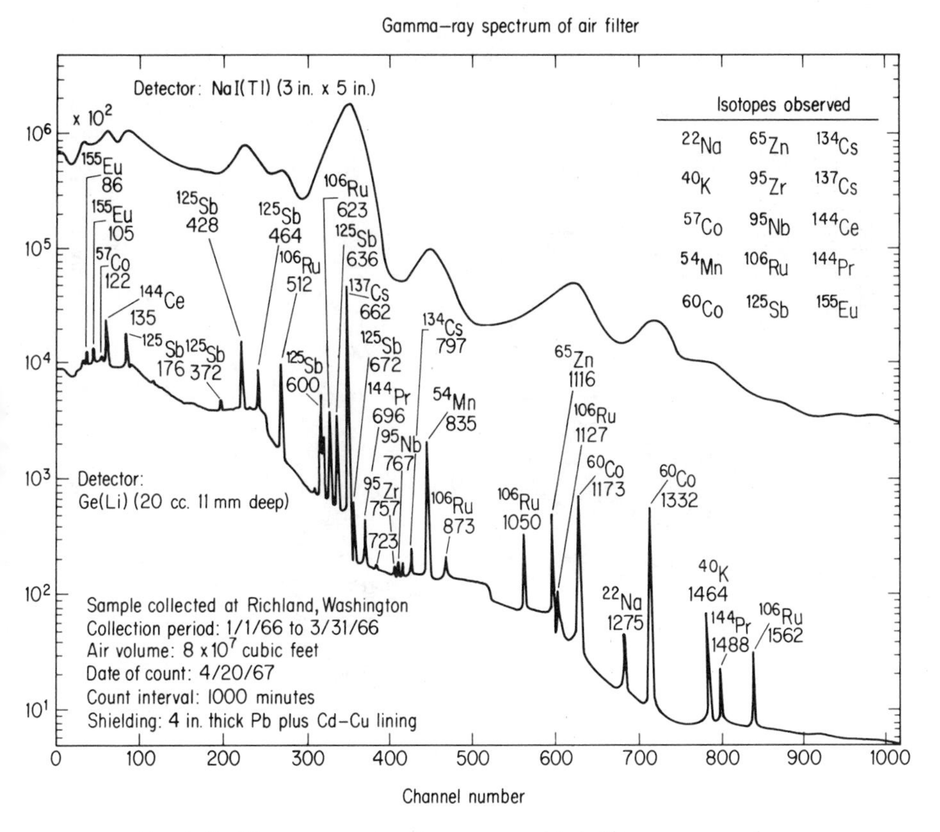

Fig. 8-13. The γ-ray spectra of a one-year old air filter as measured on a 5 in. diameter by 3 in. thick NaI(Tl) detector (top curve) and a 20 cm³ Ge(Li) detector (bottom curve). From Cooper (7).

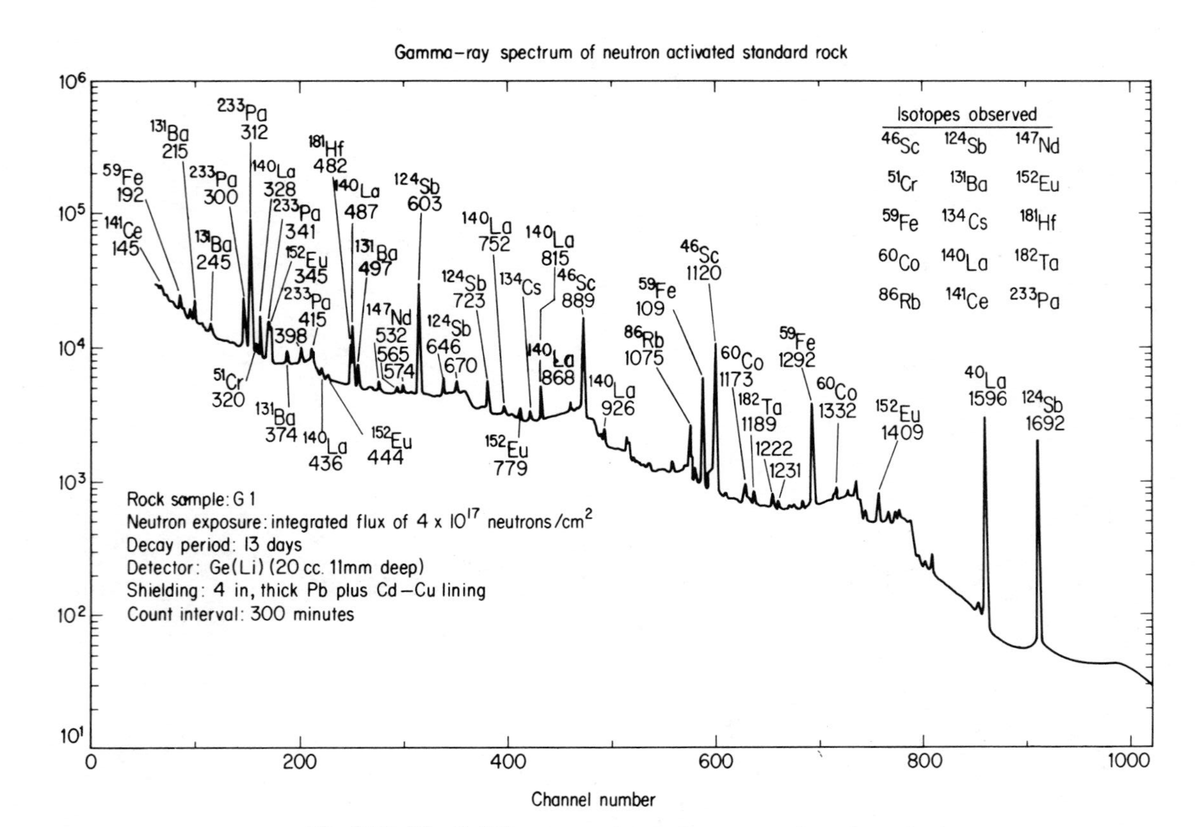

Fig. 8-14. The Ge(Li) γ-ray spectrum of neutron-activated standard rock. From Cooper (7).

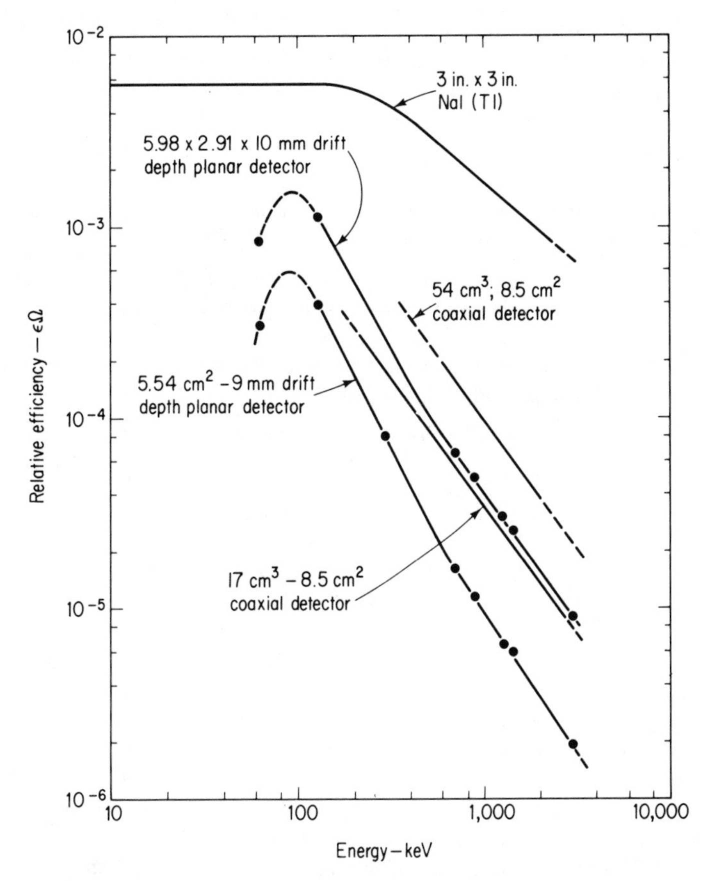

Fig. 8-15. Full-energy peak efficiency to be expected for various detectors located 25.4 cm from a source. The solid angle factor is included. From Camp (6).

3-in. NaI(Tl) detector. Thus samples must be counted for longer periods or increased activities must be used with Ge(Li) detectors as compared with NaI(Tl) detectors.

In summary, we see several significant advantages in the use of semiconductor radiation detectors:

1. A linear energy response that is independent of particle type.
2. Extremely good energy resolutions.
3. Due to the use of a solid as the stopping material, small detector sizes are possible.
4. Detectors have thin, usually negligible, windows and are insensitive to magnetic fields.

5. Extremely fast charge collection times ($\sim 10^{-8}$ to 10^{-9} sec) in the detector. This means that the detector resolving time is negligible for almost all applications.

The primary disadvantages of these detectors are as follows:

1. Small output signal necessitates high-quality, expensive electronic instruments to amplify, shape, and analyze the detector signals.
2. Some loss in counting efficiency is usually found in β^- and γ-ray counting. No loss in efficiency is usually found in α-particle counting.

The use of semiconductor radiation detectors has led to several significant advances in nuclear chemistry and physics. Hopefully, their increased use in tracer research will produce more important findings.

BIBLIOGRAPHY

1. PRICE, W. J. *Nuclear Radiation Detection.* 2nd ed. New York: McGraw-Hill, 1964. A very good introduction to semiconductor radiation detectors, although slightly dated.
2. ADAMS, F., and R. DAMS. *Applied Gamma Ray Spectrometry.* Oxford: Pergamon, 1970.
3. *ORTEC Instruction Manual for Surface Barrier Detectors,* ORTEC. (Oak Ridge, 1965) Many useful nomographs of particle ranges in Si and Ge and detector properties.
4. BROWN, W. L. "Introduction to semiconductor particle detectors," *ORTEC Laboratory Manual A,* November 1968. A classic introduction to semiconductor detectors.
5. POENARU, D. N., and N. VILCOV. *Measurement of Nuclear Radiation with Semiconductor Detectors.* New York: Chemical Publishing Company, 1969.
6. CAMP, D. C. "Applications and optimization of the lithium-drifted germanium detector system," *UCRL-50156,* March 1967.
7. COOPER, J. A. "Applied Ge(Li) gamma ray spectroscopy," *BNWL-SA-3603,* January 1971.
8. BERTOLINI, G., and A. COCHE (Eds.). *Semiconductor Detectors.* Amsterdam: North-Holland, 1968. An excellent compilation of review articles on various aspects of semiconductor detectors.
9. BROWN, W. L., W. A. HIGINBOTHAM, G. L. MILLER, and R. L. CHASE (Eds.). *Semiconductor Nuclear Particle Detectors and Circuits.* Publication 1593, National Academy of Sciences, Washington, D.C., 1969. An authoritative compilation of papers on various aspects of semiconductor detectors.
10. TAVERNDALE, A. J. "Semiconductor nuclear radiation detectors," *Ann. Rev. Nucl. Science* **17**, 73(1967). A good bibliography of recent works.

9

Measurement of Radioactivity by the Liquid (Internal-Sample) Scintillation Method

Liquid scintillation counting owes its origin to the independent discovery by Reynolds (96) and Kallman (58) in 1950 that certain organic solutions fluoresce noticeably upon interaction with high-energy radiation. This fluorescence can then be readily converted to a burst of photoelectrons in a photomultiplier and measured as an electronic pulse, as in the case of solid fluor detectors. Liquid fluors, however, offer a significant advantage over solid fluors. The radioactive sample and the fluor are mixed intimately in a medium (internal-sample), either dissolved or suspended in a suitable solvent. With this system, the sample-to-detector relation is equivalent to 4π detection geometry. Moreover, the detection sensitivity is such that low-energy β^--emitters can be assayed with quite respectable efficiency. This latter application has been most widely exploited and is emphasized in this chapter. In the late 1960's liquid scintillation counters have been used for Cerenkov counting of β^--emitters with an E_{max} above 1 MeV, using simple sample preparation procedures (29, 38, 82, 100), and to assay for adenosine triphosphate (ATP) (115), flavin mononucleotide (FMN), or reduced nicotinamide adenine dinucleotide (NADH) (111, 116) by measuring the bioluminescence associated with the luciferase enzymic reaction.

The use of liquid scintillation counters has become popular in radiotracer laboratories in recent years. The reason is primarily because of the improved reliability of the complete counting assemblies that are commercially available. Whereas elaborate liquid scintillation counters equipped with automatic sample changers and built-in computing capability continue to be expensive, reliable low-cost counters are now available.

A number of symposia have been held in recent years to cover fundamental aspects as well as recent advances in liquid scintillation counting (16, 56). The proceedings of these symposia, and those held earlier (7, 24, 57, 102–105), constitute the major sources of information regarding techniques of counting and sample preparation. Schram's monograph (110) remains the classic document on earlier developments. There are a number of brief reviews of the detection method, such as the one by Hodgson *et al.* (51). The more thorough coverages by Davidson and Feigelson (25), Hayes (42, 43), Rapkin (93), and particularly Horrocks (52–55) are to be recommended. Various instrument manufacturers, notably Packard Instrument Company, Nuclear-Chicago Corporation, Beckman Instruments, Inc., and Intertechnique, include brief descriptions of the methodology in their instrument manuals. They also provide a series of technical bulletins covering various facets of liquid scintillation detection. In addition, a fine laboratory manual covering 11 experiments on liquid scintillation counting (79) was published in 1971 by Beckman Instruments. Because of this situation, the purpose of this chapter and that part of Chapter 11 devoted to the preparation of samples for liquid scintillation detection is to provide a fundamental, although brief, survey of liquid scintillation methodology and a comprehensive list of references to its literature.

A. MECHANISM OF LIQUID (INTERNAL-SAMPLE) SCINTILLATION DETECTION

The overall sequence in liquid scintillation detection might appear superficially to be identical with that described for external-sample scintillation detection in Chapter 6. Fluor molecules are directly or indirectly excited by ionizing radiation, resulting in the emission of photons. These photons, in turn, interact with the photocathode of a photomultiplier to yield photoelectrons from the photocathode surface. The photoelectrons pass through a dynode series, which results in the production of a greatly amplified electron pulse at the photomultiplier anode. Some differences between the two detection sequences do exist, primarily in the mechanism of energy transfer from radiation to fluor. This mechanism will be briefly discussed in this chapter. [For a more detailed discussion of the mechanism of energy transfer in fluor solutions, see the papers by Birks (11, pp. 12–36;), Horrocks (55), Furst and Kallman (37, pp. 3–22, 28–29, 52–54), and Swank and Buck (121)].

1. Energy-Transfer Steps in the Fluor Solution

In liquid scintillation detection, a small amount of the fluor substance (the primary solute) is typically dissolved in a much larger quantity of an organic solvent so as to form the fluor solution. The radioisotope sample may be either dissolved or suspended in this solution. Because of the low concentration of fluor in the solvent (usually less than 1 %), the energy of the radiation particles is not transferred directly to the fluor molecules to any extent but rather indirectly by way of a complicated sequence.

The interaction of energetic electrons (or other particles) with a solute and solvent system involves a number of primary processes. Solvent, in this case, generally refers to aromatic hydrocarbons, such as toluene and xylene. The major processes are (a) excitation of molecules; (b) ionization to form ion pairs consisting of ionized molecules and secondary electrons (delta rays) that are sufficiently energetic to produce more excited molecules; (c) conversion of solvent molecules to short-lived quenching centers by a yet unknown mechanism; (d) recombination of ion pairs, leading to the formation of excited molecules; and (e) local heating. These processes have been summarized by Horrocks (55) as shown in Figure 9-1. Inasmuch as the solvent molecules are so much more abundant in comparison to the solute molecules, the bulk of the primary interaction processes occurs with solvent molecules.

Excitation of solvent molecules (aromatic hydrocarbons) may involve both σ- and π-electrons, but only the latter will lead to fluorescence. Some

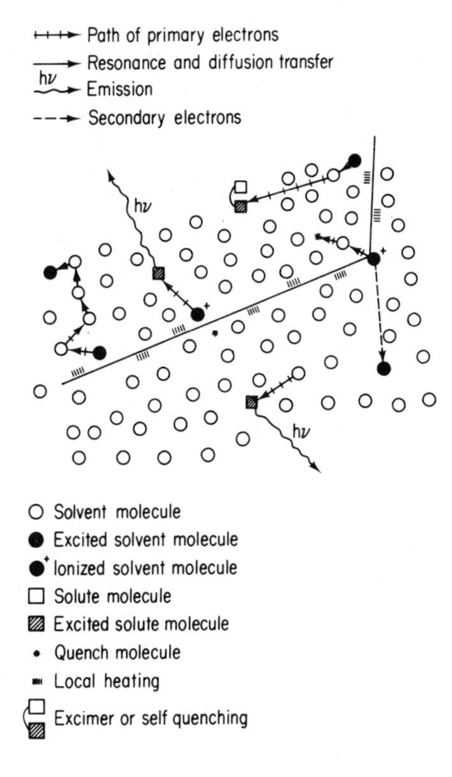

Fig. 9-1. Some processes of energetic electron excitation of a two-component scintillator solution (solvent and solute) (55).

of the excited solvent molecules are quenched through a short-lived (10^{-10} sec) process; the remaining excited molecules will transfer their energy to other solvent molecules via both resonance and diffusion-collision processes, eventually transferring energy to solute molecules mainly by a long-distance (50 Å) but efficient (90 to 100%) resonance-interaction process known as Förster-type transfer (34) and thus resulting in excited solute molecules. Only a very small fraction of interactions result in the excitation of solute molecules. The excited solute molecules return to the ground state by emitting photons with a wavelength in the visible or ultraviolet region. In the case of *p*-terphenyl in toluene, the wavelength of the fluorescence peak is near 3500 Å.

The emission peak of the primary solute may not match the most sensitive range of photocathodes of earlier vintage that were used to detect the photons. Hence the presence of a secondary solute (also a fluor) may be beneficial. This substance serves to absorb energy from the primary solute (presumably also by the Förster transfer process) by excitation and remit it as light of a longer wavelength. For this reason, the term *wave shifter* is often applied to the secondary solute. POPOP [1, 4-bis-2-(5-phenyloxazolyl)-benzene] is a commonly used wave shifter. The wavelength of the fluorescence peak of POPOP is at about 4200 Å. The use of a secondary solute often results in greatly improved detection efficiency. Figure 9-2 graphically illustrates this relation between fluorescence peaks of primary and secondary solutes and photocathode spectral sensitivity. It should be noted that in newer phototubes, such as the RCA type 8575, the wavelength response of the photocathode is such that a wavelength shifter may not be necessary.

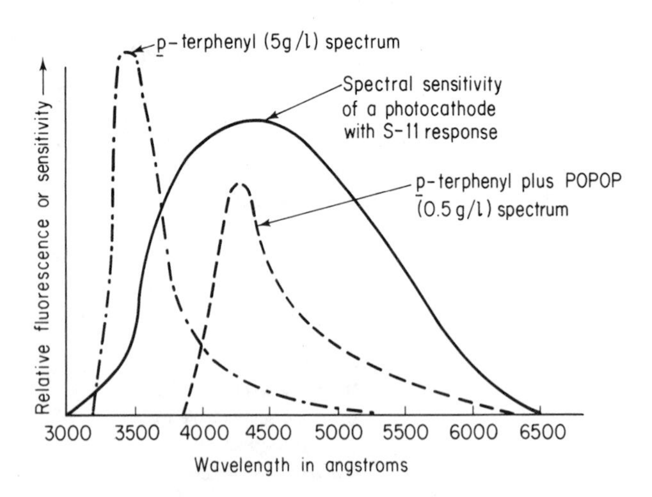

Fig. 9-2. Fluorescence spectra of primary and secondary solutes (in toluene solvent) and photocathode sensitivity.

2. Energy-Transfer Efficiency for Very Soft Beta Particles

It is even more difficult to make a generalized estimation of the efficiency in the energy-conversion processes in liquid scintillation detection than in external-sample scintillation. The variable composition of the radioactive sample and the diversity of fluor solutions that may be utilized can affect the efficiency of the energy-transfer processes and thus make any such quantitative statements meaningless. As an example of the energy-transfer process, however, idealized calculations are presented for the detection of a single β^--particle from ^{14}C with mean emission energy (0.050 MeV or 50 keV) in a *p*-terphenyl-POPOP-toluene solution.

1. Because of the short range of such a weak β^--particle, almost all its kinetic energy would be dissipated in the fluor solution. According to Hayes and Ott (43, p. 9), the fluor solution just specified will yield about 7 photons per keV of β^--particle energy dissipated. The 50-keV β^--particle under consideration would then yield approximately 350 photons.
2. A fraction of these fluorescence photons do not reach the adjacent photocathode surface. Instead they are lost by absorption in the solution, the counting vial, or the reflector, or by scattering at the air-glass interfaces. Assuming that such loss approximates 15%, about 300 photons would then impinge on the photocathode.
3. Current makes of photomultipliers have conversion efficiencies of 15 to 40% for the wavelength range of 3500 to 4500 Å. It follows that 300 photons would give rise to 45 to 120 photoelectrons.
4. A photomultiplier with an amplification factor of 10^6 would then produce 4.5 to 12.0 $\times$ 10^7 electrons at the output, which is equivalent to a charge of about 7.2 to 19.2 $\times$ 10^{-12} coul. This charge could then be transformed by a preamplifier circuit with a capacitance of 30 pF into a pulse of about 240 to 640 mV. This figure should be compared with the value calculated for a 1.17-MeV γ-ray, which was 1800 mV (see p. 128).

It should also be emphasized that the foregoing calculations are based on a ^{14}C β^--particle having an energy equivalent to its E_{mean}. The situation is more serious in reality, since half of the β^--particles from a ^{14}C sample would be emitted with lower energies. By comparison, following the calculations just given, a β^--particle from ^3H emitted at E_{mean} (5.5 keV) would yield an output pulse from the photomultiplier of only about 38 to 101 mV. The difficulty of detecting a pulse of such small size makes it necessary to use an external amplifier of higher gain than required for γ-ray detection with crystal fluors. Lukens (71) presents further discussion of these quantitative relationships.

Depending on the photocathode sensitivity and the degree of absorption loss, a certain minimum number of photons will be required in order to result in the ejection of one photoelectron from the photocathode. Beta particles of such low kinetic energy that they result in the emission of less than that minimum number of photons from the fluor solution will not result in an output pulse. In the circumstances specified in the preceding example, this minimum energy level would be over about 500 eV, even with a very efficient photomultiplier. This requirement will naturally be doubled with liquid scintillation counters equipped with coincidence circuitry, inasmuch as it takes a minimum of two photons striking two photocathodes simultaneously before a legitimate pulse can be formed. Thus a sizable fraction of the β^--particles from a tritium-labeled sample would have insufficient energy to be detected by the liquid scintillation process. In addition, many of the pulses derived from low-energy tritium β^--particles would be of the same energy as thermionic noise pulses of the photomultiplier and would be indistinguishable from them. Note, however, that even weak β^--particles lead to the production of several photons, whereas the thermionic noise from a photomultiplier involves single events. If coincidence circuitry is used, it is possible to distinguish between single and dual events, thereby eliminating a portion of the background counts. Swank (7, pp. 23–28) discusses this and other factors that limit sensitivity in liquid scintillation counting.

B. EVALUATION OF THE LIQUID SCINTILLATION METHOD

1. Advantages of Liquid Scintillation Counting

The most significant application of liquid scintillation detection is for the assay of *β^--emitting samples*, especially those with low β^--energies. Because of the intimate relation of radioactive sample and fluor solution, nearly all disintegrations within the sample are detected, except in the case of tritium samples. The complicating factors of self-absorption, window absorption, geometry, and backscatter that plague traditional G-M assay of weak β^--emitting samples are all but eliminated (see Chapter 13). Hence the major current application of liquid scintillation detection involves assay of ^3H, ^{14}C, ^{35}S, or ^{45}Ca. Under optimal conditions, ^{14}C detection efficiency may reach 98%. Because of the large proportion of extremely low-energy β^--particles it emits, tritium, in routine counting samples, is seldom detected with an efficiency exceeding 60%. Counting efficiencies are markedly affected by the nature of the sample-fluor solution and the mode of counter operations, as will be described later.

Although scintillation solutions have chiefly been used for detecting low-energy β^--particles, they have detection efficiencies approaching 100% for *α-particles*. Unfortunately, nearly all α-emitters are nuclides of high atomic weight and, as such, are insoluble in the aromatic hydrocarbon solvents used. Thus unusual sample-preparation procedures must be employed (Chapter 11). By contrast, medium- to high-energy *γ-rays* are poorly detected by small-volume fluor solutions because of the much lower energy-absorbing capability of liquids as compared to solids. On the other hand, an isotope that emits only weak X rays, such as ^{55}Fe, would be easy to assay by liquid scintillation methods. High-energy γ-radiation may, however, be detected with reasonable efficiency by large-volume ("giant") liquid scintillators, such as the whole-body counters designed for clinical use, although such practice is not generally recommended.

In liquid scintillation detection, as in external-sample scintillation detection, the size of the photomultiplier output pulse is directly related to the magnitude of energy associated with the radiation interacting with the scintillation medium. This relationship allows an approximate analysis of the energy spectrum of β^--emitters. An advantage of this proportionality of pulse size to particle energy is the possibility of differentiating between radiations from two different β^--emitting radionuclides in the same sample (a *double-labeled sample*). In order for this step to be possible, the maximum β^--ray energies of the two nuclides must be sufficiently different so as to be amenable to separation by pulse height analysis. Samples labeled with both ^{14}C and ^{3}H are most commonly encountered. For further information, see the discussions by Okita *et al.* (81), Hendler (48), Horrocks (55), and Kobayashi and Maudsley (68). The last named also cover counting techniques for the pairs $^{3}H—^{35}S$, $^{3}H—^{32}P$, $^{14}C—^{36}Cl$, and $^{36}Cl—^{32}P$.

Another difference in using fluor solutions rather than inorganic crystalline fluors for detection lies in the very much faster *fluorescence decay times* of the former, usually of the order of a few nanoseconds. Fluorescence in most inorganic crystalline fluors decays relatively slowly. Harrison (40) found up to 8% of the peak fluorescence in NaI(Tl) crystal persisting at 200 μsec following the initial γ-ray interaction. By contrast, only 1% of the fluorescence in a toluene-*p*-terphenyl solution remained after 0.3 μsec. This characteristic of scintillator solutions allows their use with "fast" amplifiers, thus avoiding coincidence loss at even very high counting rates.

Since the late 1960's liquid scintillation counters have been used to measure high-energy β^--emitters via the Cerenkov phenomenon (29, 38, 76, 82, 100). Since Cerenkov radiation is produced when a charged particle enters a transparent medium, such as water, at a velocity greater than the speed of light in the same medium, this type of counting is generally carried out with aqueous solution of radionuclides. Sample preparation is therefore simple and economical. Complete sample recovery can be easily made. However, counting efficiencies are usually low, inasmuch as threshold energy for Cerenkov emission from electrons in water is 263 keV. In addition, the Cerenkov photons have a continuous spectral distribution (Figure 9-3) (29) and a defined geometrical configuration (i.e., the photons are not emitted in all directions), thereby resulting in a noticeable reduction in coincidence counting efficiency. Cerenkov counting is not affected by chemical quenching, but it is highly vulnerable to color quenching.

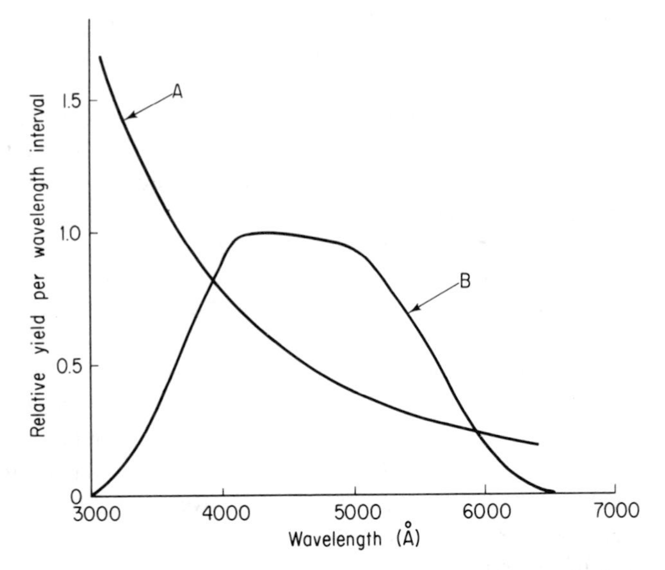

Fig. 9-3. Curve *A*. Energy distribution of Cerenkov radiation. Curve *B*. Photocathode efficiency for equal incident energy interval (29). Reprinted with permission from R. H. Elrick and R. P. Parker, *Intern. J. Appl. Radiation Isotopes* **19**, 265 (1968).

Since the threshold energy for Cerenkov emission varies inversely with the index of refraction of the solvent, various liquid compounds have been tested for Cerenkov counting of low-energy β^--emitters. However, the efficiency observed [e.g., 12.3 % for ^{14}C counting in γ-bromonaphthalene (index of refraction = 1.6582)] (101) is rather low in comparison to that of conventional liquid scintillation counting.

Cerenkov counting efficiencies for some common β^--emitters have been tabulated by Elrick and Parker (29) and are given in Table 9-1. It can readily be seen that Cerenkov counting is the technique of choice for high-energy β^--emitters such as ^{24}Na and ^{32}P, commonly used as radiotracers.

TABLE 9-1

Cerenkov Counting Efficiencies for Some β^--Emitters

β^--Emitters	E_{max} (MeV) (% of disintegrations giving rise to β^-)		Cerenkov Counting Efficiency (% of disintegration)	
			Aqueous Sample	Wavelength Shifter
^{204}Tl	0.77 (98%)		1.3	2.6
^{137}Cs	0.52 (93.5%)	1.18 (6.5%)	2.1	—
^{36}Cl	0.71 (98.1%)		2.3	4.7
^{198}Au	0.96 (99%)		5.4	—
^{47}Ca	0.65 (82%)	1.98 (18%)	7.5	14.8
^{40}K	1.31 (89%)		14	31
^{24}Na	1.40 (100%)		18	40
^{86}Rb	0.70 (8.8%)	1.78 (91.2%)	23	46
^{32}P	1.71 (100%)		25	50
^{144}Ce—^{144}Pr	2.99 (97.7%)		54	75
^{42}K	2.0 (18%)	3.52 (82%)	60	85
^{106}Ru—^{106}Rh	2.1 (1%)	2.5 (11%)	62	85
	3.1 (8%)	3.6 (79%)		

2. Problems Inherent in Liquid Scintillation Counting

a. Counting Sample Preparation. The most commonly encountered complication of liquid scintillation counting concerns sample preparation. Scintillation solvents are usually aromatic hydrocarbons which are not readily miscible with aqueous samples. However many of the common biological compounds are polar in nature and will not dissolve in such nonpolar solvents as toluene. Hence a certain amount of "witchcraft" is called for in concocting a suitable scintillation solvent mixture. In general, each new compound to be assayed must be investigated for its behavior in any given solvent system, and the proper solvent mixture must be determined by previous experimentation. Procedures covering various ways of preparing samples for liquid scintillation counting can be found in a number of review papers presented in recent symposia and published by counter manufacturers. Details in this regard are provided in Chapter 11 of this book. Nevertheless, once a proper scintillation solvent is found, the preparation of samples for liquid scintillation counting is a relatively simple procedure. Ease of routine sample preparation is, in fact, a significant advantage of liquid scintillation detection. In addition, in some cases, the sample material may be readily recovered from the scintillation solution following the counting operation, if desired (97).

b. Photomultiplier Performance and Refrigeration. The photomultipliers used in liquid scintillation counters of earlier vintage (e.g., S-11 photocathode) had high, temperature-dependent noise. Thus even a so-called quiet photomultiplier gave rise to about 10,000 cpm at 0°C and 30,000 cpm at 20°C. In order to reduce the noise level, in addition to the use of low-level discriminator circuits, refrigeration of the entire counting assembly was an inherent feature of these earlier counters. Moreover, noise level was further drastically reduced by the use of two photomultipliers operated in a coincidence mode. Thus even with limited performance of earlier coincidence circuitries (i.e., a 400-nsec coincidence time), photomultiplier noise at 20°C was reduced from 30,000 cpm to 6 cpm (i.e., the chance coincidence rate as given by Equation (4-3) is

$$[3 \times 10^4] \times [3 \times 10^4] \times \left[\frac{4 \times 10^{-7}}{60}\right] = 6.$$

With the advent of modern, improved electronics, coincidence resolving time can be reduced to as low as 20 nsec, and thus the noise level of S-11 type photomultipliers at 20°C can be reduced to 0.3 cpm, i.e.,

$$[3 \times 10^4] \times [3 \times 10^4] \times \left[\frac{2 \times 10^{-8}}{60}\right] = 0.3.$$

Another important criterion in evaluating the performance of a photomultiplier is, naturally, the photocathode sensitivity. With earlier makes, such as the type S-11, S, and Super S-11, the photocathodes were manufactured by vacuum deposition of cesium and antimony on the interior glass or quartz surface of the photomultiplier. Detection efficiencies with coincidence for ^{3}H ranged from 25 to 35% for the S-11 type (late 1950s) to 40 to 45% for the improved S-11 type (1961–1963) to 50 to 60% for the Super S-11 type (1964–1975).

Later, bialkali photomultipliers, such as the RCA type 8575 or EMI type 9634 ϕB, were used in several makes of liquid scintillation counters. This type of photomultiplier has a photocathode coating of potassium, in addition to cesium and antimony, and a superb performance. It is capable of providing a ^{3}H-counting efficiency of 55 to 60% with a noise level of 10,000 to 20,000 cpm at 40°C that can be readily eliminated by means of a fast coincidence circuitry. Moreover, the spectral response of this type of photomultiplier (Figure 9-4) is such that it eliminates the need to use a wave shifter in the counting sample (55).

As a result of the advances made in photomultiplier technology and electronics, it seems unnecessary to refrigerate liquid scintillation counting assemblies. In fact, many room-temperature models are now on the market. However, it should be noted that the photocathode sensitivity of a photomultiplier is somewhat temperature dependent (11), with the direction of change dependent on incident photon wavelength. Such variation in photomultiplier performance may be unimportant in the counting of single radioisotopes with moderately high β^--energy, but it may create problems when

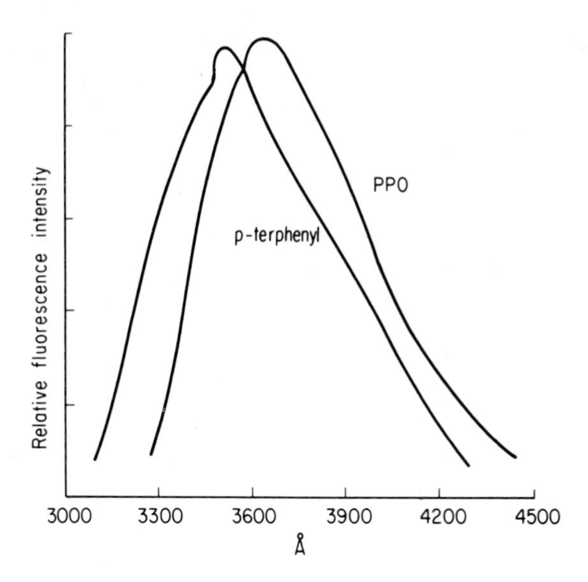

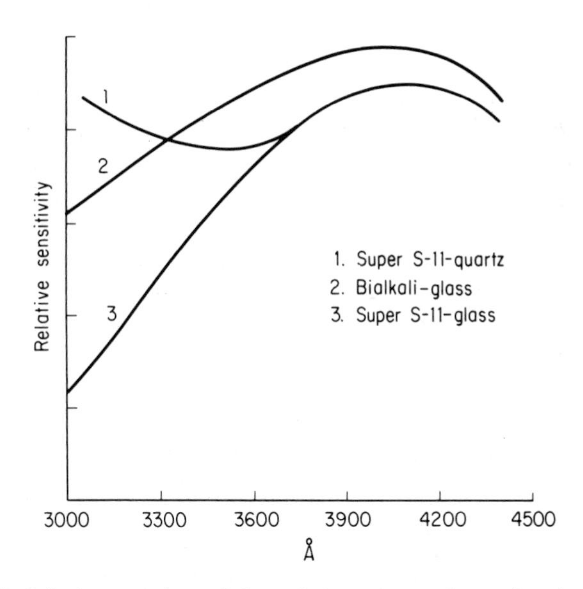

Fig. 9-4. A comparison of the emission spectra of *p*-terphenyl and PPO with the response curves for several contemporary photomultipliers. Courtesy of Intertechniques Instruments.

counting single low-energy β^--emitters, such as ^3H and EC-decaying ^{125}I, or samples containing two radioisotopes, one of which is a low-energy β^--emitter. Consequently, current high-performance liquid scintillation counters are still equipped with refrigeration or similar features to allow counting at constant temperature.

Insofar as counting samples are concerned, refrigeration of counting assembly and samples does offer a definite advantage by providing better reproducibility of data when emulsified samples are counted. There is ample evidence that emulsion-counting efficiency is higher with chilled samples (9) and is highly temperature dependent above 0°C (130). The claim that liquid scintillation counting at ambient temperatures permits one to employ higher concentrations of solute is only applicable to a few solutes, such as *p*-terphenyl and PBD. Another study indicates that the counting rate of an external ^{226}Ra source is temperature dependent. The situation is applicable to dioxane samples (27) but not to toluene samples, and it points to the need for constant-temperature counting but not refrigeration counting.

For a comprehensive analysis of the issue of temperature control in liquid scintillation counting, readers are referred to the excellent paper by Rapkin (94).

c. Fluorescence Quenching. A persistent problem in liquid scintillation detection is the effect known as quenching. This phenomenon is not to be confused with quenching in G-M detectors (see Chapter 5). Broadly defined, *quenching* is any reduction of efficiency in the energy transfer process in the scintillation solution. Thus quenching results in a decreased light output per β^--particle and, consequently, the production of a smaller output pulse in the photomultiplier or even failure to yield a detectable pulse. The net effect of quenching is, therefore, to reduce detection efficiency. Furthermore, since the extent of quenching varies considerably with different sample materials, direct comparison of activity determinations with varied sample types is almost impossible. Thus a determination of just how much the counting efficiency of every sample has been decreased by quenching must be made (see p. 217).

Quenching may occur in several ways. The substance being counted (most commonly), or other components of the scintillation solution, may interact with the excited molecules before they can emit the excitation as photons (*chemical quenching*). Any nonfluorescent dissolved molecules, particularly polar compounds, are potentially quenching agents, for they may absorb energy from the excited solvent molecules without emitting photons. On the other hand, quenching may occur simply as a result of dilution of the fluor solution by the counting sample and hence a reduction of the probability of scintillation events (*dilution quenching*). If the concentration of primary solute is too high, certain types of solutes, when excited, may associate with an unexcited solute molecule and the energy will be dissipated in nonphoton-producing events (*self-quenching*). Even if energy transfer from solvent to fluor is not reduced, colored sample materials will absorb some of the fluorescence photons before they leave the counting vial (*color quenching*) (26). Herberg's study (49) of this phenomenon is particularly valuable. Ross and Yerick (99) have investigated the relation between color quenching and the absorption spectrum of the quenching agent.

Both chemical and color quenching are associated with the molecular structure of the sample material, whereas dilution quenching is related mainly to sample concentration. Certain other effects may also result in photon quenching, such as separation of the fluor solution into two liquid phases, partial freezing of the solution, or fogging of the outside of the sample vial (*optical quenching*).

Certain substances exert a more pronounced quenching action than others. Oxygen, water, halogenated compounds, and polar compounds, in general, are severe quenchers. It is the variable presence of these agents in most biological samples that explains the somewhat empirical selection of proper scintillation solutions. Kerr *et al.* (64) and Funt and Hetherington (36) have investigated and tabulated the quenching effects of a large number of organic compounds. Helmick (47) described the quenching effect of benzoic acid. The last study is important, since labeled benzoic acid is often used as an internal standard in liquid scintillation counters in order to determine counting efficiency (see p. 217).

The quenching effect of dissolved oxygen in the scintillation solution was first pointed out by Pringle *et al.* (89). Seliger *et al.* (113) and Berlman (10) subsequently investigated the mechanism of *oxygen quenching* and found it to result from collision transfer of energy from the solvent to oxygen molecules. Inasmuch as the solubility of O_2 is such that there is one O_2 molecule dissolved in about 2000 molecules of solvent, the collision probability is greater than trillions per cubic centimeter per second. Since oxygen is an incidental contaminant and not a part of the sample, various means have been devised to remove it from the fluor solution (132). These methods include bubbling argon gas, carbon dioxide, or nitrogen through the solution, or, in addition, using ultrasonic degassing (21). As a means of stabilizing chemical quenching by ethanol, ultraviolet irradiation has been suggested (70). In general, the problem of quenching is solved not by avoiding nor eliminating the quenching agent but by devising a reliable method to estimate the precise extent of quenching.

Despite the drawbacks mentioned, the liquid scintillation method of detection is the most generally effective means for assaying low-energy β^--emitting samples, especially ^3H and ^{14}C (7, pp. 288–292). Because of the ubiquitous occurrence of the latter two elements in living organisms, this detection method is of paramount importance to the biologist using radiotracers. The relative ease of sample preparation, the versatility of sample form possible, 4π detection geometry, and the commercial availability of reliable counting systems with high-capacity automatic sample changers make this detection method the most useful where large numbers and types of samples must be assayed.

C. COMPONENTS OF A LIQUID SCINTILLATION COUNTER

A liquid scintillation counter is best visualized as consisting of two basic units: the detector assembly and the electronics. In addition, the more expensive, current commercial models provide automatic sample changers to handle a variable number of samples and built-in computers for data reduction.

1. The Detector Assembly

With regard to the liquid scintillation detector assembly alone, there are certain basic differences between its components and those of an external-sample scintillation detector as discussed in Chapter 7. These differences center primarily on the composition of the fluor (solution versus crystal). Different optical relations must also be considered, since the photomultiplier is exposed to a succession of fluor solutions, rather than remaining optically coupled to an enclosed crystal fluor.

a. Optical Components. Generally two photomultipliers are mounted in a light-tight shield so that they view the sample vial (Figure 9-5). A mechanical system must be incorporated in order to prevent light from reaching the photocathode when samples are changed. Optical coupling of the sample vial and the surface of the photomultiplier is sometimes accomplished by the use of a light guide made with such plastics as polymethylacrylate of the ultraviolet-transmissive type. Sample chamber as well as light guide are normally coated with a light-reflecting surface so as to direct the maximum light to the photocathode. This step is essential inasmuch as the light is emitted isotropically from the liquid scintillator. Various substances have been used as coating materials, of which vacuum-deposited aluminum and titanium dioxide appear to be the most effective (4). However, when the newer type of photomultipliers, such as the RCA type 8575, are used in a counter, wavelength shifters may not be required in sample preparation under certain circumstances, and the light emitted is then in the wavelength range of 3850 Å. In this case, MgO may be used as reflector material instead of TiO_2.

The *sample holding vial* is normally made of glass. Ordinary glass, however, contains a considerable quantity of the natural radioisotope ^{40}K, which contributes significantly to the background count rate. For this reason, it is common practice to employ sample vials made from special low-potassium glass. Quartz glass has also been used (1), but its cost is prohibitive except where extremely low backgrounds are essential and few samples are involved. Medium-density polyethylene vials have been suggested (66) as both inexpen-

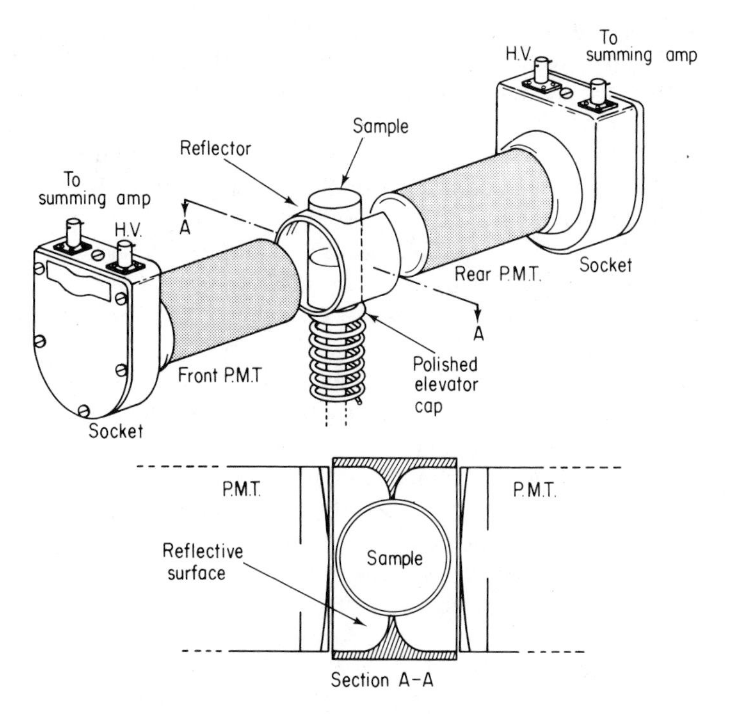

Fig. 9-5. Detector assembly in a typical liquid scintillation counter Courtesy of Beckman Instruments, Inc.

sive and not contaminated with ^{40}K. Rapkin and Gibbs (92) reported an increase in ^{3}H detection efficiency with such vials but also found them permeable to toluene (the common fluor solvent) and thus of limited usefulness. Calf (20) reported that the use of Teflon vials for tritium counting resulted in a fourfold reduction in background count rate as compared to that of low-potassium glass vials. Teflon vials appear to be impermeable to solvents, do not distort after prolonged storage with aromatic hydrocarbons, and show no phosphorescence.

b. Fluor Solution Components. The actual preparation of samples will be described in Chapter 11, but the components of the scintillation solution, excluding the sample itself, will be discussed at this point. The bulk of the scintillation solution consists of a primary solvent, with small quantities of primary solute and even smaller quantities of secondary solute (the wavelength shifter). Both solutes are fluors. Depending on the nature of the sample in question, a secondary solvent may also be employed.

Primary Solvents: The primary solvent must absorb energy by means of excitation from the β^--particles, α-particles, or weak γ-rays and transfer it to the fluor. Consequently, an efficient scintillation solvent must have good energy-transfer characteristics, as well as a small absorption coefficient for the light emitted by the fluor. Another important attribute of the solvent is that it have a good chemical solvent power for a wide variety of sample materials. Since the majority of biological samples are polar compounds and soluble in water, it is highly desirable that the primary solvent be water-miscible. Unfortunately, few solvents have been found with both characteristics. Furthermore, the solvent should not freeze at counting temperatures if the counter is refrigerated. The purity of the scintillation solvent used cannot be overemphasized (122, 123). Any degree of quenching from contaminants in this component will greatly reduce detection efficiency.

The most efficient compounds investigated as scintillation solvents are aromatic hydrocarbons of alkylbenzene structure (xylene, toluene, and so on). Of these, *toluene* is the most widely used. Unfortunately, it suffers from the twin disadvantages of low flash point and poor miscibility with water. Various substitutes for toluene have been proposed (30, 74), with either lessened fire hazard or better solvent powers. For one reason or another, none has replaced it as yet. Toluene made of carbon of recent origin contains ^{14}C and may give rise to a background count rate of as many as 20 dpm/ml of toluene. Consequently, for low-level counting, toluene derived from ancient carbon sources, such as petroleum, should be used.

A second group of compounds, which are generally less efficient but very useful for certain sample types, consists of ethers. *Anisole* and *p-dioxane*, with respective energy-transfer efficiencies of 80 and 70% relative to toluene, are commonly employed. *p*-dioxane is especially useful because it is water-miscible. A problem of freezing exists with this solvent, however, for its freezing point is 11°C, and the freezing point of water-*p*-dioxane eutectics is 4°C. However, this is no problem with ambient liquid scintillation counters. It should also be noted that *p*-dioxane undergoes degradation on standing, thus giving rise to products that display high quenching characteristics. Hence freshly purified *p*-dioxane should always be used.

Secondary Solvents: Solvent mixtures may be used to make soluble radioactive samples that are not soluble in a good primary solvent or to increase the energy-transfer efficiency of an inferior primary solvent that has desirable solvent properties. As an example, the addition of small amounts of *ethanol* (or methanol) to a toluene primary solvent considerably improves its ability to mix with aqueous samples, without causing excessive loss of scintillation efficiency. By contrast, where *p*-dioxane has been used as a primary solvent because of its miscibility with water, its lower energy-transfer efficiency may be improved by the addition of *naphthalene*. Furst and Kallman (37) have evaluated the factors that affect energy transfer in such scintillation solvent mixtures.

Secondary solvents of another important class serve as *specific binding agents* for various inorganic substances. Such solvents fall into several categories. Metallic ions, for example, may be incorporated into metal-organic complexes that are readily soluble in aromatic hydrocarbons. To date, the best organic complexing agents to be used with liquid scintillator systems are the acidic esters of orthophosphoric acid—that is, dibutyl phosphate, dioctyl phosphate, and the like. Metallic ions may also be converted to salts of 2-ethylhexanoic acid (octoic acid), which are reasonably soluble in aromatic hydrocarbons.

Of particular importance to the biologist are the binding agents for CO_2. Several organic compounds have been employed to trap the gas and convert it to a toluene-soluble carbonate. A methanolic solution of the hydroxide converted from "Hyamine 10-X" [*p*-diisobutylcresoxethoxyethyl)-dimethyl-benzyl-ammonium chloride] (Rohm and Haas, Inc.) has been the most widely used for this purpose. In addition, at a 1 molar concentration, this reagent is known to be capable of making proteins, tissues, and similar biological specimens soluble. Similarly, NCS (Nuclear-Chicago), "Primene 81-R" (Rohm and Haas, Inc.), Soluene, ethanolamine, and ethylene diamine have been utilized for economy and to reduce quenching. Specific scintillation-solvent formulations involving these substances are shown in Tables 11-2, 11-3, and 11-4. In some cases, the use of a basic solubilizing agent may cause chemical luminescence. Other sample preparation methods related to secondary solvents, including gel counting and emulsion counting, were developed to accommodate significant amounts of aqueous solution in the counting sample. Details are given in Chapter 11.

Primary Solutes: This component of the scintillation solution is the fluor, which must efficiently convert excitation energy to light quanta. During the earlier years of liquid scintillation counting, a most extensive search was conducted (largely by the Biomedical Research Group at the Los Alamos Scientific Laboratory) for compounds with suitable fluor qualities. In 1958 Hayes *et al.* (44) listed 483 different compounds that had been investigated.

A common characteristic of the most efficient fluors is that they contain three, and preferably four, aromatic or heterocyclic rings linked together in a linear manner that allows continuous conjugation throughout the molecule. In addition to being an efficient light emitter, it is also important that the light emitted by the fluor have a spectrum matched to the photocathode sensitivity. Needless to say, it is desirable that the fluor be economical and soluble in a variety of solvents, particularly at low temperature (if refrigerated counters are to be used).

The most efficient fluors fall into the classes of oligophenylenes, oxazoles, or oxadiazoles (7, pp. 101–107). Because of the complexity of standard chemical nomenclature for these classes of compounds, an abbreviated system has been devised to describe them. The letter P is used for phenyl, N for naphthyl, B for biphenyl, O for oxazole, and D for the oxadiazole group.

Despite the large number of potential fluors investigated, only a few have proved worthy of consideration for practical use. The first substance so employed, *p*-terphenyl, is particularly attractive because of its relatively low cost. For many applications, however, its solubility in toluene is too low, particularly at low temperatures. Nevertheless, it is useful in counting energetic β^--emitters, including lightly quenched ^{14}C samples, in newer ambient counters. PPO (2, 5-diphenyloxazole), because of its good solubility characteristics at low temperatures, has become the most popular primary solute for counting with refrigerated coincidence counters, despite its high cost. Photons derived from PPO give rise to a relative pulse height 3% greater than *p*-terphenyl. A third compound, PBD (phenylbiphenyloxadiazole), is the most efficient primary solute, giving rise to a relative pulse height 24% greater than *p*-terphenyl. It has the disadvantage of limited solubility. In such fluor-efficiency comparisons, *p*-terphenyl is arbitrarily assigned a relative pulse height value of 1.0. The photons derived from the three previously discussed fluors display emission peaks in the neighborhood of 3460, 3800, and 3700 Å, respectively. This fact implies that it would be advantageous to use them along with a secondary solute (i.e., a wavelength shifter) when counters employing photomultipliers of older types are used.

Other newly discovered primary solutes include butyl-PBD (2-(4′-*t*-butylphenyl)-5-(4″-biphenylyl)-1,3,4-oxadiazole, and BBOT (2,5-bis[2-(5-*t*-butylbenzoxazolyl)]-thiophene. The solubility of butyl-PBD in toluene at 0°C is 61 g/liter, well above the recommended concentration of 6 to 8 g/liter in unquenched samples. It is a stable compound and resistant to quenching or self-quenching. BBOT is less attractive as a primary solute compared to butyl-PBD, considering all factors.

In general, for a given solvent, the counting efficiency increases with increasing fluor concentration up to a certain point. Some fluors show the desirable characteristic of reaching a concentration that gives constant detection efficiency before exceeding their solubility limit in a given solvent. Obviously, whenever possible, the concentration of a fluor should be maintained at a level higher than that required to compensate effect of quenching derived from added quenching agent and thereby provides constant detection efficiency. The relationship between detection efficiency and fluor concentration for PPO in toluene is shown in Figure 9-6.

Secondary Solutes: This component of a scintillation solution acts as a secondary fluor whose effect is to shift the wavelength of the light emitted by the primary fluor to a region of greater photocathode sensitivity (see Figure 9-2). The secondary solutes are usually too insoluble, too expensive, or have too great a degree of self-quenching to function as primary solutes. They are useful because addition of only minute quantities to the scintillation solution may give impressive increases in counting efficiency. In general, their concen-

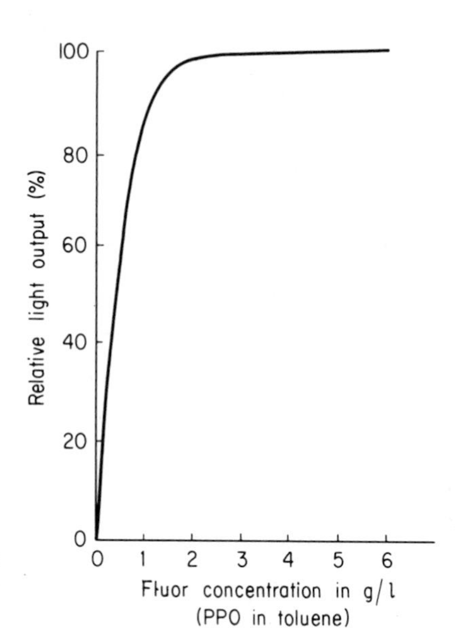

Fig. 9-6. Relative light output in relation to primary solute concentration in detection of ^{14}C.

trations in the scintillation solution are only about $\frac{1}{100}$ those of the primary solutes. Hayes *et al.* (41) have evaluated the merits of a large number of compounds for use as wavelength shifters.

Several secondary solutes have found common use. POPOP [1,4-bis-2-(5-phenyloxazolyl)-benzene] is by far the most widely utilized at present. Dimethyl-POPOP [1,4-bis-2(4-methyl-5-phenyloxazolyl)-benzene] became available later and has the advantages of better solubility and longer wavelength in photon emission peak compared to POPOP. A third compound, α-NPO [2-(α-naphthyl)-5-phenyloxazole], has limited popularity, but it is not as desirable as the other two compounds. Relative to PPO, these secondary solutes have pulse heights of 1.45, 1.45 and 1.21, respectively for a given detection assembly.

New secondary solutes introduced in the 1970s include bis-MSB (*p*-bis-*o*-methylstyryl benzene) and PBBO [2-(4-biphenylyl)-6-phenylbenzoxazole]. Both are more soluble in toluene than POPOP and show promising performance.

A good review on solvents and solutes for liquid scintillation counting has been written by Rapkin and published by Intertechnique (95).

2. The Electronics

The electronics of a typical liquid scintillation counter are illustrated by the simplified block diagram of Figure 9-7, which represents all counters made by Beckman Instruments. It can be seen that two photomultipliers, with the appropriate high-voltage power supply, view the sample vial from opposite sides. The pulses from these two tubes that are in coincidence are fed through a summing amplifier to individual single-channel analyzers and then to a fast scaler. However, a logarithmic-summing amplifier is used in

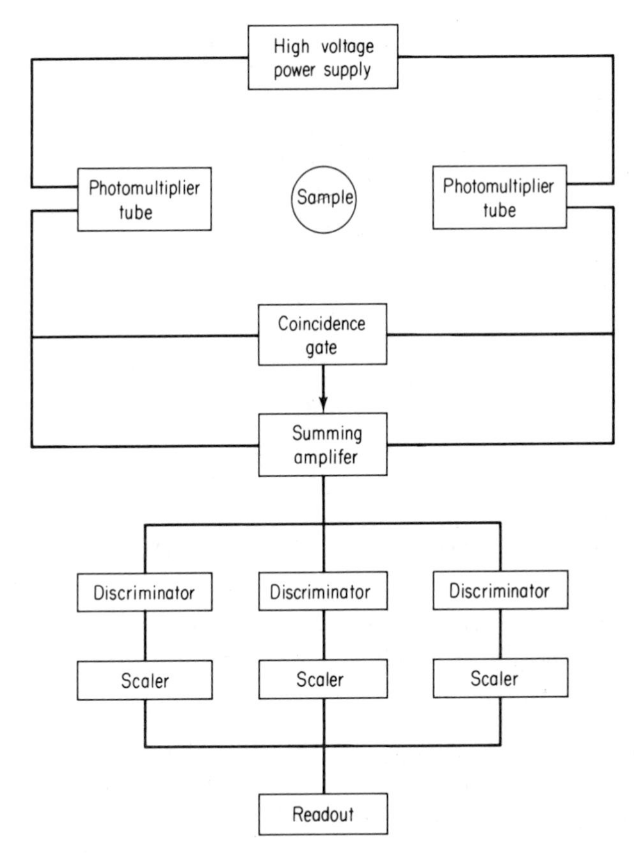

Fig. 9-7. Simplified block diagram of a liquid scintillation counter
Courtesy of Beckman Instruments, Inc.

expensive models and plug-in, single-channel analyzers with digital readout are incorporated in the "β-Mate," room-temperature, bench-top model.

Most commercially available counters are equipped with three single-channel analyzers (SCA) for sample analysis and two additional SCAs so that the channels-ratio method can be used (see p. 219) on events produced by an external standard used for quench correction (see p. 221). There are a number of manufacturers of liquid scintillation counters, notably Packard Instruments, Inc., Nuclear-Chicago, Beckman Instruments, Inc. and Intertechnique Instruments, Inc. In the discussion that follows, we shall review the basic features of several commercially available liquid scintillation counters. We do so not to endorse particular models (whose features will change with time) but to point out the basic principles behind many modern instruments.

Several series of liquid scintillation counters are currently supplied by Packard Instruments, Inc. Model 3390 is shown in Figures 9-8 and 9-9.

Fig. 9-8. Model 3390 liquid scintillation counter manufactured by Packard Instruments, Inc.

Counter performance is improved through the use of RCA 450/V4 (bialkali) photomultipliers. The counter is equipped with three SCAs for sample counting and two SCAs for external-standard ratio purposes; 300-sample automatic sample changer; a background subtraction device; a low-level activity rejection device; an on-line electronic calculator for data presentation; a normalization control to facilitate restoration of phototube gains to optimal conditions; temperature control; and an absolute activity analyzer (Model 544). The last is a device that converts raw data in counts per minute to disintegrations per minute for up to three radioisotopes in intermixed single-labeled samples or for two radioisotopes in double-labeled samples. This conversion makes use of counting-efficiency information derived from external standard ratio data. The instrument is calibrated by using a series of known quench standards, thereby providing a limited number of defined data points for correlating external-standard ratios and counting efficiencies. When an unknown sample is counted, the external-standard ratio pertaining to that sample is determined. If the ratio does not correspond to one of the preselected data

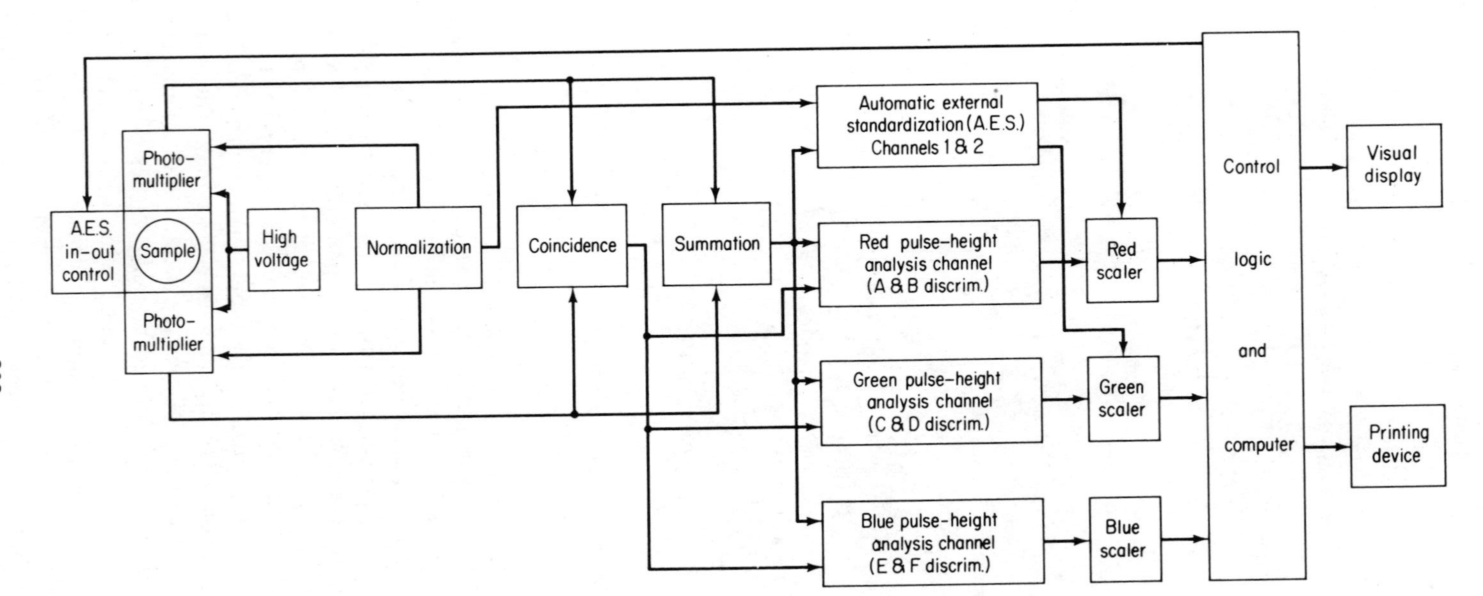

Fig. 9-9. Simplified block diagram of the liquid scintillation spectrometer system, Model 3000 series, manufactured by Packard Instruments, Inc.

points, a first-order correction is made by defocusing the two photomultipliers via a defocusing coil in which a magnetic field is generated when current is passed through. The result is a defined reduction of photomultiplier efficiency. The external-standard ratio is redetermined, and the defocusing processes can be repeated a few more times until the measured external-standard ratio of the sample is identical to one of the prestored data points of the quench correction curve. The counting efficiency is then electronically read off the curve for a calculated correction of the observed cpm to dpm. The principle underlying the external-standard ratio method is presented later in this chapter.

The model LS-255 liquid scintillation counter manufactured by Beckman Instruments, Inc. is shown in Figure 9-10. The basic electronics of this counter are essentially the same as in Figure 9-7, except that a logarithmic-summing amplifier is used in the LS-100 and LS-200 series of counters. In fact, Beckman counters were the first line of instruments utilizing two additional SCAs to provide channels ratio information on the external-standard counting data, in addition to three SCAs for sample counting. The use of RCA-developed,

Fig. 9-10. Model LS-255 liquid scintillation counter manufactured by Beckman Instruments, Inc.

bialkali, 12-stage, head-on types of photomultipliers enables these counters to give good counting performance at ambient temperature.

Beckman counters of the LS-200 series are also equipped with the Automatic Quench Calibration (AQC) device. This device provides automatic compensation for varying levels of quenching among samples through an electronic adjustment of gain for an individual sample. A given sample is first counted and the external-standard ratio pertaining to the sample is determined in order to provide necessary information for the readjustment of gain by varying the degree of amplification. Once this step is accomplished, the sample is recounted and the data registered. With ^{14}C and more energetic β^--emitters, quenching shifts a portion of the β-spectrum toward the low-energy end and below the setting of the low-level discriminator of the SCA, thus resulting in reduced counting efficiency. The AQC device has the net effect of shifting the β-spectrum toward the more energetic end and hence shifting most of the otherwise lost counts above the low-level discriminator so that they can be counted by increasing the amplifier gain. As reported by Wang (125) for ^{14}C, when the AQC device is used to provide high amplification of pulses (gain overestimation), it will be possible to compensate for the quenching effect of even severe quenching agents and to count variably quenched samples at practically constant counting efficiency (Table 9-2).

TABLE 9-2

Counting of (^{14}C) at Constant Efficiency by Gain Restoration

Concentration of Quenching Agent	Counting Rate and Efficiency			
	Without Gain Restoration		With Gain Restoration	
	(observed cpm)	(counting efficiency, %)	(observed cpm)	(counting efficiency, %)
None	61,176	60	60,851	60
0.05 ml $CHCl_3$	52,151	51	61,895	61
0.1 ml $CHCl_3$	42,509	42	60,672	60
0.2 ml $CHCl_3$	35,683	35	62,383	61
0.3 ml $CHCl_3$	27,792	27	60,613	59
0.4 ml $CHCl_3$	21,676	21	60,248	59
0.5 ml $CHCl_3$	15,769	16	59,468	58

Sample composition: Toluene and PPO (6.0 g/liter) containing (^{14}C) with total radioactivity of 102,000 dpm.

Counting procedure: Beckman Model 250 system with discriminator set at 320—∞ divisions and AQC set to provide overrestoration of (^{14}C) spectrum to the extent that restored endpoint is 80 divisions beyond endpoint of unquenched spectrum. Counting was carried out in sufficient time to provide data with no greater than 1% standard deviation.

Beckman counters also include an on-line computer ("DPM Controller") for the conversion of observed cpm to dpm, making use of quench-curve coefficient data in the form of punched paper tapes.

Intertechnique Instruments, Inc. manufactures several models of liquid scintillation counters. These models all have the same detector assembly, basic electronics, and automatic sample changer but different capability in data processing. Thus Model SL-30 only records data; Model SL-31 does simple calculations with a programmable electronic calculator; Model SL-36 is equipped with a hard-wired calculator, which, in addition to calculations for counts per minute, external-standard ratio, channels ratio, and background subtraction, can provide users with data in disintegration per minute for constant efficiency counting with cross-contribution subtraction for dual radioisotope samples; and Model SL-40 is equipped with a programmable core memory computer for the storage of external-standard ratio or channels-ratio quench correction curves. Computed dpm values are presented for variably quenched samples and with SL40-4K repetitive counts may be averaged and histograms plotted. One to four Model SL-30 counters can also be operated on line with the "Multi-8 Minicomputer" also manufactured by Intertechnique.

The Intertechnique Model SL-40 is shown in Figure 9-11 and its electronic block diagram in Figure 9-12. The front-panel controls for this counter are seen in Figure 9-13.

Fig. 9-11. Model SL-40 liquid scintillation counter manufactured by Intertechnique Instruments, Inc.

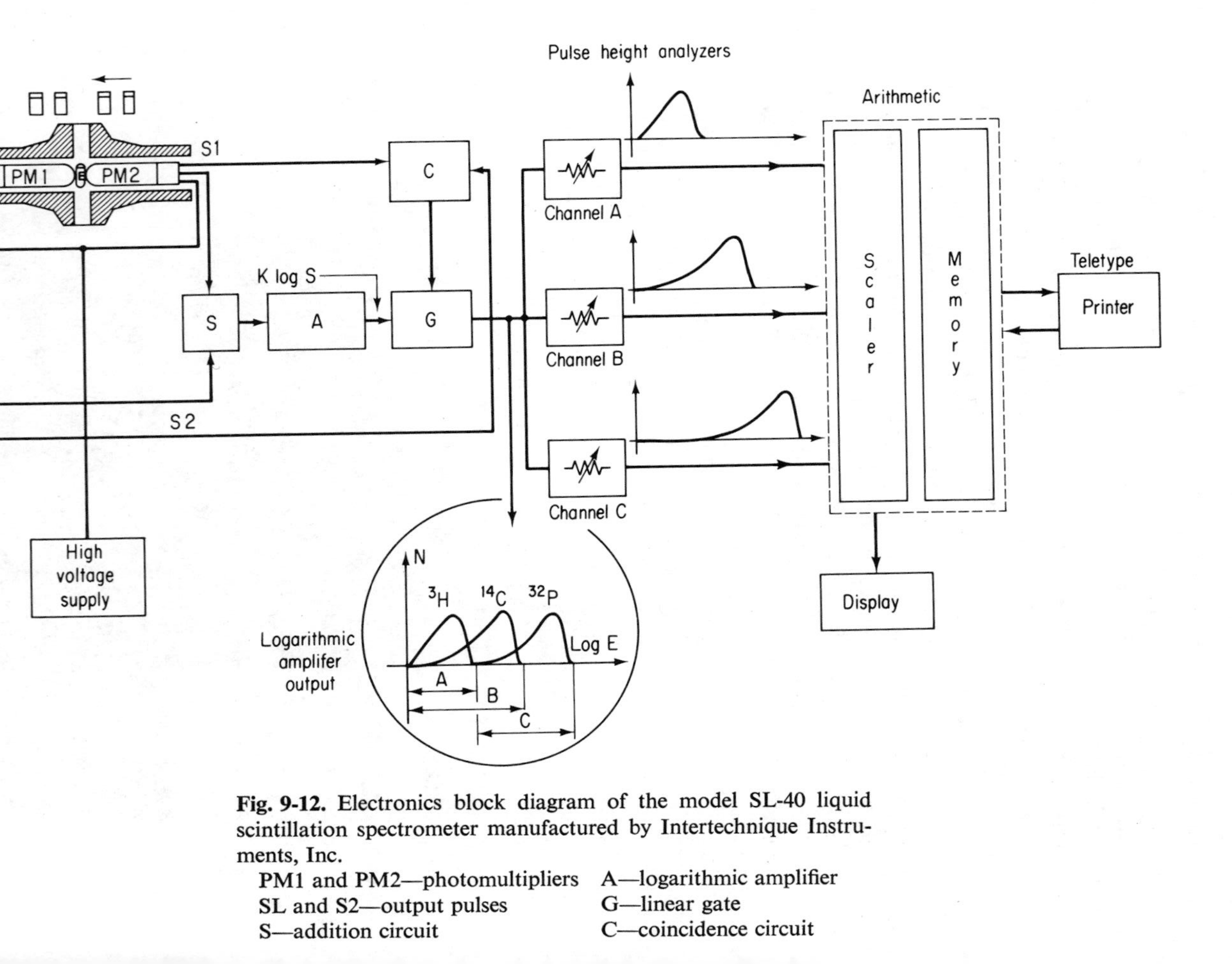

Fig. 9-12. Electronics block diagram of the model SL-40 liquid scintillation spectrometer manufactured by Intertechnique Instruments, Inc.

PM1 and PM2—photomultipliers	A—logarithmic amplifier
SL and S2—output pulses	G—linear gate
S—addition circuit	C—coincidence circuit

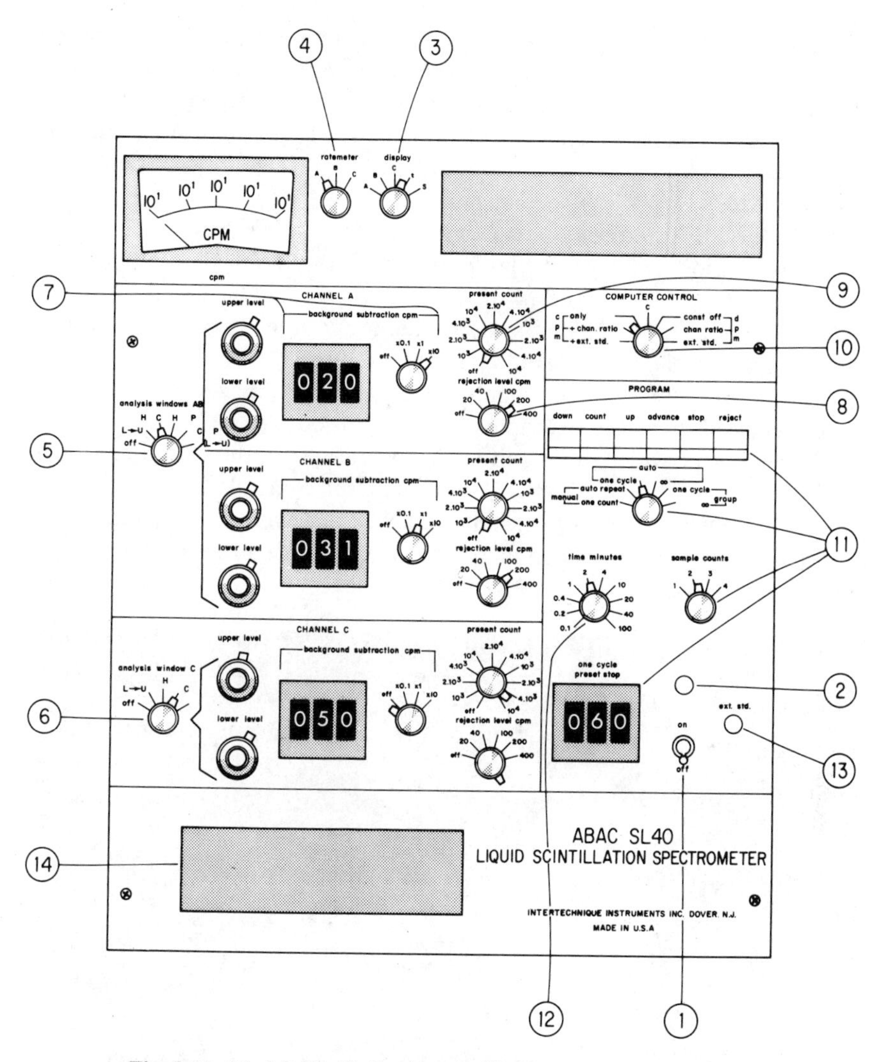

Fig. 9-13. Model SL-40 liquid scintillation counter front-panel controls (Intertechnique Instruments, Inc.).

The unique feature of Intertechnique counters is the use of a linear gate (G, Figure 9-12) to eliminate noncoincident pulses prior to pulse height analysis. The dynamic range reduction accomplished by the use of a logarithmic amplifier made it possible to incorporate such a gate, which minimizes the background noise contributed by the photomultipliers. It should be noted here that, of all current models of liquid scintillation counters, only those manufactured by Intertechnique Instruments and Beckman Instruments are equipped with a logarithmic amplifier.

As with counters of other makes, the Intertechnique Model SL-40 counter is equipped with an automatic sample changer (200-sample capacity), three channels for sample counting, and two additional channels for external-standard ratio, and it is available with or without refrigeration. A logarithmic rate meter, which may be switched to any of the three counting channels, is also provided.

The Mark II liquid scintillation counter manufactured by Nuclear-Chicago Corporation is shown in Figure 9-14, and its simplified block diagram is shown in Figure 9-15. The Mark II counter is a highly modular system, equipped with a 300-sample automatic changer in a temperature-controlled chamber. Three single-channel analyzers are provided, each capable of handling six selectable, preset counting channels. A ^{133}Ba source is used for quench correction by the external-standard method. An automatic calibration circuit is provided to monitor the operation of the photomultiplier periodically (once every 24 hours), plus analyzer channels to ensure stable system gain of the system when significant drift is detected. The module will generate a very small correction to the high voltage proportional to the amount of gain variation. This system is similar to the one used in γ-ray

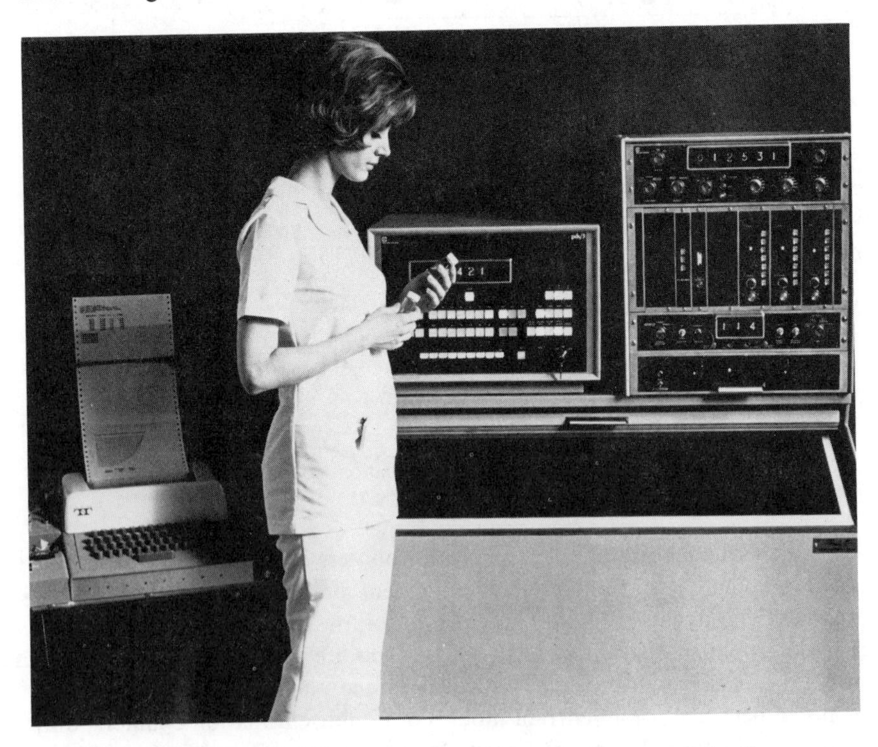

Fig. 9-14. The Mark II liquid scintillation counter manufactured by Nuclear-Chicago Corporation.

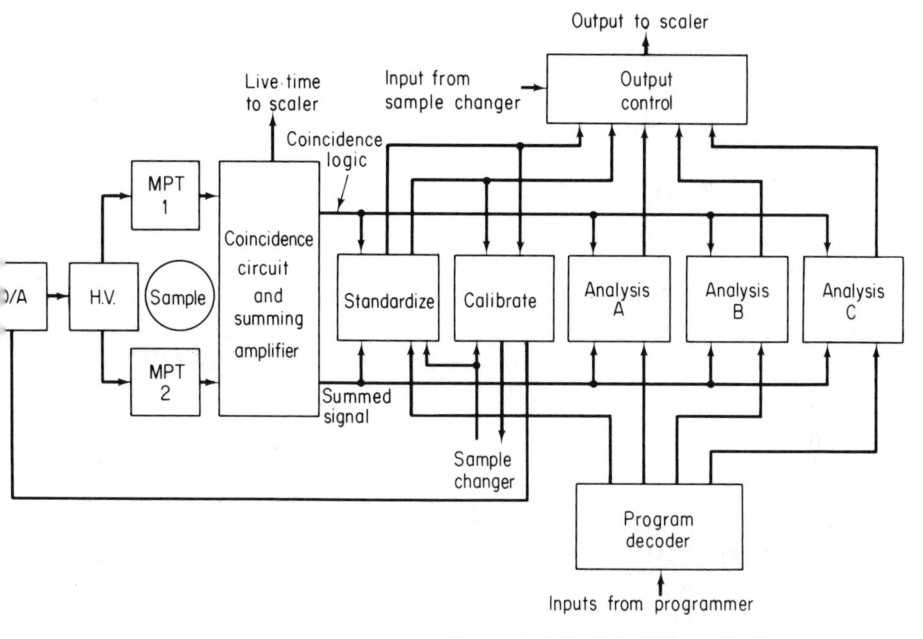

Fig. 9-15. Photomultiplier package and pulse-height-analyzer block diagram of Mark II liquid scintillation counter manufactured by Nuclear-Chicago Corporation.

pulse height analysis. The Mark II counter is also uniquely equipped with a photon-monitoring device that is capable of distinguishing multiphoton events (as derived from nuclear disintegration) from single photon events (typically resulting from chemiluminescence, phosphorescence, photomultiplier thermal noise, etc.). When the accidental coincidence count rate, resulting from a high rate of single photon events, is about 5% of the sample counting rate and greater than 10 cpm, normal counting will not proceed, and so the collection of erroneous counting data will be avoided.

In the past few years both Beckman Instruments and Intertechnique Instruments have produced low-cost, single-sample, bench-top liquid scintillation counters. The performance of these counters is reasonably good, primarily because of the use of bialkali photomultipliers. Their low cost makes them ideal for instructional programs. The Model SL-20 counter manufactured by Intertechnique is an ambient two-channel instrument equipped with coincidence circuitry and external standardization. The Beckman "β-Mate" is an ambient single-channel instrument with two Beckman 568641 photomultipliers in coincidence. The counter is equipped with an external standard (^{137}Cs) and two additional SCAs for quench correction by means of the external-standard channels-ratio method. The counter is equipped to handle a flow cell attachment and can be used for bioluminescence measurement.

D. SPECIAL TYPES OF LIQUID SCINTILLATION DETECTORS

1. Large-Volume External-Sample Detectors

The liquid scintillation detection method is not limited to internal samples, although that is the major emphasis of this chapter. The size restrictions imposed on gas-ionization and solid-crystal scintillation detectors do not apply to liquid scintillation detectors. Very large volumes of scintillation solution may be used with a number of photomultipliers in order to produce "giant" detectors (7, pp. 246–257). For numerous reasons, large-volume detectors have been constructed for use with external sources of radiation. Radiation from γ-rays, protons, neutrons, and neutrinos has been detected in this manner.

Of greatest biological interest are the liquid scintillation *whole-body counters* that are used to determine total body radioactivity in humans or animals (22). Such measurements give information on both the natural ^{40}K content of the body and the content of internal radioactive material accidentally acquired by an individual. The Los Alamos human counter will serve as an example of the immensity of this detector type. It holds 140 gallons of scintillation solution (toluene-terphenyl-POPOP) monitored by 108 photomultipliers and is surrounded by 20 tons of lead shielding (2, 7, pp. 211–219; 24, pp. 344–370).

Most such whole-body counters are one-of-a-kind models developed by individual research teams. The only commercially manufactured unit is the Packard Instrument Armac detector. This unit has a cylindrical counting-chamber volume of 1800 cc surrounded by a tank of liquid scintillation solution (Figure 9-16). Gamma rays emitted by samples in the counting chamber undergo Compton interaction with the scintillation solution to produce photons that are "seen" by one or more of the six photomultiplier tubes optically coupled to the rear of the tank. Pulses from these tubes are summed and passed to a typical γ-ray spectrometer.

Little or no sample preparation is required, and the Armac detector is particularly suitable for counting emissions from bulk samples of meat, milk, grain, water, blood, feces, urine, or soil. Thus it is most useful in assaying environmental samples with moderate-to-high levels of radioactivity. More important, living animals may be introduced directly into the counting chamber and assayed for total body radioactivity without harm. Living rats, mice, guinea pigs, chickens (and other small birds), reptiles, frogs, and even fish (using special flowing-water containers) have been used. The system is commonly utilized when repeated assays of a group of animals are required over

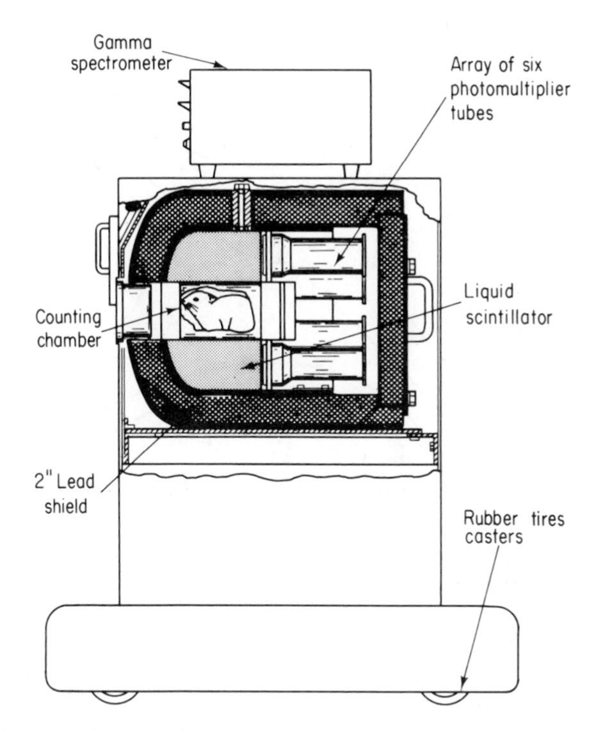

Fig. 9-16. Cross-sectional view of an Armac small-animal, whole-body, liquid scintillation counting system. Courtesy of Packard Instruments, Inc.

an extended period of time in connection with studies of the uptake, retention, or excretion of gamma-emitting radionuclides.

The Armac detector is also designed for use in clinical diagnosis where radioisotopes are used in blood-clearance studies. It permits in vivo measurements of emissions from vascular beds in a patient's forearm. Thus the trade name "Armac" is derived from the term "*arm* and *a*nimal *c*ounter."

The location of the sample in such a large-volume counting chamber affects the counting efficiency significantly, as seen in Figure 9-17. Sample holders are usually developed for each type of sample to ensure accurate repositioning with successive assays. Counting efficiency varies with the energy of the γ-rays being detected. Iodine-131, for example, can be detected with an efficiency of about 40%. Of importance in environmental studies, a nanocurie (10^{-9} Ci) of ^{137}Cs can be detected with an accuracy of $\pm 3\%$ in 4-min counting time. However, the system is not suitable for very low level samples.

In order to increase the efficiency of large liquid scintillators for *neutron detection*, the solution is best "loaded" with heavy metals. This process

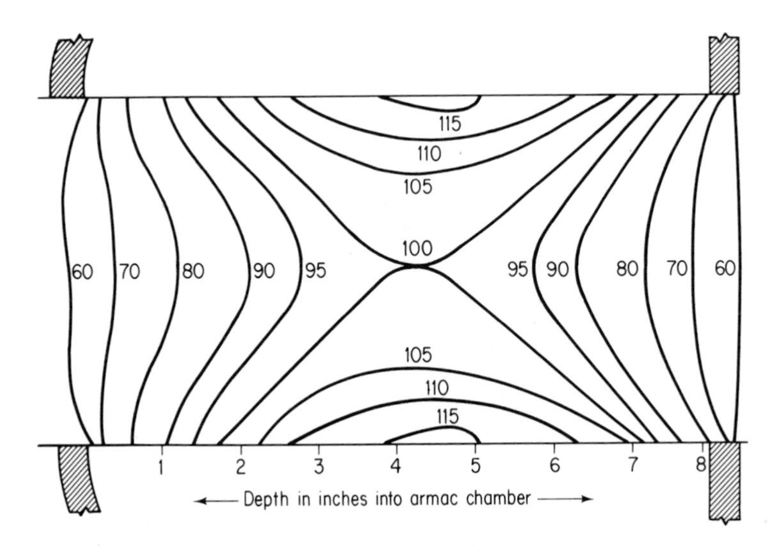

Fig. 9-17. Diagram of variations in counting efficiency in the Armac whole-body counting chamber as a function of position.

requires the use of unusual solvents (18). At various times lead, bismuth, lithium (46), uranium, iron, cadmium, and boron, in a variety of chemical forms, have been employed, but the last two appear most desirable. In the case of boron, the reaction $^{10}B(n, \alpha)^{7}Li$ occurs, and it is the resultant alpha particle that is detected by the scintillator (14). It was by means of such a cadmium-loaded scintillator that the neutrino (see Chapter 3) was first experimentally detected (98). Using paired detectors, only one of which was sensitive to neutrons, Williams and Hayes (129) developed a scintillation rate meter for differentiating mixed γ-ray-neutron radiation from a nuclear reactor.

2. Continuous-Flow Scintillation Detectors

An entirely different approach to liquid-scintillation-type assay has resulted from the work of Steinberg (117). In 1958 he introduced an efficient method for scintillation counting in a two-phase system, which consisted of a solid insoluble fluor in intimate contact with an *aqueous solution* containing a radioactive sample. At first a plastic fluor of Pilot Scintillator B (Pilot Chemicals, Inc.) was used in the form of either fine filaments or beads (117). Aqueous sample solutions were pipetted into counting vials packed with the fluor and counted in the normal manner with quite a respectable efficiency. Later (118, 119), Steinberg found that the substitution of blue-violet fluorescence-grade anthracene crystals as the fluor gave more reproducible results and higher counting efficiencies. Myers and Brush (77) have used Steinberg's technique to assay with high efficiencies ^{45}Ca, ^{90}Sr, ^{32}P, ^{210}Po, and other nuclides.

Although Steinberg's method involved the assay of individual aqueous samples, several investigators have employed his idea to construct continuous-flow detectors. In these detectors, a stream of aqueous or gaseous samples containing radioactive material is passed through a cell loosely packed with fine anthracene crystals. The cell is optically coupled to a pair of photomultipliers, which view the fluorescence of the anthracene crystals resulting from β^--ray interaction (24, pp. 222–226). Counter data are displayed by means of a rate meter and recorder. The primary application of this detection method has been for monitoring *effluents* from gas-liquid chromatographs or amino acid analyzers (86). Schram and Lombaert (108, 109), Rapkin and Gibbs (90), and Karmen *et al.* (60) have used such a flow counter for ^{14}C and ^{3}H determinations; Scharpenseel and Menke (106) have utilized it with a ^{35}S-containing effluent as well. Surprisingly, retention of most forms of activity on the surface of the anthracene crystals is minimal, and resolution of a typical column chromatograph is unaltered by passage through the cell. Counting efficiency is heavily dependent on flow rate. Efficiencies of better than 50% for ^{14}C and 3 to 6% for tritium can be realized in routine assays by using commercially available detectors. Unfortunately, the monitoring technique cannot be employed whenever organic solvents are used in the chromatography operation, inasmuch as the fluor is readily soluble in organic solvents. Karmen *et al.* (63) later employed detectors packed with silicone-coated *p*-terphenyl crystals for such flow monitoring.

Alternatively, Popják *et al.* (88) have described a system in which gaseous chromatographic effluents are trapped directly in a heated fluor solution that is then cooled and continuously circulated between two facing photomultipliers. It is possible to assay ^{14}C and ^{3}H simultaneously with good counting efficiencies by means of this procedure.

A variant type of scintillation flow detector utilizes a cell that is itself the plastic fluor. Schram and Lombaert (107) have described such a detector, in which the flowing liquid is passed through a spiral chamber machined in a plastic fluor block. The fluor block was optically coupled to two photomultipliers to complete the detector. Alternatively, the plastic scintillator can be in the form of a tight spiral of tubing, coupled to the photomultipliers. Funt and Heatherington (35) have assayed ^{32}P, ^{22}Na, and ^{14}C solutions with a detector of this type with counting efficiencies of 76, 51, and 5.7%, respectively; Kimbel and Willenbrink (65) have assayed ^{14}C and ^{35}S sample flows with a similar arrangement. Boyce *et al.* (15) have even monitored tritiated hydrogen gas from a chromatography column in this type of detector.

Another way of monitoring the radioactivity in a continuous stream of aqueous solution, such as column chromatography effluent, is to mix the effluent with liquid scintillator in a cell and count by means of a conventional liquid scintillation counter. Thus Beckman has developed a discrete sampling flow cell that is particularly useful in conjunction with an amino acid analyzer for effluent monitoring (80). Any one of several liquid scintillators can be used. A typical one consists of 4 to 8 g PPO and 0.2 to 0.4 g dimethyl POPOP in one liter of toluene, with the addition of surfactant to ensure the formation of a homogeneous solution with aqueous effluent. Counting efficiencies are said to be ~50% for ^{3}H and ~90% for ^{14}C when counting is carried out with the low-cost "β-Mate" counter. The use of an effluent stream-splitting flow system also makes it possible to preserve a portion of the effluent for sample recovery. However, as noted by Schram in his excellent review paper (111), continuous counting with liquid scintillation generally suffers from such problems as precipitation of salts from aqueous samples and destruction of the effluent samples.

E. OPERATING CHARACTERISTICS OF LIQUID SCINTILLATION COUNTERS

Scintillation counting is essentially a proportional counting method; that is, the magnitude of the output signal from the detector is proportional to the energy dissipated by the radiation particle in the scintillation solution. Since beta particles are emitted over a continuous energy spectrum and, in the case of weak betas, nearly all the particle energy is given up to the scintillation solution, a pulse height analyzer is almost universally used with a liquid scintillation detector. Pulse height analysis makes it possible to select the optimal counting condition for a given sample, to determine the extent of quenching in a sample, and simultaneously to count the activity in a sample labeled with two or even three different isotopes (131). Most commercially available counting assemblies have three or four single-channel analyzers allowing multichannel operation (16). A number of computer programs for the automated analysis of the masses of data provided by such counting systems have been prepared and described.

1. Selection of Optimal Counter Settings (amplifier gain and window setting)

The size of a detector pulse resulting from a given β^--ray interaction in a scintillation solution is directly related to both the energy of the β^--particle and the gain of the electronic circuitry. Previously, pulse amplification could be altered by varying the potential gradient applied across the dynode chain of the photomultiplier. With newer liquid scintillation counters, photomultiplier voltage is generally fixed by the manufacturer, but provisions are made to allow change of the amplifier gain. Some new counters are equipped with a logarithmic amplifier (Beckman and Intertechnique), while others use linear amplifiers (Packard and Nuclear-Chicago). With the former, it is easier to select an optimal gain setting to cover spectra of β^--particles of different E_{max} in a single wide "window"—that is, an SCA with two predetermined discriminators. This situation is shown in Figure 9-18.

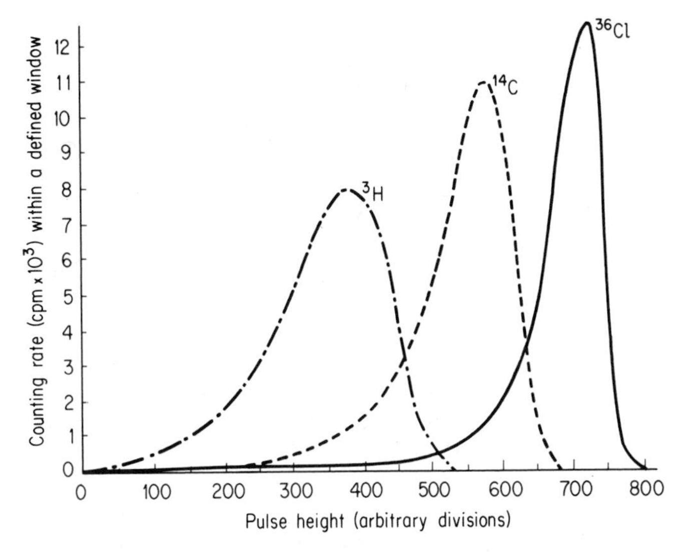

Fig. 9-18. Spectra β^--ray particles from ^3H ($E_{max} = 0.018$ MeV), ^{14}C ($E_{max} = 0.156$ MeV), and ^{36}Cl ($E_{max} = 0.714$ MeV) obtained by means of a liquid scintillation counter equipped with a logarithmic amplifier.

However, for single radioisotope counting, it is best to select an optimal gain specifically suited for the radioisotope in question, regardless of what type amplifier is involved. In order to do so, first it is important to understand the effect of gain on the shape of the beta spectrum. As amplification decreases, the upper edges (or endpoints) of the pulse height spectrum gradually shift toward the low-energy side and the spectra display increasingly sharper peaks (see Figure 9-19).

In selecting an optimal gain setting, several aspects should be considered. First, as shown in Figure 9-19, when gain is low, the β^--ray pulse height spectrum is condensed toward the low energy and shows a more pronounced peak. Any moderate quenching will shift a portion of the spectrum below the low-level discriminator, thereby resulting in a significant reduction of counting efficiency. Moreover, any slight variation in the setting of the low-level discriminator will seriously affect the reproducibility of counting data. On the other hand, when amplifier gain is high, the β^--ray spectrum is flattened somewhat and spreads out. Hence the SCA window would have to

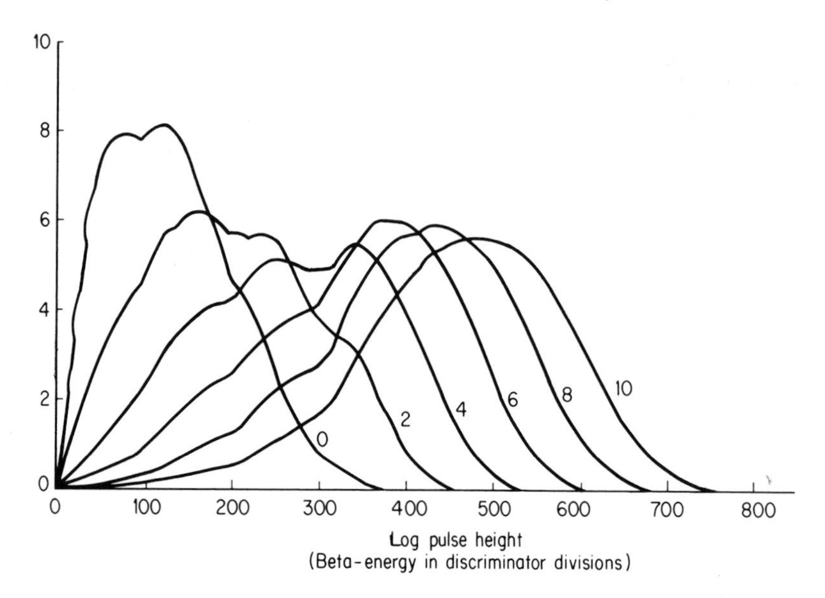

Fig. 9-19. Carbon-14 spectrum at various pulse amplifications as measured with a Beckman LS-250 liquid scintillation counter (logarithmic amplifier). Numerals in increasing order refer to increasing pulse amplification (125).

be set wider in order to encompass the entire spectrum. If the low-level discriminator stays at the same spot, the background count rate will increase significantly at higher gains. The reason is that some of the background counts (noise) cut off by the lower-level discriminator would have been shifted above the discriminator as a result of the higher gain. The situation can be remedied easily by slightly increasing the low-level discriminator setting.

It is now clear that selection of an optimal counter setting for the counting of a given radioisotope in a given type of counting sample requires the proper setting of both amplifier gain and SCA discriminators. Since the majority of counting samples are quenched by the presence of various quenching agents, it is desirable to adjust the amplifier gain so that the β^--spectrum maximum is not too close to the low-level discriminator. Needless to say, the high-level discriminator of the SCA should be set far enough to cover the upper end of the β^--spectrum and the low-level discriminator should be set high enough to minimize noise count, without cutting too much into the low-energy edge of the β^--spectrum. It should be recognized that, in the case of tritium counting, low-energy tritium emissions are of the same magnitude as noise, and therefore extraordinary care must be taken in setting the low-level discriminator.

In practice, amplifier gain and the window of the SCA are set to achieve the highest possible value of the term, "figure of merit," which is represented by S^2/B, where S is the counting rate of the counting sample in question and B is background count rate. This factor is of primary importance when one is counting samples having a very low level of radioactivity. However, in most cases, when an experiment is properly designed, the level of radioactivity in a counting sample is usually of the magnitude of a few hundred to a few thousand counts per minute. Selection of optimal counter setting can be readily accomplished by the use of moderate amplifier gain, so that the unquenched β^--spectrum is not too close to the low-level discriminator, and thus is less vulnerable to quenching, and the SCA window is set to cover the entire unquenched β^--spectrum. When counting large numbers of ^{14}C samples that contain a wide range of concentration of quenching agents, high amplifier gain can be used to flatten out the beta spectrum (*flat spectrum counting*), thereby minimizing the effect of variation in the level of quenching. Under this condition, and when the SCA window is wide open, constant counting efficiency is approached for samples having minor variations with regard to level of quenching (102, pp. 285–289). Similar consideration also applies to the Beckman counters equipped with the AQC device (16).

For the counting of dual or multiple radioisotope samples, the same principle can be applied to the selection of optimal counter setting, although several SCA windows will be involved.

2. Determination of Counting Efficiency

The efficiency of a liquid scintillation counter depends on several variable factors. Whisman *et al.* (127) have surveyed and evaluated these variables. The effect of temperature on photomultiplier thermal noise has been previously discussed. Seliger and Ziegler (114) and, more recently, Rapkin (94) have examined temperature effects on counting efficiency in general. A study of the relationship between background count rate and efficiency by Domer and Hayes (28) shows a nonlinear relation. In order to correlate the counting rate to the disintegration rate under a given set of conditions, it is essential to carry out a reliable determination of the counting efficiency.

Quenching has by far the most significant effect on counting efficiency; because of that effect, a determination of counting efficiency for every prepared sample is required. By employing the flat-spectrum counting procedure, one can, to some extent, ignore the effect of varied quenching. Nevertheless, it is necessary to determine the counting efficiency of specific groups of samples.

Methods for determining counting efficiency have been reviewed by Horrocks (55). Some important ones are given below.

a. Internal-Standardization Method. The classical method of efficiency determination is known as *internal standardization*, or *spiking* (25, 42, 81,

127). It consists of adding a precisely known amount of the same isotope (an "internal standard"), in a form that does not change the level of quenching, to a previously counted sample. Next, the resulting mixture is re-counted. The counting efficiency of the quenched sample can then be determined as follows:

$$\text{Efficiency (expressed as a fraction of unity)} = \frac{\text{net count rate (cpm)}_{\text{(internal standard+sample)}} - \text{net count rate (cpm)}_{\text{(sample)}}}{\text{disintegration rate (dpm)}_{\text{(internal standard)}}} \quad (9\text{-}1)$$

The method, when applied correctly, is very reliable and applicable to both chemically quenched and color-quenched samples. In fact, it is the primary method for standardization.

Numerous compounds have been employed as internal standards. Toluene-^{14}C, *n*-hexadecane-^{14}C, benzoic acid-1-^{14}C, tritiated toluene, tritiated water, and tritiated *n*-hexadecane are all commercially available for this purpose. Williams *et al.* (128) and Marlow and Medlock (72) have described techniques for preparing the benzoic acid-^{14}C standard. Unfortunately, each standard has a drawback. Accurate pipetting of small amounts of labeled toluene is not a simple task; benzoic acid and water are themselves quenching agents; consequently, it is necessary to use these compounds at a very low concentration. Moreover, in the case of heterogeneous samples, such as gel preparations (see Chapter 11), it is important to utilize an internal standard that can stay in the same phase as the radioactive compound in the counting sample (13).

Moghissi and Carter (75) in 1968 described the *internal standard with identical system properties method* (ISISP), by which two sample aliquots that are identical in their chemical and quenching properties are used. For spiking, a labeled standard is added to one aliquot and an unlabeled standard having the same chemical composition is added to the other aliquot. The samples are then counted consecutively to minimize the effect of any instrumental drift. This method is reported to produce excellent results in the determination of counting efficiency with samples containing both ^3H and ^{14}C.

The internal-standardization method suffers from several disadvantages. It is time consuming in that two counting processes are required for each sample and the sample vial must be opened between the counts. Furthermore, the accuracy of the determination often depends on accurate measurement of small volumes of liquid containing the labeled compound used as the standard. In addition, the original sample is contaminated by the spike and may never be counted again.

b. Dilution Method. Another means of determining efficiency involves counting a series of solutions that have different concentrations of the same radioactive sample in the fluor solution. This process can be accomplished by successive dilution of the sample solution with the fluor solution. Since the radioactivity of the counting sample remains the same, successive dilution results in stepwise reductions of the quenching-agent concentration and stepwise increases in the observed counting rate. The observed counting rate is plotted against sample (or quenching-agent) concentration, and the plot is extrapolated to zero-sample concentration, where, theoretically, sample quenching is nil. The intersection on the observed counting-rate axis will then represent the activity without sample quenching. This method, too, is time consuming and subject to considerable manipulative error. Peng (7; 24, pp. 260–275; 83) has presented an excellent evaluation and comparison of the internal-standard and dilution methods of determining counting efficiency.

c. Channels-Ratio (Pulse Height Shift) Method. When quenching occurs, the average pulse height of the spectrum decreases; consequently, the entire pulse height spectrum shifts toward a lower-energy level in a typical count rate versus pulse height plot (see Figure 9-20). Baillie (3), in 1960, first sug-

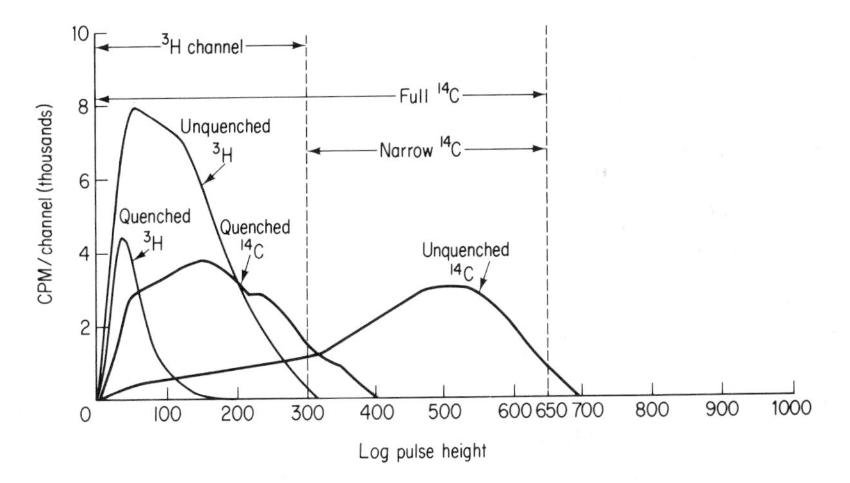

Fig. 9-20. Effect of quenching on ^{14}C and ^{3}H spectra (125).

gested a practical means for relating the degree of this pulse height shift to the extent of quenching. Bruno and Christian (17) subsequently refined the technique. The general applicability of this method has been reviewed by Bush (19). Currently the terms *pulse height shift* and *channels-ratio* method are used *interchangeably* in respect to the procedure.

Several practical techniques can be employed; the most common are outlined briefly: (a) On a dual-channel liquid scintillation spectrometer, the beta spectrum of an unquenched standard sample above the noise level is first ascertained [Figure 9-21(a)]. (b) The discriminators are now set to give

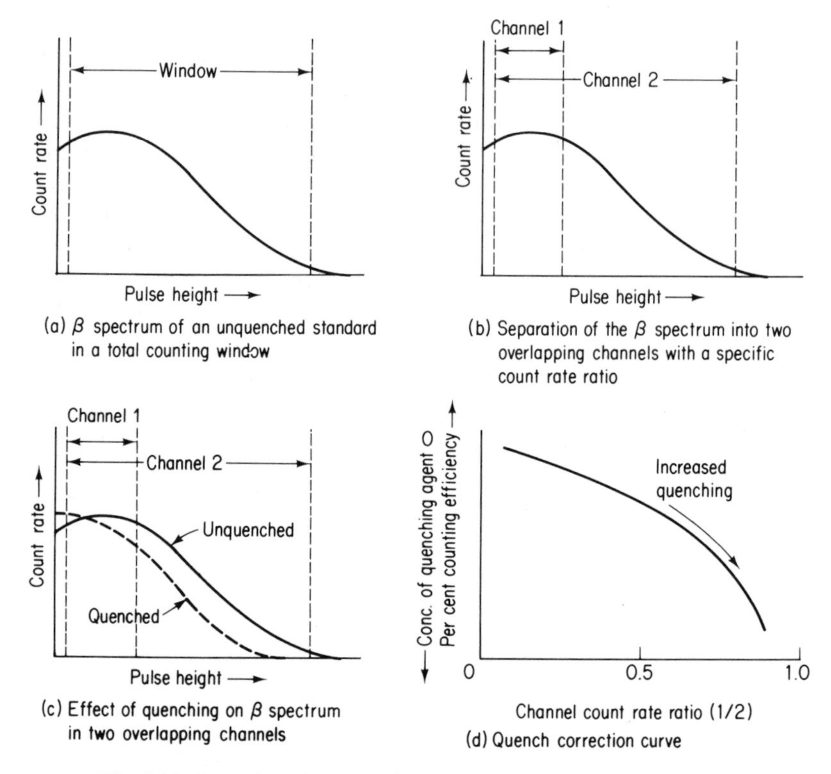

Fig. 9-21. Procedure in preparing a quench correction curve by the pulse height shift method.

two pulse windows (channels). Channel 1 covers only the first one-third of the energy spectrum, whereas channel 2 encompasses nearly the whole spectrum [Figure 9-21(b)]. Under these conditions a defined ratio will be observed between the net counting rates (i.e., sample counting rate minus background counting rate) from the two channels. (c) Increasing amounts of a quenching agent dissolved in a minimum amount of solvent are added to the counting sample, and the counting rate is observed for each of the concentrations of the quenching agent. In accordance with the quenching phenomena, the net count rate ratio between the two channels for these samples will increase as the concentration of quenching agent in the samples increases [Figure 9-21(c)] and as the counting efficiency decreases. The plot of counting efficiency versus the channels net count ratio thus provides one with a standard *quench correction curve* [Figure 9-21(d)]. It has been reported that such a plot is essentially independent of the nature of the quenching agent. Consequently, from this plot it is possible to ascertain the counting efficiency of a given sample of a specific radioisotope once the channels net count ratio is known. Obviously, if the experimenter wants to cross-compare radioactivity in a number of samples, this plot can be used to convert all the counting data to a definite counting efficiency—for example, the counting efficiency of the sample without the quencher.

The advantages of this method are evident (19, 85). A single quench correction curve for either ^{14}C or ^{3}H has been found sufficient to cover many of the common chemical quenching agents, although it will not account for color quenching (78). Such a curve, however, is strictly valid only for a specified counter and its associated settings, such as photomultiplier potential and discriminator settings. The channels-ratio method is not too dependent on the sample volume over a reasonably wide range. More recently, this method has been applied to dual radioisotope counting by Hendler (48) and Weltman and Talmage (126). Stubbs and Jackson (120) reported that the method can be readily applied to the Cerenkov counting of color-quenched samples.

The method does not apply too well to samples having a low counting rate nor to highly quenched samples. Elrick and Parker indicated that the channels-ratio method does not apply in Cerenkov counting (29); however, Moir (76) reported that the method works well in Cerenkov counting of ^{42}K. Practical suggestions concerning the channels-ratio method can be found in literature supplied by instrument manufacturers (e.g., 79).

d. External-Standard Method. Fleishman and Glazunov (31) and Higashimura *et al.* (50) have described the use of an external standard, a *γ-ray emitter*, for determining the counting efficiency associated with a given liquid scintillation counting sample. Use is made of the Compton electrons derived from the interaction of γ-rays with the counting glass vial and the scintillation solution. The γ-ray source, such as ^{137}Cs or ^{226}Ra, is stored in a shielded chamber and can be brought to a defined position immediately next to a counting sample vial on command. The energy spectrum of such Compton electrons resembles the β^--ray energy spectrum of soft beta emitters in shape. It follows that the effect of quenching agents on energy transfer from the Compton electrons to the fluor, leading eventually to the formation of photoelectrons in the photomultiplier, should be somewhat similar to that on the energy-transfer process with β^--particles. This fact makes it possible to correlate the counting efficiency of a given β^--emitter (such as ^{14}C), as determined by the internal-standard method, with the counting rate of a given γ-ray source. Such correlation curves for the ^{14}C- and ^{3}H-counting efficiencies versus the count rates of the source are shown in Figure 9-22. The observed definite correlation makes it possible to use the external γ-ray source counting rate to calibrate counting efficiency as affected by chemical quenching, color quenching, or dilution. In this method, it is essential that the exact position

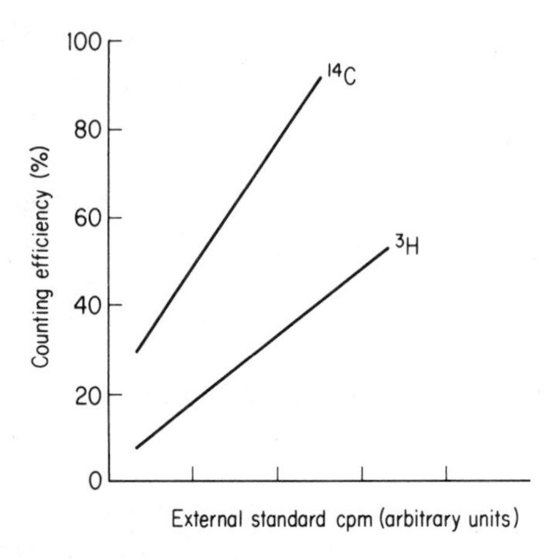

Fig. 9-22. Counting efficiency of ^{3}H- and ^{14}C-containing samples with varying amounts of added quench as a function of counting rate of an external γ-ray standard source (idealized). (55).

of the external source relative to the counting sample be reproducible in order to ensure a constant flux of γ-rays in the scintillation solution. It is equally important to establish a correction curve for each of the radioisotopes in a given type of sample as affected by a specific type of quenching agent. Even with these precautions, errors can result from variations in the thickness of counting vials or in the material used for sample-vial housing or from the presence of heavy metal atoms in the sample.

Another way of using an external standard for counting-efficiency determination is by means of the *Compton edge* (see p. 000) *technique* (32–33, 52–54). A multichannel pulse height analyzer is used to determine the Compton edges of the electron spectra derived from scattered γ-rays of the external source for both unquenched and quenched samples. As shown in Figure 9-23, the pulse height at which the Compton edge will occur is lowered by the presence of quenching agents in the counting sample to the extent proportional to the extent of quenching. This technique, as well as the related technique that measures photopeaks by stopping γ-rays from an external γ-ray or X-ray source, is tedious in operation and so is not commonly used in liquid scintillation counting practices. However, the method does offer such advantages as independence of source positioning, independence of half-life of external source, and independence of electron density charge of counting samples as derived from nonuniformity of counting vial structure.

Today, following the lead of Beckman Instruments and Nuclear-Chicago, almost all current makes of liquid scintillation counters are equipped to perform counting-efficiency determination by means of the *external-standard channels-ratio method.* Essentially this method simply applies the channels-ratio method in examining the shift of pulse height of electrons in the scintillation solution as derived from an external γ-ray source. The spectrum of the electrons is examined in two regions by the use of two SCAs (as shown

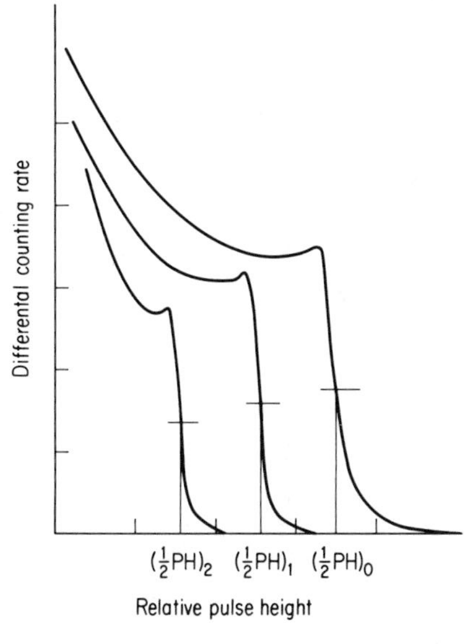

Fig. 9-23. Compton edge spectra for three samples showing the decrease in pulse height value of half-height as amount of quench is increased (55).

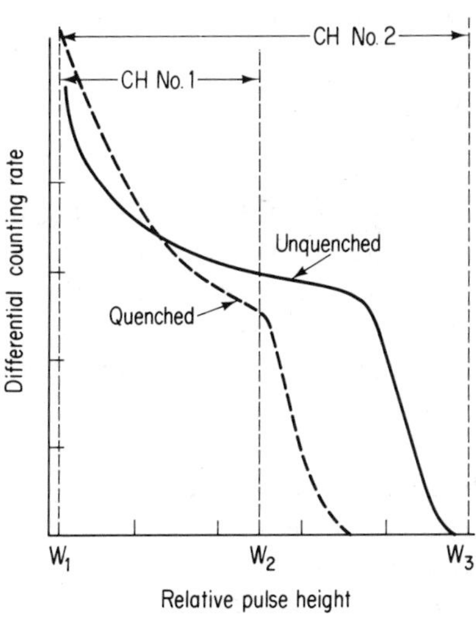

Fig. 9-24. Differential pulse spectra for quenched and unquenched samples showing relation of counts and channel settings for using dual-channels ratio technique with external γ-ray source to measure quenching (55).

in Figure 9-24), and the ratio of external-standard counts in two channels is computed and related to the extent of quenching and hence the counting efficiency. In most current counters, two independent SCAs are used exclusively to obtain the external-standard ratios. Use of the method makes the efficiency determination independent of variation in sample volume, counting vial thickness or composition, and positioning of the external source. A typical plot of counting efficiency versus external-standard ratio is shown in Figure 9-25.

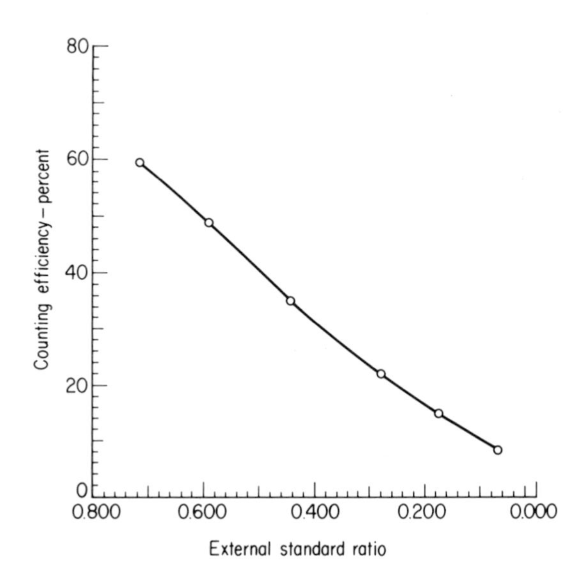

Fig. 9-25. Quench correction curves using the external-standard ratio method. Courtesy of Nuclear-Chicago.

F. SUMMARY

Internal-sample scintillation counting has the inherent advantages of high counting efficiency (particularly for low-energy β^--emitters), spectral analysis, relative ease of sample preparation, and accommodation of relatively large amounts of sample material. Because of its widespread use for assaying samples containing carbon-14 and tritium, it is of special importance to biologists and environmental scientists.

Table 9-3 summarizes pertinent information on the characteristics of the detectors discussed thus far: the ion chamber with a vibrating-reed electrometer, the proportional detector, the Geiger-Müller detector, the semiconductor detector, and both external-sample and internal-sample scintillation detectors. Although the data listed are, of necessity, general, they are based on those forms most commonly used in radiotracer assays.

BIBLIOGRAPHY

1. Agranoff, Bernard W. "Silica vials improve low-level counting," *Nucleonics* **15**(10), 106 (1957).
2. Anderson, Ernest C., et al. "The Los Alamos human counter," *Nucleonics* **14**(1), 26–29 (1956).
3. Baillie, L. A. "Determination of liquid scintillation counting efficiency by pulse height shift," *Intern. J. Appl. Radiation Isotopes* **8**, 1–7 (1960).

TABLE 9-3

Summary of Radiation Detector Characteristics

Detector Type	Energy Dis-crimi-nation	Detec-tion Medium	Repro-duci-bility	Detector Amplifi-cation Factor	Relative Electronic Amplifi-cation Required	Resolving Time (μsec)	Relative Back-ground	Relative Detection Efficiency			Most Commonly Used Sample Form	Relative Degree of Difficulty in Preparing Sample
								α	β	γ		
Ion chamber with vibrating-reed electrometer	Yes	Gas	Excel-lent	1	Very high	Not applicable	Low	Good	Good	Poor	Gas	High[a]
Proportional detector	Yes	Gas	Good	10^2–10^4	High	1–2	Low	Good	Good	Poor	Solid	Medium[b]
Geiger-Müeller detector	No	Gas	Fair	10^7	Low	100–300	Medium	Good	Good	Poor	Solid	Medium[b]
Solid (external-sample) scintillation detector	Yes	Solid	Good	10^6	Low	0.3–>1	High	Fair	Fair	Very good	Solid or liquid	Medium[b] or low[c]
Semiconductor detector	Yes	Solid	Good	1	High	0.05	Low	Very good	Fair	Fair	Solid ($\alpha\beta\gamma$) Liquid (γ)	Low
Liquid (internal-sample) scintillation detector	Yes	Liquid	Fair	10^6	Medium	0.05–<1	Low	Very good	Very good	Fair	Solution	Low

[a]Usually involves individual combustion of samples to gaseous form.
[b]Usually involves preparation of planchet-mounted samples.
[c]Where well-type detector is used for liquid samples.

4. Bannerman, D. E., and R. J. Lanter. "Durable white coating improves liquid scintillation counters," *Nucleonics* **14**(2), 60–61 (1956).

5. Barnett, Martin D., et al. "Liquid scintillators. VI. 2-aryl- and 2,7-diaryl-fluorenes." *J. Am. Chem. Soc.* **81**, 4583–4586 (1959).

6. Barnett, Martin D., et al. "Liquid scintillators. XI. 2-(2-fluorenyl)-5-aryl-substituted oxazoles and 2-(2-fluorenyl)-5-phenyl-1,3,4-oxadiazole." *J. Am. Chem. Soc.* **82**, 2282–2285 (1960).

7. Bell, Carlos G., Jr., and F. Newton Hayes (Eds.). *Liquid Scintillation Counting. Proceedings of the Conference of Northwestern University, Evanston, Ill., 1957.* New York: Pergamon, 1958.

8. Benson, Royal H., and Robert L. Maute. "Liquid scintillation counting of tritium: Improvements in sensitivity by efficient light collection," *Anal. Chem.* **34**, 1122–1124 (1962).

9. Benson, R. "Limitations of tritium measurements by liquid scintillation counting of emulsions," *Anal. Chem.* **38**, 1353–1356 (1966).

10. Berlman, Isadore B. "Luminescence in a scintillation solution excited by α and β particles and related studies in quenching," *J. Chem. Phys.* **34**, 598–603 (1961).

11. Birks, J. B. *The Theory and Practice of Scintillation Counting.* New York: Macmillan, 1964.

12. Blau, Monte. "Separated channels improve liquid scintillation counting," *Nucleonics* **15**(4), 90–91 (1957).

13. Bloom, Ben. "Use of internal scintillation standards in heterogeneous counting systems," *Anal. Biochem.* **6**, 359–361 (1963).

14. Bollinger, Lowell M., and George E. Thomas. "Boron-loaded liquid scintillation neutron detectors," *Rev. Sci. Instruments* **28**, 489–496 (1957).

15. Boyce, I. S., J. F. Cameron, and K. J. Taylor. "A simple plastic scintillation counter for tritiated hydrogen," *Intern. J. Appl. Radiation Isotopes* **9**, 122–123 (1960).

16. Bransome, Edwin D., Jr. (Ed.). *The Current Status of Liquid Scintillation Counting.* New York and London: Grune and Stratton, 1970.

17. Bruno, Gerald A., and John E. Christian. "Corrections for quenching associated with liquid scintillation counting," *Anal. Chem.* **33**, 650 (1961).

18. Buck, Warren L., and Robert K. Swank. "Use of isopropylbiphenyl as solvent in liquid scintillators," *Rev. Sci. Instruments* **29**, 252 (1958).

19. Bush, Elizabeth T. "General applicability of the channels ratio method of measuring liquid scintillation counting efficiencies," *Anal. Chem.* **35**, 1024–1029 (1963).

20. Calf, G. E. "Teflon vial for liquid scintillation counting of tritium samples," *Intern. J. Appl. Radiation Isotopes* **20**, 611–612 (1969).

21. Chleck, D. J., and C. A. Ziegler. "Ultrasonic degassing of liquid scintillators," *Rev. Sci. Instruments* **28**, 466–467 (1957).

22. Christian, J. E., W. V. Kessler, and P. L. Ziemer. "A 2π liquid scintillation counter for determining the radioactivity of large samples, including man and animals," *Intern. J. Appl. Radiation Isotopes* **13**, 557–564 (1962).

23. Continuous scintillation counting of weak beta emitters in flowing aqueous streams," *Nuclear-Chicago Tech. Bull. No. 15.*, Des Plaines, Ill., 1963.

24. Daub, Guido H., F. Newton Hayes, and Elizabeth Sullivan (Eds.). *Proceedings of the University of New Mexico Conference on Organic Scintillation Detectors, Albuquerque, 1960.* TID-7612. U.S. Government Printing Office, Washington, D.C., 1960.

25. DAVIDSON, JACK D., and PHILIP FEIGELSON. "Practical aspects of internal-sample liquid-scintillation counting," *Intern. J. Appl. Radiation Isotopes* **2**, 1–18 (1957).

26. DEBERSAQUES, J. "Relation between the absorption and the quenching of liquid scintillation samples," *Intern. J. Appl. Radiation Isotopes* **14**, 173–174 (1963).

27. DEWACHTER, R., and W. FIERS. "External standardization in liquid scintillation counting of homogeneous samples labeled with one, two, or three isotopes," *Anal. Biochem.* **18**, 351–374 (1967).

28. DOMER, FLOYD R., and F. NEWTON HAYES. "Background vs. efficiency in liquid scintillators," *Nucleonics* **18**(1), 100 (1960).

29. ELRICK, R. H., and R. P. PARKER. "The use of Cerenkov radiation in the measurement of β-emitting radionuclides," *Intern. J. Appl. Radiation Isotopes* **19**, 263–271 (1968).

30. FAISSNER, HEIMUT, et al. "New scintillation liquids," *Nucleonics* **21**, 50–55 (1963).

31. FLEISHMAN, D. G., and V. V. GLAZUNOV. "An external standard as a means of determining the efficiency and background of a liquid scintillator," *Instruments Exp. Tech.* (a translation) No. 3, 472–474 (1962).

32. FLYNN, K. F., and L. E. GLENDENIN. "Half-life and beta spectrum of Rb^{87}," *Phys. Rev.* **116**(3), 744–748 (1959).

33. FLYNN, K. F., L. E. GLENDENIN, E. P. STEINBERG, and P. J. WRIGHT. "Pulse height-energy relations for electrons and alpha particles in a liquid scintillator," *Nucl. Instr. Methods* **27**, 13–17 (1964).

34. FÖRSTER, TH. "Zwischenmolekulare energiewanderung und fluoreszenz," *Annalen der Physik.* 6 Folge, Band 2, p. 55 (1948).

35. FUNT, B. L., and A. HETHERINGTON. "Spiral capillary plastic scintillation flow counter for beta assay," *Science* **129**, 1429–1430 (1959).

36. FUNT, B. L., and A. HETHERINGTON. "The kinetics of quenching in liquid scintillators," *Intern. J. Appl. Radiation Isotopes* **13**, 215–221 (1962).

37. FURST, MILTON, and HARTMUT KALLMAN. "Fluorescent behavior of solutions containing more than one solvent," *J. Chem. Phys.* **23**, 607–612 (1955).

38. HABERER, K. "Messung von beta-aktivitäten an Wässrigen proben auf grund der Cerenkov-Strahlung," *Atom Wirtshaft* **10**(1), 36–43 (1965).

39. HARLEY, JOHN H., NAOMI A. HALLDEN, and ISABEL M. FISENNE. "Beta scintillation counting with thin plastic phosphors," *Nucleonics* **20**(1), 59–61 (1962).

40. HARRISON, F. B. "Slow component in decay of fluors," *Nucleonics* **12**(3), 24–25 (1954).

41. HAYES, F. NEWTON, DONALD G. OTT, and VERNON N. KERR. "Liquid scintillators. II. Relative pulse height comparisons of secondary solutes." *Nucleonics* **14**(1), 42–45 (1956).

42. HAYES, F. NEWTON. "Liquid scintillators: attributes and applications," *Intern. J. Appl. Radiation Isotopes* **1**, 46–56 (1956).

43. HAYES, F. NEWTON, and DONALD G. OTT. *The Small-Volume Internal-Sample Liquid Scintillation Counter.* U.S. Atomic Energy Commission, LA-2095, 1957.

44. HAYES, F. NEWTON, et al. *Survey of Organic Compounds as Primary Scintillation Solutes.* U.S. Atomic Energy Commission, LA-2176, 1958.

45. HAYES, F. NEWTON. "Solutes and solvents for liquid scintillation counting," *Tech. Bull. No. 1.* Rev. ed. Downer Grove, Ill.: Packard Instrument Company, 1962.

46. HEJWOWSKI, J., and A. SZYMANSKI. "Lithium loaded liquid scintillator," *Rev. Sci. Instruments* **32**, 1057–1058 (1961).

47. HELMICK, MARIE. "The quenching effect of benzoic acid in a liquid scintillation system," *Atomlight* (New England Nuclear Corp.), pp. 6–7 (February 1960).

48. HENDLER, RICHARD W. "Procedure for simultaneous assay of two β-emitting isotopes with the liquid scintillation technique," *Anal. Biochem.* **7**, 110–120 (1964).

49. HERBERG, R. J. "Backgrounds for liquid scintillation counting of colored solutions," *Anal. Chem.* **32**, 1468–1471 (1960).

50. HIGASHIMURA, T., et al. "External standard method for the determination of the efficiency in liquid scintillation counting," *Intern. J. Appl. Radiation Isotopes* **13**, 308–309 (1962).

51. HODGSON, T. S., B. E. GORDON, and M. E. ACKERMAN. "Single-channel counter for carbon-14 and tritium," *Nucleonics* **16**(7), 89–94 (1958).

52. HORROCKS, D. L., and M. H. STUDIER. "Low level plutonium-241 analysis by liquid scintillation techniques," *Anal. Chem.* **30**, 1747–1750 (1958).

53. HORROCKS, D. L. "Alpha particle energy resolution in a liquid scintillator," *Rev. Sci. Instr.* **35**, 334–340 (1964).

54. HORROCKS, D. L. "Measurement of sample quenching of liquid scintillator solutions with x-ray and gamma-ray sources," *Nature* **202**, 78–79 (1964).

55. HORROCKS, D. L. "Liquid scintillation counting," *Surv. Progr. Chem.* **5**, 185–235 (1969). See also HORROCKS, D. L. *Applications of Liquid Scintillation Counting*. London: Academic Press, 1974.

56. HORROCKS, D. L., and C. T. PENG (Eds.). *Organic Scintillators and Liquid Scintillation Counting*. New York and London: Academic, 1971, pp. 607–620.

57. INTERNATIONAL ATOMIC ENERGY AGENCY. *Proceedings of a Symposium on the Detection and Use of Tritium, Phys. Biol. Sci. Vienna, 1961*. Vol. 1. 1962.

58. KALLMAN, HARTMUT. "Scintillation counting with solutions," *Phys. Rev.* **78**, 621–622 (1950).

59. KARMEN, A., and H. R. TRITCH. "Radioassay by gas chromatography of compounds labelled with carbon-14," *Nature* **186**, 150–151 (1960).

60. KARMEN, ARTHUR, IRMGARDE MCCAFFREY, and ROBERT L. BOWMAN. "A flow-through method for scintillation counting of carbon-14 and tritium in gas-liquid chromatographic effluents," *J. Lipid Res.* **3**, 372–377 (1962).

61. KARMEN, ARTHUR, LAURA GIUFFRIDA, and ROBERT L. BOWMAN. "Radioassay by gas-liquid chromatography of lipids labeled with carbon-14," *J. Lipid Res.* **3**, 44–52 (1962).

62. KARMEN, ARTHUR, IRMGARDE MCCAFFREY, and BERNARD KLIMAN. "Derivative ratio analysis: A new method for measurement of steroids and other compounds with specific functional groups using radioassay by gas-liquid chromatography," *Anal. Biochem.* **6**, 31–38 (1963).

63. KARMEN, ARTHUR, et al. "Measurement of tritium in the effluent of a gas chromatography column," *Anal. Chem.* **35**, 536–542 (1963).

64. KERR, VERNON N., F. NEWTON HAYES, and DONALD G. OTT. "Liquid scintillators. III. The quenching of liquid scintillator solutions by organic compounds." *Intern. J. Appl. Radiation Isotopes* **1**, 284–288 (1957).

65. KIMBEL, K. H., and J. WILLENBRINK. "Fortlaufende Messung schwacher β-Strahler in Flüssigkeiten mit Szintillatorschlauch," *Naturwissenschaften* **45**, 567 (1958).

66. KIMBEL, K. H., and J. WILLENBRINK. "An inexpensive disposable sample container for single phototube liquid scintillation counting," *Intern. J. Appl. Radiation Isotopes* **11**, 157 (1961).

67. KOBAYASHI, YUTAKA. *Liquid Scintillation Counting and Some Practical Considerations.* Waltham, Mass.: Tracerlab, 1961.

68. KOBAYASHI, Y., and D. V. MAUDSLEY. "Practical aspects of double isotope counting." In E. D. Bransome, Jr. (Ed.), *The Current Status of Liquid Scintillation Counting.* New York and London: Grune and Stratton, 1970, pp. 273–282.

69. "Liquid scintillation counting," *Nuclear-Chicago Tech. Pub. No. 711580.* Des Plaines, Ill., 1962.

70. LOHMANN, W., and W. H. PERKINS. "Stabilization of the counting rate by irradiation of the liquid scintillation counting solutions with UV-light," *Nuclear Instruments Methods* **12**, 329–334 (1961).

71. LUKENS, H. R., JR. "The relationship between fluorescence intensity and counting efficiency with liquid scintillators," *Intern. J. Appl. Radiation Isotopes* **12**, 134–140 (1961).

72. MARLOW, W. F., and R. W. MEDLOCK. "A carbon-14 beta-ray standard, benzoic acid-7-C^{14} in toluene, for liquid scintillation counters," *J. Res. Nat. Bur. Standards* **64A**, 143–146 (1960).

73. MEINERTZ, HANS, and VINCENT P. DOLE. "Radioassay of low activity fractions encountered in gas-liquid chromatography of long-chain fatty acids," *J. Lipid Res.* **3**, 140–144 (1962).

74. MIRANDA, H. A., JR., and H. SCHIMMEL. "New liquid scintillant," *Rev. Sci. Instruments* **30**, 1128–1129 (1959).

75. MOGHISSI, A. A., and M. W. CARTER. "Internal standard with identical system properties for determination of liquid scintillation counting efficiency," *Anal. Chem.* **40**, 812–814 (1968).

76. MOIR, A. T. B. "Channels ratio quench correction using Cerenkov radiation for the assay of ^{42}K in biological samples," *Intern. J. Appl. Radiation Isotopes* **22**, 213–216 (1971).

77. MYERS, L. S., JR., and A. H. BRUSH. "Counting of alpha- and beta-radiation in aqueous solutions by the detergent-anthracene scintillation method," *Anal. Chem.* **34**, 342–345 (1962).

78. NEARY, M. P., and A. L. BUDD. "Color and chemical quench." In E. D. Bransome, Jr. (Ed.), *The Current Status of Liquid Scintillation Counting.* New York and London: Grune and Stratton, 1970.

79. NEARY, M. P. *Nuclear Laboratory Experiment Manual.* Fullerton, Calif.: Beckman Instruments, Inc., 1971.

80. NEARY, M. P. "The use of a new Beckman discrete sampling flow cell with an amino acid analyzer and liquid scintillation counter," *Beckman Technical Report No. 561.* Fullerton, Calif.: Beckman Instruments, Inc., 1971.

81. OKITA, GEORGE T., et al. "Assaying compounds containing H^3 and C^{14}," *Nucleonics* **15**(6), 111–114 (1957).

82. PARKER, R. P., and R. H. Elrick. "Cerenkov counting as a means of assaying β-emitting radionuclides." In E. D. Bransome, Jr. (Ed.), *The Current Status of Liquid Scintillation Counting.* New York and London: Grune and Stratton, 1970, pp. 110–122.

83. PENG, C. T. "Quenching of fluorescence in liquid scintillation counting of labeled organic compounds," *Anal. Chem.* **32**, 1292–1296 (1960).

84. PENG, C. T. "Correction of quenching in liquid scintillation counting of homogeneous samples containing both carbon-14 and tritium by extrapolation method," *Anal. Chem.* **36**, 2456–2461 (1964).

85. PENG, C. T. "A review of methods of quench correction in liquid scintillation counting." In E. D. Bransome, Jr. (Ed.), *The Current Status of Liquid Scintillation Counting.* New York and London: Grune and Stratton, 1970, pp. 283–292.

86. PIEZ, KARL A. "Continuous scintillation counting of carbon-14 and tritium in effluent of the automatic amino acid analyzer," *Anal. Biochem.* **4**, 444–458 (1962).

87. POPJÁK, G., et al. "Scintillation counter for the measurement of radioactivity of vapors in conjunction with gas-liquid chromatography," *J. Lipid Res.* **1**, 29–39 (1959).

88. POPJÁK, G., A. E. LOWE, and D. MOORE. "Scintillation counter for simultaneous assay of H^3 and C^{14} in gas-liquid chromatographic vapors," *J. Lipid Res.* **3**, 364–371 (1962).

89. PRINGLE, R. W., et al. "A new quenching effect in liquid scintillators," *Phys. Rev.* **92**, 1582–1583 (1953).

90. RAPKIN, E., and J. A. GIBBS. "A system for continuous measurement of radioactivity in flowing streams," *Nature* **194**, 34–36 (1962).

91. RAPKIN, E. "Liquid scintillation counting with suspended scintillators," *Tech. Bull. No. 11.* Downer Grove, Ill.: Packard Instrument Company, 1963.

92. RAPKIN, E., and J. A. Gibbs. "Polyethylene containers for liquid scintillation spectrometry," *Intern. J. Appl. Radiation Isotopes* **14**, 71–74 (1963).

93. RAPKIN, E. "Liquid scintillation counting 1957–1963: A review," *Intern. J. Appl. Radiation Isotopes* **15**, 69–87 (1964).

94. RAPKIN, E. "Temperature control in liquid scintillation counting," *Digitechniques* Intertechnique. Dover, N.J., France, U.K., Sweden, and Israel, 1969, pp. 1–7.

95. RAPKIN, E. "Sample preparation for liquid scintillation counting: part 1 and part 2," *Digitechniques*, 1970. Technical review published by Intertechnique.

96. REYNOLDS, GEORGE T., F. B. HARRISON, and G. SALVINI. "Liquid scintillation counters," *Phys. Rev.* **78**, 488 (1950).

97. RIVLIN, RICHARD S., and HILDEGARD WILSON. "A simple method for separating polar steroids from the liquid scintillation phosphor," *Anal. Biochem.* **5**, 267–269 (1963).

98. RONZIO, A. R., C. L. COWAN, JR., and F. REINES. "Liquid scintillators for free neutrino detection," *Rev. Sci. Instruments* **29**, 146–147 (1958).

99. ROSS, HARLEY H., and ROGER E. YERICK. "Quantitative interpretation of color quenching in liquid scintillator systems," *Anal. Chem.* **35**, 794–797 (1963).

100. ROSS, H. H. "Measurement of β-emitting nuclides using Cerenkov radiation," *Anal. Chem.* **41**, 1260–1265 (1969).

101. ROSS, H. H. "Cerenkov radiation: photon yield application to ^{14}C assay." In E. D. Bransome, Jr. (Ed.), *The Current Status of Liquid Scintillation Counting.* New York and London: Grune and Stratton, 1970, pp. 123–126.

102. ROTHCHILD, SEYMOUR (Ed.). *Advances in Tracer Methodology.* Vol. 1. New York: Plenum Press, 1963.

103. ROTHCHILD, SEYMOUR (Ed.). *Advances in Tracer Methodology.* Vol. 2. New York: Plenum Press, 1965.

104. ROTHCHILD, SEYMOUR (Ed.). *Advances in Tracer Methodology*. Vol. 3. New York: Plenum Press, 1966.

105. ROTHCHILD, SEYMOUR (Ed.). *Advances in Tracer Methodology*. Vol. 4. New York: Plenum Press, 1968.

106. SCHARPENSEEL, H. W., and K. H. MENKE. "Radiochromatographie mit schwachen β-Strahlern (^{35}S, ^{14}C, ^{3}H). II. Radiosaulenchromatographie mit Hilfe des Flussigkeits-Scintillations-Spektrometers." *Z. Anal. Chem.* **182**, 1–10 (1961).

107. SCHRAM, E., and R. LOMBAERT. "Determination continue du carbone-14 et du soufre-35 en mileu aqueux par un dispostif a scintillation. Application aux effluents chromatographiques." *Anal. Chim. Acta.* **17**, 417–422 (1957).

108. SCHRAM, E., and R. LOMBAERT. "Dosage continu du carbone-14 dans les effluents chromatographiques au moyen de poudres d'anthracene," *Arch. Intern. Physiol. Biochim.* **68**, 845–846 (1960).

109. SCHRAM, E., and R. LAMBAERT. "Determination of tritium and carbon-14 in aqueous solution with anthracene powder," *Anal. Biochem.* **3**, 68–74 (1962).

110. SCHRAM, E. *Organic Scintillation Detectors*. Amsterdam: Elsevier, 1963.

111. SCHRAM, E. "Flow-monitoring of aqueous solutions containing weak β-emitters." In E. D. Bransome, Jr. (Ed.), *The Current Status of Liquid Scintillation*. New York and London: Grune and Stratton, 1970, pp. 95–109.

112. SCHRAM, E. ,R. CORTENBOSCH, E. GERLO and H. ROOSENS. In D. L. Horrocks and C. T. Peng (Eds.), *Organic Scintillators and Liquid Scintillation Counting*. New York and London: Academic, 1971, pp. 125–135.

113. SELIGER, H. H., C. A. ZIEGLER, and I. JAFFE. "Role of oxygen in the quenching of liquid scintillators," *Phys. Rev.* **101**, 998–999 (1956).

114. SELIGER, H. H., and C. A. ZIEGLER. "Liquid-scintillator temperature effects," *Nucleonics* **14**(4), 49 (1956).

115. STANLEY, P. E., and S. G. WILLIAMS. "Use of the liquid scintillation spectrometer for determining adenosine triphosphate by the luciferase enzyme," *Anal. Biochem.* **29**, 381–392 (1969).

116. STANLEY, P. E. "The use of the liquid scintillation spectrometer for measuring NADH and FMN by the photobacterium luciferase and ATP by the firefly luciferase." In D. L. Horrocks and C. T. Peng (Eds.), *Organic Scintillators and Liquid Scintillation Counting*. New York and London: Academic, 1971, pp. 607–620.

117. STEINBERG, DANIEL. "Radioassay of carbon-14 in aqueous solutions using a liquid scintillation spectrometer," *Nature* **182**, 740–741 (1958).

118. STEINBERG, DANIEL. "Radioassay of aqueous solutions mixed with solid crystalline fluors," *Nature* **183**, 1253–1254 (1959a).

119. STEINBERG, Daniel. "A new approach to radioassay of aqueous solutions in the liquid scintillation spectrometer," *Anal. Biochem.* **1**, 23–39 (1959b).

120. STUBBS, R. D., and A. JACKSON. "Channel ratio color quenching correction in Cerenkov counting," *Intern. J. Appl. Radiation Isotopes* **18**, 857–858 (1967).

121. SWANK, ROBERT K., and WARREN L. BUCK. "Spectral effects in the comparison of scintillators and photomultipliers," *Rev. Sci. Instruments* **29**, 279–284 (1958).

122. TANIELIAN, C., et al. "Influence de la purification du solvant sur le rendement des scintillateurs liquides—I," *Intern. J. Appl. Radiation Isotopes* **15**, 11–15 (1964a).

123. TANIELIAN, C., et al. "Influence de la purification du solvant sur le rendement des scintillateurs liquides—II," *Intern. J. Appl. Radiation Isotopes* **15**, 17–23 (1964b).

124. Tracerlab, Inc. *Quenching in Liquid Scintillation Counting*. Waltham, Mass., 1961.

125. Wang, C. H. "Quench compensation by means of gain restoration." In E. D. Bransome, Jr. (Ed.), *The Current Status of Liquid Scintillation Counting*. New York and London: Grune and Stratton, 1970, pp. 305–312.

126. Weltman, J. K., and D. W. Talmage. "A method for the simultaneous determination of H^3 and S^{35} in samples with variable quenching," *Intern. J. Appl. Radiation Isotopes* **14**, 541–548 (1963).

127. Whisman, M. L., B. H. Eccleston, and F. E. Armstrong. "Liquid scintillation counting of tritiated organic compounds," *Anal. Chem.* **32**, 484–486 (1960).

128. Williams, D. L., et al. "Preparation of C^{14} standard for liquid scintillation counter," *Nucleonics* **14**(1), 62–64 (1956).

129. Williams, D. L., and F. N. Hayes. *Liquid Scintillator Radiation Rate Meters for the Measurement of Gamma and Fast Neutron Rates in Mixed Radiation Fields.* U.S. Atomic Energy Commission, LA-2375, 1960.

130. Williams, P. H., and T. Florkowski. "Radioactive dating and methods of low level counting." Vienna: International Atomic Energy Agency, 1967, p. 703.

131. Wu, Ray. "Simultaneous studies of phosphate transport and glycolysis by a simple liquid scintillation counting procedure with P^{32}, C^{14}, and H^3 compounds," *Anal. Biochem.* **7**, 207–214 (1964).

132. Ziegler, C. A., H. H. Seliger, and I. Jaffe. "Three ways to increase efficiency of liquid scintillators," *Nucleonics* **14**(5), 84–86 (1956).

133. Zutshi, P. K. "Low-level beta counting and absorption measurement with liquid scintillators," *Nucleonics* **21**(9), 50–53 (1963).

10

Measurement of Radioactivity by Emulsion and Track Detectors

One Picture Is Worth More Than Ten Thousand Words.

ANCIENT CHINESE PROVERB

A. INTRODUCTION

In this chapter some of the radiation detection methods and techniques in which a permanent visual record of the path of ionizing radiation in matter is made are discussed. In particular, the detection of radioactivity by using photographic emulsions and solid-state nuclear track detectors is covered. In discussing photographic emulsion detectors, a distinction is made between those applications in which the radioactivity of a sample is measured or located (*autoradiography*) and those in which the attenuation of external radiation by a sample is measured (such as X-ray and *neutron radiography*). Autoradiography is the only photographic emulsion-based technique described here.

Ionizing radiation acts on a photographic emulsion to produce a latent image much as visible light does. The production of the latent image and its subsequent development have been described by Fitzgerald *et al.* (8):

> *The formation of metallic silver from the silver bromide crystal is a complex matter.* The alpha or beta particle or photon ionizes as it passes through the emulsion and sets electrons free in the ionic lattice of the silver bromide crystal. The electrons migrate to the crystal sensitivity specks, either of metallic colloidal silver or silver sulfide, and the specks become negatively charged. The electrostatic field set up in the grain attracts the mobile silver ions and they migrate to the speck.

Upon reaching the speck, the interstitial silver ions neutralize the electrons and metallic silver is formed. Thus, the specks act as loci for the growth of nuclei of metallic silver, called *latent images*. One speck of silver initially present may grow by one silver atom for each light quantum or photon absorbed. After the production of latent images in the silver bromide crystals, developers (reducing agents) cause the reduction of the grains with latent images to metallic silver.* It is these metallic silver grains that give the darkening in photographic emulsions. The grains of silver bromide that contain no latent images are reduced at such a slow rate compared to those with latent images that judicious selection of developers and development times may reduce the latter to metallic silver, whereas the former are relatively unaffected by the developer. The hypo solutions preferentially dissolve the grains of silver bromide, leaving the metallic silver as a marker of the production of latent images and hence the trail of ionizing radiation.

In autoradiography, a sample containing radioactivity, such as a tissue section, is placed in close contact with a sensitive emulsion. After a period of exposure, the film is developed and the precise location of the radioactive matter in the sample can then be determined from the pattern of darkening on the film. This method of radiation detection appeals particularly to the biologist because no electronic equipment is needed.

Autoradiography is primarily a means of determining the *location* of radioisotopes in a given tissue section, gross sample, or chromatogram. For example, the sites of ^{45}Ca concentration in growing bone tissue, the relative distribution of ^{32}P as the phosphate through a bean plant, or the localization of thymidine-^{3}H in the DNA of cell nuclei can all be readily demonstrated by this technique. Use of the technique for precise quantitative measurements of radioactivity, however, is fraught with considerable difficulty and uncertainty (6, 14). Where gross samples are used, the relative blackening of various areas may be roughly measured with a densitometer. This is the principle behind the film badge dosimeter, as will be described in Chapter 16. Where microscopic sections are to be assayed, the tedious task of counting the individual blackened grains of silver is required (23).

To the biologist, perhaps the chief advantage of autoradiography, particularly with very soft β^--ray emitters, is that it permits a study of metabolic function at the level of the individual cell. In addition, a permanent record is produced on film for later examination. Note, however, that optimal results with this technique require considerable skill and experience. In short, autoradiography is as much an art as a science.

In this chapter we will present some general concepts of autoradiography and then discuss specific techniques briefly. A detailed description of the many and varied techniques in autoradiography is beyond the scope of this

*The metallic silver present in the latent image catalyzes the attack of the developer on adjacent silver bromide crystals.

book. (As of 1969 there were over 10,000 publications on autoradiography in the literature.) The bibliography at the end of the chapter contains some important general references.

B. GENERAL PRINCIPLES OF AUTORADIOGRAPHY

1. Resolution and Radioisotope Characteristics

Since the primary aim in using the autoradiographic technique is to determine the location of radioactivity, it is essential to achieve a high degree of resolution on the autoradiographs. *Resolution* may be defined as the minimum distance between two point sources of activity that still allows them to be distinguished from each other on the developed film. This is a particularly critical factor where intracellular localization is under investigation.

The degree of resolution is affected by a number of factors. The maximum resolution attainable in a given situation is limited by the specific emulsion used. Since radiation is emitted from a point source in all directions, the greater the distance between the radiation source and the emulsion, the more diffuse the film image produced. For optimal resolution, use the thinnest feasible sample and emulsion in the closest possible contact. A sketch of the effect of these geometrical factors on the resolution is shown in Figure 10-1.

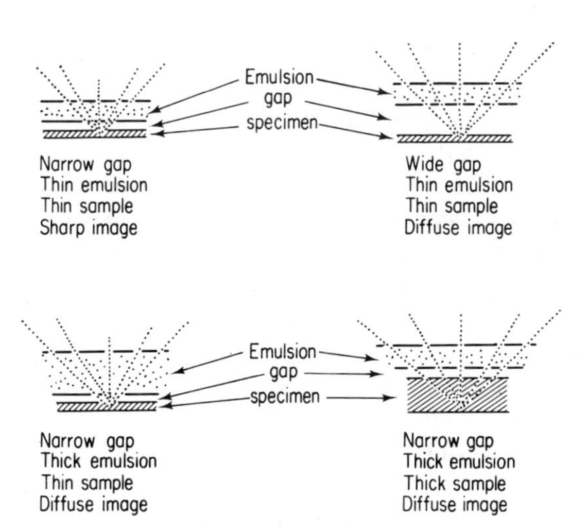

Fig. 10-1. Geometry factors in autoradiography. From Chase and Rabinowitz, *Principles of Radioisotope Methodology*, 3rd ed., 1967, Burgess Publishing Company.

One of the most significant factors affecting resolution is the character of the radiation being detected. The intensity of film blackening is directly related to the specific ionization of the ionizing particle. Alpha particles produce such intense ionization along their short paths through the emulsion that they are readily distinguished. Beta particles, with their lower specific ionization, traverse longer, more irregular paths through the emulsion and result in a somewhat more diffuse image. Gamma rays are seldom involved in autoradiographic studies because of their extremely low specific ionization.

Since the ion density along a particle track is inversely related to the kinetic energy of the beta particle, it follows that low-energy β^--emitters will create autoradiographs of high resolution. This relation is seen from the average path length in emulsion of β^--particles from the following isotopes: $^{3}\mathrm{H} = <2\ \mu$, $^{14}\mathrm{C}$ and $^{35}\mathrm{S} = <100\ \mu$, and $^{32}\mathrm{P} = <3200\ \mu$. Clearly, tritium should give exceptionally high resolution in autoradiography. Since, in addition, ^{3}H-labeled compounds can usually be prepared at much higher specific activities than the corresponding ^{14}C-labeled compounds (see Chapter 15), there has been a considerable increase in the use of tritium-labeled compounds in autoradiography in recent years (1, 15, 21, 22, pp. 291–301; 26). By contrast, high-energy β^--particles, as from ^{32}P, produce such diffuse images that highly precise intracellular localization of ^{32}P-labeled compounds is not feasible. The same is generally true for X and γ-rays. The soft X rays resulting from electron capture in such nuclides as ^{55}Fe can, however, be used to produce reasonably good images on especially sensitive emulsions.

2. Film Emulsion and Sensitivity

Film used in autoradiography is composed of three major components. The sensitive agent, usually grains of silver halide, is dispersed through a gelatin medium to form the emulsion. This emulsion is backed by a sheet of cellulose acetate or glass. The size and concentration of the silver halide grains govern both the sensitivity and the resolution of the emulsion. Grain size is directly related to the degree of resolution possible. Unfortunately, increased emulsion sensitivity to ionizing radiation necessarily increases sensitivity to background radiation "fogging." The various emulsion types available represent different compromises with regard to these factors.

Autoradiographic emulsions may be generally divided into two groups: those used only for *gross autoradiography* and those suitable for *microscopic applications*. The former group includes the various X-ray sensitive films, of which Eastman Kodak's "No-Screen" film is a typical example. This film has relatively poor resolution and a high-background sensitivity. Alternatively, Eastman Kodak's single-coated, blue-sensitive X-ray film is most often chosen for use with radiochromatograms. The so-called nuclear emulsions are used for microscopic autoradiography. They include emulsions that are especially sensitized to α- or β^--particles, such as Eastman Kodak's NTA or NTB series. The nuclear emulsions contain ~3 to 4 times more silver bromide than ordinary film and have small (~0.1 to 0.6 μ) grains that are spatially well separated. In addition, a wide variety of specialized emulsions is available for specific autoradiographic purposes. Stripping film is one of those most commonly used. It features high resolution but low sensitivity and is the emulsion of choice for intracellular localization studies. Nuclear emulsions are also available in liquid form, thus allowing autoradiographic techniques in which samples are dipped in the emulsion to form the closest possible sample-emulsion contact. (For further details, see descriptions of specific emulsion characteristics available from the various manufacturers who usually supply directions for developing and processing.)

3. Determination of Exposure Time

It must be stated frankly that the determination of exposure time in autoradiography is largely empirical. It is suggested, therefore, that in each experiment a series of duplicate samples be prepared and exposed for varying time intervals. By maintaining careful records of sample characteristics and exposure times for latter reference, it is possible to improve one's precision of determination with cumulative experience. It has been estimated that 10^6 to 10^8 β^--particles must strike each square centimeter of X-ray film in order to produce optimal blackening; detectable blackening may occur as a result of 10^5 to 10^6 beta interactions per square centimeter. On this basis, a rough estimate of exposure time can be made by measuring the activity of the sample per square centimeter with a thin end-window G-M detector. Assuming that the count rate registered in the detector is approximately equivalent to the rate of particle interaction with the emulsion, the length of time required to accumulate the aforementioned number of β^--particle interactions can be crudely estimated.

Other factors affecting exposure time are particle energy, section thickness, and background radiation. Just as background radiation produces a count rate in a G-M counter in the absence of a radioactive sample, it also produces a general fogging of film emulsions; hence it is always preferable to use fresh emulsions. Although it may be possible to decrease exposure time by the simple expedient of using higher-activity levels, one must take care not to reach specific activities that could cause radiation damage to the tissues under investigation (18).

4. Sample Preparation and Artifacts

In preparing either gross samples or tissue sections for autoradiography, extreme care must be exercised to avoid treatment that would leach out or move about the radioactive material. Each sample and each isotope poses an individual problem. For example, inorganic ^{32}P is easily leached out of many tissues, whereas organically bound ^{32}P is seldom so affected. Where tissue sections are to be prepared, the radioactive material in the tissue section must be "fixed" to avoid diffusion during exposure and development. Various chemicals, such as 95% ethanol, are used as fixing agents. In some cases, the freeze-drying technique is highly favored (17). Cosmos (5) has even found that incineration of tissue sections containing ^{45}Ca can be accomplished without displacement of the label. In autoradiographic techniques where tissue sections and emulsions are permanently mounted, the subsequent staining of the tissue may result in a reaction with the emulsion. The common hematoxylin and eosin stain combination, for example, is strongly absorbed by most emulsions.

Artifacts may result from numerous causes, and they constitute the chief difficulty in interpreting autoradiographs. An *artifact* is any nonradiation-induced darkening of the film. Vapors from volatile agents in the sample, mechanical pressure, extraneous light during film processing, dust or debris, fingerprints, and shrinkage or expansion of either sample or film have all been known to result in artifacts on the developed film (27). To recognize and control this problem, it is suggested that parallel samples without radioactivity be processed for comparsion.

C. SPECIFIC AUTORADIOGRAPHIC TECHNIQUES

Two basic methods of autoradiography are utilized. One involves contact between the emulsion and the sample only during the exposure time, after which the emulsion is removed and developed. This method is most applicable to gross samples. The second method entails permanent contact between emulsion and sample and is used exclusively with thin tissue sections.

1. Temporary Contact Method (Apposition)

The sample under study is placed in contact with the film emulsion (usually with a thin protective sheet intervening to prevent chemical fogging) and held in place by pressure. After the exposure period, the sample is removed and the film developed. This method is most suitable for use with chromatograms of labeled materials (20), leaf and whole plant tissues, gross bone sections, and tissue sections that have well-defined outlines (to allow subsequent superposition). It has the advantage that little pretreatment of the sample is required, and any subsequent tissue staining cannot affect the film. Generally poor contact between emulsion and sample results in mediocre resolution (10 to 30 μ), so that the method normally cannot be used for cellular localization studies. It is frequently difficult to superimpose the sample and the developed autoradiograph for comparison. Furthermore, the required pressure may result in artifacts on the film. The tissue sample often undergoes shrinkage, rendering interpretation of the finished autoradiograph open to some doubt. Since resolution is limited by the nature of the technique, more sensitive emulsions can be utilized for shorter exposure times.

Several unique adaptations of this method have been proposed. Hoecker *et al.* (11) have described a technique that involves clamping a flexible coverslip that holds a tissue section against the emulsion, mounted on a glass slide. After exposure, the coverslip can be bent away from the emulsion to allow development. Later, the section can be superimposed again on the developed autoradiograph without ever having been in contact with the developing solutions. This process reduces the occurrence of artifacts. Sudia and Linck (24) suggest the use of individually packaged X-ray film for gross autoradiography of plant tissues. Techniques for preparing serial cross sections of undecalicified bone for autoradiography are described by Marshall *et al.* (16). Many other adaptations appear in the literature.

2. Permanent Contact Method

For better resolution, and to avoid the problem of realignment, the tissue section may be mounted permanently in contact with the emulsion. After exposure and development of the film, the section may be stained and viewed simultaneously with the autoradiograph, or they may be examined directly by phase contrast microscopy. Four modifications of this method are in common use.

a. Mounting Method. In the mounting method, the tissue section is floated on water and the emulsion, on a glass slide, is brought up underneath it, so that the tissue lies on the film. Following the exposure, the film is developed and the tissue stained. This is a relatively simple technique and it results in reasonably high resolution (5 to 7 μ). Disadvantages are that the development may be spotty, because of nonuniform penetration of the developer through the sample, and that the film images must be viewed through the tissue sections, which may be too darkly stained. Figure 10-2 shows an autoradiogram made by the mounting method.

b. Coating Method. To avoid the problems inherent in the mounting method, the tissue section can be mounted directly on a glass slide and the emulsion applied over it. The coating method involves melting the emulsion and pouring it onto the tissue section, where it spreads and hardens, or dipping the tissue section in melted emulsion with subsequent hardening. A higher resolution is obtained than in the preceding method. The necessary handling of the emulsion results in a considerable increase in fogging, however, and it is difficult to secure a uniform and reproducible thickness of

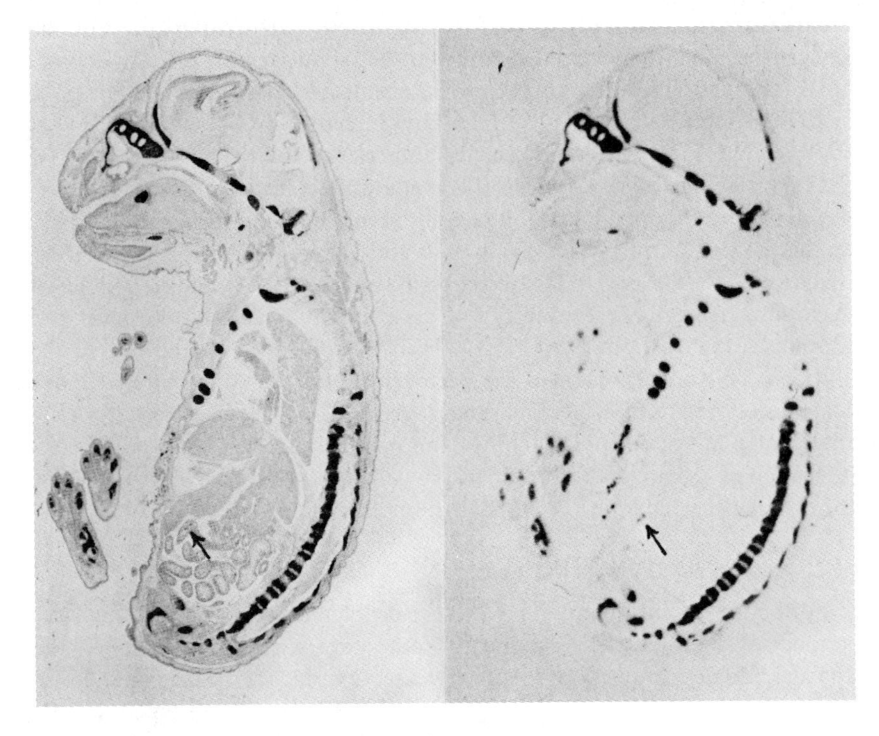

Fig. 10-2. Autoradiographic appearance of S^{35} in the 20-day-old rat embryo: (a) stained section; (b) autoradiogram showing selective deposition in cartilage. From Dominic D. Dziewiatkowski. "Sulfate-sulfur metabolism in the rat fetus as indicated by sulfur-35," *J. Exptl. Med.* **93**, 119–128 (1953).

emulsion on the sample. Baserga and Malamud (38) give an excellent discussion of the detailed experimental procedures used in the "dip-coating" method.

c. ***Stripping-Film Method.*** The stripping-film method is a less tedious means of applying an emulsion on a sample. It involves the use of films that allow the emulsion to be stripped off the base and applied directly onto the sample by means of water flotation. Because it gives better resolution than the two previous methods (1 to 3 μ), the stripping-film technique is usually the method of choice for determination of intracellular localization. The one disadvantage is the low sensitivity of the film available. Kisieleski *et al.* (22, pp. 302–308) have investigated the detection efficiency of the stripping-film method for tissue sections containing tritiated thymidine. A number of variations of the method have been developed in recent years (9, 13, 19). The stripping-film method has also been used to study the diffusion of tritium in metals and the location of metallic corrosion, fractures, inclusions, or any other feature that concentrates radioactivity.

d. "High-Resolution" Methods. Later techniques allow the application of a monolayer of silver halide crystals on a ultrathin tissue section embedded in methacrylate. Following development of the silver halide crystals, the tissue section is stained with uranyl or lead stains and examined by means of an electron microscope. Resolutions on the order of 0.05 μ are obtainable, which are obviously most desirable for studies of intracellular localization. The resolution is limited by the size of the silver halide crystals in the emulsion (0.03 to 0.05μ). Caro and van Tubergen (3) have clearly described this method and some of its specific applications. In addition, Caro (4) has discussed the problem of resolution using such monolayer preparations with tritium. Some variations of this "high-resolution" method have been termed *molecular autoradiography*. Levinthal and Thomas (12) describe such an application to the autoradiography of separated strands of DNA. A typical electron microscope autoradiograph is shown in Figure 10-3.

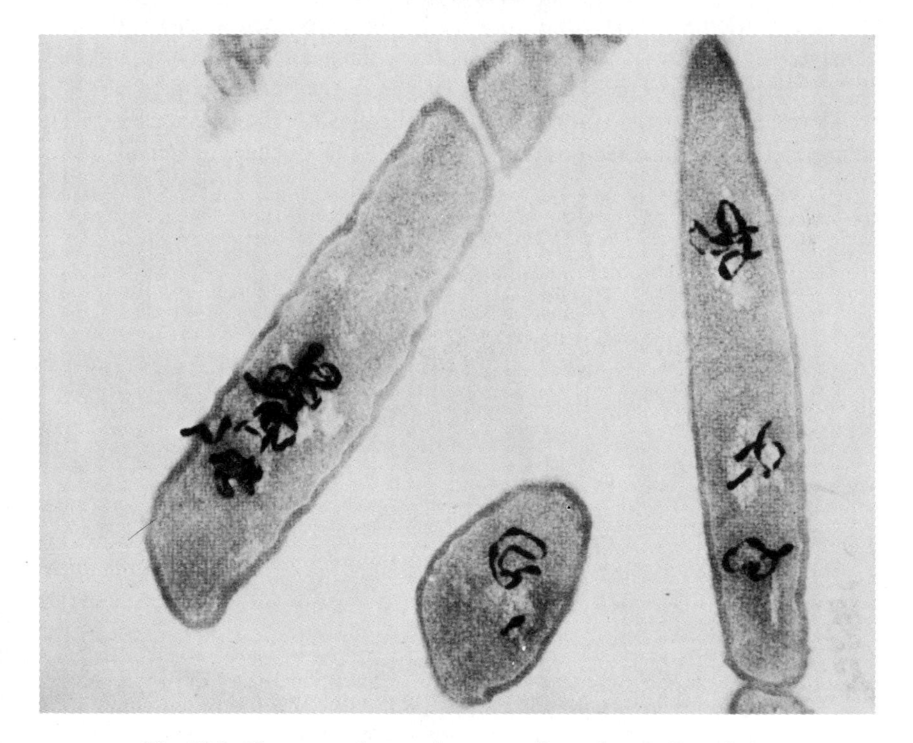

Fig. 10-3. Electron microscopic autoradiograph of *B. subtilis* labeled with thymidine-^{3}H. Silver grains are localized over the bacterial nucleoid. Approximately $\times$ 40,000. From L. G. Caro, and R. P. van Tubergen, *J. Cell Biol.* **15**, 173 (1962).

D. Nuclear Track Detectors

One of the most important modern developments in nuclear radiation detection methods has been the emergence of solid-state nuclear track detectors as a means of detecting heavy charged particles. A solid-state nuclear track detector is an homogeneous, electrically insulating solid that "stores" a permanent radiation damage "track" of the incident particle. Upon "development" of the detector, the number of tracks, their spatial distribution, length, and similiar features, can be determined. Such detectors have found widespread applications in studies of geochronology, neutron dosimetry, fission of heavy nuclei, cosmic-ray studies, and in environmental science as a means of detecting the presence of radioactive heavy nuclei in the environment.

When a charged particle impinges on a track detector, it causes a cylindrical radiation-damaged region in the detector along the particle path. This radiation-damaged region, or "track" as it is called, is formed by ionization and atom displacement in the crystal. (See Figure 10-4 for a sketch of track formation.) Under conditions to be discussed below, the radiation-damaged regions are nearly continuous along the particle trajectory and may be "developed" in order to be viewed with an optical microscope.

Development of the track consists of chemically etching and selectively enlarging the damaged regions of the crystal. Etching occurs selectively along

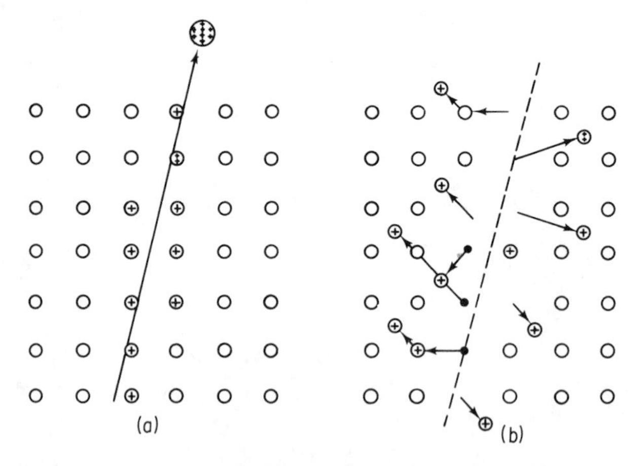

Fig. 10-4. Track formation in a simple crystalline solid: (a) the atoms have been ionized by the massive charged particle that has just passed; (b) the mutual repulsion of the ions has separated them and forced them into the lattice. Reproduced, with permission, from "Solid State Track Detectors," Annual Review of Nuclear Science, Volume 15, page 2. Copyright 1965 © by Annual Reviews Inc. All rights reserved.

the damage tracks because of the higher free energy associated with the disorder in the structure. An example of an etched track appears in Figure 10-5, where a replica of the track made by a cosmic-ray nucleus in the helmet of an astronaut is shown. Typical detector materials used include Lexan polycarbonate (etchant 6 *N* NaOH for 20 min at 50°C), mica (etchant 20% HF for 2 hr at 23°C), and cellulose nitrate (etchant 6 *N* NaOH for 2 min at 70°C). A detailed compilation of the properties of various track-detecting materials and etching conditions is given by Fleischer and Hart (34).

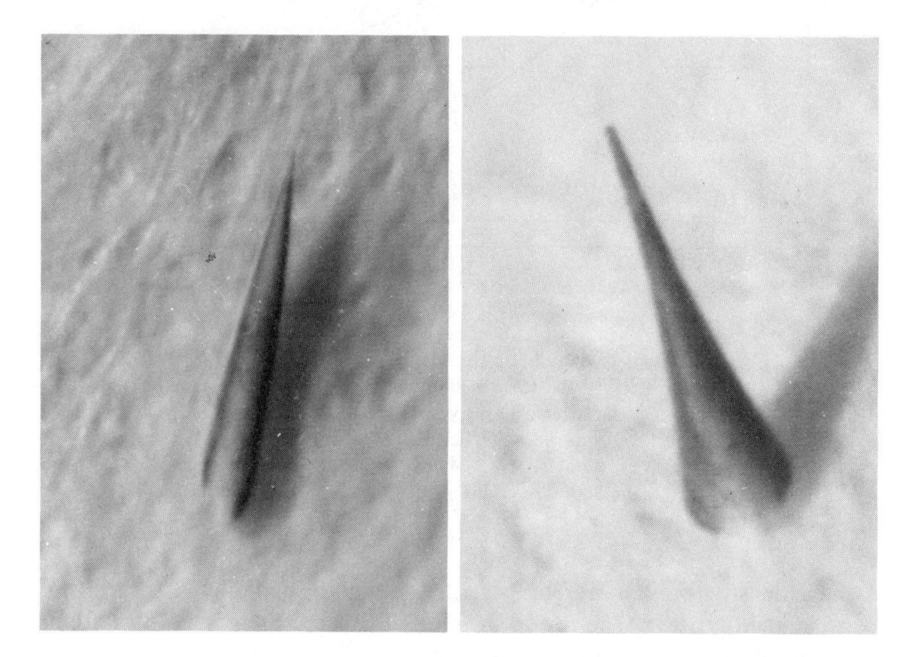

Fig. 10-5. Photographs of etched cosmic ray tracks on the inside of an Apollo space helmet made of Lexan. (Left) A track from a particle entering the helmet. (Right) An ending track from a particle that passed through the interior of the helmet and came to rest in the opposite side. The tracks are 0.5 and 0.7 mm in length. Reproduced, with permission, from "Solid Dielectric Track Detectors: Applications," Annual Review of Nuclear Science, Volume 21, page 310.

Whether a given particle will cause an etchable track to be formed in a material depends on the radiation-damage density near the particle path. Figure 10-6 shows the minimum damage density that will cause a track to register in many common detector materials. One can see, for example, that Lexan will be sensitive to almost any fission fragment but will only detect alpha particles with energies below 0.2 MeV. This dependence of radiation-

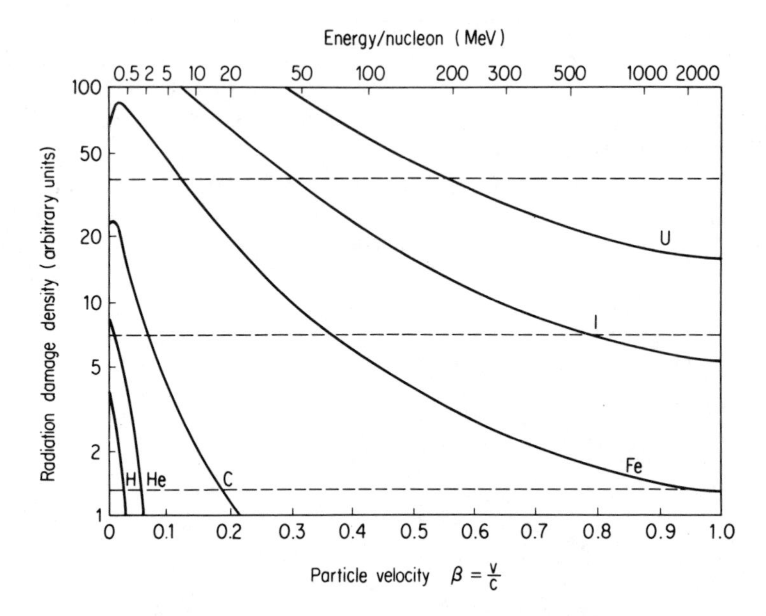

Fig. 10-6. Density of radiation damage (or ionization rate J) as a function of velocity for various bombarding nuclei. Approximate thresholds for track recording by several solids are indicated by dashed lines. Reproduced, with permission, from "Solid Dielectric Track Detectors: Applications," *Annual Review of Nuclear Science*, Volume 21, page 300.

damage density (or etch rate) on particle type can be exploited as a means of particle identification. The review article of Price and Fleischer (35) contains an extensive discussion of this technique.

One very important feature of nuclear track detectors is that they may be made in almost any size and shape for very low cost. In a typical experimental arrangement, a source (either natural or man-made) of heavy charged particles is totally surrounded by detector material, thus allowing the particle detection efficiency to approach 100%. The relative simplicity of use of such detectors (requiring only a few chemicals and a microscope) and their low cost have made them quite attractive—not to mention the previously discussed selectivity for heavy charged particles (even in immense fluxes of light particles).

One of the chief uses of this detection principle has been its application to geochronology. Uranium will fission spontaneously with a half-life for ^{238}U of $(1.01 \pm 0.03)(10^{16})$ yr. Thus any mineral, meteorite, or glass that contains uranium will also contain tracks because of the spontaneous fission of uranium. By knowing the spontaneous fission half-life of ^{238}U, the ^{238}U

content of a meteorite or mineral, and the track detection efficiency of the material, one can, by measuring the number of tracks in the material, determine the age of the material. This dating method can give reliable ages in the time range from 0 to 10^9 yr. Fleischer and Hart (34) critically review this use of nuclear track detectors.

Nuclear track detectors are beginning to find important applications in environmental studies. They can be used to measure the gross concentrations of fissile nuclides in the atmosphere. Air particulate samples are collected and then irradiated with track detectors in contact with them in a nuclear reactor. The track detectors measure the fission fragments emitted during the neutron-induced fission of the fissile nuclides and, with proper calibration, can give a quantitative estimate of the heavy-element concentrations in the air. Myers and White review the use of these detectors in environmental studies (39).

BIBLIOGRAPHY

1. ADAMIK, EMIL R. *Laboratory Procedures for Tritium Autoradiographs.* Orangeburg, N.Y.: Schwarz Bioresearch. (n.d.)
2. BOYD, GEORGE A. *Autoradiography in Biology and Medicine.* New York: Academic, 1955.
3. CARO, LUCIEN G., and ROBERT P. van TUBERGEN. "High-resolution autoradiography. I. Methods." *J. Cell Biol.* **15**, 173–188 (1962a).
4. CARO, LUCIEN G. "High-resolution autoradiography." II. The problem of resolution. *J. Cell Biol.* **15**, 189–199 (1962b).
5. COSMOS, ETHEL. "Autoradiography of Ca^{45} in ashed sections of frog skeletal muscle," *Anal. Biochem.* **3**, 90–94 (1962).
6. DOMINGUES, F. J., A. SARKO, and R. R. BALDWIN. "A simplified method for quantitation of autoradiography," *Intern. J. Appl. Radiation Isotopes* **1**, 94–101 (1956).
7. FICQ, A. "Autoradiography." In J. Brachet (Ed.), *The Cell-Biochemistry, Physiology, Morphology.* New York: Academic, 1959, Vol. 1, pp. 67–90.
8. FITZGERALD, PATRICK J. "Radioautography: theory, technic, and applications," *Lab. Invest.* **2**, 181–222 (1953).
9. GOMBERG, HENRY J. "A new high-resolution system of autoradiography," *Nucleonics* **9**(4), 28–43 (1951).
10. HERZ, R. H. "Photographic fundamentals of autoradiography," *Nucleonics* **9**(3), 24–39 (1951).
11. HOECKER, FRANK E., PAUL N. WILKINSON, and JACK E. KELLISON. "A versatile method of micro-autoradiography," *Nucleonics* **11**(12), 60–64 (1953).
12. LEVINTHAL, CYRUS, and CHARLES A. THOMAS. "Molecular autoradiography: the β-rays from single virus particles and DNA molecules in nuclear emulsions," *Biochim. Biophys. Acta* **23**, 453–465 (1957).
13. LOTZ, W. E., and P. M. JOHNSTON. "Preparation of microautoradiographs with the use of stripping film," *Nucleonics* **11**(3), 54 (1953).
14. MAMUL, YA V. "Quantitative autoradiography using a radioactive wedge," *Intern. J. Appl. Radiation Isotopes* **1**, 178–183 (1956).

15. MARKMAN, BORJE. "Autoradiography of tritium chromatograms," *J. Chromatog.* **11**, 118–119 (1963).

16. MARSHALL, J. H., V. K. WHITE, and J. COHEN. "Autoradiography of serial cross sections of undecalcified bone," *Intern. J. Appl. Radiation Isotopes* **1**, 191–193 (1956).

17. NOVEK, J. "A high-resolution autoradiographic method for water-soluble tracers and tissue constituents," *Intern. J. Appl. Radiation Isotopes* **13**, 187–190 (1962).

18. PELC, S. R. "Radiation dose in tracer experiments involving autoradiography." In *Ciba Foundation Conference on Isotopes Biochem.* Philadelphia: Blakiston, 1951, pp. 122–137.

19. PELC, S. R. "The stripping-film technique of autoradiography," *Intern. J. Appl. Radiation Isotopes* **1**, 172–177 (1956).

20. Radiochromatography. *SBR Technical Brochure 64D1*. Orangeburg, N.Y.: Schwarz Bioresearch. (n.d.)

21. ROBERTSON, J. S., V. P. BOND, and E. P. CRONKITE. "Resolution and image spread in autoradiographs of tritium-labeled cells," *Intern. J. Appl. Radiation Isotopes* **7**, 33–37 (1959).

22. ROTHCHILD, SEYMOUR (Ed.). *Advances in Tracer Methodology*. Vol. 1. New York: Plenum Press, 1963.

23. STILLSTROM, J. "Grain count corrections in autoradiography," *Intern. J. Appl. Radiation Isotopes* **14**, 113–118 (1963).

24. SUDIA, T. W., and A. J. LINCK. "Method for autoradiography using individually packaged x-ray film," *Intern. J. Appl. Radiation Isotopes* **10**, 55 (1961).

25. TAYLOR, J. HERBERT. "Autoradiography at the cellular level." In G. Oster (Ed.), *Physical Techniques in Biological Research*. New York: Academic, 1956, Vol. 3, pp. 545–576.

26. WEINSTEIN, JERRY. "Radioautography with tritium," *Atomlight* (New England Nuclear Corp.) pp. 1-2 (June 1958).

27. WILLIAMS, AGNES I. "Method for prevention of leaching and fogging in autoradiographs," *Nucleonics* **8**(6), 10–14 (1951).

28. YAGOD, HERMAN. *Radioactive Measurements with Nuclear Emulsions*. New York: Wiley, 1949.

29. FISCHER, H. A. and G. WERNER. *Autoradiography*. Berlin: DeGruyter, 1971. An important discussion of modern applications of autoradiography.

30. GUDE, W. D. *Autoradiographic Techniques*. Englewood Cliffs, N.J.: Prentice-Hall, 1968. A fine "nuts and bolts" description of many practical details of autoradiography.

31. PRICE, W. J. *Nuclear Radiation Detection*. 2nd ed. New York: McGraw-Hill, 1964. Chapter 9 contains important general information on nuclear emulsions and their use.

32. FEINENDEGEN, L. E. *Tritium Labelled Molecules in Biology and Medicine*. New York: Academic, 1967. An excellent review of the autoradiography of tritium-containing materials.

33. FLEISCHER, R. L., P. B. PRICE, and R. M. WALKER. "Solid state track detectors—Applications to nuclear science and geophysics," *Ann. Rev. Nucl. Sci.* **15**, 1 (1965). An introductory survey of applications of track detectors.

34. FLEISCHER, R. L., and H. R. HART, Jr. "Fission track dating: Techniques and problems," *G. E. Report 70-C-328*, September 1970.

35. PRICE, P. B., and R. L. FLEISCHER. "Identification of energetic heavy nuclei with solid dielectric track detectors: Applications to astrophysical and planetary studies," *Ann. Rev. Nucl. Sci.* **21**, 295 (1971).

36. Fleischer, R. L., P. B. Price, R. M. Walker, and E. L. Hubbard. "Track registration in various solid state nuclear track detectors," *Phys. Rev.* **133**, A1443 (1964). An important early paper about which particle tracks will register in track detectors.

37. Chase, G. D., and J. L. Rabinowitz. *Principle of Radioisotope Methodology*. 3rd ed. Minneapolis: Burgess, 1967.

38. Baserga, R., and D. Malamud. *Autoradiography, Techniques and Application.* New York: Harper and Row, 1969. The best single reference on the techniques of autoradiography and examples of its use.

39. Myers, W. G., and F. A. White. "Mass spectrometry and fission track analysis in nuclear environmental measurements." In *Nuclear Methods of Environmental Analysis.* Columbia, Mo.: University of Missouri, 1972.

11

Preparation of Counting Samples

Proper preparation of counting samples is as important in precise radioactivity assay as the counting operation. Samples collected in radiotracer experiments may be in such diverse forms as blood, urine, water, milk, plant or animal tissues, or respiratory gases. In most cases, such varied samples cannot be assayed directly but must be converted to a suitable form for assay. If the counting sample is to be in the form of a solid, it must also be mounted in a uniform and reproducible manner before assay is feasible. This latter feature is of particular importance for alpha-emitting or low-energy beta sources. Where liquid scintillation counting is to be used, the sample material must be suitably incorporated into the fluor medium.

This chapter outlines the various methods used to convert biological samples into chemical and physical forms more suitable for assay by the common detection methods. Cases where direct assay of the original sample is possible are cited. Finally, we consider the specific procedures necessary to introduce the sample to the detector in either a gaseous, liquid, or solid form.

In the field of liquid scintillation counting, there is currently a bewildering array of information on suitable methods of introducing the sample into the fluor medium. Because of this situation and the importance of the assay method, a major portion of this chapter is devoted to the preparation of liquid scintillation counting samples. An extensive bibliography is given at the end of the chapter to show the present state of the literature in this field. A number of papers survey sample preparation methods for a specific isotope of importance. Those pertaining to liquid scintillation assay will be cited later, but those dealing with preparation of samples for the other detection methods should be noted here. The general preparation of ^{14}C-counting samples has been most widely described (33, 141, 163); of these references, the classic work of Calvin *et al.* deserves special notice. Somewhat less general consideration has been given to other radionuclides.

A. FACTORS AFFECTING CHOICE OF SAMPLE FORM FOR COUNTING

The choice of what form the sample will take for use in counting depends on several interrelated factors. Of these, the type and energy of radiation emitted by the sample radionuclide are the most important. In the case of solid or liquid (blood, urine, etc.) samples containing α- or low-energy β^--activity (such as ^{14}C or ^3H), the problem of self-absorption leads to severely decreased counting efficiency and poor reproducibility. As a result, usually the sample must be converted to a standard and more suitable form. On the other hand, samples containing γ-ray-emitting radionuclides can generally be assayed directly with a minimum of pretreatment (such pretreatment, in any case, will probably consist of reducing the sample size to be compatible to that of the detector, plus the use of low Z materials to hold or mount the sample).

The type of detector used, the activity of each sample, the number of samples to be assayed, and the available time to do the assay will also contribute to the choice of optimum counting sample form. A last factor of importance is that the ease of sample preparation is related to the chemical and physical form of the original sample. Tables 11-1(a) and 11-1(b) summarize these factors in the form of general guidelines for the preparation of samples for α- and β-assay, respectively.

A few words of explanation and caution concerning the general guidelines in Tables 11-1(a) and 11-1(b) are necessary. First, these tables are to be regarded as *general* guidelines, not specific rules applying to every case. There may be cases where the unusual form of the initial sample precludes the processing indicated in the tables. The exact choice of counting form will depend on the particular sample in question. Tables 11-1(a) and 11-1(b) illustrate the general rules for sample preparation.

The use of the terms low, medium, and high with respect to sample activity in the tables may also raise questions. The basic idea behind such designation is that some detectors, such as the liquid scintillation counter, are poorly suited for the assay of high-level radioactivity ($>$1 mCi). Although dilution techniques may be utilized in preparing the sample for counting, it is simply easier to choose another detector. All the "external-sample" detectors (G-M, proportional, semiconductor, solid scintillation, and nuclear track detectors)

TABLE 11-1(a)

Sample Preparation Required for Alpha Assay

Original Sample Form	Detector	Sample Activity	Sample Preparation	Sample Preparation Time
Gas	Ionization chamber	Low-High	Count directly in gaseous form by introduction into detector in suitable mixture with counting gas.	Moderate
Liquid	Semiconductor	Low-High	Convert to solid thin source and assay.	Moderate-Slow
	Liquid scintillation counter	Low-Medium	Dissolve or suspend in suitable scintillator-solvent mixture and assay.	Quick
Solid	Semiconductor	Low-High	If sample is very thin source, count directly. If not, dissolve sample and redeposit as thin source.	Moderate-Slow
	Liquid scintillation counter	Low-Medium	Dissolve in suitable solvent and heat as liquid, or count directly as a suspension in a gel mixture.	Quick-Moderate
	Nuclear track detectors	Low	Same as semiconductor detector.	Moderate-Slow

could, by means of changes in sample-counter geometry, accommodate low (<1 pCi) to high-level samples (>1 mCi). However, because of its large resolving time, the G-M counter is not recommended for assay of samples with activities >1 μCi.

At least some stage of a radiotracer experiment usually requires the pipetting of radioactive solutions. Most frequently, this process occurs during preparation of the sample for counting. Pipetting of radioactive solutions follows the same general procedures as used for nonradioactive solutions, with a few important differences. The first is that *pipetting by mouth is absolutely forbidden with radioactive solutions* (and is not generally a good idea anyway). A number of commercial suction devices are available to control the pipet, one of which is shown in Figure 11-1. The use of this device is shown in Figure 11-2. Care must be taken in pipetting radioactive solutions to minimize handling the pipet during the process. Such handling greatly increases the radiation dose to the hands and fingers (see Chapter 16) and increases the possibility of spreading radioactive material to the hands. Care must also be taken to avoid spattering of liquid. In general, the usual precautions (Chapter 16) of wearing protective clothing, using gloves, and so on are to be followed during all pipetting operations.

Because the solution volumes in radiotracer experiments are frequently

TABLE 11-1(b)

Sample Preparation for Beta Assay

Original Sample Form	Detector	Sample Activity[a]	Sample Preparation	Sample Preparation Time
Gas	Geiger-Müller Proportional counter	Low-Medium Low-High	Sample (1) trapped in suitable solution and converted to solid form, or (2) may be counted directly in gaseous form by introduction into suitable detector chamber with counting gas.	Moderate
	Semiconductor Solid scintillation detector	Low-High Medium-High	Trap sample in solution and convert to thin solid source.	Slow
	Liquid scintillation counter	Low-Medium	Must be trapped in suitable solution and treated as liquid sample.	Moderate-Slow
Liquid	Geiger-Müller Proportional counter	Low-Medium Low-High	Sample (1) preferably converted to solid form for counting, or (2) may be counted directly with suitable detector if particle energy is sufficiently high.	Moderate-Slow
	Semiconductor Solid scintillation detector	Low-High Medium-High	Same as Geiger-Müller counter.	Moderate-Slow
	Liquid scintillation counter	Low-Medium	May be dissolved or suspended in a suitable scintillation solvent mixture and assayed directly without pretreatment.	Quick
Solid	Geiger Müller Proportional counter	Low-Medium Low-High	May be counted directly unless low energy of particles requires pretreatment of sample.	Quick
	Semiconductor Solid scintillation counter	Low-High Medium-High	Same as Geiger-Müller counter.	Quick
	Liquid scintillation counter	Low-Medium	(1) Should be dissolved in suitable solvent and treated as liquid sample, or (2) may be counted directly as suspension in a suitable gel mixture, or (3), in the case of ^3H and ^{14}C, samples can be burned to H_2O and CO_2, which can be dissolved or trapped as in (1).	Moderate-Slow

[a]Low $\approx <1$ pCi
Medium ≈ 1 μCi
High $\approx >1$ mCi

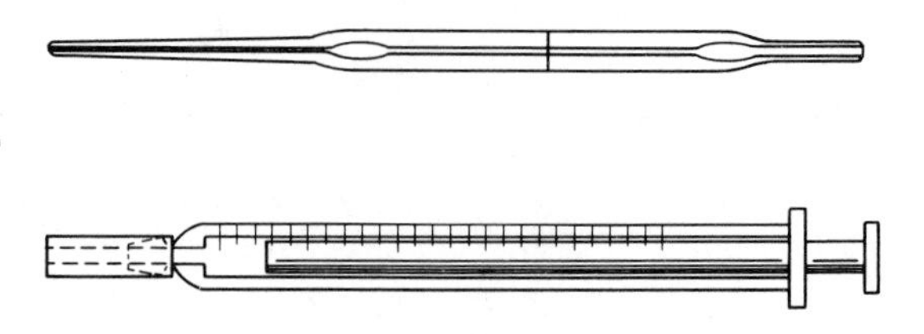

Fig. 11-1. Micropipet and control.

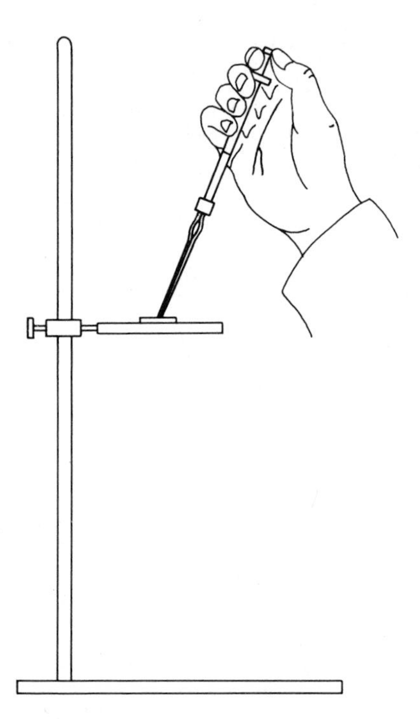

Fig. 11-2. Use of micropipet for source preparation.

quite small, special techniques for handling small quantities of liquid are appropriate. The use of micropipets or special small-volume syringes (such as those made by the Hamilton Company) is recommended. Volumes as small as 1 μl (1λ) can be reproducibly dispensed with an accuracy of $\pm 1\%$. Care must be taken to rinse the pipet several ($\sim$3 to 4) times after each use and to add the washings to the original volume pipetted to ensure complete transfer of material during the operation. A reduction in solution absorption to the pipet walls is obtained by coating the inside of the pipet with a silicone layer (such as Dri-Film) prior to use. After each use, the pipet should be thoroughly washed in dilute acid, distilled water, and acetone. Automatic pipetting devices to reproducibly dispense predetermined small quantities of liquid have become commercially available recently.

B. CONVERSION OF BIOLOGICAL SAMPLES TO SUITABLE COUNTING FORM

The pretreatment of biological samples before radioactivity assay is normally concerned with measures to reduce bulk and produce a more uniform material. Depending on the assay method to be used, it may be necessary to convert the sample to a more suitable physical state. It may also be desirable to convert the chemical form of the isotope to one more suitable for both radioactivity measurement and parallel chemical analysis. Such is often the case with liquid scintillation samples, where the original sample form is a severe quenching agent, either because of its color or its molecular structure. Frequently, it is necessary to isolate an active compound from a large volume of sample material (tissue, feces, blood, soil, water, etc.).

The specific treatment utilized for a given sample depends on its original form, particle energy, and the desired method of detection. It should be emphasized, however, that the various sample conversion methods are not unique to any given assay method. For example, $^{14}CO_2$ produced by Schöniger flask oxidation can just as easily be assayed directly in an ion chamber-electrometer system, or trapped in NaOH, converted to $BaCO_3$, and counted on a planchet using a G-M detector. Or it can be trapped in "Hyamine" and assayed by a liquid scintillation counter. Various standard chemical methods are available for converting biological samples to a more suitable form for assay. Since such methods are well described in the analytical chemical literature, emphasis here is on adaptations specific to radioactive sample conversions.

1. Ashing Methods

When the radionuclide to be assayed is metallic or mineral in nature, ashing of the sample is often the most practical method of concentration. This removal of the liquid component may be accomplished by either dry or wet ashing. *Dry ashing* involves burning the sample in a furnace, whereas *wet ashing* involves digestion of the sample with a strong oxidizing agent. Since the *dry ashing* process produces a friable residue and is often accompanied by volatilization or fusion of the sample with the crucible, wet ashing is generally preferred. Gleit and Holland (71), however, have proposed a useful modification of dry ashing for blood samples. A number of *wet ashing* procedures have been described, but most are variations of the conventional Kjeldahl method. Various combinations of nitric (146), sulfuric, hydrochloric, or perchloric acid and hydrogen peroxide have been used successfully. Wet ashing is particularly applicable to blood or tissue containing radioiron (50, 113, 166).

Animal tissues may be converted to a homogeneous solution for assay as a liquid by various forms of *pseudo wet ashing*. Pearce *et al.* (163) have pointed out the use of formamide to effect such solution. An entire animal or organ is homogenized or cut up with scissors and placed in hot formamide. Complete solution of all but bone results in 1 to 2 hr.

2. Combustion Methods

Sample material containing the low-energy beta emitters ^{14}C or ^{3}H usually requires complete oxidation. The tritium is recovered as water. The $^{14}CO_2$ produced may be assayed in the gaseous state or trapped in an alkaline solution before conversion to a suitable counting form for other types of activity measurement. Where the samples are to be assayed in a solid form, such treatment is essential in order to concentrate the isotopic material and reduce sample self-absorption. For assay by the liquid scintillation method, complete oxidation generally has the beneficial effects of improving sample solubility in the solvent and reducing quenching. Jeffay has reviewed succinctly the various oxidation techniques for preparation of liquid scintillation samples (104).

a. Wet Combustion. Several wet combustion methods have been employed, with variations of the *Van Slyke-Folch* being the most common. The apparatus for one previously undescribed modification is illustrated in Figure 11-3. In this procedure, the ^{14}C-containing sample is placed in flask A and the NaOH absorbent solution (carbonate-free variety, such as Acculute, manufactured by Anachemia Chemicals Ltd., Montreal, Canada—Champlain, New York) in flask B. (Note that a saturated NaOH solution is also relatively carbonate-free.) The system is assembled, evacuated, and sealed off. A mixture

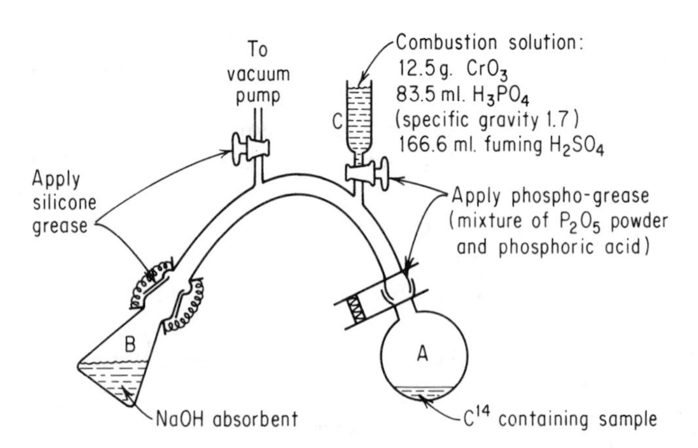

Fig. 11-3. A modified Van Slyke-Folch apparatus for wet combustion of ^{14}C-labeled samples.

of fuming sulfuric acid, phosphoric acid, and chromic trioxide (33) is introduced through stopcock C. Flask A is then heated cautiously with a low flame for 10 min to promote the combustion, and the $^{14}CO_2$ generated is allowed to come to equilibrium with the absorbent (approximately $\frac{1}{2}$ hr). Subsequent addition of $BaCl_2$-NH_4Cl solution (1 *M* in strength) to the absorbing solution yields $Ba^{14}CO_3$, which is then collected by filtration or centrifugation for planchet mounting. The production of voltatile products, such as aldehydes or chromyl chlorides, from chlorine-containing compounds sometimes limits the usefulness of this method. The method is not useful for tritium assay.

Persulfate oxidation of carbon compounds is also quite convenient (233). Samples, however, must be soluble in water, which somewhat restricts its application. This method uses a closed Erlenmeyer flask with a center well containing a carbon dioxide absorbent. The aqueous solution of the sample is placed around the well, and sulfuric acid and solid potassium persulfate are added. Silver nitrate solution is dropped in to catalyze the reaction and the flask is capped immediately. Heating accelerates the oxidation, and the CO_2 produced diffuses to the absorbent well. Unfortunately, volatile intermediate products (such as acetaldehyde) may also diffuse into the center well, giving unpredictable results.

Belcher (11) has proposed wet oxidation of tritium-labeled samples using a *nitric-perchloric acid* mixture. The resultant tritiated water is distilled off and the distillate assayed with a liquid scintillation counter.

b. Dry Combustion. Procedures involving dry combustion in an oxygen atmosphere are of particular value in preparing ^{14}C and tritium samples for liquid scintillation or ion chamber assay. The classical technique of oxidizing the sample by means of an inorganic oxidizing agent (CuO or NiO) in a *combustion furnace* with a flow of oxygen has been adapted by Peets et al. (164) for liquid scintillation counting. The tritiated water produced is trapped as ice in a trap immersed in dry ice-acetone; the $^{14}CO_2$ is taken up in a suitable alkaline-absorbing solution. Recoveries of about 96% were obtained from samples of up to 1.5 g (187, pp. 185–191). Others have adapted this combustion train method for preparation of samples for proportional counter assay (99).

In an alternative method of dry oxidation, the sample and oxide catalyst are placed in a Pyrex or Vycor tube, which is then sealed and heated. The *sealed tube* is subsequently broken and the oxidation products removed by vacuum-line transfer either to an absorbing solution or directly into an internal-sample detector. Wilzbach *et al.* (245–246) have employed this method for gas assay in an ion chamber; Buchanan and Corcoran (28) have used it with a proportional detector, and Steel (214) and Jacobson *et al.* (100) have adapted it to liquid scintillation assay of tritium. The major disadvantages of this procedure are the small sample size that can be accommodated (less than 25 mg) and the time involved per sample.

The *Schöniger oxygen-flask* combustion originally proposed for halogen determinations (195) has been applied by Kallberer and Rutschmann (107) for sample preparation in liquid scintillation counting. In this procedure, the sample material is placed in a cellophane or filter paper bag and dried. The bag is suspended in a platinum basket, which is attached to an ignition head. This assembly is placed in a gas-tight flask containing an oxygen atmosphere. When current is applied, the oxidation takes place rapidly (232, 187, pp. 185–191). Combustion products, such as ^3H_2O, may be frozen out and $^{14}CO_2$ may be trapped in a suitable alkaline absorbent introduced through a side arm. Baxter and Senoner (9) have modified the combustion apparatus to attain 96 to 100% recovery of ^{14}C activity. MacDonald (129) has reviewed this method and its specific uses for ^{14}C sample preparation; Dobbs has even applied it to ^{35}S-labeled compounds (51) and ^{14}C-labeled halogenic compounds (52). In fact, the method is also suitable for handling samples containing both ^3H and ^{14}C.

A major problem of the Schöniger flask technique is the danger of explosion if even traces of organic solvents are present in the flask. Martin and Harrison (136) have attempted to circumvent this problem by designing the top of the flask to act as a safety outlet in case of excessive internal pressure. A modification of this technique achieves ignition by focusing an external infrared beam on the sample wrapped in black paper (149). Because of the simultaneous combustion of the cellophane or paper bag with the sample, specific activity of the sample cannot be determined by this method. Only total sample activity is measurable. Sample size is limited to less than 300 mg according to Kelly *et al.* (109). Excessive quenching as a result of dissolved oxygen in combustion products has been noted. However, this condition can be taken care of by nitrogen or air flushing (106).

The *Parr oxygen bomb* (commonly used in calorimetry) offers the advantage of allowing specific activity determination on larger liquid or tissue samples (130) and has been used on even such volatile compounds as acetone. This metal bomb can be loaded with up to 1 g of lyophilized tissue sample or an equivalent amount of wet tissue. It is then filled with oxygen to 25 atm pressure. Ignition produces rapid combustion, and the combustion products are removed through a collecting train. Sheppard and Rodegker (187, pp. 192–194; 204) have reported detection of as little as 4×10^{-4} μCi of tritium and 1×10^{-4} μCi of carbon-14 in 3 g of fresh tissue.

3. Combustion of Samples for Liquid Scintillation Counting

Several unique methods that are specifically designed for combustion of samples for liquid scintillation counting exist. Specific factors in applying combustion techniques for liquid scintillation counting have been reviewed by Jeffay (104).

a. Wet Oxidation. The method developed by Cameron (34) for wet oxidation of blood samples has been applied to preparation of samples for liquid scintillation counting by Makin and Lofberg (133, 134). The method calls for digestion of 100 mg of biological samples in 0.2 ml of 60% $HClO_4$ and 0.4 ml of 30% H_2O_2 at 70 to 80°C for 30 to 60 min in a tightly capped counting vial. The resulting solution is generally clear and colorless and can be readily mixed with 6 ml of Cellosolve (2-ethoxyethanol) and 10 ml of toluene-PPO scintillation solution for counting. The chemical processes involved in wet oxidation differ from those of wet combustion in that the biological compounds in the tissue are generally not oxidized to CO_2. The method is applicable to the assay of ^{45}Ca, ^{55}Fe, ^{57}Co, ^{32}P, ^{35}S, and ^{3}H. Care should be taken when attempting to apply the wet solution method for ^{14}C assay, for a portion of the ^{14}C contained in the labeled sample could be lost as $^{14}CO_2$ during the oxidation process.

b. Automated Combustion Apparatus. The Schöniger oxygen-flask combustion technique (195) as adapted for preparation of samples for liquid scintillation counting (107, 109) has been refined by Oliverio *et al.* (149). Details of these and other developments have been reviewed by Davidson *et al.* (48). In the same paper, two types of automated combustion apparatus are also clearly reviewed.

The first one, developed by Peterson *et al.* (167, 168), is schematically shown in Figure 11-4. The basic principle underlying the Peterson apparatus is essentially the same as that of the Schöniger technique. A sample, 500 mg in weight, is encapsulated in a gelatin or Lexan polycarbonate capsule. Upon introduction of the capsule into the furnace tube, combustion is carried out under oxygen atmosphere, and the combustion products (CO_2 and H_2O) are swept out by a stream of oxygen into a condenser [Figure 11-4(a)]. There H_2O is dissolved in a stream of the scintillation solution of choice and eventually flows into a counting vial. For efficient collection of CO_2, the condenser is replaced by a CO_2 absorption column [Figure 11-4(b)], and a scintillation solution containing phenylethylamine is used as the CO_2 absorbent. Naturally, in this case, the H_2O derived from the combustion process is also dissolved. The Peterson apparatus is useful in preparing counting samples for both ^{14}C and ^{3}H assays with reproducibility reported to be approximately 3%. The performance of Peterson's apparatus has been reviewed by Tyler *et al.* (229).

The second type of automated combustion apparatus was developed by Kaartinen for tritium assay and subsequently marketed by the Packard

Instrument Co. (106). Here, again, the basic process is analogous, in principle, to that of Schöniger's oxygen flask technique. The sample is placed on a platform and ignited by means of an electrically heated, platinum sample holder under an oxygen atmosphere. Combustion products (CO_2 and H_2O) are swept by a stream of nitrogen through a condenser, mixed with a prescribed amount of scintillation solution and delivered into a counting vial. This apparatus, designed specifically for ^3H assay, provides the unique advan-

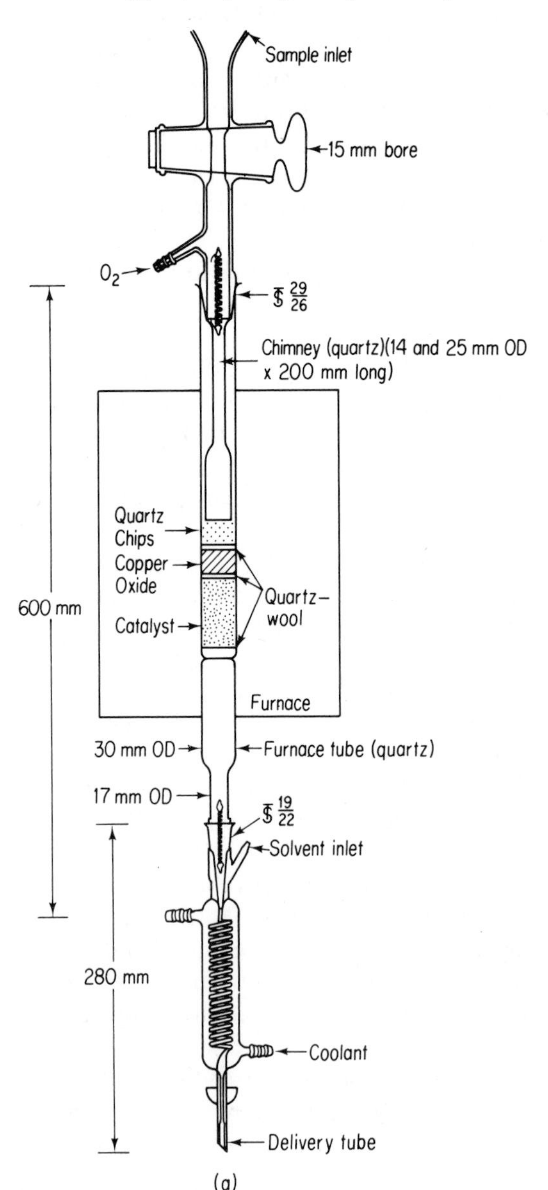

Fig. 11.4a. Diagram of Peterson tritium combustion apparatus (Courtesy of Davidson *et al.* (48).

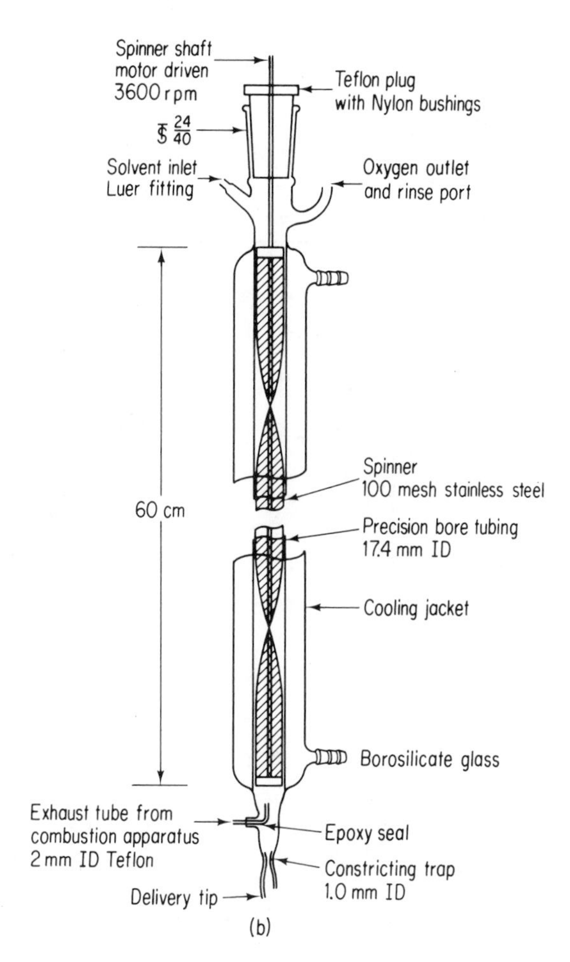

Fig. 11.4b. Diagram of Peterson carbon-tritium absorption apparatus.

tage that scintillation solution is not exposed to oxygen, and thus the quenching effect by dissolved oxygen is minimized.

Later, the Kaartinen apparatus was further modified to accommodate both ^3H and ^{14}C assays. The apparatus, under the market name of Packard Model 305 Sample Oxidizer (205), is schematically depicted in Figure 11-5. The basic process underlying Model 305 is essentially the same as that of the tritium-only oxidizer. However, the collection of combustion products has been modified to consist of two consecutive stages. First, the water produced in the combustion is condensed at 2°C in a condenser and collected in a counting vial. The combustion chamber is then purged with steam and the added condensate is collected. Finally, the condenser is rinsed with scintillation solution, thereby completing the water collection process. Second, the CO_2 produced in the combustion passes through the condenser and the tritium vial into an absorption column that is partially filled with ethanolamine and kept at 50°C. Upon completion of the absorption process, the con-

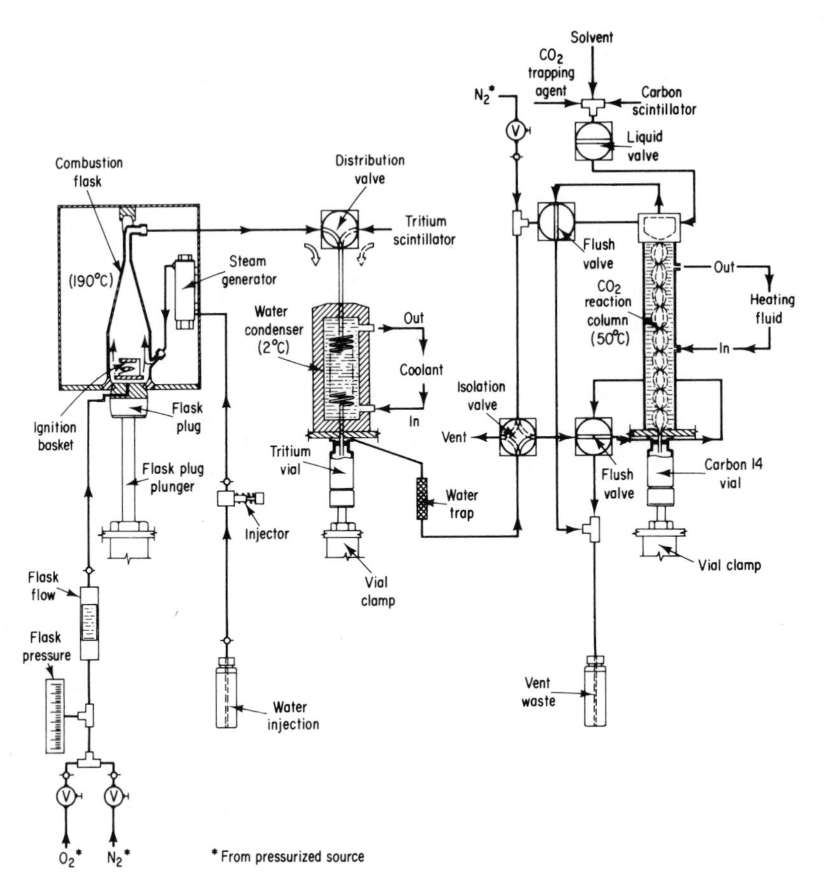

Fig. 11-5. Flow diagram of the Packard Model 305 Sample Oxidizer. Courtesy of Packard Instruments.

tent of the absorption column is drained and rinsed with methanol into a counting vial. Nitrogen gas is used to purge the entire system after each combustion and to provide pressure for the delivery of reagents. All operations, except for combustion control and steam purging, are done automatically. An elapsed time of 3 min is required for each operation that accommodates samples weighing as much as 1000 mg. Tritium and ^{14}C recoveries are reported to be good, and the retention of residual activity in the apparatus is acceptably low.

4. Miscellaneous Methods

With certain samples, the labeled compounds may be chemically extracted without the need for ashing or oxidation. Schulze and Long (197) have reported extraction by ultrasonic treatment of the sample directly in the liquid scintillation counting vial. Tritiated water samples may be prepared from biological fluids (plasma, urine) by distillation. Simpson and Greening (209) have investigated the magnitude of isotopic effect on such distillations of tritiated water from urine and found it negligible.

C. ASSAY OF SAMPLES IN VARIOUS COUNTING FORMS

The three physical states of counting samples (gaseous, liquid, solid) can be assayed by a variety of detectors, with, of course, different counting efficiencies. In general, solid counting samples require the most careful attention to mounting in order to ensure reproducibility of results. The preparation of counting samples for liquid scintillation assay has been singled out for more extensive discussion because it involves the unique feature of actually introducing the sample into the fluor medium.

1. Assay of Gaseous Counting Samples

Samples to be assayed in a *gaseous form* are usually introduced directly into the detector (ion chamber, G-M detector, or proportional detector) following their preparation from the original biological material (187, pp. 167–184). Vacuum-transfer technique is involved in this operation. The significant advantage of gas counting is that self-absorption is largely circumvented. Thus gas counting is primarily restricted to the low-energy β^--emitters ^{14}C, ^{3}H, and ^{35}S. These nuclides are most commonly assayed in the chemical form of CO_2 or C_2H_2, H_2 or CH_4, and SO_2 or H_2S, respectively. Wilzbach and Sykes (246) have described methods of $^{14}CO_2$-activity determinations in ion chambers. Although counting efficiencies may be reasonably high for soft β^--emitters and the precision of the measurement may be greater compared to other assay methods, the overall sample preparation is rather tedious and time consuming. It is necessary to calibrate gas counters with standard gaseous sources of the same nuclide that is being assayed. It is also possible to make continuous measurements of flowing gaseous samples, such as the use of the flow ion chamber for measuring respiratory $^{14}CO_2$ or the effluent from gas chromatography separations.

2. Assay of Liquid Counting Samples

If a *liquid sample* is to be assayed directly, a minimum of further sample preparation is required. Except for the internal-sample scintillation counting method, liquid sample counting is mainly restricted to γ-ray emitters and high-energy β^--emitters. G-M tubes of specialized design are available (see Figure 5-12) for the latter sample type.

NaI(Tl) scintillation detectors or Ge(Li) semiconductor detectors are normally employed for γ-ray assay. Well crystals offer the additional advantage of high counting efficiency. Whether the liquid, when placed in a polyethylene vial, is assayed in a well crystal or on top of a NaI(Tl) or Ge(Li) detector, it is important that the same volume of sample be used in all assays. Otherwise, as shown in Chapter 6, large variations in detection efficiency will take place. As in the case of gas counters, it is possible to arrange these detectors to make continuous-flow measurements on liquid samples, subject to the foregoing limitation on emitter type. Methods for assay of stationary liquid samples in a coiled tubing mounted against the solid fluor have also been described (170).

3. Assay of Solid Counting Samples

Solid counting samples are satisfactory for most radionuclides and can be more easily handled and stored than other types. The general requirement for a solid counting sample for α- or β-counting is that it consist of a relatively thin, uniform deposit of the active material supported, if possible, by a low Z (i.e., low atomic number) backing material. Solid samples of this type may be assayed by ionization chambers, G-M counters, proportional counters, solid scintillation detectors, and semiconductor detectors. As will be seen later, solid samples may also be assayed in a heterogeneous counting system by using a liquid scintillation counter.

a. Sample Preparation and Sample Backing Materials. The more commonly employed backing materials are aluminum, various plastics, nickel, carbon, platinum, paper, and stainless steel. The backing materials can be divided into two classes, those that are very thick and those that are very thin. For most routine analyses, very thick sample backings are used to ensure that the backing thickness exceeds that required for saturation backscattering (see Chapter 13). Furthermore, since the backscattering of β^--particles is related to the density of the material, it is important to use the same backing material for all samples in an experiment, plus the lowest Z material possible. The form of the thick backing material is usually that of a flat plate or a small cup called a *planchet*. It is common to coat the backing material with a thin plastic spray prior to use so as to increase its resistance to chemical attack.

The use of thin backing materials is restricted to cases where the energy spectrum of the β^--particles emerging from a source is of interest. In this case, commonly used backing materials include Mylar, Formvar (ethylene dichloride), Zapon, carbon, nickel (available down to thicknesses of 0.000004 in.), and VYNS (a polyvinylchloride-acetate copolymer consisting of $\sim 85\%$ chloride and 15% acetate). VYNS is especially useful, for it is readily available in thicknesses down to ~ 1 μg/cm and shows good tensile strength and resistance to chemical attack. Pate and Yaffee (161) detail the use of VYNS.

The necessity to prepare counting samples of uniform thickness is most stringent when dealing with low-energy β^-- or α-emitters, since *self-absorption* in solid sources of such nuclides is a significant factor. A number of generally applicable techniques for producing counting samples of uniform thickness have been developed over the years. Those most commonly used will be discussed briefly.

b. Direct Evaporation From Solution. Perhaps the simplest means of converting a dissolved sample to a solid counting source is by direct evaporation of the solvent (usually water or alcohol). The sample solution, or an aliquot of it, is pipetted onto a planchet and gently evaporated to dryness under a heat lamp. Accuracy in pipetting is a major factor in this technique. A microtransfer pipet is normally used. Application of the evaporation technique is limited because it is virtually impossible to secure reproducible sample thicknesses by its use. Variations of a factor of 100 in source activity across the sample have been observed. A number of refinements to improve uniformity of sample deposit have been developed, such as the use of wetting agents on the planchet surface (e.g., 5% insulin in water), the addition of thin sheets of absorbent material (lens tissue) to the bottom of the cupped planchets, the use of dilute agar solutions (36, 98, 128), and the rotation of the mount during evaporation.

c. Filtration of Precipitates. Larger sample amounts can be obtained from suspensions of precipitates, such as $Ba^{14}CO_3$, than from direct evaporation. Direct filtration of the suspension, leaving the precipitate as a uniform layer on the filter medium, is the simplest procedure. Filter paper or fritted disks are generally used for this purpose and subsequently serve as the source mount (17). A typical filtration apparatus is shown in Figure 11-6. Care must

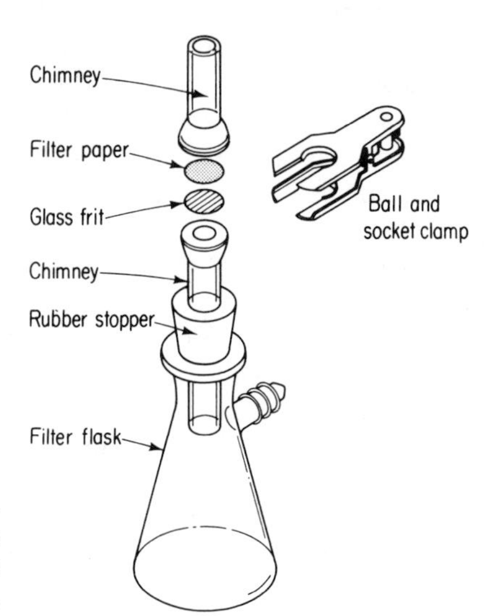

Fig. 11-6. Convenient filter apparatus for the preparation of radioactive samples for measurement. Courtesy of Brookhaven National Laboratory.

be taken to obtain crystals of $Ba^{14}CO_3$ that are both small and uniform. When filter paper is used, there is some difficulty in securing reproducible results because of variable paper texture and damage to the thin sample layer when it is subsequently transferred to a mounting block and dried. With fritted glass or sintered metal disks, this problem is minimized, but their cost is too great to allow holding large numbers of samples for permanent reference. The use of Millipore filters has been suggested by Jervis (105). Bronner and Jernberg (23) have proposed a centrifugal filter assembly for mounting ^{45}Ca and ^{35}S precipitates. A self-absorption correction curve should be made for each type of counting sample that contains weak β^--emitters. (Details are given in Chapter 13.)

d. Settling or Centrifugation of Slurries. Precipitates that have been separated from excess solvent by centrifugation and decantation may be mounted in slurry form by two different techniques. The precipitate may be resuspended in a volatile solvent and poured into a cup planchet. Slow drying of the slurry with occasional tapping sometimes produces a uniform sample layer. Although this technique gives good reproducibility, it is tedious. Alternately, the resuspended slurry may be centrifuged again, using an arrangement whereby a removable cupped planchet serves as the false bottom of the centrifuge tube. Such an apparatus is commercially available. Following centrifugation, the solvent is decanted and the planchet removed from the tube and dried before counting. This procedure is relatively simple and can be easily used with large numbers of samples. It gives reasonable reproducibility.

e. Mounting Dry Powdered Samples. When a sample consists of ash or dried soil, it may be most convenient to weigh it directly into a cup planchet. The particles may be compressed or the planchet may be tapped so as to produce a more even distribution. It is highly desirable to treat such a mounted sample with a binder, such as a solution of collodion, in order to prevent disturbing or spilling the deposit in handling. In general, this method is not applicable to samples containing low-energy β^--emitters.

f. Electroplating. Uniform, thin films of many metals may be prepared by electrodeposition. The sample is usually plated out from solution in the form of an elemental deposit, which allows a maximum specific activity to be obtained (see Figure 11-7 for a sketch of a typical electrodeposition cell). The thin, uniform films have the two advantages of minimizing self-absorption and ensuring reproducibility. Since the early 1960s a variation on the electroplating technique, called "molecular plating," has received widespread use (156, 157, 158). In molecular plating, a very small quantity of inorganic material is dissolved in an organic solvent and forms a positively charged

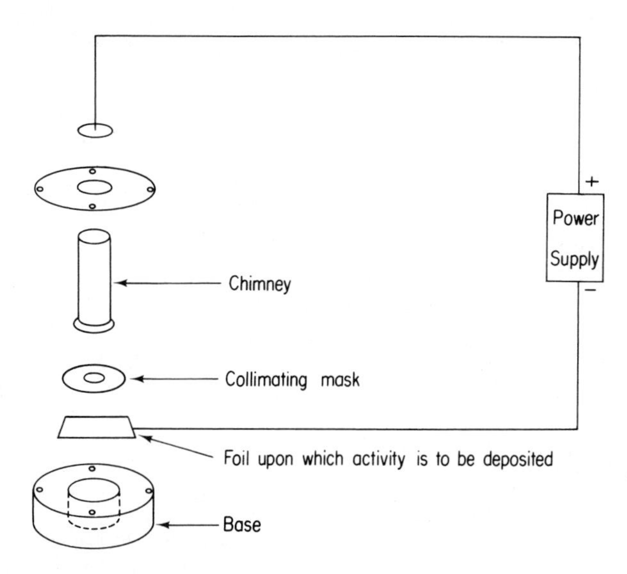

Fig. 11-7. Details of an electrodeposition cell for the preparation of α- and β-activities.

complex in which the inorganic molecule is surrounded by a cluster of solvent molecules. The positively charged complex is attracted to the high negative potential on the cathode of the plating cell and is implanted in the cathode (which acts as the source background). The method has the advantages that any water-soluble compound can be deposited; the deposits are thin and uniform; no chemical attack of the backing occurs; the nuclide is tightly bound to the backing and will not rub off or be pulled off by tape; and the time for deposition is not long (<1 hr). Efficiencies of deposition of $>50\%$ have been achieved for the actinides, the rare earths, ^{137}Cs, ^{60}Co, ^{45}Ca, ^{55}Fe, ^{65}Zn, and ^{90}Sr. Details of the procedure are found in the literature (157, 158).

g. Vacuum Evaporation. Vacuum evaporation is a technique for forming thin, uniform films of any nonrefractory material on a backing material. The procedure is to deposit the material to be evaporated on a thin, metallic filament (see Figure 11-8). The space inside the bell jar is evacuated and the filament is heated electrically. Under the reduced pressure, most materials will rapidly melt and then evaporate. The vapors from the material being evaporated are condensed on a cold collector plate. The area of the source on the backing plate may be defined by collimating apertures. Extensive tables exist in the literature, showing how this technique may be used to form a thin film of almost any element (159).

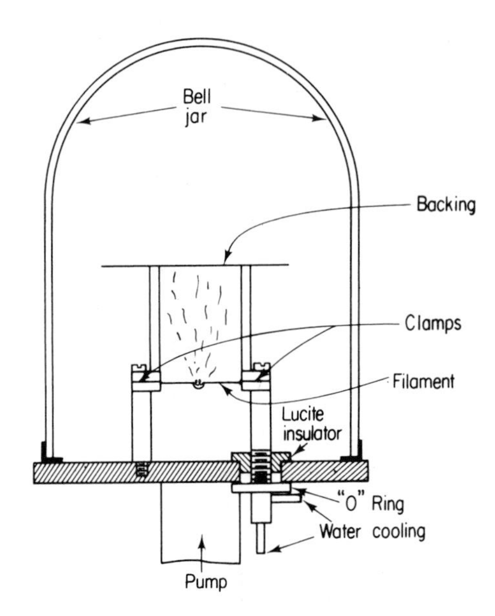

Fig. 11-8. Diagrammatic picture of a simple evaporation apparatus.

4. Radiochromatogram scanning

Paper chromatography is so widely used to separate complex mixtures of labeled compounds that it is not surprising to find that much effort has been expended on developing methods of directly assaying radiochromatograms (96). In general, paper chromatograms containing ^{14}C, ^{35}S, or, to a lesser extent, ^{3}H lend themselves fairly well to direct assay, despite the problem of self-absorption. Since the typical paper-strip chromatogram may have a series of active spots at varying distances from the point of origin, it is highly desirable to have some means of scanning the length of the strip and recording both the extent of the activity and its location.

Many types of radiochromatogram scanners have been described (37, 124, 151, 176, 185, 193, 207), and several models are commercially available. Basically, they consist of a detector (usually a windowless G-M tube), a count rate meter, a strip chart recorder, and a drive mechanism to move the chromatogram past the collimated detector at a fixed speed. The detector window is equipped with a set of slits with widths ranging from $\frac{1}{16}$ to $\frac{1}{4}$ in. or even wider. The selection of slit width depends on the desired resolution with regard to the location of radioactive substances, in the form of spots or bands, on a given radiochromatogram. For radiochromatograms having a very low counting rate, it may be desirable to use a pair of detectors with appropriate slits, such as windowless G-M detectors, mounted end to end, thus attaining nearly 4π geometry. Alternatively, the strip may be cut into serial sections, which are assayed individually. Such a procedure, using liquid scintillation counting, is described on p. 280. Two-dimensional radiochromatograms pose a most difficult problem, but a few semiautomatic scanners developed for this purpose are of some value (138). Svendsen (221) has suggested a method of eluting the spots onto planchets and counting as solid samples.

Operational procedures for scan counting of radiochromatograms are dependent on the features and the control devices unique to each of the commercial types of scanners. However, several general considerations can be summarized.

a. Preliminary Scanning. It is generally advisable to carry out a fast scanning count of the entire radiochromatogram by using a wide detector window slit. Doing so will provide information on the counting times needed to establish optimal conditions for final counting.

b. Selection of Detector Window Slit Widths. Naturally, the narrower the width of the slit, the better resolution one obtains in separating closely located spots or bands on the radiochromatogram. In this process, the counting rate of the small fraction of the radiochromatogram exposed to the detector window through the slit is naturally low. Hence it may be necessary to compromise in selecting proper slit width in order to have reasonably good resolution, together with reasonably high counting rate, so as to minimize uncertainty in the counting data.

c. Scanning Speed. The driving mechanism to move the radiochromatogram across the detector window is usually equipped with a control to permit the use of several speeds. For samples having a low counting rate, the radiochromatogram moving speed must be low. Here the selection of adequate speed depends on the desired reliability of the counting data and the counting rate of background radiation.

d. Time Constant of Rate Meter. Data from the radiochromatogram scanning counter is generally presented via a rate meter equipped with several time constants. For radiochromatograms having a low counting rate, as noted above, the moving rate of the radiochromatogram is necessarily low; hence the time constant should be set as long as necessary in order to provide adequate counting time for a given exposed fraction of a sample. Judgment, in this case, naturally depends on statistical factors.

e. Recording Chart Speed. Output of the counting rate meter is normally presented on a chart recorder. The chart speed, therefore, should be adjusted to correspond with the scanning speed. In recent models, the scanning counter is also equipped with electronic devices to digitize and integrate data from the rate meter as recorded on the chart records. Such a device is useful in allowing one to obtain numerical values of total counts accumulated for a given spot or band on the radiochromatogram.

The more recently developed procedure of *thin-layer chromatography* (TLC) finds widespread usage in the resolution and analysis of mixtures of radioisotope-labeled compounds. TLC, utilizing a glass plate coated with a uniform, thin layer of absorbent material, generally offers a more clean-cut separation of fractions in considerably less time than either paper or column chromatography. An excellent technical bulletin from the Nuclear-Chicago Corporation (97) describes practical techniques in the use of radioisotopes with TLC.

D. PREPARATION AND ASSAY OF LIQUID SCINTILLATION COUNTING SAMPLES

1. Basic Considerations in the Choice of Scintillation Solutions

a. Sample Solubility. The unique feature of liquid scintillation counting is that the sample is in intimate contact with the scintillation solution. The components of such solutions have been discussed previously (Chapter 9). Clearly, the best contact between sample and fluor is achieved when both are dissolved in the same solvent mixture. Unfortunately, simple solution is not always feasible for many biological samples. Various alternative methods have been developed to improve *sample solubility* or at least to bring the sample material and the scintillation solution into defined proximity. Because of these considerations and the numerous biological samples encountered, suitable scintillation solutions have been prepared on a rather empirical basis. We can cite several general references on the preparation of liquid scintillation counting samples (12, 46, 64, 187, 250), of which the review by Funt (64) and the handbook published by Nuclear-Chicago (250) are particularly recommended. The Packard Technical Bulletins (47, 68, 92, 178–181) and Intertechnique Technical Review Series (182–184) treat specific aspects of sample preparation for liquid scintillation counting thoroughly.

b. Quenching. A second technical problem affecting the choice of a suitable scintillation solution is quenching. Quenching is commonly caused by the dissolved sample and is a consequence of the sample being polar in nature or colored. *Color quenching* may be avoided by using pretreatment procedures (adsorption or combustion) that decolorize the original biological sample (208). Alternatively, Fales (55) suggests the direct addition of ethanolic sodium borohydride solution to the sample in the counting vial to effect decoloration. *Chemical quenching* can often be partially compensated for by the addition of naphthalene to the scintillation solution. The extent of quenching should be routinely determined as discussed in Chapter 9.

c. Homogeneous Versus Heterogeneous Counting Systems. In cases where the sample is, or can be made, soluble in the fluor solution, a *homogeneous* counting system can be used. Where this is not possible or where the solution leads to excessive quenching, the sample and the scintillation solution can still be brought into intimate contact by various means in a *heterogeneous* system. Since the sample types that may be assayed by liquid scintillation counting are so diverse, obviously no single scintillation solution is suitable for all samples. Of course, some fluor solutions do have wider application than others. The remainder of this chapter attempts to point out generally successful means of incorporating various sample types into fluor solutions and to list in tabular form many of the liquid scintillation solutions that have been proposed. These tables are presented as a guide to the choice of a suitable scintillation solution, but not as a substitute either for the discussion in the references cited or for individual experience.

As an added practical point, note that a variety of causes, aside from β-particle interaction, can produce fluorescence of the scintillation solution. Such fluorescence results in spurious pulses. An easily overlooked cause of such induced fluorescence is the light from fluorescent lamps. For this reason, liquid scintillation samples should be prepared in rooms illuminated by incandescent lamps, and the counting samples should be kept in the dark before being assayed.

Equally important is the interference with counting from *chemiluminescence*. For example, there is the chemiluminescence resulting from the oxidation of an unsaturated compound by molecular oxygen, a reaction readily catalyzed by a base, such as sodium hydroxide or hyamine hydroxide. Note that the emulsifying agents used in the preparation of gel counting samples (see Section 3a of this chapter) are unsaturated in nature. The problems arising from chemiluminescence can often be handled by allowing the enclosed counting sample to stand for an hour or so before actual counting. By then the oxygen in the sample will be completely depleted.

2. Homogeneous Counting Systems

a. Direct Solution in the Scintillation Solvent. Where sample solubility in the most efficient scintillation solvents (toluene, xylene, etc.) is no problem, the most urgent factor is to choose a fluor solution in which the sample will cause a *minimum quenching* effect. Relatively few biological samples fall into this category, since most are polar compounds and not readily soluble in toluene. Table 11-2 lists scintillation solution "recipes" that are known to contribute little to pulse quenching. The sample recipes marked with an

TABLE 11-2

Composition of Minimum-Quenched Scintillation Solutions

Solvent(s)	Primary Fluor	Secondary Fluor (wave shifter)	Antiquencher	References
Toluene*	PPO, 0.4–0.6%	POPOP, 0.01%		(80, 82, 108)
Toluene*	*p*-terphenyl, 0.5%	POPOP, 0.05%		(12, pp. 101–107)
Methoxybenzene (Anisole)	PPO, 0.3%			(12, pp. 88–95)
Methoxybenzene (Anisole)	PPO, 0.3%			(80)
Xylene	PPO, 0.3%			(152)
Xylene	PBD, 1.0%			(12, pp. 101–107)
Phenylcyclohexane	*p*-terphenyl, 0.5%			(108)
Phenylcyclohexane	PPO, 0.3%			(80)
Triethylbenzene	PPO, 0.3%			(80)
Triethylbenzene	*p*-terphenyl	POPOP		(44)
1, 3-dimethoxybenzene	PPO, 0.3%			(80)
Paraldehyde, 72.3%* Xylene, 4.4%	PPO, 0.4%	POPOP, 0.01%	Naphthalene, 23.3%	(12, pp. 261–267; 115, 169)
p-cymene*	*p*-terphenyl, 0.3%			(81, 84, 108)
Hexane-Octane*	*p*-terphenyl, 0.1%			(81, 108)
Toluene, 98% Ethanol, 2% Acetylene*	PPO, 0.3–0.4%			(12, pp. 288–292; 74)
Toluene Ethanol*	PPO, 0.4%	DPHT, 0.002% (diphenylhexatriene)		(2)
Benzene*	*p*-terphenyl, 0.45%			(80, 108)
Benzene*	PPO, 0.3%			(80)
Ethylbenzene*	PPO, 0.3%			(213)
Xylene, 90% Methanol,* 10%	*p*-terphenyl, 0.4%	POPOP, 0.01%		(173, 174)
Trimethylborate, 47% Xylene, 31%	PPO, 0.3%		Naphthalene, 22%	(66)
Xylene, 83%	PBD, 0.3%		Naphthalene, 7%	(173)
Benzene*	PPO, 0.4%	POPOP, 0.01%		(116)
Benzene-^{3}H, 85% Toluene, 15%	PPO, 0.4%	POPOP, 0.01%		(225)

asterisk are those used in carbon-14 dating studies. In this case, maximum detection efficiency is achieved by adding the sample in the same chemical form as the solvent. This step usually involves elaborate chemical or biological syntheses, starting with $^{14}CO_2$ (224). The advent of this technique made possible a considerable backward extension of the period of time that could be accurately measured by the ^{14}C-dating method. Several reviews of the procedures involved are available (12, pp. 129–134, 261–267; 68, 81, 173). Later, Tamers and Bibron (225) converted tritiated rain water to benzene for direct incorporation in the scintillation solvent.

b. Indirect Solution in the Scintillation Solvent. In general, biological samples are not soluble in toluene, the most efficient scintillation solvent in common use. Thus it is necessary to take an indirect approach in order to achieve a homogeneous sample-fluor solution (180). One can use a mixture of solvents that will accommodate small quantities of *aqueous solutions*. Inevitably, this step increases quenching and reduces counting efficiency. The mere addition of ethanol to the common toluene solvent results in a mixture that can hold up to 3% of aqueous sample material in solution. Dioxane, as a solvent, allows the presence of much larger percentages of water (up to 29%), although it is less efficient in energy transfer than toluene. Dioxane solutions usually require the addition of naphthalene as an anti-quenching agent. Dioxane has the additional disadvantage of a relatively high freezing point (+12°C), and it often needs the addition of an antifreeze component (120). Table 11-3 lists "recipes" for scintillation solutions with varying degrees of water tolerance. These solutions have particular significance in tritium counting, since it is usually assayed as tritiated water (4, 35, 100). In an attempt to avoid the disadvantage of dioxane as a solvent, Avinur and Nir (3) have used an acid-catalyzed tritium exchange between tritiated water (THO) and toluene. Much-improved counting efficiency resulted, but the procedure is quite time consuming.

The measurement of $^{14}CO_2$ by the liquid scintillation counting method poses the problem of finding a suitable absorption medium. This trapping agent should combine with CO_2 to produce a toluene-soluble salt without introducing additional quenching effect. The first such trapping agent proposed was a methanolic or ethanolic solution of the hydroxide of *Hyamine* 10-X (Rohm and Haas, Inc.) (160). The original technique was later refined by Frederickson and Ono (60). Rapkin has reviewed the uses of this substance in liquid scintillation counting (178). "Hyamine" continues to be the most widely used trapping agent for $^{14}CO_2$ (210). It suffers, however, from the disadvantage of a relatively high cost, derived from the tedious operations involved in converting the commercially available Hyamine chloride to the hydroxide of Hyamine by using either silver oxide or Dowex 1-OH resin (12, pp. 123–125). Moreover, the use of an alcoholic solution of the

TABLE 11-3

Scintillation Solvent Mixtures to Make Aqueous Samples Soluble

Solvent(s)	Primary Fluor	Secondary Fluor (wave shifter)	Antiquencher	Water Tolerance	References
Toluene, 8.5–12.5 ml Ethanol or methanol, 2.5–6 ml	PPO, 0.6%	POPOP, 0.01%		0.15–0.50 ml	(8, 147, 249)
p-dioxane. 76–80%	PPO, 0.7%	POPOP, 0.005%	Naphthalene, 7–12%	20–24%	(31, 237)
Xylene, 5 ml *p*-dioxane, 5 ml Ethanol, 3 ml	PPO, 0.5%	α-NPO, 0.005%	Naphthalene, 6%	1 ml	(110)
p-dioxane, 60% Anisole, 10% 1, 2-dimethoxyethane, 10%	PPO, 1.2%	POPOP, 0.005%	Naphthalene, 6% (with ethylene glycerol)	20%, or in place of water, ethylene glycol, 20%	(12, pp. 88–95; 100, 164)
p-dioxane, 88% Methanol, 10% Ethylene glycol, 2%	PPO, 0.4%	POPOP, 0.02%	Naphthalene, 6%	10%	(20, 21)
Xylene, 14% *p*-dioxane, 43% Ethylene glycol monoethyl ether, 43%	PPO, 1.0%	POPOP, 0.08%	Naphthalene, 8%	16.1%	(27)
p-dioxane, 83%; Ethylene glycol monoethyl ether, 17%	PPO, 1.0%	POPOP, 0.05%	Naphthalene, 5%	29.2%	(27)
Toluene, 50% Cyclohexene, 50%	PPO, 0.5%	POPOP, 0.01%			(1)
Toluene, 5 ml *p*-dioxane, 5 ml Ethanol, 3 ml	PPO, 0.5%	POPOP, 0.005%	Naphthalene, 10%	6%	(51)
Toluene, 10 ml Ethylene glycol monomethyl ether, 6 ml	PPO, 1.5%	POPOP, 0.005%		0.5 ml	(175)

hydroxide of Hyamine in a liquid scintillation solution results in significant fluorescence quenching. Fortunately, no additional quenching occurs when, upon absorbing CO_2, the hydroxide of Hyamine is converted to the carbonate form.

Other $^{14}CO_2$-trapping agents have been suggested. *Primene* 81-R (Rohm and Haas, Inc.) is cheaper, may be used directly as supplied, and shows little quenching effect in moderate concentrations (150). Unfortunately, it does not absorb CO_2 as readily as "Hyamine" and it requires the use of multiple traps. *Ethanolamine* is relatively inexpensive and shows a tolerable quenching effect (187, pp. 113–114). It, however, requires the use of ethylene glycol monomethyl ether (a rather toxic compound) or ethanol to facilitate solubility of the resulting ethanolamine carbonate in toluene. *Ethylene diamine* is also usable but, like ethanolamine, can accommodate only limited amounts of CO_2 (up to 1 mmole with 10 ml of an ethanol-ethylenediamine [2:1] solution) (234). *Phenylethylamine* has much greater trapping capacity and less quenching effect than "Hyamine." Rapkin (34, 181) has reviewed in detail the various $^{14}CO_2$ trapping agents proposed. The first portion of Table 11-4 lists several common "recipes" used in solution counting of $^{14}CO_2$.

The method of Gordon *et al.* (73) for liquid scintillation assay of $H_2{}^{35}S$ should also be mentioned. Finding "Hyamine" an unsatisfactory trapping agent, they modified the caps of their counting vials to allow direct injection of the gas through a silicone rubber insert into the air space over a toluene-PPO-POPOP solution. They found very linear counting results with increasing quantities of $H_2{}^{35}S$. It is suggested that this method may have application to other gaseous samples.

In an attempt to overcome the insolubility of most *metallic ions* in the organic scintillation solvents, various specific solubilizing agents have been investigated (53, 92). In general, these agents act either by forming an organic complex with the ions or by converting them to the salt of an organic acid. To date, the best organic complexing agents to be used with liquid scintillation systems are the acidic esters of orthophosphoric acid—that is, dibutyl phosphate, tributyl phosphate, dioctyl phosphate, and so on. Other organic complexing agents, however, have been used in specific cases to good advantage. Metallic salts of 2-ethylhexanoic acid (octoic acid) are readily soluble in toluene and have high scintillation efficiencies even at quite high concentrations. Table 11-4 lists solvent systems and binding agents that are useful in incorporating a large variety of inorganic ions into a homogeneous counting system.

Considerable attention has been directed in recent years toward the development of techniques allowing liquid scintillation counting of labeled *whole tissues*, *blood*, and *urine* with minimal sample pretreatment (12, pp. 223–229; 25, 89, 146). "Hyamine" in its hydroxide form has been used to make soluble animal tissues (5, 26), bacterial cell debris (79), and blood

TABLE 11-4

Scintillation Solutions to Accommodate Inorganic Ions

Solvent(s)	Primary Fluor	Secondary Fluor (wave shifter)	Binding Agent	Inorganic Ions	References
Toluene, 55% Ethylene glycol monomethyl ether, 39%	PPO, 0.6%		Ethanolamine, 6%	CO_2	(102)
Toluene, 93%	PPO, 0.3%		Methanolic primene, 7%	CO_2, sulfate, phosphate, chloride, organic acids	(150, 177)
Toluene, 67%	*p*-terphenyl, 0.3%	POPOP, 0.003%	Methanolic or ethanolic hyamine-OH, 33%	CO_2	(12, pp. 108–114, 123–125; 88, 160, 226, 231)
Toluene, 67% Ethylene glycol, 3% 2-methoxyethanol, 27%	PPO, 0.6%		Ethylene diamine, 3%	CO_2	(234)
Toluene, 46% Methanol, 27%	PPO, 0.5%	POPOP, 0.01%	2-phenylethylamine, 27%	CO_2	(247)
Toluene, 80%	PPO, 0.6%		2-ethylhexanoic acid, 20%	Hg, Cd, Ca, K	(12, pp. 88–95)
Toluene, 70% Methanol, 30%	PPO, 0.3%			KOH, 0.2%, with plasma, urine, and so on.	(69)
Toulene, 58.8% *N*, *N*-dimethyl-formamide, 26.5% Glycerol, 5.9% Ethanol, 8.8%	PPO, 0.3%	POPOP, 0.01%		Sulfates	(101)
Phenylcyclohexane	*p*-terphenyl, 0.2%		Methyl isobutyl ketone	Uranium	(6)
Toluene, 90% *p*-dioxane, 10%	PPO, 0.3%		Methyl isobutyl ketone	Uranium	(92)
Xylene	*p*-terphenyl, 0.4–0.5%	POPOP, 0.01%	Dibutyl phosphate	Pu, Zr, Nb	(90, 122)
Toluene	*p*-terphenyl 0.5%	POPOP, 0.005%	Dibutyl phosphate	Y, Sr	(58)
Toluene, 78% Isoamyl alcohol, 12%	PPO, 0.545%	POPOP, 0.018%	Orthophenanthroline	Fe	(113, 114)
Toluene	*p*-terphenyl, 0.4%	POPOP, 0.01%	Dioctyl phosphate	Th	(72)
Xylene	*p*-terphenyl, 0.4%	POPOP, 0.01%	Dioctyl phosphate	Pu, Sm, Ni	(91)

Solvent(s)	Primary Fluor	Secondary Fluor (wave shifter)	Binding Agent	Inorganic Ions	References
Xylene	*p*-terphenyl, 0.4%	POPOP, 0.01%	*p*-toluidine	Ru	(91)
Toluene, 29% Methanol, 29% Phenethylamine, 29%	PPO, 0.5%	POPOP, 0.005%	2-pentene, 6.5% naphthalene, 10% (anti-quencher)	Cl^-, I^-, Br^-	(52)
Xylene	*p*-terphenyl, 0.4%	POPOP, 0.01%	2-ethylhexylhydrogen 2-ethyl-hexyl Phosphonate	Pu	(91)
Toluene	*p*-terphenyl, 0.04%	POPOP, 0.01%	Octoic acid in Ethanol	Rb, Ca, K, Na, Sm, Pb, Cd, Bi, U	(10, 57, 72, 186, 248)
p-dioxane, 75% 1,2-dimethoxyethane, 12.5% Anisole, 12.5%	PPO, 0.7%	POPOP, 0.005%		Na_2SO_4, $NiCl_2$ in water	(91)
Toluene, 95% Ethanol, 5%	PBD, 0.8%	POPOP, 0.01%		Po, Cs, Ba, Ca in HCl H_3PO_4 $CoCl_2$, NaI in water	(126, 199, 217)
p-dioxane	PPO, 5%			At in water	(7)
p-dioxane, 95% Naphthalene, 5%	PPO, 0.6%	POPOP, 0.05%		P, Cl, Ca, S in water	(165)
Toluene, 50%	PPO, 0.3%		Tri-*n*-butyl phosphate, 50%	U, Th	(4)
Toluene, 87% Ethanol, 8.7%	PPO, 0.4%	POPOP, 0.01%	*n*-caproic acid, 4.3%	Ni	(70)
p-dioxane, 90% Naphthalene, 10%	PPO, 0.4%	POPOP, 0.005%	Di (2-ethylhexyl)-orthophosphoric acid (HDEHP)	U in water	(22)
Toluene, 44.5% Ethanol, 44.5% Ethylene glycol, 11%	PPO, 0.53%			Ca in nitric acid	(194)
1,2-dimethoxyethane, 91% Naphthalene, 9%	PPO, 0.64%	POPOP, 0.0046%		U in water	(117)

serum (39, 223). At present several types of "solubilizers" are available from commercial suppliers. The NCS solubilizer sold by Nuclear-Chicago is a dimethyl-dialkyl (chain length from C_6 to C_{20}) quaternary ammonium base that is soluble in toluene and furnished in a 0.6 *N* solution (172). Three types of Bio-Solv solubilizers are sold by Beckman (127). BBS-1 and BBS-3 are general-purpose solubilizers, and BBS-2 is a solubilizer for alkaline materials (21, 172). Another solubilizer marketed by the Packard Instrument Company is Soluene-100, which is a quarternary ammonium hydroxide of a molecular weight of 386. The compound is of the type $R_2R'R''$ NOH, where R is methyl, R′ is a C_{11} straight chain, and R″ is a C_{12} straight chain alkyl group. The marketed product is an 0.5 *N* solvent in toluene. The use of Soluene-100 for the preparation of whole blood samples for liquid scintillation counting has been reported by Laurencot and Hempstead (112). The solubilization process can generally be facilitated by heat (up to 60°C) or sonication. The solubilizing power of Hyamine, NCS, and alcoholic KOH has been comparatively examined by Hansen and Bush (78), who concluded that NCS is the most useful solubilizing agent and that alcoholic KOH has little value in sample preparation. Hyamine, in its chloride form, and Triton-X-100 (Rohm and Haas, Inc.) have also been suggested for emulsifying tissues or body fluids (137). The problems of poor solubility and color quenching have plagued the use of solubilizing agents, but decreased counting efficiency can sometimes be tolerated as the price for more simplified sample preparation. The direct liquid scintillation assay of untreated urine samples containing tritium has been attempted (31, 111, 147), but decolorizing and centrifuging are usually necessary. Dioxane is typically used as the solvent in such cases. Such urine assays are of particular importance as a health physics practice for investigators using high levels of certain tritiated compounds. An excellent review has been written by Rapkin (182) on sample preparation procedures with urine, whole blood, plasma, serum, and so on. Later, Bray summarized the problem of counting aqueous samples in an excellent review paper (21). The problem of solubilization of animal tissues for liquid scintillation counting has been reviewed extensively by Pollay and Stevens (172).

3. Heterogeneous Counting Systems

Radioactive samples that cannot be dissolved in some suitable solvent system may still be assayed by the liquid scintillation process in a heterogeneous system. The samples may be in the form of a gel or an emulsion, or they may be dried on filter paper and the paper strips placed in a fluor solution. Rapkin (179, 184) and Greene (77) have reviewed these systems in detail. In addition, an aqueous sample may be introduced into a counting vial containing a solid, insoluble fluor (anthracene) as described in Chapter 9. It should be noted that the counting of heterogeneous systems commonly entails much poorer reproducibility despite good counting efficiencies.

a. Gel Suspension Counting. Hayes *et al.* (83) reported a technique involving the liquid scintillation counting of finely ground sample material that was suspended by agitation immediately before assay. The technique was applied to a wide variety of sample materials but had two disadvantages: it required repeated counting of the same sample and it lacked precision. The technique is not widely used today. The significant point was that the presence of opaque materials, either as dispersed water droplets or fine white solids, did not reduce the counting efficiency significantly, compared to a homogeneous system. Self-absorption appeared to be only a minor problem when the particulate material was finely ground, except for tritium-labeled samples.

Funt (61) was the first to propose the use of a gel agent to suspend particulate samples for liquid scintillation counting. He used *aluminum stearate* (5%) as the gelling agent and injected the sample material throughout the already formed gel by means of a fine hypodermic needle. This technique proved particularly suitable for assay of finely ground $Ba^{14}CO_3$ (62). Gel counting is important because samples that would cause severe quenching in solution may be counted in suspension with only minimal quenching. It is essential, however, that the sample material be completely insoluble in the scintillation solvent used, so that variable counting efficiencies from sample to sample do not result.

A modification of this technique using the gelling agent, *Thixcin* (ricinoleic acid) (Baker Castor Oil Co., Inc.), was subsequently introduced by White and Helf (238). They suggested a gel system prepared by adding 25 g Thixcin to 1 liter of scintillator solution (toluene-0.4% PPO-0.01% POPOP) and blending (12, pp. 96–100; 87); the result is a pourable solution that quickly sets to a gel. This gel system has been used in assaying organic nitrocompounds (45), ground barium carbonate (144), and aqueous solutions of carbon-14 and tritium compounds (201).

Later, Ott *et al.* (153) used *Cab-O-Sil* M-5 (Godfrey L. Cabot, Inc.), a pure silica of extremely fine particle size, as a gelling agent. It is particularly advantageous, for it requires neither blending nor heating, can be used with either toluene or dioxane, and forms an almost transparent gel merely by shaking in the counting vial. Cab-O-Sil gels will support about twice as much sample material as the same weight of Thixcin gel and also give higher counting efficiencies. Gordon and Wolfe (75) have used Cab-O-Sil gels in assaying aqueous solutions (up to 6.5% by volume) with counting efficiencies reported to be 65% for ^{14}C and 14% for ^{3}H, whereas Snyder and Stephens (211) and Brown and Johnston (24) have used such gels to count ^{14}C- and ^{3}H-labeled scrapings from thin-layer chromatograms. Cluley (42) has used a silica gel to hold up to 1 g of $Ba^{14}CO_3$ per 10 ml of gel with reasonable counting efficiency. Cab-O-Sil, however, being finely powdered silica, poses a possible health hazard to its users if proper precautions are not taken. Cab-O-Sil has also been utilized in reducing the extent of absorption of certain polymer samples to the walls of the counting vials, which would result in decreased counting efficiency (16). Gel counting has also been applied to samples containing ^{32}P and ^{36}Cl. Thus Erdtmann and Hermann (54) mixed an aqueous solution of these nuclides with toluene/PPO/POPOP, silica, and a small amount of nonylphenoxyethanol by aeration to form a gel for counting at practically theoretical efficiency. When high concentrations of salts are present, a dioxane/ethanol/naphthalene/xylene mixture is used as the scintillation solution.

Other gelling agents have also been suggested, such as dissolved polystyrene and methyl methacrylate (Plexiglas). Shakhidzhanyan *et al.* (200) have used gels of the latter substance in assaying ^{40}K-containing ash from human organs. The use of Poly-Gel-B as a gelling agent has been reported by Benakis (13). This new agent consists of polyolefine resins that are the result of the polymerization of ethylene, propylene, butylene, and similar products, as well as the copolymerization of two or more of these olefines with low molecular weight. In practice, the agent, in the form of opalescent white pellets and containing fluors (e.g., PPO/POPOP or butyl-PBD/POPOP), is dissolved in warm (60 to 70°C) toluene at a concentration of 10% and a stable gel is formed upon cooling. This new, low-cost gelifying agent is reported to remain stable without any settling of suspended material for several weeks, to cause no quenching effect, and to give rise to good counting efficiency for both ^{14}C- and ^{3}H-containing samples.

Later, Spencer and Baneyi (212) prepared a 26% solution of ^{14}C-labeled plasma in dimethyl sulfoxide (DMSO), and up to 2 ml of the solution was mixed with 1.0 ml of a scintillation mixture containing Cab-O-Sil and dioxane/naphthalene as solvent for counting at a remarkable 75% efficiency. Because of the superb solvent characteristics of DMSO for a variety of compounds, this procedure may have wide applications.

Chemiluminescence is often a problem in gel suspension counting when the counting sample is alkaline in nature.

b. Emulsion Counting. Meade and Stiglitz (137) were the first to introduce the idea of counting aqueous samples in the form of an emulsified preparation. A mixture of 0.5 ml of an aqueous solution, 2.0 ml of methanolic Hyamine 10-X chloride (1.5 *M*), and 17.5 ml of a toluene-based scintillation solution containing 8 g PPO and 50 mg POPOP was emulsified by blending and counted for ^{14}C or ^{3}H at 80% and 10% efficiencies, respectively. Later, the use of Triton X-100 (isooctyl-phenoxy-polyethoxyethanol) at high concentrations (25 to 50%) was reported as a good emulsifier by Patterson and Greene (162). Over 40% of water can be accommodated if one uses a very high concentration of six parts Triton-X-100 to seven parts of toluene, although counting efficiency is reduced to 5% for ^{3}H and 55% for ^{14}C. However, it should be noted that counting efficiencies of emulsified samples vary with the water content in a nonlinear manner (14). When water content gets higher, up to a concentration of 15%, counting efficiency decreases accordingly. From that point, counting efficiency actually increases with high water content over a narrow range and then decreases again. This phenomenon appears to be associated with the transition from true solution to emulsion and then to the formation of mircoscopic water droplets. For the same reason, when determining counting efficiency by means of the internal-standard method, one must keep in mind the partition of spiking standard between the organic and the aqueous phases to ensure that the distribution approximates that of the labeled compound being counted (230). If a good grade of Triton-X-100 is available, it can be used directly without purification. The importance of sample cooling in emulsion counting has been reported by Benson (14). Triton and toluene mixtures should be kept in the dark under refrigeration. Emulsification of aged mixtures with alkaline solution often causes a rapidly decaying luminescence.

The advantage of using emulsified samples instead of homogeneous aqueous samples has been thoroughly discussed by Turner (227), Mostafa *et al.* (140), Williams and Florkowski (243), and Williams (242). These authors also reported a surprisingly great variation in counting efficiency at a given water content when different proportions of toluene and Triton were employed, pointing again to the importance of achieving optimum phase partitioning in emulsion counting. A high order of reproducibility of sample preparation is indeed one of the most important factors in using emulsion counting for routine analysis.

Several other emulsifiers related to Triton-X-100 but differing principally in the lengths of their polyethoxyethanol moieties have been examined for emulsion counting by Greene, Patterson, and Istes (76). They concluded that all emulsifiers tested gave rise to essentially the same counting efficiencies, which were also independent of the visual appearance of the emulsion. They also reported that refrigeration greatly enhances counting performance, presumably because of the fact that the water content of the toluene-scintillator phase is minimized as the temperature decreases. Liebermann and Moghissi (118) also made a comparative study on the use of various emulsifiers and solvents for HTO counting. A mixture of *p*-xylene and Triton-N-101 (2.75 to 1) with 7 g PPO and 1.5 g bis-MSB per liter of mixture is judged to be the best system. At room temperature, 15 ml of this mixture can be readily emulsified with 10 ml of water and will provide a ^{3}H-counting efficiency of 22 to 24%. The chemiluminescence problem can be avoided by preparing samples under red lights. A similar study has also been carried out by Lupica (125).

Fox (59) reported that Triton-X-100/toluene can be used to prepare emulsions (or rather colloids, as preferred by Fox) for counting with such aqueous solutions as 2 *M* NaCl, 8 *M* urea, 5% trichloroacetic acid, 5%

$HClO_4$, and 5% sucrose. Counting efficiency with these types of samples can be determined by means of the channels-ratio method (228) or the internal-standard method using a standard that distributes itself between phases in the same manner as the substance being counted. Emulsion counting samples can also be made with an alkaline solution containing trapped $^{14}CO_2$ (142) or dissolved amino acid or protein (131) added to Triton/toluene. However, in the case of $^{14}CO_2$ samples, counting must be delayed for at least 18 hr to allow the induced chemiluminescence to decay, inasmuch as the alkaline emulsion cannot be neutralized to avoid loss of CO_2. Similarly, emulsion counting has been applied to tissue and allied samples solubilized with NCS (119) or Hyamine 10-X hydroxide (240). Here, again, care should be taken to avoid interference from chemiluminescence by allowing the counting samples to stand for several hours before counting.

The application of emulsion counting for monitoring a flowing stream of sucrose gradients has been discussed by Schram (196). Emulsion counting has also been applied to the counting of ^{51}Cr (203) and ^{45}Ca (143, 155).

c. Filter-Paper Counting. Another effective means of liquid scintillation counting of heterogeneous systems involves direct counting of toluene-insoluble samples on filter paper (235). This method has been applied most successfully to counting *paper chromatogram sections* but it has also been used with pieces of filter paper on which samples have been added and dried in place. Davidson (47) has reviewed the various techniques proposed in this type of system.

BIBLIOGRAPHY

1. ANBAR, M., P. NETA, and A. HELLER. "The radioassay of tritium in water in liquid scintillation counters—the isotopic exchange of cyclohexene with water," *Intern. J. Appl. Radiation Isotopes* **13**, 310–312 (1962).
2. ARNOLD, JAMES R. "Scintillation counting of natural radiocarbon. I. The counting method." *Science* **119**, 155–157 (1954).
3. AVINUR, P., and A. NIR. "Tritium exchange between toluene and aqueous sulphuric acid," *Bull. Res. Council Israel* **7A**, 74–77 (1958).
4. AXTMANN, R. C., and LECONTE CATHEY. "Liquid scintillators containing metallic ions," *Intern. J. Appl. Radiation Isotopes* **4**, 261 (1959).
5. BADMAN, H. G., and W. O. BROWN. "The determination of ^{14}C and ^{32}P in animal tissue and blood fractions by the liquid-scintillation method," *Analyst* **86**, 342–347 (1961).
6. BASSON, J. K., and J. STEYN. "Absolute alpha standardization with liquid scintillators," *Proc. Phys. Soc.* **67**, 297–298 (1954).
7. BASSON, J. K. "Absolute alpha counting of astatine-211," *Anal. Chem.* **28**, 1472–1474 (1956).

8. BATEMAN, JEANNE C., *et al.* "Investigation of distribution and excretion of C^{14}-tagged triethylene thiophosphoramide following injection by various routes," *Intern. J. Appl. Radiation Isotopes* **7**, 287–298 (1960).

9. BAXTER, CLAUDE F., and ILSE SENONER. "Liquid scintillation counting of C^{14}-labeled amino acids on paper, using trinitrobenzene-1-sulfonic acid, and an improved combustion apparatus," *Atomlight No. 33.* (New England Nuclear Corp.), pp. 1–8 (November 1963).

10. BEARD, G. B., and W. H. KELLY. "The use of a samarium loaded liquid scintillator for the determination of the half-life of Sm^{147}," *Nuclear Phys.* **8**, 207–209 (1958).

11. BELCHER, E. H. "The assay of tritium in biological material by wet oxidation with perchloric acid followed by liquid scintillation counting," *Phys. Med. Biol.* **5**, 49–56 (1960).

12. BELL, CARLOS G., JR., and F. NEWTON HAYES (Eds.). "Liquid scintillation counting," *Proceedings of the Conference of Northwestern University, Evanston, Ill., 1957.* New York: Pergamon, 1958.

13. BENAKIS, A. "A new gelifying agent in liquid scintillation counting." In D. L. Horrocks and C. T. Peng, (Eds.), *Organic Scintillators and Liquid Scintillation Counting.* New York and London: Academic, 1971, pp. 735–745.

14. BENSON, R. H. "Limitations of tritium measurements by liquid scintillation counting of emulsions," *Anal. Chem.* **38**, 1353–1356 (1966).

15. BLAIR, ALBERTA, and STANTON SEGAL. "Use of filter paper mounting for determination of the specific activity of gluconate-C^{14} by liquid scintillation assay," *Anal. Biochem.* **3**, 221–229 (1962).

16. BLANCHARD, F. A., and I. T. TAKAHASHI. "Use of submicron silica to prevent count loss by wall absorption in liquid scintillation counting," *Anal. Chem.* **33**, 975–976 (1961).

17. BLOOM, BEN. "Filter paper support for mounting and assay of radioactive precipitates," *Anal. Chem.* **28**, 1638 (1956).

18. BLOOM, BEN. "The simultaneous determination of C^{14} and H^3 in the terminal groups of glucose," *Anal. Biochem.* **3**, 85–87 (1962).

19. BOUSQUET, WILLIAM F., and JOHN E. CHRISTIAN. "Quantitative radioassay of paper chromatograms by liquid scintillation counting: Application to carbon-14-labeled salicylic acid," *Anal. Chem.* **32**, 722–723 (1960).

20. BRAY, G. A. "A simple efficient liquid scintillator for counting aqueous solutions in a liquid scintillation counter," *Anal. Biochem.* **1**, 279–285 (1960).

21. BRAY, G. A. "Determination of radioactivity in aqueous samples." In E. D. Bransome, Jr. (Ed.), *The Current Status of Liquid Scintillation Counting.* New York and London: Grune and Stratton, 1970.

22. BRITT, R. D., JR. "The radiochemical determination of promethium-147 in fission products," *Anal. Chem.* **33**, 602–604 (1961).

23. BRONNER, FELIX, and NILS A. JERNBERG. "Simple centrifugal filtration assembly for preparation of solid samples for radioassay," *Anal. Chem.* **29**, 462 (1957).

24. BROWN, JERRY L., and JOHN M. JOHNSTON. "Radioassay of lipid components separated by thin-layer chromatography," *J. Lipid Res.* **3**, 480–481, (1962).

25. BROWN, W. O., and H. G. BADMAN. "Liquid-scintillation counting of ^{14}C-labeled animal tissues at high efficiency," *Biochem. J.* **78**, 571–578 (1961).

26. Bruno, Gerald A., and John E. Christian. "Note on suitable solvent systems usable in the liquid scintillation counting of animal tissue," *J. Am. Pharm. Assoc. Sci. Ed.* **49**, 560–561 (1960).

27. Bruno, Gerald A., and John E. Christian. "Determination of carbon-14 in aqueous bicarbonate solutions by liquid scintillation counting techniques: Application to biological fluids," *Anal. Chem.* **33**, 1216–1218 (1961).

28. Buchanan, Donald L., and Betty J. Corcoran. "Sealed tube combustions for the determination of carbon-14 and total carbon," *Anal. Chem.* **31**, 1635–1638 (1959).

29. Buhler, Donald R. "A simple scintillation counting technique for assaying $C^{14}O_2$ in a Warburg flask," *Anal. Biochem.* **4**, 413–417 (1962).

30. Burr, William W., Jr., and Donald S. Wiggans. "Direct determinations of C^{14} and S^{35} in blood," *J. Lab. Clin. Med.* **48**, 907–911 (1956).

31. Butler, Frank E. "Determination of tritium in water and urine—liquid scintillation counting and rate-of-drift determination," *Anal. Chem.* **33**, 409–414 (1961).

32. Cahn, Arno, and R. M. Lind. "An improved procedure for plating uniform $BaCO_3$ precipitates," *Intern. J. Appl. Radiation Isotopes* **3**, 44–45 (1958).

33. Calvin, Melvin, *et al.* "Isotopic carbon." *Techniques in Its Measurement and Chemical Manipulation.* New York: Wiley, 1949.

34. Cameron, B. F. "Determination of iron in heme compounds. II. Hemoglobin and myoglobin." *Anal. Biochem.* **11**, 164–169 (1965).

35. Cameron, J. F., and I. S. Boyce. "Liquid scintillation counting of tritiated water," *Intern. J. Appl. Radiation Isotopes* **8**, 228–229 (1960).

36. Campbell, H., H. A. Glastonbury, and Margaret D. Stevenson. "A direct-plating method for the assay of radioactive isotopes in aqueous and alcoholic samples," *Nature* **182**, 1100 (1958).

37. Carleton, F. J., and H. R. Roberts. "Determination of the specific activity of tritiated compounds on paper chromatograms using an automatic scanning device," *Intern. J. Appl. Radiation Isotopes* **10**, 79–85 (1961).

38. Carr, T. E. F., and B. J. Parsons "A method for the assay of calcium-45 by liquid scintillation counting," *Intern. J. Appl. Radiation Isotopes* **13**, 57–62 (1962).

39. Chen, Philip S., Jr. "Liquid scintillation counting of C^{14} and H^3 in plasma and serum," *Proc. Soc. Exp. Biol. Med.* **98**, 546–547 (1958).

40. Chiriboga, J. "Radiometric analysis of metals using chelates labeled with carbon-14 and liquid scintillation counting procedures," *Anal. Chem.* **34**, 1843 (1962).

41. Chiriboga, J., and D. N. Roy. "Rapid method for determination of decarboxylation of compounds labeled with carbon-14," *Nature* **193**, 684–685 (1962).

42. Cluley, H. J. "Suspension scintillation counting of carbon-14 barium carbonate," *Analyst* **87**, 170–177 (1962).

43. Comar, C. L. *Radioisotopes in Biology and Agriculture.* New York: McGraw-Hill, 1955. An excellent discussion of the conversion of biological samples to proper counting form.

44. Cowan, C. L., Jr., *et al.* "Detection of the free neutrino: A confirmation," *Science* **124**, 103–104 (1956).

45. Cuppy, Dianna, and Lamar Crevasse. "An assembly for $C^{14}O_2$ collection in metabolic studies for liquid scintillation counting," *Anal. Biochem.* **5**, 462–463 (1963).

46. Daub, Guido H., F. Newton Hayes, and Elizabeth Sullivan (Eds.). *Proceedings of the University of New Mexico Conference on Organic Scintillation Detectors, Albuquerque, 1960.* U.S. Government Printing Office (TID-7612), Washington, D.C., 1961.

47. Davidson, Eugene A. "Techniques for paper strip counting in a scintillation spectrometer." Rev. ed. *Packard Instrument Co. Tech. Bull.* No. 4. Downers Grove, Ill., 1962.

48. Davidson, J. D., V. T. Oliverio, and J. I. Peterson. "Combustion of samples for liquid scintillation counting." In E. D. Bransome, Jr. (Ed.), *The Current Status of Liquid Scintillation Counting.* New York and London: Grune and Stratton, 1970, pp. 222–235.

49. Dern, Raymond J., and Willie Lee Hart. "Studies with doubly labeled iron. I. Simultaneous liquid scintillation counting of isotopes Fe^{55} and Fe^{59} as ferrous perchlorate." *J. Lab. Clin. Med.* **57**, 322–330 (1961).

50. Dern, Raymond J., and Willie Lee Hart. "Studies with doubly labeled iron. II. Separation of iron from blood samples and preparation of ferrous perchlorate for liquid scintillation counting." *J. Lab. Clin. Med.* **57**, 460–467 (1961).

51. Dobbs, Horace E. "Oxygen flask method for the assay of tritium-, carbon-14-, and sulfur-35-labeled compounds," *Anal. Chem.* **35**, 783–786 (1963).

52. Dobbs, Horace E. "Determination of carbon-14 in halogenic compounds by an oxygen flask method," *Anal. Chem.* **36**, 687–689 (1964).

53. Erdtmann, G., and G. Herrmann. "Über die Zählung von Radioisotopen metallischer Elemente in flüssigen Szintillatoren," *Z. Elektrochem.* **64**, 1092–1098 (1960).

54. Erdtmann, G., and G. Hermann. "Wasseureiche emulsionen zu szintillation wäbrigen lösunger von *β*-strahlen," *Radiochimica Acta* **1**, 98–103 (1963).

55. Fales, Henry M. "Discoloration of samples for liquid scintillation counting," *Atomlight* (New England Nuclear Corp.), **25**, 8 (January 1963).

56. Feinendegen, Ludwig E. *Tritium-labeled Molecules in Biology and Medicine.* New York: Academic, 1967. A comprehensive discussion of sample preparation of tritium-containing compounds.

57. Flynn, K. F., and L. E. Glendenin. "Half-life and beta spectrum of Rb^{87}," *Phys. Rev.* **116**, 744–748 (1959).

58. Foreman, H., and M. B. Roberts. "Determination of $strontium^{90}$ in bone." In *Biological and Medical Research Group* (H-4 of the Health Division, Los Alamos Scientific Laboratory—semiannual report). U.S. Atomic Energy Commission. LAMS-2455, January-June 1960 pp. 61–70.

59. Fox, B. W. "Liquid scintillation counting of aqueous H^3 and C^{14}-protein solutions at room temperature," *Intern. J. Appl. Radiation Isotopes* **18**, 223–230 (1968).

60. Frederickson, Donald S., and Katsuto Ono. "An improved technique for assay of $C^{14}O_2$ in expired air using the liquid scintillation counter," *J. Lab. Clin. Med.* **51**, 147–151 (1958).

61. Funt, B. Lionel. "Scintillating gels," *Nucleonics* **14**(8), 83–84 (1956).

62. Funt, B. Lionel, and Arlene Hetherington. "Suspension counting of carbon-14 in scintillating gels," *Science* **125**, 986–987 (1957).

63. Funt, B. Lionel, and Arlene Hetherington. "Scintillation counting of beta activity on filter paper," *Science* **131**, 1608–1609 (1960).

64. Funt, B. Lionel. "Scintillation counting with organic phosphors," *Canad. J. Chem.* **39**, 711–716 (1961).

65. FURLONG, N. B. "Liquid scintillation counting of samples in solid supports." In E. D. Bransome, Jr. (Ed.), *The Current Status of Liquid Scintillation Counting.* New York and London: Grune and Stratton, 1970, pp. 201–206.

66. FURST, MILTON, and HARTMUT KALLMAN. "Enhancement of fluorescence in solutions under high-energy irradiation," *Phys. Rev.* **97**, 583–587 (1955).

67. GEIGER, JOHN W., and LEMUEL D. WRIGHT. "Liquid scintillation counting of radioautograms," *Biochem. Biophys. Res. Com.* **2**, 282–283 (1960).

68. GIBBS, JAMES A. "Liquid scintillation counting of natural radiocarbon." Rev. ed. *Packard Instrument Co. Tech. Bull. No.* **8**, Downers Grove, Ill., 1962.

69. GJONE, EGIL, HUGH G. VANCE, and DAVID ALAN TURNER. "Direct liquid scintillation counting of plasma and tissues," *Intern. J. Appl. Radiation Isotopes* **8**, 95–97 (1960).

70. GLEIT, C. E., and J. DUMOT. "Liquid scintillation counting of nickel-63," *Intern. J. Appl. Radiation Isotopes* **12**, 66 (1961).

71. GLEIT, C. E., and W. D. HOLLAND. "Retention of radioactive tracers in dry ashing of blood," *Intern. J. Appl. Radiation Isotopes* **13**, 307–308 (1962).

72. GLENDENIN, L. E. "Present status of the decay constants," *Ann. N.Y. Acad. Sci.* **91**, 166–180 (1961).

73. GORDON, B. E., H. R. LUKENS, JR., and W. TEN HOVE. "Liquid scintillation counting of H_2S^{35}," *Intern. J. Appl. Radiation Isotopes* **12**, 145–146 (1961).

74. GORDON, B. E., and R. M. CURTIS. "The anti-quench shift in liquid scintillation counting," *Anal. Chem.* **40**, 1486–1493 (1968).

75. GORDON, CHARLES F., and ARTHUR L. WOLFE. "Liquid scintillation counting of aqueous samples," *Anal. Chem.* 32, 574 (1960).

76. GREENE, R. C., M. S. PATTERSON, and A. H. ISTES. "Use of alkylphenol surfactants for liquid scintillation counting of aqueous tritium samples," *Anal. Chem.* **40**, 2035–2037 (1968).

77. GREENE, R. C. "Heterogeneous systems: Suspensions." In E. D. Bransome, Jr. (Ed,.) *The Current Status of Liquid Scintillation Counting.* New York and London: Grune and Stratton, 1970, pp. 189–200.

78. HANSEN, D. L., and E. T. BUSH. "Improved solubilization procedures for liquid scintillation counting of biological materials," *Anal. Biochem.* **18**, 320–332 (1967).

79. HASH, JOHN H. "Determination of tritium in whole cells and cellular fractions of *Bacillus megaterium* using liquid scintillation techniques," *Anal. Biochem.* **4**, 257–267 (1962).

80. HAYES, F. NEWTON, BETTY S. ROGERS, and PHYLLIS C. SANDERS. "Importance of solvent in liquid scintillators," *Nucleonics* **13**(1), 46–48 (1955).

81. HAYES, F. NEWTON, ERNEST C. ANDERSON, and JAMES R. ARNOLD. "Liquid scintillation counting of natural radiocarbon." In *Proc. Intern. Conf. Peaceful Uses of Atomic Energy, Geneva*, **1955**. New York: United Nations, 1956, pp. 188–192.

82. HAYES, F. NEWTON, DONALD G. OTT, and VERNON N. KERR. "Liquid scintillators. II. Relative pulse height comparisons of secondary solutes." *Nucleonics* **14**(1), 42–45 (1956).

83. HAYES, F. NEWTON, BETTY S. ROGERS, and WRIGHT H. LANGHAM. "Counting suspensions in liquid scintillators," *Nucleonics* **14**(3), 48–51 (1956).

84. HAYES, F. NEWTON, ELIZABETH HANSBURY, and V. N. KERR. "Contemporary carbon-14: The *p*-cymene method," *Anal. Chem.* **32**, 617–620 (1960).

85. HELF, SAMUEL, et al. "Radioassay of tagged sulfate impurity in cellulose nitrate," *Anal. Chem.* **28**, 1465–1468 (1956).

86. HELF, SAMUEL, and CECIL WHITE. "Liquid scintillation counting of carbon-14-labeled organic nitrocompounds," *Anal. Chem.* **29**, 13–16 (1957).

87. HELF, SAMUEL, C. G. WHITE, and R. N. SHELLEY. "Radioassay of finely divided solids by suspension in a gel scintillator," *Anal. Chem.* **32**, 238–241 (1960).

88. HERBERG, R. J. "Phosphorescence in liquid scintillation counting of proteins," *Science* **128**, 199–200 (1958).

89. HERBERG, R. J. "Determination of carbon-14 and tritium in blood and other whole tissues," *Anal. Chem.* **32**, 42–46 (1960).

90. HORROCKS, D. L., and M. H. STUDIER. "Low-level plutonium-241 analysis by liquid scintillation techniques," *Anal. Chem.* **30**, 1747–1750 (1958).

91. HORROCKS, D. L., and M. H. STUDIER. "Determination of the absolute disintegration rates of low energy beta emitters in a liquid scintillation spectrometer," *Anal. Chem.* **33**, 615–620 (1961).

92. HORROCKS, D. L. "Liquid scintillation counting of inorganic radioactive nuclides." Rev. ed. *Packard Instrument Co. Tech, Bull. No.* **2**. Downers Grove, Ill., 1962.

93. HORROCKS, D. L. "Liquid scintillation counting," *Surv. Prog. Chem.* **5**, 185–235 (1969).

94. How to prepare radioactive samples for counting on planchets. Part I. *Nuclear-Chicago Tech. Bull. No.* **7**. Des Plaines, Ill., 1961.

95. How to prepare radioactive samples for counting on planchets. Part 2. *Nuclear-Chicago Tech. Bull. No. 7B*. Des Plaines, Ill., 1961.

96. How to use radioactivity in paper chromatography. *Nuclear-Chicago Tech. Bull. No. 4*. Des Plaines, Ill., 1959.

97. How to use radioisotopes with thin-layer chromatography. *Nuclear-Chicago Tech. Bull. No. 16*. Des Plaines, Ill., 1963.

98. ISBELL, HORACE S., HARRIET L. FRUSH, and RUTH A. PETERSON. "Tritium-labeled compounds. I. Radioassay of tritium-labeled compounds in 'infinitely thick' films with a windowless, gas-flow, proportional counter." *J. Res. Nat. Bur. Standards* **63A**, 171–175 (1959).

99. ISBELL, HORACE S., and JOSEPH D. MOYER. "Tritium-labeled compounds. II. General-purpose apparatus, and procedures for the prepation, analysis and use of tritium oxide and tritium-labeled lithium borohydride." *J. Res. Nat. Bur. Standards* **63A**, 177–183 (1959).

100. JACOBSON, H. I., *et al.* "Determination of tritium in biological material," *Arch. Biochem. Biophys.* **86**, 89–93 (1960).

101. JEFFAY, HENRY, FUNSO O. OLUBAJO, and WILLIAM R. JEWELL. "Determination of radioactive sulfur in biological materials," *Anal. Chem.* **32**, 306–308 (1960).

102. JEFFAY, HENRY, and JULIAN ALVAREZ. "Liquid scintillation counting of carbon-14: Use of ethanolamine-ethylene glycol monomethyl ether-toluene." *Anal. Chem.* **33**, 612–615 (1961).

103. JEFFAY, HENRY, and JULIAN ALVAREZ. "Measurement of C^{14} and S^{35} in a single sample." *Anal. Biochem.* **2**, 506–508 (1961).

104. JEFFAY, HENRY. "Oxidation techniques for preparation of liquid scintillation samples." *Packard Instrument Co. Tech. Bull. No.* **10**. Downers Grove, Ill., 1962.

105. JERVIS, R. E. "The use of molecular filter membrane in mounting and assaying of radioactive precipitates," *Talanta* **2**, 89–91 (1959).

106. KAARTINEN, NIILO. "A new oxidation method for the preparation of liquid scintillation samples." *Packard Instrument Co. Tech. Bull. No. 18*. Downers Grove, Ill., 1969.

107. KALBERER, F., and J. RUTSCHMANN. "Eine Schnellmethode zur Bestimmung von Tritium, Radiokohlenstoff und Radioschwefel in beliebigem organischem Probenmaterial mittels des Flüssigkeits-Scintillations-Zahlers," *Helv. Chim. Acta* **44**, 1956–1966 (1961).

108. KALLMAN, HARTMUT, and MILTON FURST. "Fluorescent liquids for scintillation counters," *Nucleonics* **8** (3), 32–39 (1951).

109. KELLY, R. G., *et al.* "Determination of C^{14} and H^3 in biological samples by Schöniger combustion and liquid scintillation techniques," *Anal. Biochem.* **2**, 267–273 (1961).

110. KINARD, FRANK E. "Liquid scintillator for the analysis of tritium in water," *Rev. Sci. Instruments* **28**, 293–294 (1957).

111. LANGHAM, W. H., *et al.* "Assay of tritium activity in body fluids with use of a liquid scintillation system," *J. Lab. Clin. Med.* **47**, 819–825 (1956).

112. LAURENCOT, H. J., and J. L. HEMPSTEAD. "Liquid scintillation counting of biological materials. I. Solubilized whole blood." In D. L. Horrocks and C. T. Peng (Eds.), *Organic Scintillators and Liquid Scintillation Counting*. New York and London: Academic, 1971, pp. 635–657.

113. LEFFINGWELL, T. P., G. S. MELVILLE, JR., and R. W. RIESS. "A semi-microtechnic for iron-59 determination in biologic systems, using beta counting in a liquid scintillator." U.S. Air Force, School of Aviation Medicine, Randolph AFB, Texas. *AF-SAM-58-93*, 1958.

114. LEFFINGWELL, T. P., R. W. RIESS, and G. S. MELVILLE, JR. "Liquid scintillator beta counting of iron-59 in clear and colored systems," *Intern. J. Appl. Radiation Isotopes* **13**, 75–86 (1962).

115. LEGÉR, CONCÉLE, and LOUIS PICHAT. "Utilisation du paraldéhyde pour incorporer de grandes quantités de carbone marqué dans un scintillateur liquide," *Compt. Rend. Hebdomadaires Séances Acad. Sci.* **244**, 190–192 (1957).

116. LEGER, C., and M. A. TAMERS. "The counting of naturally occurring radiocarbon in the form of benzene in a liquid scintillation counter," *Intern. J. Applied Radiation Isotopes* **14**, 65–70 (1963).

117. LEVIN, LESTER. "Liquid scintillation methods for measuring low level radioactivity of aqueous solutions: Determination of enriched uranium in urine," *Anal. Chem.* **34**, 1402–1406 (1962).

118. LIEBERMANN, R., and A. A. MOGHISSI. "Low-level counting by liquid scintillation. II. Applications of emulsions in tritium counting." *Intern. J. Appl. Radiation Isotopes* **21**, 319–327 (1970).

119. LINDSAY, P. A., and N. B. KURNICK. "Preparation of tissues for liquid scintillation radioactivity counting," *Intern. J. Appl. Radiation Isotopes* **20**, 97–102 (1969).

120. LOEWUS, F. A. "The use of bis-(2-alkoxyethyl) ethers as antifreeze in naphthalene-1, 4-dioxane scintillation mixtures," *Intern. J. Appl. Radiation Isotopes* **12**, 6–9 (1961).

121. LOFTFIELD, ROBERT BERNER, and ELIZABETH ANN EIGNER. "Scintillation counting of paper chromatograms," *Biochem. Biophys. Res. Com.* **3**, 72–76 (1960).

122. LUDWICK, J. D. "Liquid scintillation spectrometry for analysis of zirconium-95-niobium-95 mixtures and coincidence standardization of these isotopes," *Anal. Chem.* **32**, 607–610 (1960).

123. LUDWICK, J. D., and R. W. PERKINS. "Liquid scintillation techniques applied to counting phosphorescence emission: Measurement of trace quantities of zinc sulfide," *Anal. Chem.* **35**, 1230–1235 (1961).

124. LUDWIG H., *et al.* "Automatic direct quantitation of radioactivity on paper chromatograms," *Biochim. Biophys. Acta.* **37**, 525–527 (1960).

125. LUPICA, S. B. "Polyethoxylated nonionic surfactants in toluene for liquid scintillation counting of tritium in aqueous samples," *Intern. J. Appl. Radiation Isotopes* **21**, 487–490 (1970).

126. LUTWAK, LEO. "Estimation of radioactive calcium-45 by liquid scintillation counting," *Anal. Chem.* **31**, 340–343 (1959).

127. MCCLENDON, D., M. P. NEARY, M. GALASSI, and W. STEPHENS. "Study of the use of Bio-Solve^{TM} solubilizer with biologically significant samples." In D. L. Horrocks and C. T. Peng (Eds.), *Organic Scintillators and Liquid Scintillation Counting.* New York and London: Academic, 1971, pp. 587–598.

128. MCCREADY, C. C. "A direct-plating method for the precise assay of carbon-14 in small liquid samples," *Nature* **181**, 1406 (1958).

129. MACDONALD, A. M. G. "The oxygen flask method. A review," *Analyst* **86**, 3–12 (1961).

130. MCFARLANE, A. S., and K. MURRAY. "^{14}C and ^{3}H specific activities by bomb combustion and scintillation counting," *Anal. Biochem.* **6**, 284–287 (1963).

131. MADSEN, N. P. "Use of toluene/triton X-100 scintillation mixture for counting C^{14}-protein radioactivity," *Anal. Biochem.* **29**, 542–544 (1969).

132. MAIN, RAYMOND K., and E. RICHARD WALWICK. "A simplified quantitative assay for tritiated thymidine incorporated into deoxyribonucleic acid," *Biochem. Biophys. Res. Com.* **4**, 52–55 (1961).

133. MAKIN, D. T., and R. T. LOFBERG. "A simplified method of sample preparation for determination of tritium, carbon-14, or sulfur-35 in blood or tissue by liquid scintillation counting," *Anal. Biochem.* **16**, 500–509 (1966).

134. MAKIN, D. T., and R. T. LOFBERG. "Determination of several isotopes in tissue by wet oxidation." In E. D. Bransome, Jr. (Ed.), *The Current Status of Liquid Scintillation Counting.* New York and London: Grune and Stratton, 1970, pp. 212–221.

135. MANS, RUSTY J., and G. DAVID NOVELLI. "Measurement of the incorporation of radioactive amino acids into protein by a filter-paper disk method," *Arch. Biochem. Biophys.* **94**, 48–53 (1961).

136. MARTIN, L. E., and C. HARRISON. "The determination of ^{14}C- and tritium-labeled compounds in biological materials," *Biochem. J.* **82**, 18 p (1962).

137. MEADE, R. C., and R. A. STIGLITZ. "Improved solvent systems for liquid scintillation counting of body fluids and tissues," *Intern. J. Appl. Radiation Isotopes* **13**, 11–14 (1962).

138. MOSES, V., and K. K. LONBERG-HOLM. "A semiautomatic device for measuring radioactivity on two-dimensional paper chromatograms," *Anal. Biochem.* **5**, 11–27 (1963).

139. MOSS, G. "A simple device for the rapid routine liberation and trapping of $C^{14}O_2$ for scintillation counting," *Intern. J. Applied Radiation Isotopes* **11**, 47–48 (1961).

140. MOSTAFA, I. Y., E. PALLIN, and S. C. FANG. "Comparative studies of liquid scintillation counting of aqueous ^{14}C samples," *Anal. Biochem.* **36**, 238–243 (1970).

141. MOYER, JOSEPH D., and HORACE S. ISBELL. "Preparation and analysis of carbon-14-labeled cyanide," *Anal. Chem.* **29**, 393–396 (1957).

142. MURRAY, J. "Liquid scintillation counting of $^{14}CO_2$ in a toluene/Triton X-100 system," *Intern. J. Appl. Radiation Isotopes* **22**, 209–216 (1971).

143. NADARAJAH, A., B. LEESE, and G. I. JOPLIN. "Triton X-100 scintillant for counting calcium-45 in biological fluids," *Intern. J. Radiation Isotopes* **20**, 733–735 (1969).

144. NATHAN, DAVID G., *et al.* "The counting of barium carbonate in a liquid scintillation spectrometer," *J. Lab. Clin. Med.* **52**, 915–917 (1958).

145. NUNEZ, J., and CL. JACQUEMIN. "Comptage de radiochromatogrammes par scintillation liquide," *J. Chromatog.* **5**, 271–272 (1961).

146. O'BRIEN, R. D. "Nitric acid digestion of tissues for liquid scintillation counting," *Anal. Biochem.* **7**, 251–254 (1964).

147. OKITA, GEORGE T., JAMES SPRATT, and GEORGE V. LEROY. "Liquid-scintillation counting for assay of tritium in urine," *Nucleonics* **14**(3), 76–79 (1956).

148. OKUYAMA, TAUNEO, and YUTAKA KOBAYASHI. "Determination of diamine oxidase activity by liquid scintillation counting," *Arch. Biochem. Biophys.* **95**, 242–250 (1961).

149. OLIVERIO, VINCENT T., CHARLENE DENHAM, and JACK B. DAVIDSON. "Oxygen flask combustion in determination of C^{14} and H^3 in biological materials," *Anal. Biochem.* **4**, 188–189 (1962).

150. OPPERMAN, R. A., *et al.* "Use of tertiary alkyl primary C_{12}-C_{14} amines for the assay of $C^{14}O_2$ by liquid scintillation counting," *Intern. J. Appl. Radiation Isotopes* **7**, 38–42 (1959).

151. OSINSKI, P. A. "Detection and determination of tritium labeled compounds on paper chromatograms," *Intern. J. Appl. Radiation Isotopes* **7**, 306–310 (1960).

152. OTT, DONALD G., *et al.* "Argon treatment of liquid scintillators to eliminate oxygen quenching," *Nucleonics* **13**(5), 62 (1955).

153. OTT, DONALD G., *et al.* "Cab-O-Sil suspensions for liquid-scintillation counting," *Nucleonics* **17**(9), 106–108 (1959).

154. OVERMAN, RALPH T., and HERBERT M. CLARK. *Radioisotope Techniques.* New York: McGraw-Hill, 1960. An important elementary discussion of sample preparation techniques with extensive references to counting of gases.

155. OXBY, C. B., and P. A. KIRBY. "Measurement of calcium-45 in biological fluid," *Intern. J. Appl. Radiation Isotopes* **19**, 151–152 (1968).

156. PARKER, W., and R. FALK. "Molecular plating: A method for the electrolytic formation of thin inorganic films," *Nucl. Instr. and Meth.* **16**, 355–357 (1962).

157. PARKER, W., H. BILDSTEIN, and N. GETOFF. "Molecular plating." I. A rapid and quantitative method for electrodeposition of thorium and uranium. *Nucl. Instr. and Meth.* **26**, 55–60 (1964).

158. PARKER, W., H. BILDSTEIN, and N. GETOFF. "Molecular plating." III. The rapid preparation of radioactive reference sources. *Nucl. Instr. and Meth.* **26**, 314–316 (1964).

159. PARKER, W., and H. SLATIS. "Sample and window technique." In K. Siegbahn (Ed.), *α-, β-, and γ-ray Spectroscopy.* Amsterdam: North-Halland, 1966, Vol. 1, p. 379.

160. PASSMAN, JOHN M., NORMAN S. RADIN, and JOHN A. D. COOPER. "Liquid scintillation technique for measuring carbon-14-dioxide activity," *Anal. Chem.* **28**, 484–486 (1956).

161. PATE, B. D., and L. YAFFE. "A new material and techniques for the fabrication and measurement of very thin films for use in 4π-counting," *Can. J. Chem.* **33**, 15–23 (1955).

162. PATTERSON, M. S., and R. C. GREENE. "Measurement of low energy beta emitters in aqueous solution by liquid scintillation counting of emulsions," *Anal. Chem.* **37**, 854–857 (1965).

163. PEARCE, ELI M., *et al.* "Rapid determination of radiocarbon in animal tissues," *Anal. Chem.* **28**, 1762–1765 (1956).

164. PEETS, EDWIN A., JAMES R. FLORINI, and DONALD A. BUYSKE. "Tritium radioactivity determination of biological materials by a rapid dry combustion technique," *Anal. Chem.* **32**, 1465–1468 (1960).

165. PENG, C. T. "Quenching of fluorescence in liquid scintillation counting of labeled organic compounds," *Anal. Chem.* **32**, 1292 (1960).

166. PERRY, S. W., and G. T. WARNER. "A method of sample preparation for the estimation of Fe^{55} in whole blood by the liquid scintillation counting technique," *Intern. J. Appl. Radiation Isotopes* **14**, 397–400 (1963).

167. PETERSON, J. I. "A carbon dioxide collection accessory for the rapid combustion apparatus for preparation of biological samples for liquid scintillation analysis," *Anal. Biochem.* **31**, 204–210 (1969).

168. PETERSON, J. I., F. WAGNER, S. SIEGEL, and W. NIXON. "A system for convenient combustion preparation of tritiated biological samples for scintillation analysis," *Anal. Biochem.* **31**, 189–203 (1969).

169. PICHAT, L., J. CLEMENT, and C. BARET. "Synthèse du paráldehyde au départ de carbonate de baryum en vue du datage d'échantillons archéologiques carbonés par scintillation liquide," *Bull. Soc. Chim. France.* (**2**), 329–333 (1959).

170. PICKERING, DONALD E., HELEN L. REED, and ROBERT L. MORRIS. "Detection of calcium-45 in bone solutions," *Anal. Chem.* **32**, 1214–1215 (1960).

171. PINTER, KAROLY G., JAMES G. HAMILTON, and O. NEAL MILLER. "Liquid scintillation counting with glass fiber paper," *Anal. Biochem.* **5**, 458–461 (1963).

172. POLLAY, M., and F. A. STEVENS. "Solubilization of animal tissue for liquid scintillation counting." In E. D. Bransome, Jr. (Ed.), *The Current Status of Liquid Scintillation Counting.* New York and London: Grune and Stratton, 1970, pp. 207–211.

173. PRINGLE, R. W., W. TURCHINETZ, and B. L. FUNT. "Liquid scintillation techniques for radiocarbon dating," *Rev. Sci. Instruments* **26**, 859–865 (1955).

174. PRINGLE, R. W., *et al.* "Radiocarbon age estimates obtained by an improved liquid scintillation technique," *Science* **125**, 69–70 (1957).

175. PROCKOP, DARWIN J., and PAUL S. EBERT. "A simple method for differential assay of tritium and carbon-14 in water-soluble biological materials," *Anal. Biochem.* **6**, 263–271 (1963).

176. Quantitative radiochromatography. ("Selected scientific papers from the Rendiconti of the Instituto Superiore di Sanita," Rome, Vol. 1, Parts 2 and 3. Title taken from preface in Vol. 1, Part 2.) New York: Interscience, 1956, pp. 241–548.

177. RADIN, NORMAN S., and RAINER FRIED. "Liquid scintillation counting of sulfuric acid and other substances," *Anal. Chem.* **30**, 1926–1928 (1958).

178. RAPKIN, EDWARD. "Hydroxide of Hyamine 10-X." Rev. ed. *Packard Instrument Co. Tech. Bull. No.* **3**. Downers Grove, Ill., 1961a.

179. RAPKIN, EDWARD. "Liquid scintillation measurements of radioactivity in heterogeneous systems." Rev. ed. *Packard Instrument Co. Tech. Bull. No. 5.* Downers Grove, Ill., 1961b.

180. RAPKIN, EDWARD. "The determination of radioactivity in aqueous solutions." Rev. ed. *Packard Instrument Co. Tech. Bull. No. 6.* Downers Grove, Ill., 1961c.

181. RAPKIN, EDWARD. "Measurement of $^{14}CO_2$ by scintillation techniques." *Packard Instrument Co. Tech. Bull. No.* 7. Downers Grove, Ill., 1962.

182. RAPKIN, EDWARD. "Sample preparation for liquid scintillation counting. Part 1. Solubilization techniques." *Digitechniques No. 2.* Technical Review published by Intertechnique, 1969.

183. RAPKIN, EDWARD. "Sample preparation for liquid scintillation counting. Part 2. Solvents and scintillators." *Digitechniques No. 3.* Technical Review published by Intertechnique, 1970.

184. RAPKIN, EDWARD. "Gel and emulsion counting of aqueous solutions." *Digitechniques No. 5.* Technical Review published by Intertechnique, 1971.

185. ROBERTS, HENRY R., and FREDERICK J. CARLETON. "Determination of specific activity of carbon-14-labeled sugars on paper chromatograms using an automatic scanning device," *Anal. Chem.* **28**, 11–16 (1956).

186. RONZIO, A. R. "Metal loaded scintillator solutions, *Intern. J. Appl. Radiation Isotopes* **4**, 196–200 (1959).

187. ROTHCHILD, SEYMOUR (Ed.). *Advances in Tracer Methodology.* Vol. 1. New York: Plenum Press, 1963.

188. ROTHCHILD, SEYMOUR (Ed.). *Advances in Tracer Methodology.* Vol. 2. New York: Plenum Press, 1965.

189. ROTHCHILD, SEYMOUR (Ed.). *Advances in Tracer Methodology.* Vol. 3. New York: Plenum Press, 1966.

190. ROTHCHILD, SEYMOUR (Ed.). *Advances in Tracer Methodology.* Vol. 4. New York: Plenum Press, 1968. These volumes contain the technical reports presented at the Annual Symposia on Tracer Methodology sponsored by New England Nuclear Corp.

191. ROUCAYROL, JEAN-CLAUDE, ERICH OBERHAUSER, and RICHARD SCHUSSLER. "Liquid scintillators in filter paper—a new detector," *Nucleonics* **15**(11), 104–108 (1957).

192. RYVES, T. B. "Use of a liquid scintillator counter for beta particles," *J. Sci. Instruments* **37**, 201–203 (1960).

193. SALOMON, LOTHAR L. "Sensitive 4π detector for scanning radiochromatograms," *Science* **131**, 415–417 (1960).

194. SARNAT, MARLENE, and HENRY JEFFAY. "Determination of radioactive calcium by liquid scintillation counting," *Anal. Chem.* **34**, 643–646 (1962).

195. SCHÖNIGER, W. "Eine mikroanalytische Schnellbestimmung von Halogen in organischen Substanzen," *Mikrochim. Acta* (1), 123–129 (1955).

196. SCHRAM, E. "Flow monitoring of aqueous solutions containing weak β-emitters." In E. D. Bransome, Jr. (Ed.), *The Current Status of Liquid Scintillation Counting.* New York and London: Grune and Stratton, 1970, pp. 95–109.

197. SCHULZE, JANOS, and E. A. LONG. "A method for liquid scintillation counting utilizing ultrasonic extraction," *Anal. Biochem.* **4**, 99–102 (1962).

198. SELIGER, H. H., and B. W. AGRANOFF. "Solid scintillation counting of hydrogen-3 and carbon-14 in paper chromatograms," *Anal. Chem.* **31**, 1607–1608 (1959).

199. SELIGER, H. H. "Liquid scintillation counting of α-particles and energy resolution of the liquid scintillator for α- and β-particles," *Intern. J. Appl. Radiation Isotopes* **8**, 29–34 (1960).

200. SHAKHIDZHANYAN, L. G., *et al.* "Measurement of natural radioactivity in human organs," *Trans. Biol. Sci. Sec., Doklady, Akad. Nauk SSSR* **125**, 166–167 (1959).

201. Shapira, Jacob, and William H. Perkins. "Liquid scintillation counting of aqueous solutions of carbon-14 and tritium," *Science* **131**, 414–415 (1960).

202. Shapiro, Irwin L., and David Kritchevsky. "Radioassay of cholesterol-C^{14} digitonide," *Anal. Biochem.* **5**, 88–91 (1963).

203. Sheppard, G., and C. G. Marlow. "The simultaneous measurement of ^{51}Cr and ^{14}C by liquid scintillation counting," *Intern. J. Appl. Radiation Isotopes* **22**, 125–127 (1971).

204. Sheppard, Herbert, and Waldtraut Rodegker. "Determination of H^3 and C^{14} in biological materials using oxygen bomb combustion, "*Anal. Biochem.* **4**, 246–251 (1962).

205. Sher, D. W., N. Kaartinen, L. J. Everett, and V. Justes, Jr. "Preparing samples for liquid scintillation counting with the Packard sample oxidizer." In D. L. Horrocks and C. T. Peng (Eds.), *Organic Scintillators and Liquid Scintillation Counting*. New York and London: Academic, 1971, pp. 849–868.

206. Sherman, John R. "Rapid enzyme assay technique utilizing radioactive substrate, ion-exchange paper, and liquid scintillation counting," *Anal. Biochem.* **5**, 548–554 (1963).

207. Shipotofsky, S. H. "A simple sensitive Geiger counter for scanning chromatograms (A simple radiochromatogram scanner)," *Anal. Biochem.* **7**, 233–239 (1964).

208. Shneour, E. A., S. Aronoff, and M. R. Kirk. "Liquid scintillation counting of solutions containing carotenoids and chlorophylls," *Intern. J. Appl. Radiation Isotopes* **13**, 623–627 (1962).

209. Simpson, J. D., and J. R. Greening. "Preparation of tritiated water samples by distillation," *Nature* **186**, 467–468 (1960).

210. Snyder, Fred, and Paul Godfrey. "Collecting $C^{14}O_2$ in a Warburg flask for subsequent scintillation counting," *J. Lipid Res.* **2**, 195 (1961).

211. Snyder, Fred, and Nelson Stephens. "Quantitative carbon-14 and tritium assay of thin-layer chromatography plates," *Anal. Biochem.* **4**, 128–131 (1962).

212. Spencer, R. B., and M. A. Baneyi. "Rapid preparation of ^{14}C-blood plasma for liquid scintillation counting in dimethyl sulfoxide-thixotropic gel," *Intern. J. Appl. Radiation Isotopes* **21**, 431 (1970).

213. Starik, I. E., *et al.* "Liquid scintillators for radiocarbon dating in archaeology," *Intern. J. Appl. Radiation Isotopes* **9**, 193–194 (1960).

214. Steel, G. G. "A simple method of estimating the tritium content of biological samples," *Intern. J. Appl. Radiation Isotopes* **9**, 94–99 (1960).

215. Steele, Robert, William Bernstein, and Clara Bjerknes. "Single phototube liquid scintillation counting of C^{14}: Application to an easily isolated derivative of blood glucose," *J. Appl. Physiol.* **10**, 319–326 (1957).

216. Steenberg, K., and A. A. Benson. "A scintillation counter for soft-β paper chromatograms," *Nucleonics* **14**(12), 40–43 (1956).

217. Steyn, J. "Absolute standardization of beta-emitting isotopes with a liquid scintillation counter," *Proc. Phys. Soc.* **69A**, 865–867 (1956).

218. Steyn, J., and F. J. Haasbroek. "The application of internal liquid scintillation counting to a 4π beta-gamma coincidence method for the absolute standardization of radioactive nuclides." In *Proceedings of the Second UN International Conference on Peaceful Uses Atomic Energy*. Geneva: United Nations, 1958, Vol. 21, pp. 95–100.

219. Stitch, S. R. "Liquid scintillation-counting for (^{14}C) steroids," *Biochem. J.* **73**, 287–292 (1959).

220. Stitch, S. R., and R. E. Oakey. "The evaluation of radiochromatograms by liquid scintillation counting and photography," *Biochem. J.* **81**, 12P–13P (1961).

221. Svendsen, Reiner. "A method by which radioactive material may be transferred from a paper chromatogram to a planchette," *Intern. J. Appl. Radiation Isotopes* **5**, 146–147 (1959).

222. Takahashi, Hajime, Toshie Hattori, and Bunji Maruo. "Liquid scintillation counting of C^{14} paper chromatograms," *Anal. Biochem.* **2**, 447–462 (1961).

223. Takesue, E. I., *et al.* "A radiometric assay of tritiated tetracycline in serum and plasma of laboratory animals," *Intern. J. Appl. Radiation Isotopes* **8**, 52–59 (1960).

224. Tamers, M. A. "Carbon-14 dating with the liquid scintillation counter: Total synthesis of the benzene solvent," *Science* **132**, 668–669 (1960).

225. Tamers, M. A., and R. Bibron. "Benzene method measures tritium in rain without isotope enrichment," *Nucleonics* **21**(6), 90–94 (1963).

226. Toporek, Milton. "Liquid scintillation counting of C^{14} plasma proteins using a standard quenching curve," *Intern. J. Appl. Radiation Isotopes* **8**, 229–230 (1960).

227. Turner, J. C. "Triton X-100 scintillant for carbon-14-labelled materials," *Intern. J. Appl. Radiation Isotopes* **19**, 557–564 (1968).

228. Turner, J. C. "Tritium counting with the Triton X-100 scintillant," *Intern. J. Appl. Radiation Isotopes* **20**, 499–505 (1969).

229. Tyler, T. R., A. R. Reich, and C. Rosenblum. "An analytical study of the Peterson combustion apparatus." In D. L. Horrocks and C. T. Peng (Eds.), *Organic Scintillators and Liquid Scintillation Counting.* New York and London: Academic, 1971, pp. 869–877.

230. Van der Laarse, J. D. "Experience with emulsion counting of tritium," *Intern. J. Appl. Radiation Isotopes* **18**, 485–491 (1967).

231. Vaughan, Martha, Daniel Steinberg, and Jane Logan. "Liquid scintillation counting of C^{14}- and H^3-labeled amino acids and proteins," *Science* **126**, 446–447 (1957).

232. Von Schuching, Susanne, and Carl W. Karickhoff. "Low-level carbon-14 determination by improved Schöniger combustion and ionization chamber," *Anal. Biochem.* **5**, 93–98 (1963).

233. Walker, L. A., and R. Lougheed. "A simple method for the assay of carbon-14 in compounds or mixtures," *Intern. J. Appl. Radiation Isotopes* **13**, 95–97 (1962).

234. Wang, C. H. "Radiorespirometry." In David Glick (Ed.), *Methods of Biochemical Analysis.* New York: Interscience, 1967, Vol. 15, pp. 311–366.

235. Wang, C. H., and D. E. Jones. "Liquid scintillation counting of paper chromatograms," *Biochem. Biophys. Res. Com.* **1**, 203–205 (1959).

236. Weg, M. W. "Beta-scintillation counting of radioactive tracers insoluble in toluene," *Nature* **194**, 180–181 (1962).

237. Werbin, Harold, I. L. Chaikoff, and Miles R. Imada. "Rapid sensitive method for determining H^3-water in body fluids by liquid scintillation spectrometry," *Proc. Soc. Exp. Biol. Med.* **103**, 8–12 (1959).

238. White, C. G., and Samuel Helf. "Suspension counting in scintillation gels," *Nucleonics* **14**(10), 46–48 (1956).

239. White, D. R. "An assessment of the efficiencies and costs of liquid scintillation mixer for aqueous tritium sample," *Intern. J. Appl. Radiation Isotopes* **19**, 49–61 (1968).

240. Whyman, A. E. "Measurement of tritium in neat plasma and urine with a toluene/Triton X-100/Hyamine 10-X scintillant," *Intern. J. Appl. Radiation Isotopes* **21**, 81–86 (1970).

241. WILLENBRINK, J. "On the quantitative assay of radiochromatograms by liquid scintillation counting," *Intern. J. Appl. Radiation Isotopes* **14**, 237–238 (1963).

242. WILLIAMS, P. H. "Liquid scintillation counting of tritium in water with Triton emulsion systems," *Intern. J. Appl. Radiation Isotopes* **19**, 377–383 (1968).

243. WILLIAMS, P. H., and T. FLORKOWSKI. *Radioactive Dating and Methods of Low-Level Counting.* Vienna: International Atomic Energy Agency, 1967, p. 703.

244. WILSON, A. T. "Detection of tritium on paper chromatograms," *Biochim. Biophys. Acta* **40**, 522–526 (1960).

245. WILZBACH, K. E., LOUIS KAPLAN, and W. G. BROWN. "The preparation of gas for assay of tritium in organic compounds," *Science* **118**, 522–523 (1953).

246. WILZBACH, K. E., and W. Y. SYKES. "Determination of isotopic carbon in organic compounds," *Science* **120**, 494–496 (1954).

247. WOELLER, FRITZ H. "Liquid scintillation counting of $C^{14}O_2$ with phenethylamine," *Anal. Biochem.* **2**, 508–511 (1961).

248. WRIGHT, P. M., E. P. STEINBERG, and L. E. GLENDENIN. "Half-life of samarium-147," *Phys. Rev.* **123**, 205–208 (1961).

249. ZIEGLER, C. A., D. J. CHLECK, and J. BRINKERHOFF. "Radioassay of low specific activity tritiated water by improved liquid scintillation techniques," *Anal. Chem.* **29**, 1774–1776 (1957).

250. ZIMMERMAN, MARTIN E. (Ed.). *Handbook for Preparation of Samples for Liquid Scintillation Counting.* Des Plaines, Ill.: Nuclear-Chicago Corp., 1967. A very useful, loose-leaf collection of liquid scintillation recipes for a wide variety of sample types, each with a reference to the original source.

12

Nuclear Statistics

A. INTRODUCTION

Radioactive decay is a random process. The number of nuclei in a sample of radioactive material that decay in any time period is not a fixed number but will differ, in general, for various time periods. This point can be readily demonstrated by making repeated measurements of the activity of a long-lived radionuclide, each for the same time duration. The results of such an experiment might be as shown in Table 12-1. Note that in these measurements there is a large range of activity values with a clustering near the center of the range.

Given this typical situation, what do we call the average count rate? How do we compare counts from different samples? The answers to these and many other questions can be obtained from a study of the statistics of nuclear processes. Consideration of the statistical nature of radioactive decay, as undertaken in this chapter, will allow us to extract the maximum amount of correct information from radioactivity measurements.

B. BASIC STATISTICAL DISTRIBUTIONS FOR RADIOACTIVE DECAY

Suppose that we repeated the experiment that led to the data of Table 12-1 a large number of times and that we plotted the number of times a given count rate occurred versus the count rate. Such a plot is shown in Figure 12-1. The curve represents the distribution of count rate values obtained in

TABLE 12-1

Typical Sequence of Counts of a Long-Lived Sample (^{170}Tm)*

Measurement Number	Number of Counts in 0.1-min Interval, X_i	Deviation of Individual Count from Mean $X_i - \overline{X}$	(Deviation from Mean)^2 $(X_i - \overline{X})^2$
1	1880	−18	324
2	1887	−11	121
3	1915	17	289
4	1851	−47	2209
5	1874	−24	576
6	1853	−45	2025
7	1931	33	1089
8	1866	−32	1024
9	1980	82	6724
10	1893	−5	25
11	1976	78	6084
12	1876	−22	484
13	1901	3	9
14	1979	81	6561
15	1836	−62	3844
16	1832	−66	4356
17	1930	32	1024
18	1917	19	361
19	1899	1	1
20	1890	−8	64

Sum of squared deviations $= (X_i - \overline{X})^2 = 37{,}194$

$$\overline{X} = \frac{\Sigma X_i}{N} = \frac{37{,}966}{20} = 1898$$

$$\sigma = \left[\frac{\Sigma (X_i - \overline{X})^2}{N-1}\right]^{1/2} = \left(\frac{37{,}194}{19}\right)^{1/2} = 44$$

*We are indebted to Professor R. A. Schmitt for the data contained in this table.

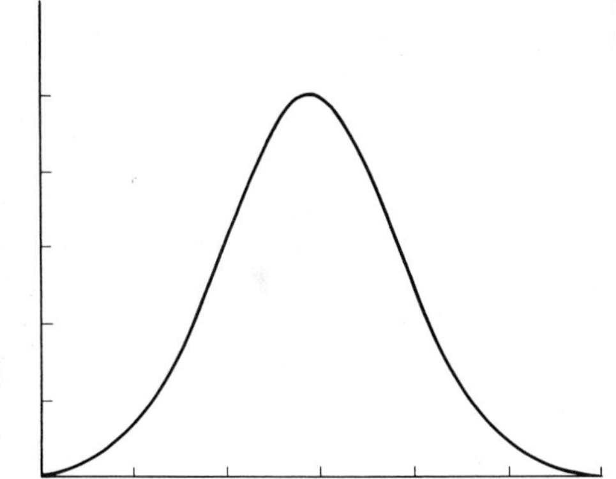

Fig. 12-1. A plot showing the frequency of occurrence of a given count rate value in a series of count rate measurements with a long-lived sample.

successive counts of a long-lived radioactive source. It is called the *normal* or *gaussian distribution** and is represented mathematically by

$$P_x = \frac{1}{\sigma\sqrt{2\pi}} \exp\left[-\frac{(x-m)^2}{2\sigma^2}\right] \tag{12-1}$$

*The actual correct distribution is the binomial distribution. The normal distribution is, however, an excellent approximation to the binomial distribution when the sample half-life is long compared to the counting time t and when the number of nuclei present is large.

where σ is a parameter, known as the standard deviation, characterizing the breadth of the distribution (see Section C for a detailed discussion of σ) and m is the true mean or average number of decays occurring in time t.*

Since the normal distribution is commonly used to interpret radiotracer counting experiments, it is important to understand some of the basic properties of the normal distribution. Figure 12-2 shows a plot of the normal distribution curve with various features outlined.

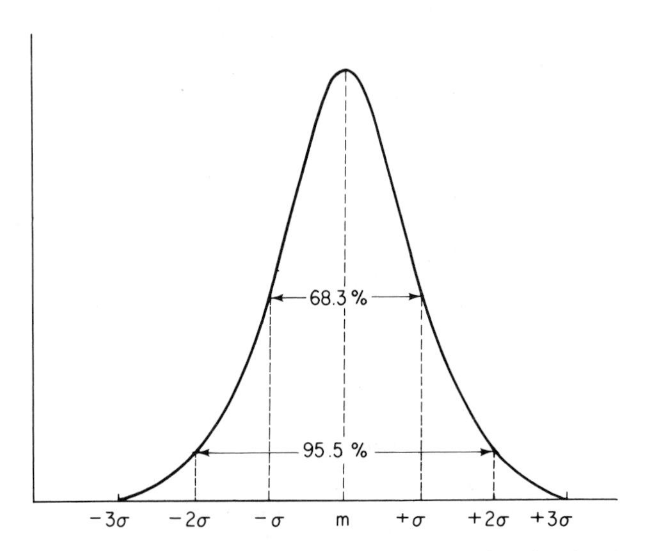

Fig. 12-2. A plot of the normal distribution function showing the 1σ and 2σ points and the position of the data mean value, m.

As noted, the standard deviation σ gives information about the width of the distribution function. Note also, as shown in Figure 12-2, that 68.3 % of the measured values of activity lies within $\pm 1\sigma$ of the mean value, m. Furthermore, 95.5 % of all measurements lies within $\pm 2\sigma$ of the mean, m, while 99.7 % of the measurements lies within $\pm 3\sigma$ of the mean value. Thus if the standard deviation of a normal distribution (see Section C on how to estimate it) is known, we are able to make a statement concerning the true mean value of the distribution from a single measurement. For example, if we know that $\sigma = 10$, and if we measure a count rate of 100, we can say,

*Consider N measurements where the number of counts in each measurement is x_i ($i = 1, \ldots, N$). The sample mean $\bar{x}$ is the arithmetic average of the N counts; that is,

$$\bar{x} = \sum_{i=1}^{N} \frac{x_i}{N}$$

As the number of measurements becomes infinite, $\bar{x}$ approaches m, the true mean of the data. In other words, we measure $\bar{x}$ for many sets of samples and we use it as an approximation of m, the true mean of the distribution if an infinite number of measurements were made.

with a 68.3% chance of being correct, that the mean count rate is between $100 - 10 = 90$ and $100 + 10 = 110$. With 95.5% certainty, we can say that m lies between $100 - 20 = 80$ and $100 + 20 = 120$. Thus the specification of activity values is a "probabilistic event"; that is, we specify that the mean activity lies somewhere in the range from (measured value $-n\sigma$) to (measured value $+n\sigma$), where n is some number between 0 to 3 that specifies the chance that the statement is correct. If $n = 1$, the statement has a 68.3 % chance of being correct; if $n = 2$, a 95.5% chance, and so on. Table 12-2 shows various n values and the associated chance of being correct when one says that the mean count rate lies in the range (measured value $\pm n\sigma$). The upper and lower limits on a quoted measurement, x, $(x + n\sigma, x - n\sigma)$ are called the *confidence limits* of the measurement or, more specifically, the C% confidence limits, where C is the value associated with n in Table 12-2. For example, $x + \sigma$ and $x - \sigma$ are referred to as the 68.3% confidence limits.

TABLE 12-2

n Value	Chance that Mean Value Lies in Range (measured value $\pm n\sigma$) C
0.6745	50%
1.0000	68.3
1.6449	90
1.9600	95
2.0000	95.5
2.5758	99
3.0000	99.7

It is standard practice to quote results of radiotracer experiments as $x \pm \sigma$—that is, using the 68.3% confidence limits. It should be noted, however, that doing so means that one is wrong 31.7% of the time; that is, the mean count rate will lie outside of $x \pm \sigma$. If this risk is not acceptable, one should pick a higher level of confidence, such as the 95% confidence limits, and quote all activity values as $x \pm 1.96\sigma$.

In addition to considering how the values of the count rate are distributed about the mean value in a radiotracer experiment, there is the question of what the distribution of time intervals between successive counts is. We know that the average time between counts is 1/count rate, but the counts may not all occur with equal time intervals between them. In fact, the distribution of time intervals between successive counts in random processes (such as radioactive decay) is given by the *interval distribution* which has the form shown in Figure 12-3. Note that, according to the figure, small time intervals between events have a higher probability of occurrence than larger or average time intervals. In other words, one should expect counts to occur in "bunches" in the detection system and not be uniformly spaced in time.

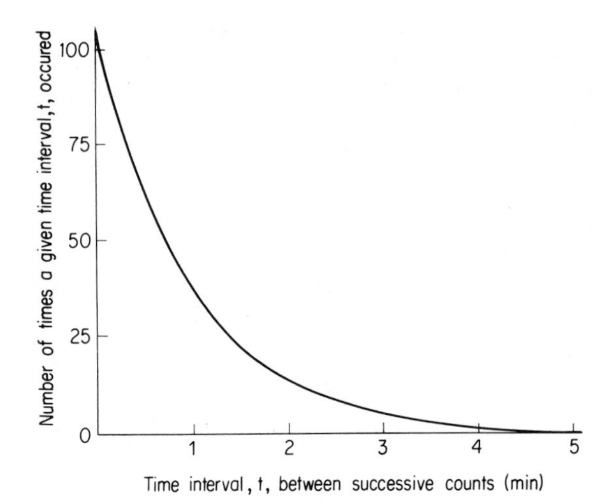

Fig. 12-3. A plot of the frequency of occurrence of a time interval t between successive counts, where the average count rate is 1 cpm.

Many people doing low count-rate experiments for the first time have confused this feature of the interval distribution with instrument malfunction or some startling new physical phenomenon.

C. PARAMETERS THAT DESCRIBE STATISTICAL DISTRIBUTION FUNCTIONS

Various parameters can be used to characterize the normal distribution of count values obtained in a sequence of measurements. In this section we discuss how to estimate these quantities and use them in characterizing radioactivity measurements. The first quantity that is used to characterize a set of N measurements of the count rate is the average or mean value, $\bar{x}$. Formally,

$$\bar{x} = \frac{\sum_{i=1}^{N} X_i}{N} \tag{12-2}$$

where X_i is the count rate in the ith measurement. Applying this formula to the data of Table 12-1, we have

$$\bar{x} = \frac{\begin{matrix} 1882 + 1887 + 1915 + 1851 + 1874 + 1853 + 1931 \\ + 1866 + 1980 + 1893 + 1976 + 1876 + 1901 + 1979 \\ + 1836 + 1832 + 1930 + 1917 + 1899 + 1890 \end{matrix}}{20}$$

$$= \frac{37{,}966}{20} = 1898 \text{ cpm}$$

As shown in Section B, the standard deviation of a set of measurements, σ, expresses something of the width of the distribution of values about the mean value. In general, we say that for a set of N measurements,*

$$\sigma = \left[\frac{1}{N-1}\sum_{i=1}^{N}(x_i - \bar{x})^2\right]^{1/2} \tag{12-3}$$

Applying this formula to the data of Table 12-1, we have

$$\sigma = \left[\frac{1}{19}\left(\begin{array}{l} 324 + 121 + 289 + 2209 + 576 + 2025 \\ + 1089 + 1024 + 6724 + 25 + 6084 + 484 + 9 \\ + 6561 + 3844 + 4356 + 1024 + 361 + 1 + 64 \end{array}\right)\right]^{1/2}$$

$$= \left(\frac{37{,}194}{19}\right)^{1/2} = 44 \text{ cpm}$$

Now suppose that we were to repeat the measurements outlined in Table 12-1 N times, getting N values of the mean, $\bar{x}$ (x_i, $i = 1, 2, 3, \ldots, N$). The many values of $\bar{x}$ obtained would also be normally distributed but with a "tighter" (i.e., more clustered) distribution than that observed for x_i (see

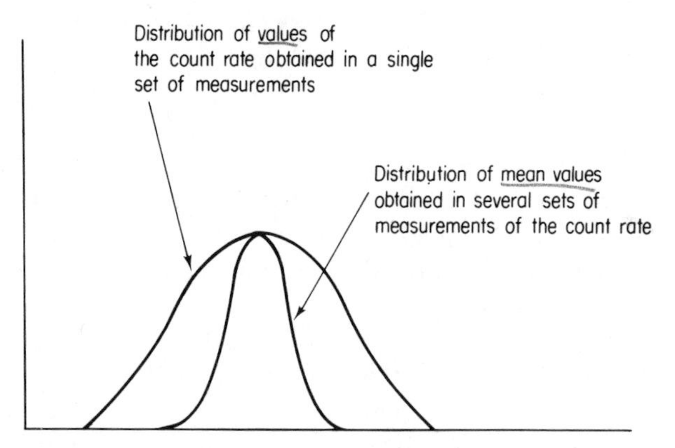

Fig. 12-4. Schematic view of distribution of values obtained in single measurements of the count rate and the distribution of values obtained in several measurements of the *mean* count rate.

Figure 12-4). We express this point formally by stating that the standard deviation of the distribution of *mean* values is given by

$$\sigma_{\bar{x}} = \left[\frac{1}{N(N-1)}\sum_{i=1}^{N}(x_i - \bar{x})^2\right]^{1/2} = \frac{\sigma}{N^{1/2}} \tag{12-4}$$

Thus for the data of Table 12-1, the mean count rate was 1898 cpm, σ was

*For purposes of calculation, it is frequently more convenient to express this equation as

$$\sigma = \left(\frac{1}{N-1}\left[\sum_{i=1}^{N} x_i^2 - N\bar{x}^2\right]\right)^{1/2}$$

44, N was 20, and the standard deviation of the mean was

$$\frac{\sigma}{N^{1/2}} = \frac{44}{20^{1/2}} = 9.8 \text{ cpm}$$

Using 68.3% confidence limits, we would quote the mean value as 1898 ± 10 cpm. In other words, we would expect that, in 68.3% of all cases, *another* measurement of $\bar{x}$ would lie in the range of 1888 to 1908.

What about the more common situation where, instead of making a set of measurements on a radioactive sample, one makes only one measurement, recording x counts. How does one proceed to estimate the standard deviation of the distribution? Knowing that the activity values are normally distributed, we can say that

$$\sigma = m^{1/2} \qquad \text{for} \quad m \gg 1 \tag{12-5}$$

That is, the standard deviation of the distribution is equal to the square root of the mean value. If we further approximate that x is our best estimate of the mean value m, we can say that

$$\sigma \approx x^{1/2} \qquad \text{for} \quad x \gg 1 \tag{12-6}$$

Thus we would say (with 68.3% confidence) that the mean count rate is in the range $x \pm \sigma$; that is, $x \pm x^{1/2}$. Applying this idea to the data of Table 12-1, we would estimate the mean value as $1898 \pm 1898^{1/2}$; that is, 1898 ± 44.

To recapitulate, two standard deviations are associated with a normal distribution of counting measurements. The first is the standard deviation of the distribution, which is given in a general form by Equation (12-3) and which can be estimated from a single measurement by using Equation (12-6). This expresses the breadth of the distribution, and *all single measurements of an activity, x, should be quoted as $x + n\sigma$, where n determines the degree of confidence in the measurement* (see Table 12-2). The second statistical parameter associated with the distribution is the standard deviation of the mean value, $\sigma_{\bar{x}}$, which is given by Equation (12-4). The standard deviation of the mean is to be used in connection with a set of activity measurements, not a single measurement, and it expresses the breadth of the distribution of mean values obtained in sets of measurements of the same quality. At the risk of being repetitious, let us repeat the most significant statement of the discussion.

From a single measurement of an activity, x, the best estimate of the mean value of the activity is to say that the mean value lies in the range from $x - n\sqrt{x}$ to $x + n\sqrt{x}$, where n is determined by using Table 12-2.

D. GENERAL RULES FOR CALCULATIONS INVOLVING NORMALLY DISTRIBUTED QUANTITIES

Consider the case where you have measured the activity of a sample (plus background) in your detector and obtained a value of 100 counts in 1 min. You would then estimate the mean counting rate as 100 ± 10 cpm. But a second measurement showed a counter background rate of 10 cpm or, more specifically, 10 ± 3.2 cpm. What is the true sample rate and its uncertainty [(sample + background) − background]? This raises the general question of calculating the uncertainty in the result of some mathematical operation on an uncertain number. If we consider two numbers and their uncertainties (standard deviations), $A \pm \sigma_a$, $B \pm \sigma_b$, we can write down, as shown in Table 12-3, some rules for the uncertainty in the result of some common mathematical operations.*

TABLE 12-3

Operation	Answer	Uncertainty
Addition	$A + B$	$(\sigma_a^2 + \sigma_b^2)^{1/2}$
Subtraction	$A - B$	$(\sigma_a^2 + \sigma_b^2)^{1/2}$
Multiplication	$A \cdot B$	$A \cdot B\left[\left(\frac{\sigma_a}{A}\right)^2 + \left(\frac{\sigma_b}{B}\right)^2\right]^{1/2}$
Division	A/B	$A/B\left[\left(\frac{\sigma_a}{A}\right)^2 + \left(\frac{\sigma_b}{B}\right)^2\right]^{1/2}$

As an example of the use of the relations outlined in Table 12-3, we would calculate that for our sample and background counting case,

$$\begin{aligned}\text{Net rate} &= (\text{sample} + \text{background}) - (\text{background}) \\ &= 100 - 10 = 90 \text{ cpm}\end{aligned}$$

$$\begin{aligned}\text{Uncertainty in net rate} &= (10^2 + 3.2^2)^{1/2} \\ &= (100 + 10)^{1/2} = 10.5 \text{ cpm}\end{aligned}$$

Thus we would quote (at the 68.3% confidence limit) a value of the net rate of 90 ± 10.5 cpm.

Using various mathematical operations on a set of counting data to extract meaningful information from it naturally raises the familiar question of *significant figures*. The measured value of any quantity is accurate only within the capabilities of the measuring system. For example, we cannot use a rough balance to weigh an object to an accuracy of 1 μg. Therefore the results of a measurement with this balance should not be quoted as 2.500000 g, because such a statement is basically a lie. It is a lie because the statement that $m = 2.500000$ g implies that we have measured the mass to be in the range between 2.499995 and 2.500004 g, when we may have only determined it to be in the range from 2.4 to 2.6 g. This entire discussion is summarized by

*Some other, less frequently used relations are as follows:

Operation	*Answer*	*Uncertainty*
Raising to a power (A^a, where a is well known)	A^a	$\simeq a\sigma_a A^{a-1}$
Exponentiation ($e^{\lambda A}$)	$e^{\lambda A}$	$\simeq e^{\lambda A}\lambda\sigma_A$
Taking a logarithm ($\ln \lambda A$)	$\ln \lambda A$	$\simeq \frac{\sigma_A}{\lambda A}$

stating that, in quoting the value of any experimental quantity, we must use the proper number of *significant figures*—that is, to make sure that the value of each digit of our number has been experimentally determined. (The last digit is always assumed to be uncertain.) Thus if our balance will weigh with an accuracy of 0.01 g, we may quote weights to three-significant-figure accuracy, such as 2.42 g. If our balance weighs to 0.001-g accuracy, we may quote results to four significant figures, such as 2.425 g. Applying this notion to radioactivity measurements, we can say that if we get a count rate of 102 cpm, we should quote it as 100 ± 10 cpm, not 102 ± 10.1 cpm. The latter number implies that the "2" of the 102 and the ".1" of the 10.1 are experimentally known, when they are not.

It is important to keep track of significant figures in common operations. In multiplication and division, we retain in the product or quotient the number of significant figures present in the least precise of the numbers being multiplied or divided.* So if we were to divide the number 8.1 by 4.596, we should quote the answer as 1.8—that is, to two-significant-figure accuracy, not as 1.7624. If we are adding and subtracting numbers, the precision is that of the least precise number in the calculation. For example, if we add $0.003 + 30 + 1.02$, the result should be quoted as 31, not 31.023.

E. NUCLEAR COUNTING STATISTICS

Now let us examine the application of some of the general principles of nuclear statistics, as outlined in Sections A to D, to specific problems involved in radioactivity measurements.

1. Average Counting Rate

Up to now we have carefully restricted our discussion of nuclear statistics to cases where 1-min counts were taken. If the number of counts recorded in 1 min was x, then the counting *rate* has been quoted as $x \pm (x)^{1/2}$ cpm. Suppose, however, that we recorded 160 counts in 5 min. What would be the standard deviation of the average counting rate (in cpm)? The best estimate of the mean number of counts in the 5-min period would by $160 \pm (160)^{1/2}$—that is, 160 ± 13 counts. The average rate would be $160/5 \pm 13/5 = 32 \pm 3$ cpm. In general, therefore, the rate R is given as

$$R = \frac{\text{number of counts recorded}}{\text{measurement time}} = \frac{x}{t}$$

*A more exact rule is to retain in the product or quotient the number of significant figures necessary to have the same relative error as the least precise of the numbers being multiplied or divided. Using this rule, $8.1/4.596 = 1.76$. That is, 8.1 is known to $\sim \pm 1\%$, as is 1.76.

The standard deviation of the rate, σ_R, is

$$\sigma_R = \frac{\sqrt{x}}{t} = \frac{\sqrt{R \cdot t}}{t} = \sqrt{\frac{R}{t}} \tag{12-7}$$

Thus for the preceding example we could have calculated directly that

$$\sigma_R = \sqrt{\frac{R}{t}} = \sqrt{\frac{32}{5}} \simeq 3$$

2. Sample and Background Counting

As the example worked in Section D illustrates, a poorly determined background counting rate can contribute significant uncertainty in the measurement of the true or net sample counting rate. This example naturally raises the question of the proper division of time in an experiment between counting the sample and counting the background. It can be shown that the lowest fractional standard deviation (σ_x/x) is obtained when the ratio of time spent counting (sample + background), t_{s+b}, to the time spent counting the background alone, t_b, is

$$\frac{t_{s+b}}{t_b} = \sqrt{\frac{R_{s+b}}{R_b}} \tag{12-8}$$

where R_{s+b} is the (sample + background) rate and R_b is the background rate. (Note that the true or net sample rate is $R_s = R_{s+G} - R_b$.) This result assumes that the total time spent counting background and (sample + background) is a fixed number. Thus if the (sample + background) rate is 40 cpm, and the background rate is 10 cpm, the ratio of (sample + background) counting time to background counting time should be

$$\sqrt{\frac{40}{10}} = \sqrt{4} = 2$$

Convenient nomographs (such as that shown in Figure 12-5) exist, which allow one to calculate quickly the optimum time spent counting sample and background in order to achieve a given percentage uncertainty in the results.

For example, consider the situation where the (sample + background) rate is known to be approximately 15 cpm, while the background rate was known to be 10 cpm. We then calculate the ratio of these rates, r, to be 1.5. Then, in order to achieve 10% uncertainty in the results, we find the intersection of the horizontal 10% line with the $r = 1.5$ diagonal line on Figure 12-5. By interpolating between the two background curves we determine that the total number of background counts should be approximately 900, while the total number of (sample + background) counts should be approximately 1600. Therefore, we should count background for 90 min. and the (sample + background) for 107 min. In examining Figure 12-5, we can also see a valuable, rather rough rule of thumb, which states that

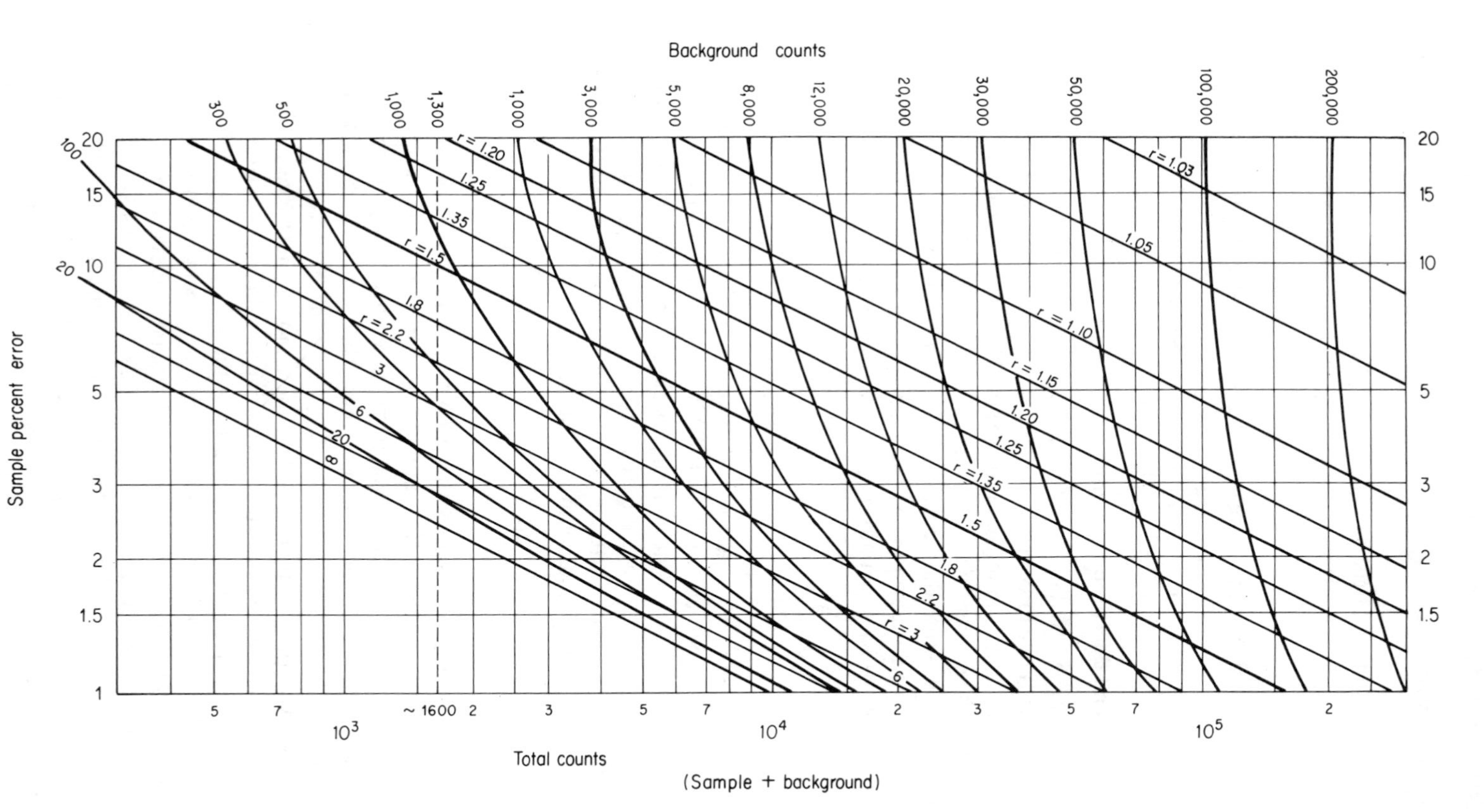

Fig. 12-5. Optimum counting time nomographs. See text for explanation. Reprinted from NUCLEONICS, July, 1951 Copyright 1951, McGraw-Hill Publishing Co., Inc.

for small counter backgrounds, a 1% error in the measured sample rate is achieved by collecting 10^4 total (sample + background) counts.

3. Optimum Choice of Detector

Frequently, we are faced with the problem of choosing the best tracer and counting technique to solve a given problem. Nuclear counting statistics can help to establish a quantitative basis for such decisions. Reynolds (6) has treated this problem and offers some significant insights into various aspects of the problem, which we shall discuss.

A useful number (*or figure of merit*) for judging tracers and counters is the ratio S/σ, where S is the sample rate and σ is the standard deviation of that rate. One tries to achieve the highest numerical value of S/σ for any situation for best performance. Clearly,

$$\frac{S}{\sigma} = \frac{S}{[(s + B) + B]^{1/2}} = \frac{S}{(S + 2B)^{1/2}} \tag{12-9}$$

where B is the background rate and $S + B$ is the sample plus background rate. If the tracer disintegration rate is D and the detector efficiency is E, we can say that $S = DE$, or

$$\frac{S}{\sigma} = \frac{S}{(S + 2B)^{1/2}} = \frac{DE}{(DE + 2B)^{1/2}} \tag{12-10}$$

Note that when $S > 10B$,

$$\frac{S}{\sigma} \cong S^{1/2} = (DE)^{1/2} \tag{12-11}$$

Also, when $S < 0.1B$,

$$\frac{S}{\sigma} \sim \frac{S}{(2B)^{1/2}} = \frac{DE}{(2B)^{1/2}} \tag{12-12}$$

If S/σ is taken as a figure of merit, note that the choice of *detectors depends on the sample disintegration rate, the detector efficiency, and the detector background.* For large disintegration rates, one chooses the detector of highest efficiency; for low disintegration rates, one chooses the detector with the highest ratio of E/B.

Consider a practical application of these ideas [taken from Reynolds (6)]. Imagine that a particular tracer experiment could be done by using ^{85}Sr and counting in a well crystal with efficiency $E = 0.35$ and background $B = 100$ cpm or by using ^{89}Sr tracer and counting in a β-counter with efficiency 0.25 and background $B = 1$ cpm. Which method should one use? The answer depends on the quantity of tracer used in the experiment. Consider two cases: (a) $D = 1000$ dpm and (b) $D = 10{,}000$ dpm. For case (a), the figure of merit of the scintillator is 14.9, whereas the figure of merit for the beta counter is 15.7. That is, the better choice would be the beta counter and the ^{89}Sr tracer.

For case (b), the figure of merit of the scintillator is ~59.2, whereas the figure of merit of the beta counter is ~50. In other words, the scintillator and ^{85}Sr tracer are the method of choice.

4. Weighted Averages

Often we wish to compute the average of two numbers, x_1 and x_2, both of which have an uncertainty denoted by their standard deviations σ_1 and σ_2, respectively. The best average of these two numbers is not $(x_1 + x_2)/2$ but weighted average $\bar{x}$, given by

$$\bar{x} = \left(\frac{x_1}{\sigma_1^2} + \frac{x_2}{\sigma_2^2}\right) \Big/ \left(\frac{1}{\sigma_1^2} + \frac{1}{\sigma_2^2}\right)$$

$$\bar{x} = \frac{x_1 + wx_2}{1 + w} \tag{12-13}$$

where

$$w = \left(\frac{\sigma_1}{\sigma_2}\right)^2 \tag{12-14}$$

In short, each number is weighted by the inverse of its standard deviation squared. For the weighted average of N values, x_1, with standard deviation, σ_i, we have

$$\bar{x} = \sum_{i=1}^{N} \left(\frac{x_i}{\sigma_i^2}\right) \Big/ \sum_{i=1}^{N} \left(\frac{1}{\sigma_i^2}\right)$$

The uncertainty or standard deviation of $\bar{x}$ is given by

$$\sigma_{\bar{x}} = \left(\frac{1}{\sum_{i=1}^{N} (1/\sigma_i^2)}\right)^{1/2}$$

which, for the case where $N = 2$, gives

$$\sigma_{\bar{x}} = \left[\frac{\sigma_1^2 + w^2\sigma^2}{(1 + w)^2}\right]^{1/2} \tag{12-15}$$

For example, suppose that we make two independent measurements of a tracer activity, obtaining results of 35 ± 10 cpm and 46 ± 2 cpm. The weighted average of the two measurements is

$$w = \left(\frac{10}{2}\right)^2 = 25$$

$$\bar{x} = \frac{(35) + (25)(46)}{(1 + 25)} = \frac{1185}{26} \simeq 46 \text{ cpm}$$

The standard deviation of the weighted average is

$$\sigma_{\bar{x}} = \left(\frac{100 + (625)(4)}{26^2}\right)^{1/2} \simeq 2.0$$

Thus we would say that the average rate was 46 ± 2 cpm. Note that because the first measurement was so uncertain relative to the second measurement, it did not affect the average significantly.

5. Rejection of Abnormal Data

In our discussions thus far we have only considered uncertainty in the experimental data due to the randomness of radioactive decay. Other factors, such as variation in sample preparation procedures and measuring errors, will also contribute to the distribution of values obtained in a radioactivity measurement. As a result, when repeated assays of similar samples are made under seemingly identical circumstances, we occasionally find that one measurement differs from the others by a large amount. Since the number of observations is usually limited, one abnormal value can introduce considerable error into the overall average, which is used as the best estimate of the "true" value. Various criteria for rejecting such suspected data are in use, but many investigators prefer "Chauvenet's criterion"; namely, an observation should be rejected if its deviation from the mean is such that the probability of occurrence of all deviations equally large or larger is equal to, or less than, $1/2N$, where N represents the number of observations involved. However, as Jaffey (3) points out, this criterion is *not* a good one; it rejects good observations in $\sim 40\%$ of the situations to which it is applied. A better criterion used by some investigators is to reject suspected values that deviate from the mean of a series by more than 2σ or 3σ. The probabilities of the occurrence of such deviations are 4.5 and 0.27%, respectively.

If a measurement is rejected, using the preceding criterion, and a second measurement is suspected of being "bad," the same criterion must be applied to the second measurement. Before applying the criterion to the second measurement, the first measurement is rejected, a new mean and standard deviation are computed, and the criterion is reapplied. If a significant fraction of the measurements is discarded using this method, then we begin to suspect that the distribution of results was not a normal distribution and the criterion should not be used. In any case, it is imperative that the experimentor report, along with his data, the particular measurement(s) rejected and the criterion for rejection.

6. Other Statistical Tests

Numerous statistical tests can be used to verify a particular experimental hypothesis that might be used in a radiotracer experiment. Although it is beyond the scope of this book to discuss such tests in detail, it does seem worthwhile to mention some of them, as well as their uses. Detailed descriptions of the tests can be found in standard references (1, 2).

A very important test is the χ^2 test (chi-square test). The χ^2 test is a test for evaluating whether a given theory fits experimental data that are normally distributed. The t test and F test are tests used to compare the means and standard deviations, respectively, of two sets of data that are normally distributed, in order to decide if the two sets have the same mean and standard deviation.

F. APPLICATION OF NUCLEAR STATISTICS TO RADIOTRACER EXPERIMENTS

In radiotracer experiments one of three situations is generally encountered, in regard to the use of nuclear statistics. The first of these situations is when the sample activities are low or the counting time is limited because some feature of the experiment requires the use of low activities or because there is an inability to predict the outcome of some experiment when designing it. For example, fear of deleterious radiation effects on an organism or some aspect of the environment may force the use of low activities. In this case, all the rules and methods outlined in previous sections of this chapter can and should be used to extract properly the maximum amount of valid information from the data. This point is particularly true in environmental radiotracer experiments.

The second situation occurs when careful planning of the experiment is possible, thus leading to sample activities and counting times that make the percent of uncertainty in the results due to statistical fluctuations in the count rate negligible (usually $< 1\%$). For a given sample and background activity, Figure 12-5 can be used to plan exactly how long counting times must be for a given percent uncertainty in the count rate. In this situation, a rough rule of thumb is that accumulation of 10,000 sample counts (with a low background rate) will lead to a 1% uncertainty in the count rate due to statistical fluctuations.

The third situation is simply the intermediate one. Either the planning has not been careful enough or some other effect prevents the achievement of negligible uncertainties in count rate due to statistical fluctuations. In this case, the methods outlined in this chapter are applicable.

BIBLIOGRAPHY

1. DIXON, W. J., and F. J. MASSEY. *Introduction to Statistical Analysis*. 3rd ed. New York: McGraw-Hill, 1969. An authoritative general treatment of statistics.
2. EVANS, R. D. *The Atomic Nucleus*. New York: McGraw-Hill, 1965. The best general reference on nuclear statistics.
3. JAFFEY, A. H. "Statistical tests for counting," *Nucleonics* **18**(11), 180 (1960). A very good, highly recommended summary of nuclear counting statistics.

4. Masterton, W. L., and E. J. Slowinski. *Mathematical Preparation for General Chemistry*. Philadelphia: Saunders, 1970. A typical example of an excellent discussion of the topic of significant figures.

5. Price, W. J. *Nuclear Radiation Detection*. 2nd ed. New York: McGraw-Hill, 1964. A good review of the basics of nuclear statistics.

6. Reynolds, S. A. "Choosing optimum methods and radiotracers," *Nucleonics* **22**(8), 104 (1964).

7. Young, H. D. *Statistical Treatment of Experimental Data*. New York: McGraw-Hill, 1962. A very simple, good introduction to the basic ideas of statistics.

13

Correction Factors in Radiotracer Assay

In ordinary radioactivity assay encountered in radiotracer experiments, the observed counting rate is not equal to the true disintegration rate of the sample, but nevertheless, it is related to it. A number of factors affect the relation between absolute disintegration rate and the observed counting rate. Some tend to increase the observed counting rate, but most tend to decrease it in relation to the disintegration rate. Of these factors, some are inherent in the detectors, others arise from the nature of the counting sample, and still others are related to the sample-to-detector arrangement. Of the many recognized factors, only a few are important in connection with radiotracer experimentation. Figure 13-1 illustrates the ones that are relatively more important in relation to an end-window G-M detector. The factors to be described in detail are primarily important in relative β^--measurements. The important efficiency factors for γ-ray assay were discussed in Chapter 6. Even though only relative measurements are being made, it is necessary to understand the origin and the effect of these factors, plus the corrections to be made for them.

A. BACKGROUND

The background count rate registered on a radiation counter can come from such varied sources as cosmic radiation, natural radioactivity in the vicinity, artifical radioactivity (fallout), nearby X-ray generators, and thermal or other circuit noise. Several means are employed to reduce the background

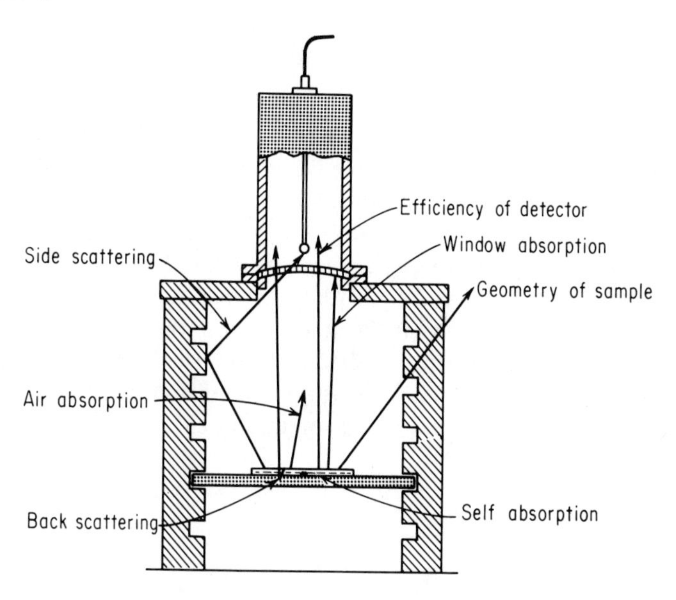

Fig. 13-1. Factors in G-M radiotracer assay.

counting rate for various counter assemblies. Detector shielding with lead or iron is most widely utilized. In addition, coincidence circuitry and photomultiplier cooling in liquid scintillation counters, plus anticoincidence circuitry in special low-background G-M counters, are exceptionally effective in reducing background count—in some cases, to the level of 10 counts per hour (24). Regardless of such background reduction, the gross sample count rate (m_g) always includes some background count rate (m_b). The net observed counting rate (m_o) due to sample activity alone is thus

$$m_o = m_g - m_b \tag{13-1}$$

Scales (21) has discussed in detail the problem of determining background counts in liquid scintillation counting where the complication of quenching also occurs.

By ignoring some minor factors, the net observed counting rate can now be related to the absolute disintegration rate (D) of a sample as follows:

$$m_o = DEf_\tau f_b f_w f_s \tag{13-2}$$

where E represents the correction for geometry and detector efficiency, f_τ stands for the coincidence loss factor, f_b is the backscatter factor, f_w represents the factor for window and air absorption, and f_s stands for the self-absorption factor.* In gamma ray measurements, the last four factors may usually be ignored.

*See Bleuler, Ernst and George L. Goldsmith. *Experimental Nucleonics*. New York: Rinehart, 1952, Chapter 3.

B. GEOMETRY

Since sample radiation is emitted equally in all directions, the placement of the detector with respect to the sample is clearly a primary limiting factor on the percentage of disintegrations that will be detected. If Ω represents the solid angle subtended by the sensitive volume of the detector, then $\Omega/4\pi$ equals the fraction of the disintegrations that would be directed toward that sensitive volume. Figure 13-2 indicates this relationship graphically. The inside radius of the detector is a, the distance from the source to the edge of the detector window is D, and the distance from the source to the detector active volume is denoted by Z. (Z is usually the distance from the source to the detector window + 4 mm.) For a point source,

$$f_g = \frac{1}{2}\left[\frac{a^2}{D(D+Z)}\right] \tag{13-3}$$

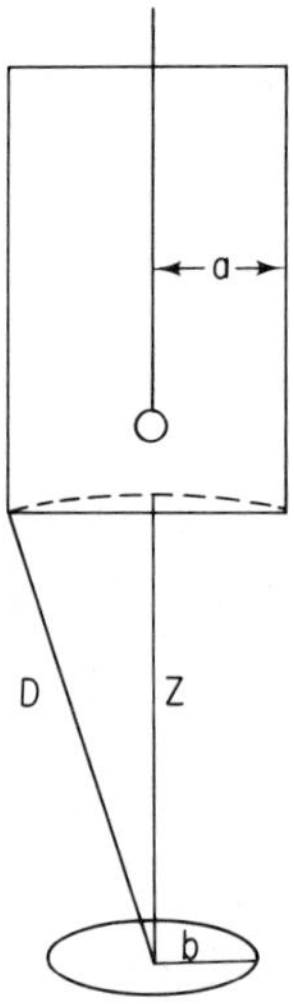

Fig. 13-2. Geometry of an end-window G-M counter.

In practice, a counting sample is spread over some finite area and is an extended source rather than a point source. If the source activity is spread uniformly over a circle of radius b, the geometry factor, f_g, can be calculated by using the formula

$$f_g \simeq \frac{1}{2}\left[\frac{a^2}{D(D+Z)}\right] - \frac{3}{16}b^2\frac{a^2 Z}{D^5} + \frac{5}{32}b^4\frac{a^2 Z}{D^9}\left(Z^2 - \frac{3}{4}a^2\right) + \cdots \tag{13-4}$$

The preceding theoretical formulas should agree to within $\pm 1\%$ with experimental values. In general, the size of the source should be as small as possible and should by no means approach the size of the detector window.

Since it is tedious to calculate the solid angle accurately for extended sources of arbitrary shapes, it is strongly recommended that, for comparative measurements, counting samples of uniform size and distribution be prepared and that they be placed at a reproducibly identical distance from the detector. A sample holding rack with several shelves usually provides for this latter feature. Where desired, the relative values of $\Omega/4\pi$ (shelf ratios) for the various shelf positions of a sample holder can be determined.

The value of Ω for G-M and proportional end-window detectors is inherently less than 2π and frequently less than 1π (7). For G-M and proportional windowless detectors, where the sample is introduced into the detector, the solid angle approximates 2π. In counting with liquid scintillation detectors and internal-sample ion chambers, the geometry involved approaches 4π. An arrangement in which two windowless G-M detectors are placed face to face and the sample is suspended in the center (a "4π counter") is widely used to standardize β^--emitting sources. A common method for greatly increasing the geometrical efficiency of NaI(Tl) crystal scintillation detectors is the use of "well" crystals, where the counting sample (usually liquid in a vial) can be inserted deep into the crystal. In this case, sample volume is limited, since geometrical efficiency decreases sharply as the sample volume reaches and exceeds the well capacity (2, 11, 27).

C. DETECTOR EFFICIENCY

The factor E in Equation (13-2) is made up of the geometry factor $\Omega/4\pi$ and the intrinsic efficiency factor (f_e) of the given detector for the radiation from a specific nuclide. Factor f_e represents the fraction of radiation within the sensitive volume of a detector that is actually detected. For G-M detectors, the values of f_e are 100%, about 96%, and 1% for α-, β^--, and γ-radiation, respectively. With a NaI(Tl) crystal scintillation detector, the γ-ray-detection efficiency varies from 0.1 to 60%, depending on the photon energy (see Chapter 6). The efficiencies of a Ge(Li) detector are a factor of 3 to 100 lower (see Chapter 8). In an ion chamber, α- and low-energy β^--particles (^{14}C, ^{35}S) would be nearly 100% detected if the chamber was sufficiently large. Detection efficiencies would decrease with increasing β^--particle energy, and γ-rays would pass through virtually undetected. The value of f_e in liquid scintillation detectors is strongly dependent on beta particle energy, degree of quenching, nature of the solvent and fluor, concentration of fluor, and the like (see Chapter 9) (See Chapter 8 for a discussion of semiconductor detector efficiencies.)

Not only do detectors show a different intrinsic efficiency for different radiations, but the sensitivity of a specific detector to a given radiation may vary over a period of time as well. This problem particularly plagues scintillation and proportional counters in which the detection efficiency is a function of detector potential. Even slight fluctuations in this potential can offset detection efficiency noticeably. For this reason, stability of the high-voltage supply for these counters is essential. In addition, G-M detectors may show a variation in radial response within the sensitive volume of the tube (1). Semiconductor detectors are usually free from these difficulties.

In organically quenched G-M detectors, failure of the quenching mechanism may occur with increasing age, thereby resulting in multiple pulses for a single ionizing event. The factor f_m (for multiple pulsing) must then be incorporated into factor *E*. Generally speaking, however, multiple pulses are not a significant factor in counting with G-M detectors. Similarly, electrons may be created in a photomultiplier tube without a photon interaction. Such a phenomenon may create a serious problem in γ-ray spectrometry.

In order to determine the absolute value of *E*, it would be necessary to calibrate the detector with a standard source of the same nuclide to be assayed later, making the measurements under physical conditions identical to those to be used later (9). In most radiotracer assays, however, only relative values of *E* need be known for the purpose of correcting counting rates for variation in efficiency with time. This process is easily accomplished by making routine measurements of the activity of a standard source, such as $Ba^{14}CO_3$, $Na_2{}^{14}CO_3$, toluene-^3H, or U_3O_8. These and many other radionuclide standards can be obtained from the National Bureau of Standards or from commercial sources. Any variation in the counting rate of this standardized source over a few days or weeks can be attributed to a change in detector efficiency and all sample counts can be correspondingly corrected.

D. RESOLVING TIME LOSSES

A radiation counting assembly requires a finite time to clear the pulse created by each radiation particle. During this period, other incident radiation is either not detected or not differentiated. The result of this failure to resolve two incident radiation particles that are closely spaced in time is coincidence loss. Note that disintegrations are not evenly spaced in time. The minimum time interval between which two events can be registered is termed the *resolving time* of the counter.

Coincidence loss has several sources in a counting assembly. Loss can occur because of the amplifier circuitry. Normally such loss is negligible, since the better-quality amplifiers have resolving times of 1 to 5 μsec. In the economical "training" scalers, however, it may be as high as 250 μsec. Another possible source of coincidence loss is any mechanical register used to record counting events. These registers, being very slow devices, must be used with electronic scaling circuits in order to avoid loss. Davidon (8) has prepared a nomogram for computing such register losses. Recently it has become a common practice to eliminate the mechanical register altogether and to substitute a series of electronic scaling units. To accommodate the very high counting rates attainable with scintillation, semiconductor, and proportional detectors, it is necessary to use scalers capable of operating at rates of >1 MHz.

For the G-M *counter*, the most important source of coincidence loss is the detector. The resolving time (τ) of G-M detectors is typically 100 to 200 μsec but may be as high as 1000 μsec. This resolving time has two components: dead time and recovery time. As seen in Figure 13-3, for a brief interval following each ionizing event, a G-M detector is insensitive or "dead." For a longer period of time, during which the tube is "recovering," it will respond to incident radiation with pulses of increasing size, up to its maximum. The total resolving time of the detector will be the dead time (τ_d), plus a portion of the recovery time (τ_r). The portion of the recovery time included will depend on the discriminator setting of the associated scaler. Typically, the coincidence loss in a G-M detector begins to be significant over about 3000 cpm. Scintillation, semiconductor, and proportional detectors, on the other hand, have such extremely short resolving times that they are not the cause of significant coincidence loss.

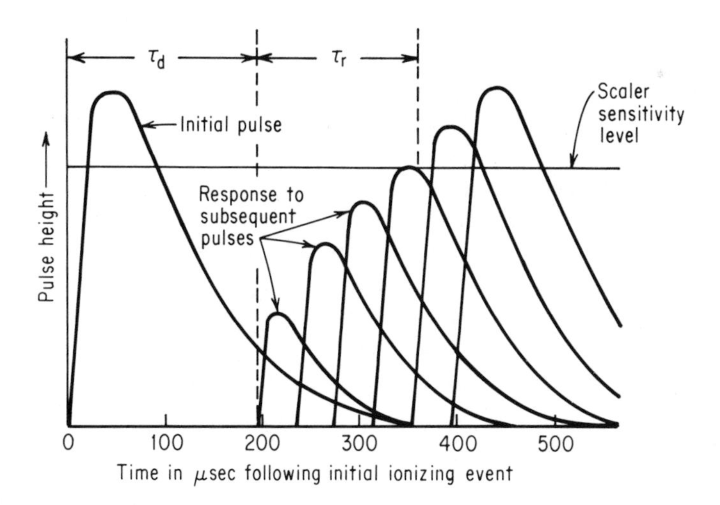

Fig. 13-3. Resolving time of a G-M detector.

If the resolving time (τ) of a counting assembly is known, the correction for coincidence loss can be calculated. When τ is expressed in units of seconds, a counting interval of τ sec (usually a small fraction of a second) is lost for each count registered. If m represents the observed counting rate per second, the total lost counting time is $m\tau$ sec for each second of counting time, and the net useful counting time per second is $1 - m\tau$ sec. Letting n represent the corrected counting rate per second, its value is found as

$$n = \frac{m}{1 - m\tau} \tag{13-5}$$

As an example of the use of this equation for coincidence loss corrections, consider a situation where a counting rate of 6000 cpm (i.e., 100 cps) was measured with a G-M counter whose resolving time was 200 μsec. The corrected counting rate would be calculated as follows:

$$n = \frac{100 \text{ cps}}{1 - 100 \times 0.0002 \text{ sec}} = 102 \text{ cps, or } 6120 \text{ cpm}$$

In contrast, where the same counting assembly observed a counting rate of 60,000 cpm, the correct count rate would be

$$n = \frac{1000 \text{ cps}}{1 - 1000 \times 0.0002 \text{ sec}} = 1250 \text{ cps, or } 75{,}000 \text{ cpm}$$

The percentage coincidence losses in these two examples were 2 and 25%, respectively. Scott (23) has designed a useful slide rule for making rapid determinations of such coincidence loss corrections.

The resolving time of G-M detectors can be determined by several methods. An oscilloscope may be used to visualize the resolving time directly, essentially as in Figure 13-3. Another method involves making a consecutive series of measurements on a nuclide of short half-life, whose initial activity is such that considerable coincidence loss occurs in the detector. As decay occurs and the count rate decreases, correspondingly less coincidence loss takes place. The actual extent of loss can be determined graphically and the resolving time calculated from this graphical procedure. As a variant of this method, one can prepare a set of counting samples of proportionately increasing and known relative activities and determine the coincidence loss from the plot of apparent counting rate versus known activity (20).

The simplest means of determining detector resolving time is the paired-source method. It involves comparing the sum of the measured activities of two individual sources (of approximately 7000 to 10,000 cpm each) with the activity of the two sources measured together. Clearly, the latter figure will be smaller because the relative coincidence loss is greater. Let n_1, n_2, $n_{1,2}$, and n_b represent the true counting rates (in cps) of source 1, source 2, both sources together, and background, respectively; and let m_1, m_2, $m_{1,2}$, and m_b equal the corresponding observed counting rates. Then

$$n_1 + n_2 = n_{1,2} + n_b \tag{13-6}$$

Substituting values of n from Equation (13-5) would yield

$$\frac{m_1}{1 - m_1\tau} + \frac{m_2}{1 - m_2\tau} = \frac{m_{1,2}}{1 - m_{1,2}\tau} + \frac{m_b}{1 - m_b\tau} \tag{13-7}$$

This result leads to a quadratic equation for τ. Normally a simplified approximation is used as follows:

$$\tau \approx \frac{m_1 + m_2 - m_{1,2} - m_b}{m_{1,2}^2 - m_1^2 - m_2^2} \quad \text{(expressed in seconds)} \tag{13-8}$$

A source kit designed for this determination can be obtained from several manufacturers.

E. BACKSCATTERING

Radiation from a sample that is emitted in a direction away from the detector into the backing material and subsequently scattered back toward the detector is said to be *backscattered*. The effect is to increase the apparent counting rate of the sample, above that anticipated for a given sample-to-detector geometry. Backscatter is primarily a phenomenon of β^--particle assay (6).

The backscatter factor (f_b) can be determined experimentally by first assaying a sample mounted on Mylar film and subsequently measuring the count rate with various thicknesses of backing material beneath it. It will be found that the observed counting rate increases directly with increased thickness of backing up to a "saturation" value (25). Theoretically, this saturation thickness would be at 0.5 of the β^--particle range, but, in practice, it occurs at 0.2 of the range value. At saturation backing, the thickness is generally called the saturation backscattering thickness for a given β^--emitter.

As Figure 13-4 shows, backscatter increases more rapidly with increasing backing thickness (d), the lower the β^--particle energy and the higher the atomic number of the backing material. But the ultimate backscatter value,

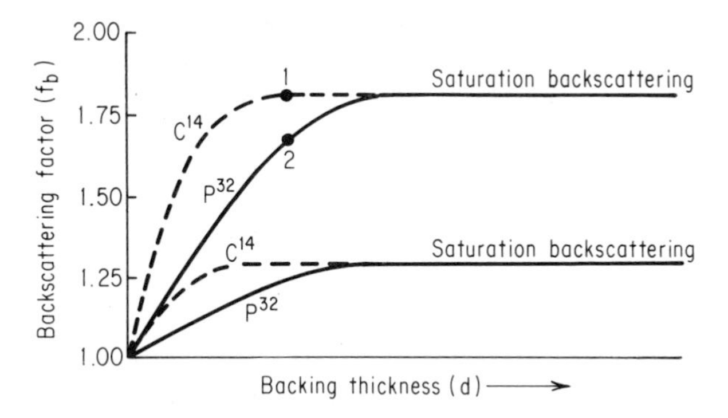

Fig. 13-4. The effect of β^--particle energy and backing thickness on backscatter.

f_b, attained (at saturation backscattering thickness) is independent of β^--particle energy within a range between 0.3 and 2.3 MeV for a given backing material. In addition, the value of the saturation backscattering thickness for a given β^--particle energy is approximately the same for all backing materials.

The value of f_b at saturation backscattering thickness is directly related to the atomic number of the backing material, as shown by Figure 13-5. This relationship can be used to advantage to increase the counting rate of weak samples. To this end, it will be noted that lead and platinum sample planchets would give f_b values of about 1.80, compared to the value of about 1.30 for the commonly used aluminum planchets.

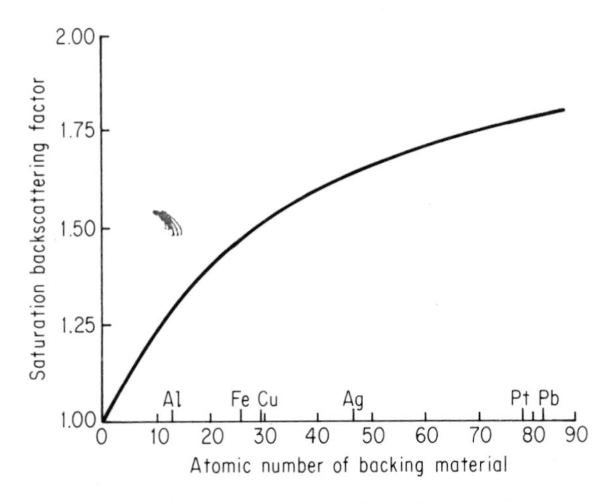

Fig. 13-5. Saturation backscattering factor in relation to the atomic number of the backing material.

In practice, because of the variation of f_b with backing thickness up to the saturation value, it is necessary to use sample planchets of saturation backscattering thickness for the radiation being assayed. It is further advisable to utilize planchets of the same material throughout a given series of experiments where comparisons will be made. Furthermore, a planchet designed to give saturation backscattering thickness for ^{14}C assay (point 1 on Figure 13-4) could be too thin for ^{32}P assay (point 2 on Figure 13-4), because only a slight variation in thickness from planchet to planchet would produce a significant difference in counting rate observed. Thus the selection of planchets depends on the nuclides to be assayed in them.

A certain amount of *side scattering* of radiation into the detector may occur (see Figure 13-1). In β^--particle assay, this scattering can increase the observed counting rate noticeably if material of high atomic number is near the source or detector. Hence sample holders are usually of lucite or other low Z material.

F. ABSORPTION

An often-considerable fraction of the radiation emitted in the direction of a detector may be absorbed before it reaches the sensitive volume of the detector. This absorption may occur in the sample itself (self-absorption), in the intervening air space, or in the window of the detector (G-M and proportional type). The last two areas of absorption are included in the correction factor f_w; self-absorption, the more important factor, is designated f_s.

1. Window and Air Absorption

Corrections for *window absorption* in the G-M detectors of earlier years with mica windows of over 2 mg/cm² thickness were often considerable. Figure 3-12 indicates the percentage transmission of β^--particles of various E_{max} through different window thicknesses (5). With the more recent development of ultrathin, aluminized, Mylar windows of about 0.15 mg/cm², window absorption becomes negligible for all nuclides but tritium. Since these Mylar windows leak slightly, such detectors must be operated with a constant gas flow through the chamber. *Air absorption* of sample radiation is usually slight except for α-particles and the very weakest β^--particles.

Both components of f_w are normally determined together by means of an *absorption curve* prepared with various thicknesses of external aluminum absorbers. In brief, the technique consists of extrapolating the linear semilog plot of observed sample activity to zero thickness by accounting for air and window absorption and thus obtaining the corrected activity at zero total absorber. Figure 13-6 illustrates such an extrapolated curve.

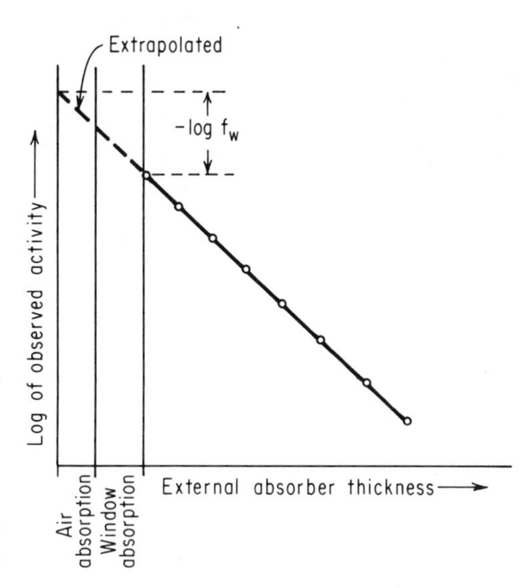

Fig. 13-6. Correction for f_w by means of an extrapolated absorption curve.

2. Self-Absorption

If radioactivity measurements are made on a series of β^--particle-emitting samples, particularly soft β^--emitters, of increasing thickness but constant specific activity, the phenomenon of self-absorption can be readily demonstrated. Without self-absorption, we would expect the count rate to increase linearly as the amount of sample is increased. In practice, this is not so. Evidently, with thick samples, β^--particles from the lower layers are absorbed to some extent by the overlying sample material. Eventually a thickness is reached where only the β^--particles from the topmost layers are being measured. A plot of such counting data (Figure 13-7) shows that the observed

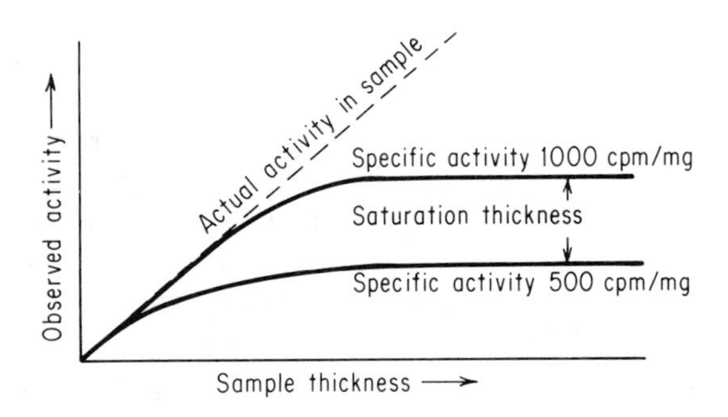

Fig. 13-7. Self-absorption curve for samples of constant specific activity.

activity curve bends away from the calculated activity curve (dotted line) and approaches a limiting value. This value is known as the *saturation* (or infinite) *thickness.* As Figure 13-7 shows, for samples of two different specific activities, the observed activity at saturation thickness is directly proportional to the respective specific activities. The self-absorption curve can be theoretically calculated for a given radioisotope; however, since, in reality, the observed self-absorption curve also reflects backscattering and other factors related to the counting assembly, it is best to determine the curve experimentally for a given counting assembly.

Another version of the self-absorption curve (Figure 13-8) results if a constant total activity is maintained in a series of samples, each diluted with increasing amounts of unlabeled material. Without interference from self-absorption, the observed activity should remain constant (dotted line). Instead the observed counting rate decreases exponentially with increasing sample thickness.* Since the specific activity in this series of counting samples

*One should be aware that in many samples a simple decrease in count rate with increasing sample thickness is *not* observed. Instead the count rate may increase with increased thickness [due to scattering within the sample (self-scattering)] and then decrease because of self-absorption.

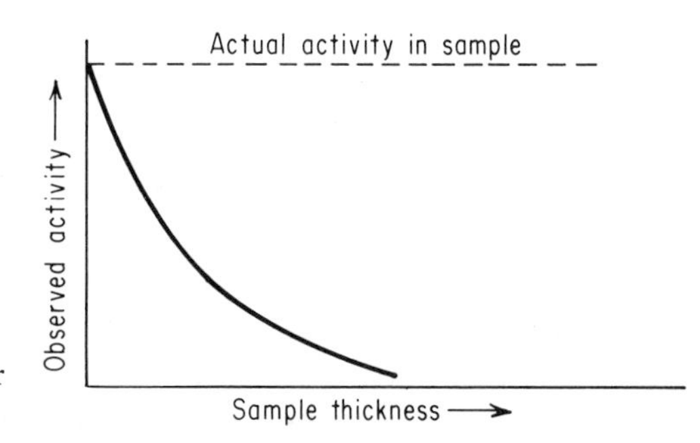

Fig. 13-8. Self-absorption curve for samples of constant activity.

decreases as the total amount of sample increases, the phenomenon of saturation thickness will not be realized.

Self-absorption correction is primarily a problem in the assay of low-energy β^--particle emitters, such as ^{3}H, ^{14}C, ^{35}S, and ^{45}Ca (4, 10, 14–18, 22, 28–30). High-energy β^--emitters suffer relatively little self-absorption with the usual sample thicknesses (less than 15 mg/cm^2). Alpha emitters, because of their typically short range, are almost always measured by using essentially weightless samples, whereas γ-radiation is easily transmitted through the thickest assay sample with almost no absorption. In assaying liquid samples of β^--emitters, it is advisable to adjust the sample thickness so that saturation thickness is exceeded. This principle applies to the use of dipping counters or counting liquid samples in cupped planchets.

The extent of self-absorption cannot be predicted on a theoretical basis because of the associated phenomena of self-scattering and backscattering. Therefore it must be determined empirically. Several methods have been developed to *measure the effect of self-absorption* on counting rate when varying thicknesses are encountered. Some workers prepare all samples to a constant thickness. Although accurate for comparative purposes, this is an extremely tedious and impractical technique. Alternatively, direct plating of samples from solution may be used to produce samples of negligible thickness, so that the interference of self-absorption can be ignored. This process requires preparing samples of less than 1 mg/cm^2 thickness for ^{14}C or ^{35}S and is quite tedious. In contrast, if sufficient sample material is available, all samples can be prepared at or beyond saturation thickness. In this case, the observed count rate will be directly proportional to the specific activity of the sample, regardless of the sample thickness. The method is quite satisfactory for comparative measurements involving soft β-emitters, although it is often impossible to secure sufficient amounts of counting samples in radiotracer experiments.

The most commonly used method for the correction of the self-absorption effect among counting samples of varied thickness is that involving the use of a *transmission* or *self-absorption graph.* Briefly, this process involves the preparation and assay of a series of samples of constant specific activity but varying thickness. A curve is prepared of the apparent specific activity of the samples plotted against the sample thickness in mg/cm² (or weight of sample, if a standardized size of planchet is to be used). The curve is extrapolated to zero-sample thickness, and the specific activity at that point is, by definition, the "true" specific activity of the samples—that is, specific activity unaffected by self-absorption. The extrapolation can be facilitated if the data are plotted on semilog paper. A second curve is then prepared of percent specific activity (apparent specific activity/"true" specific activity—i.e., percentage transmission) versus sample thickness, taking the true specific activity as the value corresponding to 100% transmission—that is, no self-absorption. Correction factors can then be read from this curve for transmission at any sample thickness. Figure 13-9 shows a schematic version of such a curve.

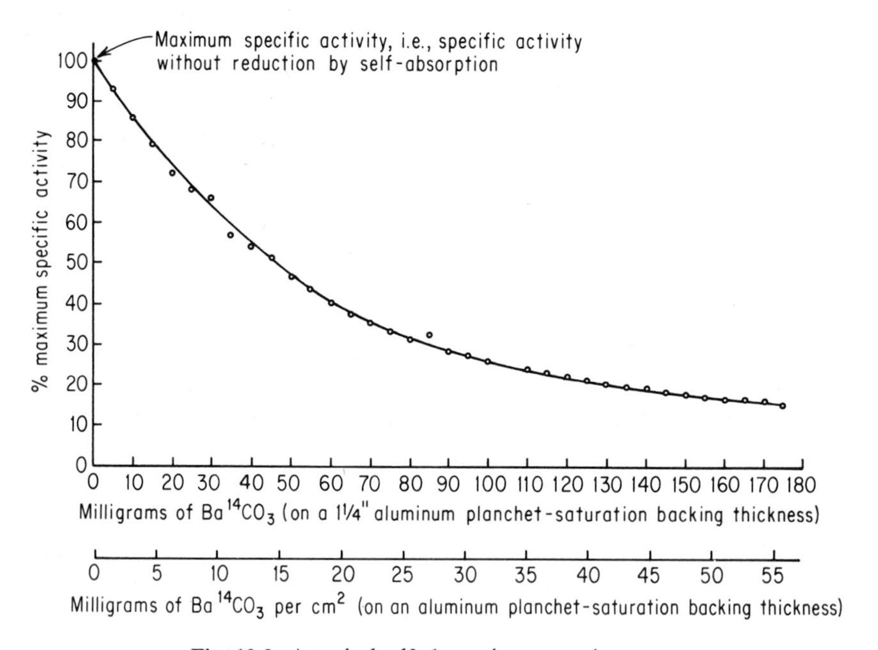

Fig. 13-9. A typical self-absorption correction curve.

Up to now we have discussed how to determine the magnitude of each individual factor that lowers (or raises) the efficiency of the detection system. Bayhurst and Prestwood (3) have published a number of standard counter-efficiency curves as a function of average beta energy and sample weight. They claim that, with the use of these curves, it is possible to estimate one's detection efficiency to ±3% without having to evaluate the various individual contributors to the counting efficiency.

G. SUMMARY

In making comparative measurements of β^--sample activity with G-M counters, correction for all the factors in Equation (13-2) (except f_τ and f_s) and background can be avoided by keeping assay conditions constant. If sample counting rates are below 3000 cpm, and if the detector has a resolving time faster than 200 μsec, f_τ may also be ignored. Correction for self-absorption is made by means of a correction curve (Figure 13-9). In contrast, when the activities of different samples of given γ-ray-emitting nuclides are being compared, all the factors in Equation (13-2) may be kept constant, thereby making it possible to cross-compare scintillation counting data directly. In liquid scintillation counting, the value of factor E—that is, detector efficiency—and particularly the subfactor f_e, is of paramount importance as a result of variation in quenching effect. Other factors can be standardized or ignored. It must be stressed that such comparisons are possible only when assays have been carried out under conditions as nearly identical as possible.

Obviously, for absolute activity determinations, each correction factor in Equation (13-2) must be individually evaluated as previously outlined. Zumwalt (31) has thoroughly described such absolute β^--counting, using an end-window G-M detector.

BIBLIOGRAPHY

1. AMITH, AVRAHAM, and W. WAYNE MEINKE. "Radial response of typical end-window G-M tubes," *Nucleonics* **11**(5), 60–61 (1953).
2. BASKIN, R., H. L. DEMOREST, and S. SANDHAUS. "Gamma counting efficiency of two well-type NaI crystals," *Nucleonics* **12**(3), 82–85 (1954).
3. BAYHURST, B. P., and R. J. PRESTWOOD. "A method for estimating beta-counting efficiencies," *Nucleonics* **17**(3), 82–85 (1959).
4. BELTZ, RICHARD E. "Self-absorption correction in the use of tritium-labeled substrates for enzyme assays," *Anal. Biochem.* **2**, 303–316 (1961).
5. CHANG, CHIA-HUA, and C. SHARP COOK. "Relative transmission of beta particles through G-M counter windows," *Nucleonics* **104**(4), 24–27 (1952).
6. CHRISTIAN, DARLEANE, WAYNE W. DUNNING, and DON S. MARTIN, JR. "Backscattering of beta rays in windowless G-M counters," *Nucleonics* **10**(5), 41–43 (1952).
7. COOK, G. B., J. F. DUNCAN, and M. A. HEWITT. "Geometrical efficiency of end-window G-M counters," *Nucleonics* **8**(1), 24–27 (1951).
8. DAVIDON, WILLIAM C. "Nomogram for computing register losses," *Nucleonics* **10**(2), 76–77 (1952).
9. GLEASON, G. I., J. D. TAYLOR, and D. L. TABERN. "Absolute beta counting at defined geometries," *Nucleonics* **8**(5), 12–21 (1951).

10. GRAF, WALTER L., C. L. COMAR, and IRA B. WHITNEY. "Relative sensitivities of windowless and end-window counters," *Nucleonics* **9** (4), 22–27 (1951).

11. GUNNINK, RAY, and A. W. STONER. "Photopeak counting efficiencies for 3 × 3 inch solid and well-type NaI scintillation crystals," *Anal. Chem.* **33**, 1311–1313 (1961).

12. LIBBY, W. F. "Measurement of radioactive tracers," *Anal. Chem.* **19**, 2–6 (1947).

13. LOEVINGER, ROBERT, and MONES BERMAN. "Efficiency criteria in radioactivity counting," *Nucleonics* **9**(1), 26–39 (1951).

14. MURAMATSU, MASAMI, and HARRIS BUSCH. "Self-absorption corrections for C^{14}-labeled protein," *Anal. Biochem.* **4**, 384–394 (1962).

15. NERVIK, W. E., and P. C. STEVENSON. "Self-scattering and self-absorption of betas by moderately thick samples," *Nucleonics* **10**(3), 18–22 (1952).

16. PODDAR, R. K. "Quantitative measurement of S^{35} in biological samples," *Nucleonics* **15**(1), 82–83 (1957).

17. POWSNER, E. R. "Self-absorption of carbon-14 samples on thin paper planchets," *Intern. J. Appl. Radiation Isotopes* **10**, 22–29 (1961).

18. REID, ALLEN F., ADELE S. WEIL, and J. R. DUNNING. "Properties and measurement of carbon 14," *Anal. Chem.* **19**, 824–827 (1947).

19. REID, J. F., W. E. BAKER, and D. M. COMPTON. "Low-counting-rate errors from photomultiplier fatigue," *Nucleonics* **20**(3), 80–82 (1962).

20. RUDSTAM, SVEN GÖSTA. "Measuring detector dead times accurately," *Nucleonics* **19**(12), 62–63 (1961).

21. SCALES, B. "Liquid scintillation counting: The determination of background counts of samples containing quenching substances," *Anal. Biochem.* **5**, 489–496 (1963).

22. SCHWEITZER, GEORGE K., and BERNARD R. STEIN. "Measuring solid samples of low-energy beta emitter," *Nucleonics* **7**(3), 65–72 (1950).

23. SCOTT, JAMES H. "Slide rule for making dead-time correction," *Nucleonics* **19**(9), 90–92 (1961).

24. SOFTKY, D. D., and R. E. NATHER. "Low-background counter for solid β-emitting samples," *Nucleonics* **15**(5), 90–93 (1957).

25. TITTLE, C. W. "How to compute absorption and backscattering of beta rays," *Nuclear-Chicago Tech. Bull. No. 8*. Des Plaines, Ill., 1961.

26. UPSON, U. L. "Beta energy dependence in end-window G-M tubes," *Nucleonics* **11**(12), 49–54 (1953).

27. VERHEIJKE, M. L. "Calculated efficiencies of a 3 × 3 in. NaI(Tl) well-type scintillation crystal," *Intern. J. Appl. Radiation Isotopes* **13**, 583–585 (1962).

28. WICK, ARNE N., HARRY N. BARNET, and NANCY ACKERMAN. "Self-absorption curves of C^{14}-labeled barium carbonate, glucose, and fatty acids," *Anal. Chem.* **21**, 1511–1513 (1949).

29. YANKWICH, PETER E., THOMAS H. NORRIS, and JOHN HUSTON. "Correcting for the absorption of weak beta-particles in thick samples," *Anal. Chem.* **19**, 439–441 (1947).

30. YANKWICH, P. E., G. K. ROLLEFSON, and T. H. NORRIS. "Chemical forms assumed by C^{14} produced by neutron irradiation of nitrogeneous substances," *J. Chem. Phys.* **14**, 131–140 (1946).

31. ZUMWALT, LLOYD R. "Absolute beta counting using end-window Geiger-Müeller counters and experimental data on beta-particle scattering effects." U.S. Atomic Energy Commission. AECU-567, 1950.

32. STEINBERG, ELLIS P. "Counting methods for the assay of radioactive samples." In A. H. Snell (Ed.), *Nuclear Instruments and Their Uses*. New York: Wiley, 1962, Vol. 1.

33. OVERMAN, R. T. and H. M. CLARK. *Radioisotope Techniques*. New York: McGraw-Hill, 1960, Chapters 6 and 7.

34. CHOPPIN, G. R. *Experimental Nuclear Chemistry*. Englewood Cliffs, N.J.: Prentice-Hall, 1961, Chapter 5.

35. FRIEDLANDER, G., J. W. KENNEDY, and J. M. MILLER. *Nuclear and Radiochemistry*. 2nd ed. New York: Wiley, 1964, Chapter 12.

14

Design and Execution of Radiotracer Experiments

It would be difficult to determine the magnitude of the contribution that radioactive tracer methodology has made to scientific investigations. The use of radioactive isotopes allows the tracing of tagged molecules in a manner hitherto impossible. This is the reason for the wide use of radiotracers. Yet radiotracer methodology has certain inherent limitations and problems that must be understood and evaluated if the method is to be used successfully. The effect of these limitations on the proper design and execution of radiotracer experiments is examined in this chapter. The following discussion is focussed on the applications of radiotracers in biological studies because of the complexity usually associated with experiments involving biological systems. However, most of the points made are equally applicable in studies on physical systems or environmental systems.

A. UNIQUE ADVANTAGES OF RADIOTRACER EXPERIMENTS

The *sensitivity of detection* when using radiotracers far exceeds that with most other chemical or physical methods. It is equivalent to the magnification of processes to an extraordinary power to allow ready detection. For example, the specific activity of carrier-free tritium is approximately 30 Ci/mmole. This fact implies that a dilution factor of 10^{12} can be tolerated without jeopardizing the detection of tritium-labeled compounds. It is thus possible to detect the occurrence of metabolic substances that are normally present in tissues at such low concentrations as to defy the most sensitive chemical methods of identification. An example is the tracing of tritiated thymidine incorporation into nucleic acids in cell nuclei.

Perhaps the most outstanding advantage of the use of radioisotopes is the opportunity offered to *trace dynamic mechanisms*. Such biological phenomena as ion transport across cell membranes, turnover, intermediary metabolism, or translocation in plants could, before the advent of radiotracer methods, be approached only indirectly. Unfortunately, the very fact that radiotracer studies so commonly involve such dynamic conditions frequently makes the interpretation of experimental results most difficult.

The use of the *isotope effect* to study *rate-determining steps* in a sequence of chemical reactions represents an additional advantage of radiotracer methodology. The term *isotope effect* (to be discussed more fully later) refers to the influence on a reaction rate of the difference in the masses of isotopes. This effect may create significant problems in the use of radioisotopes as tracers but can, nevertheless, be used to advantage in a limited number of cases in order to understand the kinetics of certain chemical reactions.

Despite such advantages, radiotracers are often *indiscriminately used* in scientific investigations. It should be emphasized that radiotracers are not magic tools that can be used to solve any problem. Rather, their use involves, *in addition to the knowledge of one's own field*, added complexity, radiation detection, safety precautions, and the like. Experimental design of a radiotracer experiment is often more complex than that of conventional tracer experiments. The problems of data analysis and data interpretation are inherently more difficult. In short, radiotracer methods should be employed only when their use can be fully justified.

The scope of radiotracer applications is so extensive that it is not feasible to attempt a detailed recital. Reynolds and Leddicotte (9), however, have prepared a comprehensive review of radiotracer uses in analytical chemistry. The current applications of ^{14}C and tritium in the biological sciences alone are so numerous as to merit annual (11) and monthly (15) bibliographies, respectively.

B. PRELIMINARY FACTORS IN DESIGN OF RADIOTRACER EXPERIMENTS

1. Basic Assumptions Underlying the Validity of Radiotracer Experiments

Experiments with radiotracers will be valid and interpretations of results meaningful only insofar as careful consideration has been given to certain basic assumptions. Such assumptions concern the behavior and nature of the radioisotope-labeled compound involved and are given below.

a. There is No Significant Isotope Effect. We have stated, as a primary assumption in all radiotracer work, that a radioactive isotope behaves chemically in a fashion identical to the stable isotopes of the same element. Actually, this assumption is not true. The difference in masses between radiotracer nuclei and stable nuclei can cause a shift in the reaction rate or equilibria (the *isotope effect*). It is true, however, that in most cases the isotope effect does not significantly jeopardize the utility of the radioisotope method. Since the degree of chemical bond stability is directly related to the square root of the masses of the isotopes involved, it is apparent that an isotope effect will be of significance only for elements of low atomic weight.

The isotopes of hydrogen present the extreme case. Thus ^{1}H, ^{2}H (D), and ^{3}H (T) could scarcely be expected to act as the same substances chemically, since the relative mass differences are so great. Hence *tritium* cannot be uncritically employed as a tracer for hydrogen in regard to reaction rates, although, of course, its use in determining hydrogen location in an organism is not precluded.

The so-called isotopic effect, in reality, should be considered from the standpoint of the two basic types: the *intramolecular* and the *intermolecular* isotopic effects. In regard to the first type, the decarboxylation of malonic acid-1-^{14}C ($HOO^{14}C—^{12}CH_2—^{12}COOH$) is a good example. Here the pyrolytic decarboxylation of malonic acid gives rise to CO_2 and acetic acid. Consequently, the reaction is concerned with the relative bond strength of $^{12}C—^{14}C$ and $^{12}C—^{12}C$. Since the former bond is relatively more stable (owing to the greater mass of the ^{14}C), we would expect that the CO_2 resulting from the reaction would be comparatively enriched in ^{12}C and that the acetic acid would be comparatively enriched in ^{14}C from the —COOH. Such is the case. Inasmuch as the competitive reactions occur within the same molecule, the isotopic effect will be observed even if the reaction goes to completion.

In the case of the *intermolecular* isotope effect, the decarboxylation of benzoic acid-7-^{14}C can be cited as an example. Here, unless carrier-free compounds (a rare case in radiotracer studies) are being used one is actually dealing with the decarboxylation of two compounds: $C_6H_5—^{14}COOH$ and $C_6H_5—^{12}COOH$. Since the $^{12}C—^{14}C$ bond is relatively more stable, we would expect that, during the initial phase of the decarboxylation reaction, there would be an enrichment of $^{12}CO_2$ compared to the specific activity of the labeled carbon in the benzoic acid. As the reaction approaches completion, however, the reactant (benzoic acid) will be relatively enriched with the ^{14}C-labeled variety, and, consequently, the decarboxylation product (CO_2) will have a specific activity higher than the labeled carbon of the original starting material. When the reaction is driven to completion, the overall specific activity of CO_2, will naturally be the same as the labeled carbon atom in the benzoic acid; that is, no isotope effect can be observed. This is an important concept, for it indicates that the significance of the isotope effect can be minimized or ignored in radiotracer studies if the intramolecular type of reaction is not involved.

A large volume of both theoretical and experimental literature concerning isotope effects exists: Eidinoff (10, pp. 222–226) has discussed isotopic effect as it relates to C—H and C—T bonds. In his review, Yakushin (17) states that the replacement of ^{1}H by ^{2}H (deuterium) in a reactive bond usually results in a decrease in reaction velocity by a factor of 3 to 8, whereas a corresponding replacement of ^{1}H by ^{3}H (tritium) gives a decrease by a factor of 6 to 20. The temperature of the reaction and the specific character of the bonds involved are important factors in determining the exact magnitude of the isotope effect for a given reaction. The magnitude of the effect for the isotopes of carbon is much less than for hydrogen. The isotope effect in the decarboxylation of malonic acid-1-^{14}C has been determined experimentally and calculated theoretically as equal to approximately 4%. Rabinowitz *et al.* (8) found that urea-^{12}C was hydrolyzed by urease at a rate about 10% faster than urea-^{14}C. An excellent survey of the problem of biological fractionation of isotopic mixtures by Protista, higher plants, Mollusca, and vertebrates has been made by Bowen (1). An ACS monograph reviews many aspects of isotope effects in chemical reactions (3).

b. There is No Significant Radiation Damage to the Experimental System. It is essential that radiation from the tracer dose does not elicit a response from the experimental system that would distort the experimental results. The amount of activity employed should be restricted to the minimum necessary to permit reasonable counting rates in the samples to be assayed. The possibility of excessive concentration of the tracer compound in certain biological tissues and the degree of radiosensitivity of these tissues must also be carefully considered, especially when α- or β^{-}-emitting tracers are used. Fortunately, the excellent sensitivity of most radioactivity assay methods minimizes the need to employ tracer doses of such a magnitude that any detectable radiation damage occurs. The possibilities of interference due to physiological response to radiation are further minimized because most studies are short term and thus completed before any latent radiation effects appear. In physical tracer studies, the radiation damage produced by the decaying nuclei must not so destroy the crystal structure and similar features, as to modify the experimental results.

c. There is No Deviation From the Normal Physiological State. If, in order to administer the required tracer activity, the chemical level of the compound given to an organism greatly exceeds the normal physiological or chemical level, the experimental results are open to question. In other words, the specific activity of the tracer compound must be high enough for the total chemical level administered to be within the normal range. As an example, ^{36}Cl might be quite useful for biological investigation, but the maximum specific activity obtainable in the inorganic form is about 100 μCi/g Cl. This situation is in contrast to specific activities of ^{14}C (as $Ba^{14}CO_3$) of up to 2.2 Ci/g of carbon.

d. The Physical State of the Radioisotope-Labeled Compound is Identical to the Unlabeled Variety. Some carrier-free tracers (usually cations) in solution behave like colloids rather than true solutions. The term *radiocolloid* has been applied to such substances. A radiocolloid behaves quite differently in experimental systems than the same element would when present in the true solute form. Stokinger (13) has examined the problem of phagocytosis of radiocolloids in organisms and the consequent effect on their tissue distribution. Thus an ill-defined physical state of the radiotracer compound leads to difficulty in interpreting its behavior. One must also be aware of the difficulties posed by adsorption and other surface effects in very dilute solution.

e. The Chemical Form of the Radioisotope-Labeled Compound is Identical to the Unlabeled Variety. Anomalous experimental results have frequently been traceable to the chemical form of the administered radiotracer. Since reactor production of radioisotopes often results in side reactions, this point is not surprising. In one case involving phosphate-^{32}P uptake in plants, the unexpected experimental results were explained by the fact that a large percentage of the tracer dose was actually in the form of phosphite-^{32}P.

The *radiochemical purity* of a compound cannot be assumed. The presence of other radioactive species in low chemical concentration but high specific activity is frequently encountered. This situation is particularly true in the labeling of compounds with ^{3}H by means of the Wilzbach direct-exposure method (see Chapter 15). Thus, for example, direct hydrogenation of a double bond with ^{3}H during the Wilzbach operation may give rise to a small amount of impurity (saturated form of the compound in question) having a specific activity many times higher than the ^{3}H-labeled compound derived from the recoil-labeling operation. Cohn (2) has examined the sources, detection, and means of removal of such radioactive contaminants.

The problem of radiochemical purity with respect to the chemical state of aged tritium or ^{14}C-labeled compounds is still more acute. Because of the short range of the soft β^--particles associated with these two isotopes, the sizable radiation dose delivered to such compounds by their own radiation leads to self-decomposition (*radiolysis*) and hence a variable concentration and number of products. This problem is more fully discussed by Lemmon (5) and Tolbert (10, pp. 64–68; 13).

f. Only the Labeled Atoms are Traced. Never assume that the appearance of the radioactive label in a given sample indicates the presence of the administered compound. It is the labeled atoms that are being followed, not the intact compound. Not only may metabolic reactions involve the cleavage of the labeled atom from the original compound, but exchange reactions may also occur, thus removing any labile atom from the labeled compound. Such chemical exchanges particularly plague many experiments with tritium-labeled compounds. The extent of chemical exchange is strongly dependent on the molecular species involved, the position of the label in the molecule, and the environmental factors (such as the pH of a biological fluid).

2. Evaluation of the Feasibility of Radiotracer Experiments

a. Availability of the Radiotracer. A primary factor is whether a radioisotope of the element to be traced is available with the proper characteristics (half-life, particle energy, and so on.) For example, although radioisotopes of oxygen and nitrogen would be highly desirable in many investigations, the longest-lived radionuclides available of these elements have half-lives of 2 and 10 min, respectively. Clearly, such short half-lives severely limit the use of such isotopes for most tracer purposes. On the other hand, for several elements, a choice of usable radioisotopes may be available, such as ^{22}Na or ^{24}Na and ^{58}Co or ^{60}Co. Of equal importance is the available specific activity of a given radioisotope. There are radioisotopes, such as ^{36}Cl, that cannot readily be made with desirable specific activity.

A second factor is whether the tagged compound desired is commercially available or can be easily synthesized. The number of labeled compounds available is large indeed, and most radiochemical suppliers will attempt custom syntheses of unstocked compounds. In some cases, however, it is not economically feasible or even possible to introduce a given radioactive atom into the molecular structure under consideration. Furthermore, the specific activity secured may be too low for the proposed experimental use of the tagged compound. This general problem will be considered in Chapter 15.

b. Limits of Detection. The degree to which the administered tracer dose will be diluted by the experimental system must be more carefully evaluated. Dilution must not be so great that the activity of the counting samples will be below the limits of detection. In situations where sample specific activity is unavoidably low, it will be necessary to consider the use of the most sensitive detecting system. Furthermore, the anticipated sample specific activity and means of detection will often dictate the choice of counting sample preparation. This choice will also be influenced by the number of counting samples to be prepared and the required precision of the experiment.

c. Evaluation of Hazard. The first item to be considered is the possibility of harm to the experimenter or to his co-workers. In the great majority of radiotracer experiments, the hazard from direct external radiation does not pose a serious problem. However, there are situations where such is not the case—for example, where high levels (millicuries) of γ-ray emitters are utilized. For instance, 10 mCi of ^{24}Na will deliver a dose of about 204 milliroentgens per hour (at 1-ft distance). One should also be quite cautious about the radiation dose delivered to the hands and fingers while handling radioactive materials. Another item of concern in the use of α- or β^--emitting tracers is the possibility of ingestion of the labeled compounds, particularly those known to have a long turnover time in the human body. This problem is made acute where the sample is in the form of an aerosol or a dry powder at some stage of the experiment. (A more complete discussion of radiotracer hazards will be found in Chapter 16.)

Radiation damage to a biological system under study may occur at two levels: the physiological and the histological. In general, higher radiation doses are required to elicit the latter type of damage. Whenever it is suspected that radiation damage is influencing the physiological response of the organism, it is advisable to repeat the experiment with lower levels of radioactivity, while maintaining the same total chemical level of the administered compound. Biological effects of radiation from radiotracer doses have been reported to occur at the following dose levels: 0.045 μCi ^{131}I/g of body weight in mice, 0.8 μCi ^{32}P/g of body weight in mice, 47 μCi ^{24}Na/g body weight in mice and rats, 0.5 μCi ^{89}Sr/g body weight in mice and rats, 0.05 μCi ^{32}P/ml of rearing solution for mosquito larvae, 2 μCi ^{32}P/liter of nutrient solution for barley plants.

Attention must also be given to the disposal of *radioactive wastes* resulting from the experiment, such as excreta, carcasses, or large volumes of solutions. The possible method of disposal will depend on the specific radioisotope present, its concentration and activity, and the nature of the waste. Disposal must conform to the Code of Federal Regulations, Title 10, Part 20. (Methods and principles of waste disposal are further discussed in Chapter 16.)

d. Evaluation of Proposed Methodology and Data Analysis. Experimental data are useless until they can be properly interpreted. This fact is particularly true in radiotracer experiments. First, a clear statement of the hypothesis to be tested must be made. Next, the sampling process to be followed must be carefully evaluated. Only then can proper methods of data analysis be selected. For further discussion of this aspect of experimental design, see Wilson's excellent general treatment (16).

The *precision* required in the proposed experiment must be clearly established. In radiotracer investigations, this feature is heavily influenced by the method of sample preparation and the assay technique employed. The inherent variability in organisms adds further complications where highly precise results are desired. If one is looking for only a small experimental difference, a high degree of precision is critical for the success of the investigation. Such a degree of precision is not always compatible with the statistical nature of radioactive decay.

C. BASIC FEATURES OF EXPERIMENTAL DESIGN

1. The Nature of the Experiment

The use of radiotracers will not allow an investigator to approach an experimental problem with a lesser degree of preparation and forethought than is necessary in nontracer experiments. Rather, the tracer method requires that the investigator be even more familiar with the specific system to be used and the general nature of the problem. In pursuing scientific studies, there is no substitute for a thorough knowledge of one's own field. Radiotracers must always be regarded as only a tool, not a panacea.

When radioisotopes are used in a dynamic biological system, one of the most common difficulties is failure to recognize the *kinetic aspects* of the system. It is of prime importance to consider the possible pathways that a labeled compound may take in the experimental organism, based on what is already known about the kinetics of the system. One must ponder the effect of such factors as the existence of alternate pathways, the extent of dilution from endogenous sources, the possibility of the reentry of degraded products into the system, the degree of chemical exchange of the labeled atoms, and the general consideration of whether one is dealing with an open or closed system. Even if the kinetics of the system under consideration are not well known, the investigator must still be aware of their possible influence on his experimental results.

In an outstanding volume Sheppard (12) has discussed extensively the expected occurrences in a kinetic system following the introduction of labeled material into a portion of it. The author mathematically analyzes multicompartmental systems, such as are found in living organisms. There is a sizable literature on the experimental design necessary in radiotracer turnover measurement studies; one might cite the articles by Zilversmit and Shore (18), Lax and Wrenshall (4), Kamen (10, pp.210–216), and Jeffay (10, pp. 217–221). Nevertheless, current understanding of the kinetics of biochemical processes is, by and large, limited to the simplest systems.

2. The Scale of Operation

A basic requirement in designing a radiotracer experiment is the calculation of the necessary quantity of labeled compound. In other words, will kilogram or milligram quantities be required? The factors affecting this calculation are discussed first; then specific examples will be cited. Putnam (7) has summarized some of these factors in a concise mathematical expression.

a. The Labeled Compound. If the labeled compound must be synthesized, the starting amount required to permit satisfactory physical or chemical manipulation and to give the desired yield must be determined. A generous allowance must be made for side reactions and for general loss in chemical manipulations. On the other hand, if the tagged compound is to be bought, then cost must be considered. It is frequently necessary to redesign an entire experiment on the basis of the cost factor alone. For example, if hydrocortisone-4-^{14}C (current commercial price over \$100/10 μCi) were being used, it would be unrealistic to design a single tracer experiment requiring mCi amounts.

b. The Experimental System. The extent to which the labeled compound will be diluted in the experimental system is a major factor in determining the quantity to be administered. This *dilution* is primarily a function of the relative size of the system chosen. Clearly, a smaller quantity of tracer compound will be needed for a given purpose in a rat than in a dog, or in a seedling plant than in a mature tree. The use of microorganisms in radiotracer investigations offers the distinct advantage of requiring very small quantities of active material. This point is most useful where the tracer compound used is available in quite limited amounts. When no reliable basis exists for calculating the extent of anticipated dilution, liberal estimates should be made to account for unexpectedly large dilution. In addition, the number of specimens necessary to achieve a given statistical validity must be determined in advance. One must also plan for some loss in adding the labeled material to the experimental system.

c. Purpose of the Experiment. The purpose of the experiment will also be a determining factor in the calculation of the scale of operation. In some cases, as in assaying respired $^{14}CO_2$, the purpose will be merely to identify a labeled compound. In other investigations, the location of a labeled compound is sought, as in the use of tritiated thymidine to trace DNA localization in chromosomes. On the other hand, the determination of the labeling distribution pattern in a compound isolated from an organism metabolizing a specifically labeled substrate, as in a study of precursor-product relationships, may be desired. Such investigations require not only isolation of the labeled compound but further chemical processing in degradation studies as well. An example would be the identification of the labeling pattern in amino acids synthesized by microorganisms metabolizing ^{14}C-specifically labeled glucose.

Where the labeled product is to be degraded in order to determine the pattern of labeling, the scale of operation must consider the necessity of diluting the end product to a desired amount with carrier for degradation studies. Such an operation often involves the dilution of micrograms or milligrams of a compound to as much as the gram level; hence the specific activity of the diluted sample goes below the limit of detection.

3. Detection Efficiency

A primary factor in the size of the initial tracer dose required is the detection efficiency of the counting system to be employed. This factor, along with the degree of counting precision desired, sets a minimum level on the required specific activity and total activity in counting samples. Furthermore, the detection efficiency of the various counting systems is dependent on the radioisotope being assayed. Thus a much higher activity of the γ-ray emitter ^{131}I would be required if the detector were to be a G-M tube than if a NaI(Tl) scintillation detector were used. (For a discussion of the relative efficiencies of the various detection systems, see Chapters 5 to 9. Other factors affecting counting efficiency, such as self-absorption and coincidence loss, were dealt with in Chapter 13.)

4. Specific Activity

Once the scale of operations for a given experiment is determined, one would know the exact amount—that is, the chemical amount in such units as milligrams, millimoles, and so on—of the labeled compound as needed for the experiment in question. The next question is, naturally: What is the desired specific activity of the labeled compound and hence the *total radioactivity* in the defined amount of the compound as determined by the scale of operation? It can be answered only by evaluating the entire experimental procedure, particularly that portion relating to the radiation detection system and procedures for the preparation of the counting samples.

It is most convenient to determine the desired specific activity of the initial labeled compound by first calculating the desired specific activity of a *counting sample* one expects to secure in the experiment. Once this feature is determined, one can then consider the anticipated dilution factors involved in all steps of the experimental procedures, the anticipated loss of radioactivity due to experimental operations, and any other pertinent factors to estimate the magnitude of the desired specific activity in the initial labeled compound. The most trobulesome problem, in fact, is the dilution of labeled compounds by unlabeled compounds during a series of processes. In many cases, only an educated guess can be made. Hence a safety factor in the magnitude of two- to tenfold is often incorporated into the experimental design to accommodate unexpected dilution factors and losses.

5. Example of Experimental Design

As an illustrative example, consider the case of an experiment that involves the study of the utilization of L-alanine-1-^{14}C by a microorganism such as *Azotobacter vinelandii*. The purpose of the experiment is to determine how much of the substrate activity is converted into respiratory CO_2. The experimental design is given below.

a. Scale of Operation. It has previously been decided that the nonsterile incubation period for such radioisotope experiments should not exceed 5 hr, in order to avoid contamination of the microbial culture. Results obtained in previous studies indicate that 10 mg (dry weight) of the microbial cells suspended in 10 ml of an aqueous incubation medium can utilize 10 mg of L-alanine in 4 hr. The amount of respiratory CO_2 given off by the microorganisms will be very small. This fact necessitates the addition of carrier carbonate ion to the CO_2-absorbing trap before sufficient amounts of barium carbonate can be isolated from the trap solution for radioactivity assay. Nevertheless, it is clear that, in this particular experiment, one would need 10 mg (or 0.11 mmole) of L-alanine-1-^{14}C, which, in fact, directly reflects the scale of operation.

b. Specific Activity Requirement. In this particular experiment, a G-M detector assembly is to be used. The overall detection efficiency of the counting assembly has been previously determined as 10%. Counting will be carried out on duplicate samples of respiratory CO_2 in the nature of 100 mg of barium carbonate mounted on aluminum planchets. Since 0.11 mmole of L-alanine is used as substrate, the maximum amount of respiratory CO_2 that can be collected in an absorption trap will be 0.33 mmole if the substrate L-alanine is completely converted to CO_2. In reality, the results of preliminary experiments reveal that the total amount of CO_2 produced in the entire experiment is only approximately 0.15 mmole. This fact implies that one has to add approximately 0.85 mmole of sodium carbonate to the CO_2-absorbing solution so that, upon addition of the $BaCl_2$-NH_4Cl precipitation solution, one can realize a yield of 200 mg of barium carbonate, which is just enough for the preparation of two counting samples.

The optimal radioactivity level of the counting sample is set as approximately 1000 cpm (apparent counting rate). Such a counting rate will be fast enough to collect sufficient counts within a short length of time for good statistical precision, yet sufficiently slow to permit the correction for coincidence loss to be ignored. Since the transmission of ^{14}C β^--particles in a 100-mg plate is approximately 25%, it follows that the desired radioactivity in each plate is 1000 cpm/0.25 = 4000 cpm (counting rate corrected for self-absorption). The total radioactivity in 200 mg of barium carbonate (for both counting samples) is therefore 8000 cpm or 80,000 dpm (since the detection efficiency is 10% in the present case).

In order to meet this requirement, one is faced with the question of how much of the substrate activity will be converted to respiratory CO_2. This determination, however, is actually the purpose of the experiment, which clearly implies that the exact extent of conversion is unknown. It is entirely possible to make an educated guess at this point for the sake of the design of the experiment. Such a guess requires that two separate factors be considered. First, the extent of oxidation of C-1 of L-alanine to CO_2 must be examined. From biochemical considerations, it is likely to be extensive; consequently, an assumption that a minimum of 50% conversion occurs is not too unrealistic.

Second, it is possible that the substrate L-alanine can be diluted by the unlabeled cellular compounds. The maximum dilution factor can be estimated if one considers that substrate L-alanine is in equilibrium with all carbonaceous compounds in the microorganism. The amount of the unlabeled compound can be crudely estimated as equivalent to 0.4 mmole of carbon. This value is derived from the fact that the mean carbon content of biological samples is approximately 40 to 50%. Since 10 mg (dry weight) of the microbial cells is involved in the experiment, the carbon content is therefore approximately 5 mg or 0.4 mmole. This sum, in comparison to the carbon in the substrate L-alanine—that is 0.11×3 (since L-alanine has three carbon atoms) $= 0.33$ mmole—constitutes a maximum dilution of approximately twofold. On the other hand, the minimum dilution factor is naturally nil; that is, there exists no equilibrium between the substrate L-alanine and cellular compounds. On the basis of the foregoing discussion, 25% of the substrate activity will not appear in CO_2 samples as a result of equilibrium with the cellular compounds.

By considering the extent of conversion and the extent of dilution, one would then estimate that (in order to realize 80,000 dpm in the CO_2 sample) the substrate L-alanine should contain approximately

80,000 × 2 (factor to cover extent of conversion) × 1.25 (factor covering the possible equilibration of substrate with several compounds) = 200,000 dpm

or

$$\frac{2 \times 10^5 \text{ dpm}}{2.2 \times 10^6 \text{ dpm}/\mu\text{Ci}} = 0.9\ \mu\text{Ci of L-alanine-1-}^{14}\text{C}$$

The preceding example illustrates the primary pattern involved in the design of a radiotracer experiment. In this particular case, the design is aimed at the radioactivity assay of the respiratory CO_2 sample. Obviously, if one desires to examine the incorporation of L-alanine into cellular constituents employing paper chromatography techniques, considerably more radioactivity must be contained in the substrate, because the amount of cellular compound that can be separated on the paper chromatogram is necessarily small (10 to 50 μg for each compound), and the specific activity of each of the compounds must be sufficiently high (from 1000 dpm/μg up) before a radioactivity assay can be conveniently made. In addition, it should be emphasized that it may also be advisable to introduce a safety factor in an experimental design of this type so that unexpected dilution can be accommodated.

6. Anticipated Experimental Findings

Certain features of experimental design concern the anticipated experimental results. If a time course is to be followed, the experimenter must decide in advance the number of samples to be collected and the collecting intervals that will give the best kinetic information. As for procedures that involve paper chromatography, he must know the optimal amount of sample that can be applied on the particular chromatograph paper to be used. It is especially important to design the experiment so that the information derived can be interpreted meaningfully. Consequently, it must be decided whether counting results will be expressed as percentage yield or specific activity (see Section E of this chapter).

Of equal importance is the *desired sensitivity* with respect to the projected findings. The following example will illustrate this point. Let us assume that an experimenter wants to study the conversion of glucose-2-^{14}C to lactic acid by a tissue culture. If the purpose of the experiment is to demonstrate the precise magnitude of such a conversion, even to the extent of only $\pm 1\%$ of the substrate glucose, then the experimental design must consider this factor. Considerable amounts of radioactivity in the glucose will probably be needed. On the other hand, if he merely wants to determine that a conversion is extensive, the precision of the extent of conversion would not be critical. In other words, it is immaterial whether the conversion is 81 or 80%. This fact makes it possible to design an experiment with much less radioactivity in the substrate glucose, since the sensitivity of the method need not be such as to differentiate one or a few percentage points of the substrate glucose conversion.

D. EXECUTION OF RADIOTRACER EXPERIMENTS

A major part of this book has dealt with the execution of radiotracer experiments (sample preparation, assay, and so on). A few specific points should be mentioned here. An initial step is the determination of the *radiochemical purity* of the labeled compound. Some degree of radiolysis in compounds labeled with ^{3}H, ^{14}C, or ^{35}S must be assumed if they have been stored for any length of time. Establishing the purity, followed by the necessary purification operations if impurities are detected, is an essential step prior to carrying out the experiment.

Never assume that the employment of radiotracer methodology allows a decreased concern for accuracy in the routine procedures of *weighing and measuring*. On the contrary, errors in these activities will profoundly affect the significance of the most precise counting data. It is essential to develop the habit of proper laboratory techniques so as to reduce errors of this type to the minimum. It should be realized that the overall accuracy of the experiment is determined by the precision associated with each step of the experimental operation, which includes physical manipulation, chemical measurement, and radioactivity measurement.

Good laboratory "housekeeping" is obviously important with regard to *safety* in the handling of radioisotopes. Careless technique rapidly results in a technically contaminated laboratory, if not one where an actual health hazard may exist. Note that technical contamination may make a laboratory unfit for sensitive radiotracer studies long before any health hazard develops.

E. DATA ANALYSIS

1. Expression of Results

The results of radioactivity determinations can be stated in several ways. The most suitable form of expression is frequently determined by the nature of the specific experiment and the assay system employed. It is not practical to use any standard form for all types of experiments. It is of utmost importance, however, that the results be presented unambiguously and with sufficient information to permit comparison. In expressing radioactivity data, since a "count" is a purely arbitrary unit, it is vital to state clearly the conditions under which the measurements were made and what, if any, corrections were applied to the observed count rates (i.e., self-absorption corrections, net or gross counts, and so on). Consequently, results are given in units of microcuries, millicuries, or similar expressions in order to facilitate cross-comparison of results. The results of radiotracer experiments are also commonly expressed as percentage yield, specific activity, and the like.

a. Percentage yield. The total radioactivity of the sample assayed can be expressed as a percentage of the labeled compound administered. This step has the advantage of presenting a straightforward comparison and is especially useful in radiorespirometric studies.

b. Specific Activity. The specific activity of a radioisotope sample is the amount of radioactivity per unit amount of material. Specific activity may be expressed in numerous ways. The activity may be stated as count rate, disintegration rate, or in microcuries; the unit amount of material may be indicated as per gram or per millimole of compound or element involved, or any other similar expression. The most convenient form for expressing specific activity of pure compounds is as microcuries per millimole (μCi/mmole). To facilitate ready comparison of results, it is obvious that uniform expressions throughout a piece of work are of primary importance.

2. Interpretation of Results

The interpretation of the results of radiotracer experiments is subject to greater pitfalls than the actual conduct of the experimental procedures. Of course, this situation is not peculiar to tracer experiments, but the opportunities for misinterpretation are probably much greater in this type of investigation than in the traditional approaches. First and foremost, the investigator must take an *open-minded, unprejudiced approach* to the interpretation. An expert knowledge of the experimental system under investigation is the best safeguard against unwarranted interpretations of tracer experiment results.

An important *historical case* involving a misinterpretation of valid radiotracer data concerns early studies of the tri-carboxylic acid (TCA) cycle mechanisms. It is generally recognized that the TCA cycle is one of the most important pathways for the biochemical oxidation of carbonaceous compounds to respiratory CO_2. Coupled with supplementary processes, such as CO_2 fixation or malate synthetase, the TCA cycle also functions as an important mechanism for the biosynthesis of the amino acids.

The reaction scheme for the TCA cycle is given in Figure 14-1. It is noted that citric acid is one of the essential intermediates in the cyclic mechanism. In the late 1930s, however, in experiments using labeled carbon dioxide, together with unlabeled pyruvate administered to liver slices, it was found that the ^{14}C label was incorporated into the α-ketoglutarate, an intermediate in the TCA cycle. The incorporation is recognized as involving a CO_2 fixation process to give rise to oxaloacetate-4-^{14}C—that is, HOOC*—CH_2—CO—COOH. Conversion of the latter compound to α-ketoglutarate via a portion of the TCA cycle (Figure 14-1) will then give rise to labeling of α-ketoglutarate. On degradation of the latter compound, however, it was found that the labeling was confined to the α-carboxyl group (C-1 of α-ketoglutarate). Since citric acid appears to be a symmetrical compound, one would expect that the α-ketoglutarate derived from the citric acid would be labeled at both the C-1 and C-5 positions, as shown here:

$$\underset{\text{(oxaloacetic acid-4-}^{14}\text{C)}}{\text{HOOC*—CH}_2\text{—CO—COOH}} + \underset{\text{(acetyl-CoA)}}{\text{CH}_3\text{—CO—CoA}} \longrightarrow$$

$$\underset{\text{(citric acid-1-}^{14}\text{C)}}{\begin{array}{l}\quad\;\; \text{CH}_2\text{—COOH} \\ \quad\;\;\, | \\ \text{HO—C—COOH} \\ \quad\;\;\, | \\ \quad\;\; \text{CH}_2\text{—C*OOH}\end{array}} \longrightarrow \longrightarrow \longrightarrow \underset{\text{(}\alpha\text{-ketoglutaric acid-1, 5-}^{14}\text{C)}}{\underset{(5)}{\text{HOOC*}}\text{—}\underset{(4)}{\text{CH}_2}\text{—}\underset{(3)}{\text{CH}_2}\text{—}\underset{(2)}{\text{CO}}\text{—}\underset{(1)}{\text{C*OOH}}}$$

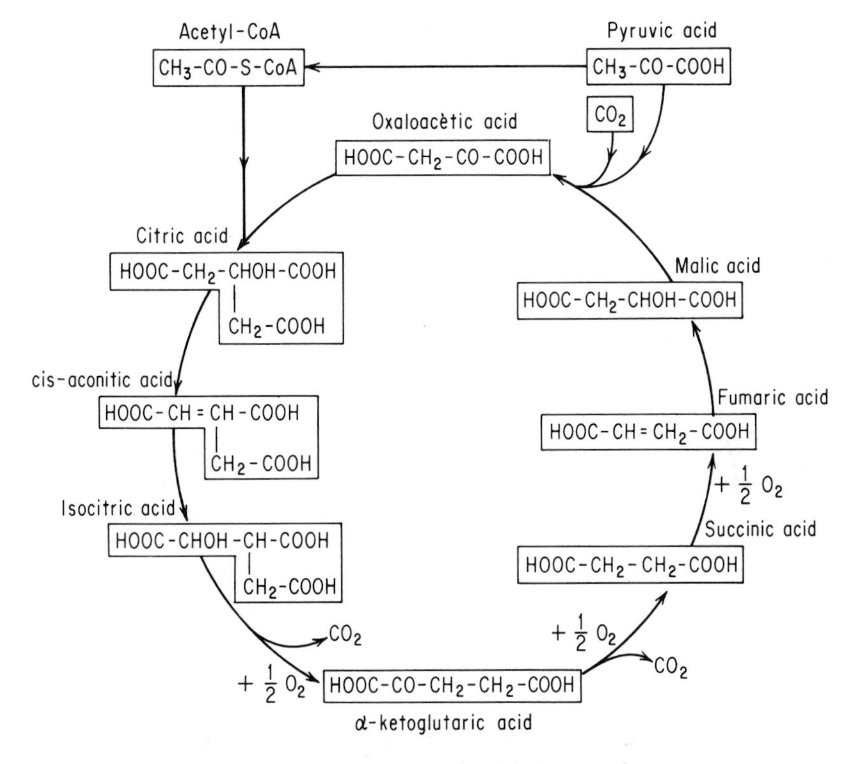

Fig. 14-1. The tricarboxylic acid (TCA) cycle.

The observed discrepancy consequently led biochemists to believe that citric acid should not be considered a direct intermediate of the TCA cycle. Soon afterward, however, Ogston postulated that if the substrate is attached to the enzyme surface at three points that are catalytically different, it might pass through the stage of a symmetrical intermediate (citric acid) and then be converted to the unsymmetrical compound without randomization of the label. This attractive postulation was later confirmed experimentally by Potter and Heidelberger. The development accounted for biochemical texts of the late 1930s having citric acid in, out, and in again in the TCA cycle scheme. This situation occurred even though the radiotracer data were unchanged throughout the period. Besides its historical interest, the case offers a good illustrative example of the importance of proper interpretation of radiotracer findings.

Another common error of interpretation is failure to recognize that, in a radiotracer experiment, one is *following the label,* not the administered compound. If, for example, one detects ^{14}C in a biochemical compound isolated from a biological system metabolizing acetate-1-^{14}C, it simply means that there is incorporation of labeling in the biochemical compound. The incorporation mechanism could be direct in nature, but it could equally well be complex, involving many intermediate steps. Obviously, one should not claim that the conversion of acetate to this compound has been realized, since such a statement implies the presence of a direct mechanism.

Equally confusing is the phenomenon of *randomization* encountered in metabolic studies with radiotracers. This problem is well illustrated in the following case of the biochemical production of fumaric acid:

$$\underset{\text{(pyruvic acid)}}{C^{*}O_2 + CH_3{-}CO{-}COOH} \rightleftharpoons \underset{\text{(oxaloacetic acid-4-}^{14}\text{C)}}{HOOC^{*}{-}CH_2{-}CO{-}COOH}$$

$$\underset{\text{(1-malic acid-4-}^{14}\text{C)}}{HOOC^{*}{-}CH_2{-}CHOH{-}COOH} \rightleftharpoons \begin{array}{c} HOOC^{*}{-}CH{=}CH{-}COOH \\ \text{and} \\ HOOC{-}CH{=}CH{-}C^{*}OOH \\ \text{(fumaric acid-4 or 1-}^{14}\text{C)} \end{array}$$

It will be seen that the fumaric acid is a symmetrical molecule. Consequently, in most textbooks the incorporation of the ^{14}C activity from CO_2 into fumaric acid is indicated as $HOOC^{*}{-}CH{=}CH{-}C^{*}OOH$. This designation implies that the labeling has been randomized to the two corresponding groups, so that each has one-half of the total radioactivity. It must, however, be recognized that although the compound appears to be doubly labeled by such a designation, it is actually composed of a mixture of two singly labeled compounds.

Finally, one must be alert to possible *artifacts* derived from operations in radiotracer experiments. These artifacts may result from such causes as chemical or radiochemical contaminants in the original labeled compounds. In liquid scintillation counting, unreasonably high count rates may result from such phenomena as chemiluminescence and therefore may be entirely unrelated to counting sample activity. Autoradiography is particularly prone to artifacts (see p. 237). Levi (6) has reported a unique artifact in the autoradiography of leaves of bean plants containing ^{32}P. He showed that the apparent accumulation of isotope in primary leaves that were oven dried before exposure to X-ray films was not metabolic but due to a drying gradient occurring within the leaves. This artifact disappeared when autoradiographs were made from leaves that were freeze dried under vacuum. This case illustrates the unexpected sources of artifacts.

It has been possible merely to highlight the basic features and problems of radiotracer experimental design here. The task of applying these generalities to particular biological experiments must lie with the reader.

BIBLIOGRAPHY

1. Bowen, H. J. M. "Biological fractionation of isotopes," *Intern. J. Appl. Radiation Isotopes* 7, 261–272 (1960).
2. Cohn, Waldo E. "Radioactive contaminants in tracers," *Anal. Chem.* **20**, 498–503 (1948).

3. COLLINS, C. J., and N. S. BOWMAN. *Isotope Effects in Chemical Reactions.* ACS Monograph 167. New York: Van Nostrand, 1970.

4. LAX, LOUIS C., and GERALD A. WRENSHALL. "Measurement of turnover rates in systems of hydrodynamic pools out of dynamic equilibrium," *Nucleonics* **11**(4), 18–20 (1953).

5. LEMMON, RICHARD M. "Radiation decomposition of carbon-14-labeled compounds," *Nucleonics* **11**(10), 44–45 (1953).

6. LEVI, A. "An artifact in plant autoradiography," *Science* **137**, 343–344 (1962).

7. PUTNAM, J. L. "An expression for source counting rate required in tracer experiments," *Intern. J. Appl. Radiation Isotopes* **13**, 99–100 (1962).

8. RABINOWITZ, JOSEPH L., *et al.* "Carbon isotope effects in enzyme systems. I. Biochemical studies with urease," *Arch. Biochem. and Biophys.* **63**, 437–445 (1956).

9. REYNOLDS, S. A., and G. W. LEDDICOTTE. "Radioactive tracers in analytical chemistry," *Nucleonics* **21**(8), 128–142 (1963).

10. ROTHCHILD, SEYMOUR (Ed.). *Advances in Tracer Methodology.* New York: Plenum Press, 1963, Vol. 1.

11. SCHARFFENBERG, R. S , and JOHN K. POLLARD. *Carbon-14. 1963.* A comprehensive annual bibliography of applications in chemistry, biology, and medicine. Los Angeles: Calbiochem., January-December 1962, Vol. 1.

12. SHEPPARD, C. W. *Basic Principles of the Tracer Method.* New York: Wiley 1962.

13. STOKINGER, HERBERT E. "Size of dose; its effect on distribution in the body," *Nucleonics* **11**(4), 24–27 (1953).

14. TOLBERT, B. M. "Self-destruction in radioactive compounds," *Nucleonics* **18**(8), 74–75 (1960).

15. *Tritium. 1(1) et seq., 1963.* A monthly bibliographic newsletter by Calbiochem., Los Angeles.

16. WILSON, E. BRIGHT, JR. *An Introduction to Scientific Research.* New York: McGraw-Hill, 1952.

17. YAKUSHIN, F. S. "Application of the kinetic isotope effect of tritium to the investigation of the mechanism of hydrogen substitution and transfer reactions," *Russ. Chem. Rev.* (a translation) **31**, 123–131 (1962).

18. ZILVERSMIT, D. B., and MORIS L. SHORE. "A hydrodynamic model of isotope distribution in living organisms," *Nucleonics* **10**(10), 32–34 (1952).

15

Availability of Radioisotope-Labeled Compounds

The investigator desiring to use radiotracer methods in his own research is first faced with the practical problem of securing suitable radioisotope-labeled compounds. Most frequently, this problem is resolved by buying labeled compounds from a commercial radiochemical supplier. In some cases, the desired tracer compound may not be commercially available and synthesis must be attempted. In either case, it is important to have some idea of the problems of primary radioisotope production and conversion of the primary product to an experimentally useful radiotracer compound. This chapter presents a brief survey of these processes, although they are primarily in the province of the nuclear chemist. For an excellent, more detailed discussion, see the 1966 book by the staff of The Radiochemical Centre at Amersham, England (22).

A. PRIMARY PRODUCTION OF RADIONUCLIDES

1. Introduction

Whether made in the experimenter's laboratory or purchased from a commercial supplier, radionuclides used in tracer studies must be produced by using nuclear reactions. As a result, we should examine and understand the basic features of nuclear reactions before proceeding. First, let us consider the terminology used to describe nuclear reactions that is similar to the terminology used for chemical reactions. In a nuclear reaction, a projectile nucleus, P, with some finite kinetic energy is caused to strike a stationary target nucleus, T. In the ensuing nuclear reaction, a smaller emitted particle, E, is given off, leaving a residual nucleus, R. In a notation similar to that employed for chemical reactions, we write

$$\underset{\text{(projectile)}}{P} + \underset{\text{(target)}}{T} \longrightarrow \underset{\substack{\text{(emitted}\\\text{particle)}}}{E} + \underset{\substack{\text{(residual}\\\text{nucleus)}}}{R} \tag{15-1}$$

This situation is shown schematically in Figure 15-1. A shorthand notation for the reaction given in Equation (15-1) is

$$T(P, E)R \tag{15-2}$$

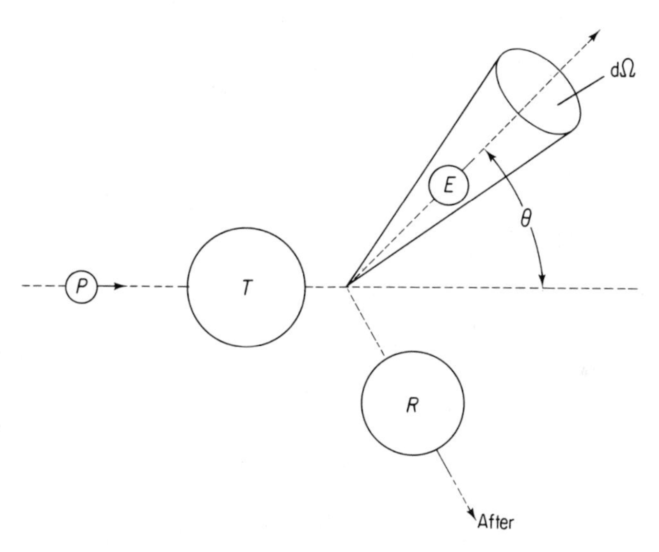

Fig. 15-1. A schematic view of a nuclear reaction showing incoming projectile a striking target nucleus A, giving rise to residual nucleus B, and emitted particle b moving to solid angle $d\Omega$.

For example, if a neutron strikes an ^{27}Al nucleus and a γ-ray is emitted in the ensuing reaction, leaving a ^{28}Al residual nucleus, we could write the reaction equation as

$$^{27}\text{Al} + n \longrightarrow {}^{28}\text{Al} + \gamma$$

or, in the commonly used shorthand,

$$^{27}\text{Al}\ (n, \gamma)\ {}^{28}\text{Al}$$

In words, we say that an "aluminum-27-n-γ-aluminum 28" reaction has occurred.

Note that in writing down nuclear reactions, just as in chemical reactions, we must conserve material. Applying this idea to nuclear reactions, it means that the number of protons and the number of neutrons in the reacting system must not change during the reaction.* In practical terms, the number of neutrons on the left-hand side of the equation must equal the number of neutrons on the right-hand side. Thus if we consider the following reaction

$$^{14}\text{N}(n, p)\ ^{14}\text{C}$$

or, writing it out longhand,

$$^{14}_{7}\text{N} + ^{1}_{0}n \longrightarrow ^{14}_{6}\text{C} + ^{1}_{1}p$$

we can see the conservation principle in action. There are seven protons on the left-hand side of the equation (in the ^{14}N nucleus) and seven protons on the right-hand side (six protons in ^{14}C and one free proton). Similarly, the eight neutrons in the projectile-target system before the reaction takes place equal the eight neutrons found in the reaction products.

Let us consider another example of the application of these ideas. Consider the reaction.

$$^{6}\text{Li}(n, \alpha)?$$

What is the nature of the residual nucleus in this reaction? Writing the reaction out in longhand, we have

$$^{6}_{3}\text{Li} + ^{1}_{0}n \longrightarrow ^{4}_{2}\alpha + ?$$

Applying our conservation principles, the nucleus denoted by ? must have $3 - 2 = 1$ proton and $3 + 1 - 2 = 2$ neutrons. Therefore it must be $^{3}_{1}$H—that is, tritium. The completed reaction is then

$$^{6}\text{Li}(n, \alpha)\ ^{3}\text{H}$$

reaction.

2. Energetics and Probability of Nuclear Reactions

The act of writing down any possible nuclear reaction and successfully balancing it is not evidence that such a reaction will occur. As with chemical reactions, one must compute the energy flow and the probability of the reaction in order to determine the extent to which it will proceed. Such computations are simple but, in general, not consistent with the main idea of our discussion—that is, that of radiotracer methodology. Most radiotracer users choose simply to purchase any necessary primary radioisotopes from a commercial supplier. For those radiotracer users who wish to produce their own primary radioisotopes, a simple discussion of the basic principles involved can be found in the Appendix. More detailed discussions may be found in the literature (9, 11).

*This statement is strictly true for nuclear reactions involving nucleons only. For nuclear reactions involving mesons, the number of neutrons and protons may change as it does in radioactive decay.

3. Practical Techniques

In Section Al of this chapter and in the Appendix, we have discussed the basic terminology of nuclear reactions, how to predict which nuclear reactions might occur and how to calculate how much radioisotope will be produced in a nuclear reaction. In this section we wish to discuss the practical techniques for carrying out isotope production.

a. Isotope Production Using Nuclear Reactors. The most common method of isotope production utilizes neutron irradiation of some material in a nuclear reactor to bring about such nuclear reactions as the (n, γ), (n, p), or (n, α) reactions. The nuclear reactor (or "atomic pile" as it was once described) serves as a massive source of neutrons for carrying out these reactions.

In brief, nuclear reactors are devices to control and sustain nuclear *fission* reactions. In the fission process, nuclei of ^{235}U, ^{239}Pu, or certain other heavy nuclides divide into two smaller nuclei of roughly unequal size following the capture of a neutron. Fission of a heavy nucleus is accompanied by the simultaneous emission of several neutrons (an average of 2.5 to 3.0) with a spectrum of energies. A great quantity of kinetic energy (averaging 172 MeV per fission) is also released. The following equation indicates a typical fission reaction:

$$^{235}_{92}\mathrm{U} + n \longrightarrow {}^{97}_{40}\mathrm{Zr} + {}^{137}_{52}\mathrm{Te} + 2n + 176\text{-MeV kinetic energy}$$

The kinetic energy release is important in reactors used for nuclear power or propulsion, but it is the neutron emission that is of value in radioisotope production.

In order to achieve a self-sustaining or *chain reaction* in a mass of fissionable material, it is necessary to conserve the neutrons emitted at each fission in such a way that at least one of them induces a successive fission in another nucleus. Such neutron-induced fission is highly energy dependent. The neutrons emitted by fissioning nuclei have kinetic energies ranging roughly from 10 keV to over 10 MeV and are termed *fast neutrons.* Unfortunately, fast neutrons are not readily captured by fissionable nuclei; hence they are quite ineffective in inducing fission. In order for these fast neutrons to be absorbed by other fissionable nuclei, they must be slowed to thermal energies by a series of elastic collisions. *Thermal neutrons*, as they are called, have velocities of approximately 2200 m/sec at room temperature or kinetic energies of about 0.025 eV—equivalent to the energy of atmospheric gas molecules. Moderating substances, such as graphite, water (or heavy water), or beryllium, are incorporated into nuclear reactor cores in order to slow the fast neutrons to thermal energies without necessarily capturing them. Typical neutron fluxes in nuclear reactors range from 10^{11} to 10^{13} neutrons/cm^2-sec.

There are many kinds of nuclear reactors, but it is beyond the scope of this book to consider these variations in detail. The essential components of all reactors are the fissionable material, a moderating substance, some mechanism for controlling the rate of fission, shielding material around the reactor core to protect operating personnel, and some means of removing the heat produced in the reactor core. In reactors used for radioisotope production, channels into the core must be provided for inserting and removing materials to be irradiated.

A specific example of reactor design may be seen in the illustration of the Oak Ridge National Laboratory's famous Graphite Reactor (Figure 15-2).

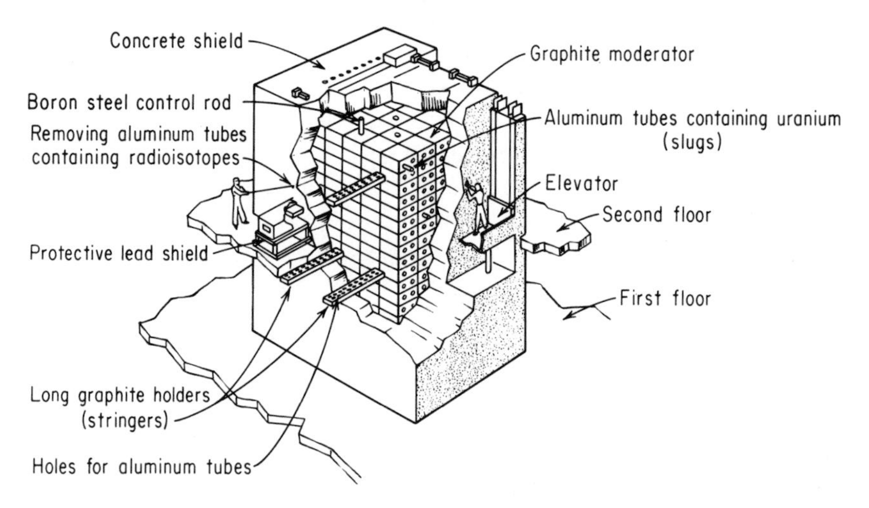

Fig. 15-2. Cutaway view of the Oak Ridge National Laboratory Graphite Reactor.

In this design, graphite is used as the moderator; normal uranium serves as the fissionable material; movable boron-surfaced rods are utilized to control the chain reaction; and the core is cooled by the circulation of air through the fuel channels. The reactor core is a cube measuring 24 ft on an edge and is surrounded by a 7-ft concrete shield. Within the core, a maximum flux of 8×10^{11} thermal neutrons/cm^2/sec is obtained. This reactor was first put into service in 1943 as part of the wartime nuclear weapons program, but it was subsequently operated as the primary production source of radioisotopes in this country, until its retirement.

If a sample material, such as ^{31}P, is placed within a reactor, it will be exposed to a massive flux of thermal neutrons. Those nuclei that capture a thermal neutron will be converted to radioactive ^{32}P by the following reaction: ^{31}P (n, γ) ^{32}P. Thus the typical result of thermal neutron absorption is the emission of γ-radiation from the excited nucleus, and the resultant nuclide is the next higher isotope of the sample material. Other similar thermal neutron reactions of importance are $^{44}Ca(n, \gamma)^{45}Ca$, $^{64}Zn(n, \gamma)^{65}Zn$, and $^{59}Co(n, \gamma)^{60}Co$.

In a few exceptional cases involving light nuclei, bombardment with thermal neutrons results in the emission of one or more nucleons rather than gamma photons. Hence the product nucleus is of a different element than the initial nucleus. Fortunately, the products of such reactions are some of the most important radionuclides for tracer purposes. The best examples of such thermal neutron reactions are $^{6}Li(n, \alpha)^{3}H$, $^{14}N(n, p)^{14}C$, and $^{35}Cl(n, p)^{35}S$.

The neutron flux in a reactor also consists of many neutrons of higher than thermal energies (epithermal). The capture of such neutrons by a sample nucleus greatly increases the energy of the nucleus. Such capture is generally followed by the emission of one or more nucleons. Some important examples of fast neutron reactions of this type are $^{32}S(n, p)^{32}P$, $^{58}Ni(n, p)^{58}Co$, $^{40}Ca(n, \alpha)^{37}Ar$, and $^{12}C(n, 2n)^{11}C$.

For any given sample material in a nuclear reactor, many competing nuclear reactions may be occurring simultaneously as a result of interaction with neutrons of various energies. Thus it is necessary to select a starting material carefully so as to minimize production of undesirable radionuclides and to maximize output of the desired radioisotope product. Furthermore, the starting compound must be sufficiently stable under the thermal and γ-radiation conditions existing in the reactor core. For example, NH_4NO_3 in solution was originally employed in the production of ^{14}C, but because of its excessive rate of decomposition, it was soon discarded and replaced by the more stable Be_3N_2 in solid form.

Besides the neutron-produced isotopes from nuclear reactors, many fission products themselves are useful radioisotopes. The most important in biological research is ^{131}I. Commonly used isotopes of strontium, cesium, and barium are also derived from fission sources.

The source of starting material presents another production problem that is important to the user of radioisotopes. If, for example, ^{32}P is produced by the reaction $^{31}P(n, \gamma)^{32}P$, it will be inseparably mixed with the unreacted stable ^{31}P. The *specific activity*—that is, the amount of radioactivity per unit weight of the element—is unavoidably low and its use as a radiotracer is limited. If, on the other hand, the route $^{32}S(n, p)^{32}P$ was used, the product ^{32}P could be readily separated by chemical means from the unreacted ^{32}S, so that almost all the phosphorus in the sample would be radioactive ^{32}P. Such a sample would be known as carrier-free—that is, free of any stable isotope of the same element. Unfortunately, for some nuclides, a carrier-free state is impossible because of the limited choice of starting materials available.

The radioactive products of neutron irradiation or fission must be chemically processed as they come from the reactor. The processed isotopic materials are generally sold as simple inorganic compounds or in elementary form. Table 15-1 lists common radiotracers in order of increasing half-life.

TABLE 15-1

Common Radioisotopes Listed According to Half-life

Half-life[a]	Isotope	Maximum Specific Activity[b]	Radiation	Production Method: Reactor	Production Method: Cyclotron
12.3 h	^{130}I	~40 Ci/g I	β, γ	(n, γ)	
12.36 h	^{42}K	>450 mCi/g K	β, γ	(n, γ)	
12.80 h	^{64}Cu	~25 Ci/g Cu	EC, β^-, β^+, γ	(n, γ)	
13.47 h	^{109}Pd	~6 Ci/g Pd	β, γ	(n, γ)	
14.12 h	^{72}Ga	~2 Ci/g Ga	β, γ	(n, γ)	
14.96 h	^{24}Na	~10 Ci/g Na	β, γ	(n, γ)	
17.4 h	^{194}Ir	>30 Ci/g Ir	β, γ	(n, γ)	
19.2 h	^{142}Pr	>10 Ci/g Pr	β, γ	(n, γ)	
23.9 h	^{187}W	~10 Ci/g W	β, γ	(n, γ)	
24 h	^{197m}Hg	~500 mCi/g Hg	EC, γ	(n, γ)	
26.4 h	^{76}As	~4 Ci/g As	β, γ	(n, γ)	
35.34 h	^{82}Br	~1 Ci/g Br	β, γ	(n, γ)	
38.7 h	^{77}As	CF	β, γ	(n, γ), $\beta\rightarrow$	
40.22 h	^{140}La	~9 Ci/g La	β, γ	(n, γ)	
46.8 h	^{153}Sm	~40 Ci/g Sm	β, γ	(n, γ)	
53.5 h	^{115}Cd	~50 mCi/g Cd	β, γ	(n, γ)	
64.0 h	^{90}Y	CF, >1 Ci/g	β	Fission, $\beta\rightarrow$ (n, γ)	
64.728 h	^{198}Au	~60 Ci/g Au	β, γ	(n, γ)	
65 h	^{197}Hg	~1 Ci/g Hg	EC, γ	(n, γ)	
66.7 h	^{99}Mo	>140 mCi/g Mo	β, γ	(n, γ)	
67.2 h	^{122}Sb	~2 Ci/g Sb	β, γ	(n, γ)	
3.15 d	^{199}Au	CF	β, γ	(n, γ)	
3.70 d	^{186}Re	~15 Ci/g Re	EC, β, γ	(n, γ)	
4.535 d	^{47}Ca	>150 mCi/g Ca	β, γ	(n, γ)	
5.013 d	^{210}Bi	~50 mCi/g Bi	β, γ	(n, γ)	
5.270 d	^{133}Xe	CF	β, γ	Fission	
6.7 d	^{177}Lu	~20 Ci/g Lu	β, γ	(n, γ)	
7.5 d	^{111}Ag	CF	β, γ	(n, γ), $\beta\rightarrow$	
8.05 d	^{131}I	CF	β, γ	(n, γ), $\beta\rightarrow$	
9.3 d	^{169}Er	~1 Ci/g Er	β, γ	(n, γ)	
11.06 d	^{147}Nd	CF	β, γ	Fission	
12.0 d	^{131}Ba	~10 mCi/g Ba	EC, γ	(n, γ)	
12.80 d	^{140}Ba	CF	β, γ	Fission	
13.59 d	^{143}Pr	CF	EC, γ	Fission	
14.28 d	^{32}P	CF	β		^{32}S$(n, p)^{32}$P
15.0 d	^{191}Os	>400 mCi/g Os	β, γ	(n, γ)	
16.0 d	^{48}V	CF	β^+, EC, γ		^{48}Ti$(p, n)^{48}$V
18.66 d	^{86}Rb	~1 Ci/g Rb	β, γ	(n, γ)	
27.8 d	^{51}Cr	~500 Ci/g Cr	EC, γ	(n, γ)	
32.5 d	^{141}Ce	~2 Ci/g Ce	β, γ	(n, γ)	
35.0 d	^{95}Nb	CF	β, γ	Fission	
35.1 d	^{37}Ar	CF	EC		^{40}Ca(n, α) ^{37}Ar
39.5 d	^{103}Ru	CF	β, γ	Fission	
42.5 d	^{181}Hf	~2 Ci/g Hf	β, γ	(n, γ)	
43 d	^{115m}Cd	~100 mCi/g Cd	β, γ	(n, γ)	
45.6 d	^{59}Fe	~20 Ci/g Fe	β, γ	(n, γ)	
46.9 d	^{203}Hg	~1 Ci/g Hg	β, γ	(n, γ)	
50.0 d	^{114m}In	~1 Ci/g In	γ	(n, γ)	
52.7 d	^{89}Sr	CF	β	Fission	
53 d	^{7}Be	CF	EC		^{7}Li(p, n) ^{7}Be
58.8 d	^{91}Y	CF	β, γ	Fission	
60.2 d	^{125}I	CF	EC, γ	(n, γ), EC$\rightarrow$	
60.4 d	^{124}Sb	~2 Ci/g Sb	β, γ	(n, γ)	
64 d	^{85}Sr	CF	EC, γ		^{85}Rb(p, n) ^{85}Sr

TABLE 15-1 (Continued)

Half-life[a]	Isotope	Maximum Specific Activity[b]	Radiation	Production Method	
				Reactor	Cyclotron
65.5 d	^{95}Zr	CF	β, γ	Fission	
72 d	^{58}Co	CF	EC, γ		^{58}Ni(p, n) ^{58}Co
75 d	^{185}W	~500 mCi/g W	EC, β, γ	(n, γ)	
83.9 d	^{46}Sc	>5 Ci/g Sc	β, γ	(n, γ)	
87.9 d	^{35}S	CF	β		^{35}Cl(n, p) S
106.7 d	^{88}Y	CF	EC, γ		^{88}Sr(p, n) Y
109 d	^{127m}Te	~500 mCi/g Te	β, γ	(n, γ)	
115 d	^{113}Sn	~300 mCi/g Sn	EC, γ	(n, γ)	
115.1 d	^{182}Ta	>500 mCi/g Ta	β, γ	(n, γ)	
120.4 d	^{75}Se	~500 Ci/g Se	EC, γ	(n, γ)	
134 d	^{170}Tm	~600 Ci/g Tm	β, γ	(n, γ)	
140 d	^{139}Ce	CF	EC, γ		^{139}La(p, n) ^{139}Ce
144 d	^{159}Dy	~1 Ci/g Dy	EC, γ	(n, γ)	
165 d	^{45}Ca	~10 Ci/g Ca	β	(n, γ)	
183 d	^{195}Au	CF	EC, γ		^{195}Pt(p, n) ^{195}Au
242 d	^{153}Gd	~3 Ci/g Gd	EC, γ	(n, γ)	
245 d	^{65}Zn	~3 Ci/g Zn	EC, β^+, γ	(n, γ)	
		CF			^{65}Cu(p, n) ^{65}Zn
255 d	^{110m}Ag	~1 Ci/g Ag	β, γ	(n, γ)	
270 d	^{57}Co	CF	EC, γ		^{56}Fe(p, γ) ^{57}Co
280 d	^{68}Ge	CF	β^+, EC, γ		^{69}Ga($p, 2n$) ^{69}Ge
284 d	^{144}Ce	CF	β, γ	Fission	
303 d	^{54}Mn	CF	EC, γ		^{54}Fe(n, p) ^{54}Mn
330 d	^{49}V	CF	EC		^{49}Ti(p, n) ^{49}V
368 d	^{106}Ru	CF	β	Fission	
453 d	^{109}Cd	~1 Ci/g Cd, CF	EC, γ	(n, γ)	^{109}Ag(p, n) ^{109}Cd
700.8 d	^{171}Tm	CF	β	(n, γ), $\beta\rightarrow$	
2.05 y	^{134}Cs	~25 Ci/g Cs	β, γ	(n, γ)	
2.6 y	^{55}Fe	~12 Ci/g Fe, CF	EC	(n, γ)	^{55}Mn(p, n) ^{55}Fe
2.62 y	^{147}Pm	CF	β, γ	Fission	
2.62 y	^{22}Na	>1 mCi/mg Na, CF	β^+, EC, γ		^{24}Mg(d, α) ^{22}Na
2.71 y	^{125}Sb	CF	β	(n, γ), $\beta\rightarrow$	
3.81 y	^{204}Tl	~1 Ci/g Tl	EC, β	(n, γ)	
5.263 y	^{60}Co	~50 Ci/g Co	β, γ	(n, γ)	
7.2 y	^{133}Ba	~1 Ci/g Ba	EC, γ	(n, γ)	
10.76 y	^{85}Kr	~21 Ci/g Kr	β, γ	Fission	
12.262 y	^{3}H	CF	β		^{6}Li(n, α) ^{3}H
12.7 y	^{152}Eu	>250 mCi/g Eu	EC, β, γ	(n, γ)	
16 y	^{154}Eu	>250 mCi/g Eu	β, γ	(n, γ)	
27.7 y	^{90}Sr	CF	β	Fission	
30.0 y	^{137}Cs	CF	β	Fission	
30.2 y	^{207}Bi	CF	EC, γ		^{207}Pb(p, n) ^{207}Bi
48 y	^{44}Ti	CF	EC, γ		^{45}Sc($p, 2n$) ^{44}Ti
87 y	^{151}Sm	25 Ci/g Sm	β, γ	Fission	
92 y	^{63}Ni	~10 Ci/g Ni	β	(n, γ)	
1.2×10^3 y	^{166m}Ho	~1 mCi/g Ho	β, γ	(n, γ)	
5730 y	^{14}C	>2 Ci/g C	β		^{14}N(n, p) ^{14}C
2×10^4 y	^{94}Nb	0.2 mCi/g Nb	β, γ	(n, γ)	
8×10^4 y	^{59}Ni	~10 mCi/g Ni	EC	(n, γ)	
2.12×10^5 y	^{99}Te	20 mCi/g Te	β	Fission	
3.08×10^5 y	^{36}Cl	~10 mCi/g Cl	β	(n, γ)	
7.4×10^5 y	^{26}Al	~1 μCi/g Al	β^+, γ		^{27}Al(p, pn) ^{26}Al
2.7×10^6 y	^{10}Be	~15 μCi/g Be	β	(n, γ)	
1.7×10^7 y	^{129}I	~260 μCi/g I	β, γ	Fission	

[a]h = hour, d = day, y = year
[b]Cf = carrier-free

Data on maximum specific activities available, type of radiation emitted, and method of production are noted. Further information on the availability of these radionuclides can be obtained from any company listed in the annual issue of *Science* magazine's *Guide to Scientific Instruments.* It should also be mentioned that many universities, as well as commercial suppliers, will perform irradiations of target materials furnished by outside individuals. A list of reactors and accelerators whose facilities can be used can be found in the ORTEC Laboratory Manual (8).

b. Isotope Production Using Cyclotrons. The cyclotron is a versatile device for accelerating charged particles to high energies, thus allowing them to participate in nuclear reactions. In a cyclotron, charged particles move in a spiral path of ever-increasing radius, receiving energy "kicks" during each revolution by means of an oscillating electric field. Typical particles accelerated range from protons to α-particles, and final particle energies range from 10 to 150 MeV. Typical particle fluxes are in the range 10^{12} to 10^{15} projectiles/sec striking the target.

The cyclotron is a versatile tool for isotope production because of the wide variety of types and energies of accelerated particles. Operating costs for cyclotrons are so much greater than those for nuclear reactors, however, that the former devices are used for radioisotope production only when they offer a significant advantage over reactors.

A major advantage of the cyclotron is that certain useful radioisotopes may be produced in it that are not produced in significant quantities in nuclear reactors. Some examples of isotopes in this category are ^{7}Be, ^{22}Na, ^{26}Al, ^{48}V, ^{49}V, ^{54}Mn, and ^{74}As. In the case of ^{26}Al, ^{48}V, ^{49}V, and ^{54}Mn, these are the only isotopes of their elements suitable for use as radiotracers. An additional advantage is that, in many cases, radioisotopes may be produced with much *higher specific activities* (frequently carrier-free) in the cyclotron than in nuclear reactors. Certain radioisotopes cannot be produced free of other radioisotopes of the same element in a reactor without resorting to extremely expensive, highly enriched target elements. By means of the cyclotron, however, it has been possible to prepare ^{55}Fe free of ^{59}Fe and ^{85}Sr free of ^{89}Sr and ^{90}Sr.

B. CONVERSION OF PRIMARY RADIOISOTOPES TO LABELED COMPOUNDS

In certain experiments, the primary radioisotopes may be employed directly, but usually the investigator wants to secure a specific labeled compound for use in radiotracer experiments. Before considering the production details of these labeled compounds, let us discuss the nomenclature and rules used in referring to them.

1. The position of a single labeled atom in a molecule is shown following the chemical name of the compound. Thus acetic-1-^{14}C acid is $CH_3^{14}COOH$, while acetic-2-^{14}C acid is $^{14}CH_3COOH$.
2. Certain terms are used to indicate the distribution of label in material with more than one labeled atom. These terms and their meanings are as follows:
   a. *Specifically labeled.* Chemicals are designated as specifically labeled when all labeled positions are included in the name of the compound and 95% or more of the radioactivity of the compound is at these positions. Thus specifically labeled aldosterone-1,2-^{3}H implies that $\geq$95% of the tritium label is in the 1 and 2 positions.
   b. *Uniformly labeled* (*U*) (sometimes called evenly labeled). Uniformly labeled compounds are labeled in all positions in a uniform or nearly uniform pattern. Thus L-valine-^{14}C(U) implies that all the carbon atoms in L-valine are labeled with approximately uniform amounts of ^{14}C.
   c. *Nominally labeled* (*N*). This designation means that some part of the label is at a specific position in the material but that no further information is available as to the extent of labeling at other positions. Thus cholestrol-7-^{3}H(N) implies that some tritium is at position 7, but it may also be at other positions in the molecule.
   d. *Generally labeled* (*G*). This designation is for compounds (usually tritium labeled) in which there is a random distribution of labeled atoms in the molecule. Not all positions in a molecule are necessarily labeled.

Since the great preponderance of labeled compounds are ^{14}C-tagged, our discussion on the preparation of labeled compounds deals mainly with them. However, many of the general principles outlined can be applied to other nuclides and molecules as well. Generally atoms of the radioisotope (^{14}C) may be introduced into the molecules to be labeled by either chemical synthesis or biosynthesis. In the case of tritium, certain unique methods of labeling are also possible.

1. Chemical Synthesis

A carbon-14 label may be introduced into a wide variety of compounds by the standard synthetic procedures of organic chemistry. (The same generally applies to ^{32}P, ^{35}S, and ^{131}I labels.) In addition, some completely new methods have been devised in order to conserve the labeling radioisotope. When chemical synthesis is at all possible, it is usually *the method of choice*. As a rule, synthetic methods give the greatest control over yield, position of labeling, and purification of the product.

It is not feasible here to list all the possible syntheses that have been used. Figure 15-3, however, illustrates a few of the synthetic pathways that might

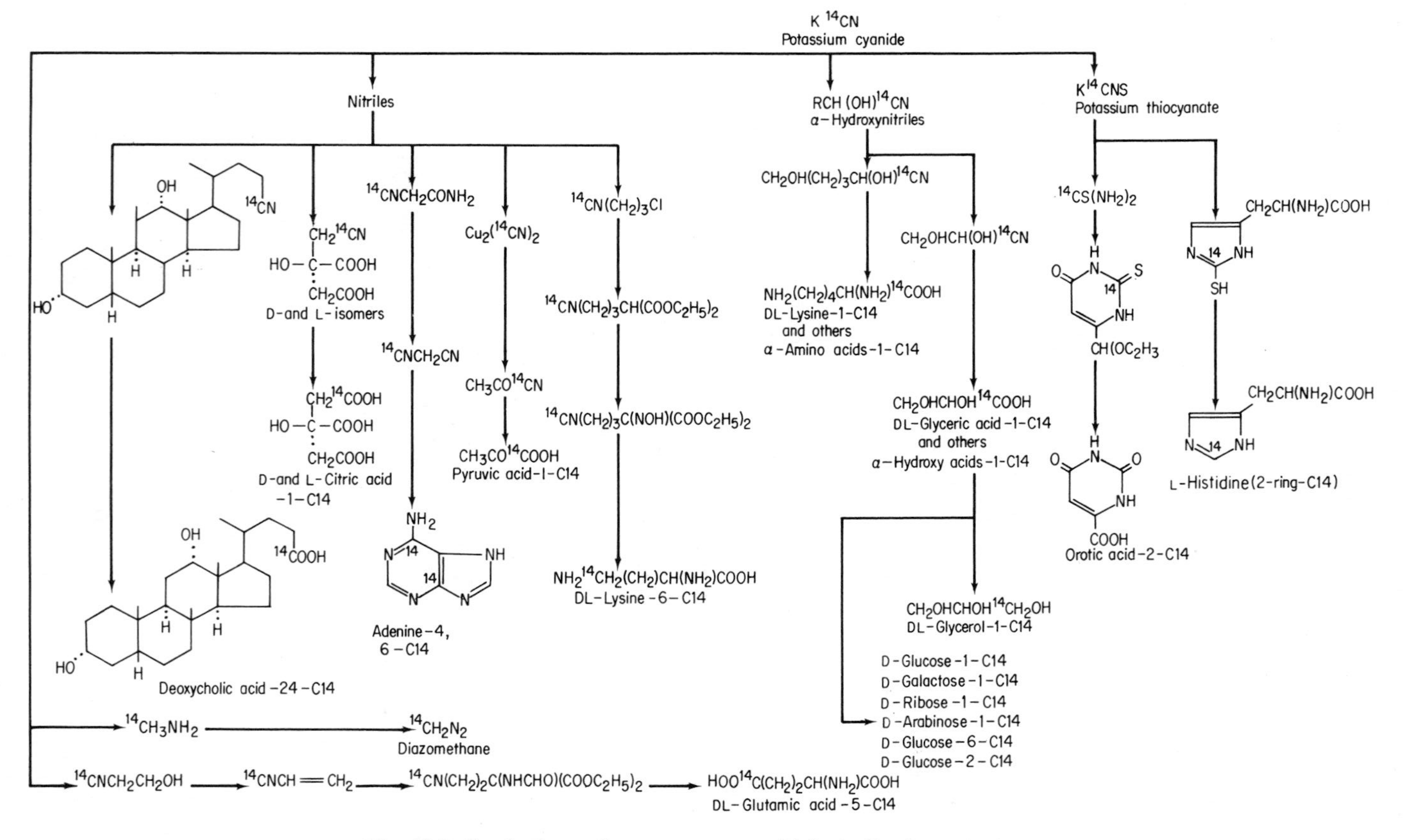

Fig. 15-3. Synthetic pathways to some biologically important ^{14}C-labeled compounds via cyanide. By kind permission of the Radiochemical Centre, Amersham.

be followed to produce certain biologically important compounds, starting with $Ba^{14}CO_3$ via $K^{14}CN$. Calvin *et al.* (2, pp. 148–239) denote a large part of their book to a general survey of the synthetic routes used in producing ^{14}C-labeled compounds. Kamen (14, pp. 311–332) has presented in tabular form an extensive list of specific syntheses for labeling with carbon-14. In addition, work by Murray and Williams (17) covers this field in an encyclopedic manner, although it does not treat biologically important compounds to any extent. Part 1 deals with carbon labeling, whereas Part 2 is concerned with halogen (I, Cl, Br), tritium, ^{32}P and ^{35}S compounds. Catch's book (3) is devoted entirely to ^{14}C compounds; the Radiochemical Manual (22) has separate chapters noting the general aspects of compounds labeled with ^{14}C, ^{3}H, ^{35}S, ^{21}P, ^{131}I, and ^{36}Cl, respectively.

Chemical synthesis of labeled compounds suffers from several *limitations* and problems. One limitation concerns the amount and cost of the isotopic starting material available. This factor necessitates devising synthetic routes to the desired compounds in which the radioisotopic label can be introduced near the end of the sequence of reactions, so as to secure as high an overall yield of labeled material as possible. Since the syntheses will normally be carried out on a micro or semimicro scale, losses due to transfer may be great and overall yield consequently low. Often high-vacuum manifold systems are required for the transfer of liquid or gaseous compounds. At present, numerous labeled compounds are available commercially as starting materials for syntheses. Still, in planning a new synthetic route, it is necessary to consider its compatability with the specific starting material available.

Another disadvantage of chemical synthesis is that when it is used to produce certain biologically important compounds, such as amino acids, a racemic mixture of D- and L-isomers results. Since organisms, by and large, metabolize the L-form selectively, as in the case of amino acids, the use of such racemates in biological investigations is somewhat unphysiological and may lead to undesirable confusion. Methods for the resolution of racemic mixtures are available. Most of the available methods are rather tedious, however, and not well suited to small-scale operations, particularly when one desires to isolate the labeled compound in a reasonably pure form.

Note that many compounds of biochemical importance cannot be synthesized in the laboratory by the classical methods of organic chemistry. Uniformly labeled L-amino acids are a good example. They are currently being prepared from hydrolysates of yeast cells grown on glucose-U-^{14}C as the carbon source. Nevertheless, it is generally recognized that chemical synthesis provides the most effective way to prepare specifically labeled compounds with the highest possible specific activity and purity, and it is capable of producing relatively large quantities of labeled compounds. Several commercial firms, such as New England Nuclear Corp., offer custom synthesis service for desired labeled compounds that are not normally available.

2. Biosynthesis

Living organisms, or active enzyme preparations, offer a biochemical means of synthesizing certain labeled compounds that are not obtainable by chemical synthesis. These include both the macromolecules (proteins, polysaccharides, nucleic acids, etc.) and many simpler molecules (vitamins, hormones, amino acids, sugars). The successful use of biosynthesis for production of a given labeled tracer compound depends on several factors. First, an organism must be selected that will synthesize and accumulate practical quantities of the desired compound. Culture conditions must be established so as to provide optimal yields of the highest specific activity. Last, and most important, procedures must be established for isolating and purifying the labeled compound, as well as determining the distribution pattern of the label, if a specific labeling is desired.

Photosynthetic methods offer the advantage of using the relatively cheap $^{14}CO_2$ (from $Ba^{14}CO_3$) as the starting material. Carbon-14-labeled starch, glucose, fructose, and sucrose can be isolated in good yields from green leaves or algal suspensions that have been exposed to a $^{14}CO_2$ atmosphere and illuminated for a prolonged period. Specific activities of up to 200 mCi/mmole of glucose-U-^{14}C have been achieved in such biosyntheses. Similarly, green algae of the genus *Chlorella* can be used to produce ^{14}C-labeled amino acids.

Microorganisms, or enzyme systems prepared therefrom, have been used to produce organic acids tagged with ^{14}C, either by direct synthesis or transformation of labeled substrates. Several species of microorganisms have been used to produce higher fatty acids by condensation. Calvin *et al.* (2, pp. 262–277) describe some of the specific biosyntheses in general use.

In general, biosynthetic procedures for preparing radioisotope-labeled compounds are likely to be laborious and are limited to small-scale operations. One often encounters purification problems when attempting to isolate specific biological compounds from a number of other biological compounds in a typical biological system.

3. Tritium Labeling

Tritium-labeled compounds have been increasingly used in recent years (30), even though the mean energy of β^--particles emitted from ^3H is only about 5.5 keV. This increased attention is due largely to the development of improved methods of tritium labeling and the application of liquid scintillation counting to tritium assay. Since carrier-free tritium gas can be obtained, very high specific activities in ^3H-labeled tracer compounds are attainable. Commercially available compounds (such as thymidine, tryptophan, and epinephrine) can be secured with specific activities of up to 13 Ci/mmole when the label is tritium, whereas ^{14}C-labeled compounds are usually available only at specific activities of the order of mCi/mmole. So important has tritium usage become that in recent years several symposia have dealt exclusively with the subject (12, 18, 19).

Compounds may be labeled with tritium by several methods (23). The classical *synthetic methods* utilizing labeled intermediates (as previously described) have the advantage of yielding products that have predictable specific activities, are specifically labeled, and have a minimum of tagged by-products. The disadvantages of synthesis are cost and the limitation on the variety of compounds that can be successfully labeled. Hence many compounds are labeled by one of several unique methods.

a. By Reduction of Unsaturated Precursors. The method of choice for labeling with tritium is the reduction of a suitable unsaturated precursor (containing a double bond, carbonyl group, and so forth) with carrier-free tritium gas or tritiated metal hydrides. The major limitation of this method is the availability of a suitable unsaturated precursor of the desired compound. It is essential to carry out the reduction in a nonhydroxylic solvent (dioxane, ethyl acetate, and the like). Reductions carried out in alcohol or water will lead to almost complete exchange of the tritium gas with the solvent. Catalytic reduction may also be complicated by the occurrence of random catalytic exchange reactions, which obviously jeopardize the specific activity of labeling. Careful purification of the product is required, since competing reactions may also occur. The advantages of this method are the very high specific activities attainable, the purity of the product, and the known position of labeling.

b. By Exchange Reactions. Random tritium labeling may be secured by simple exchange methods, with or without catalytic action. For example, the simple exchange of tritiated water (^3H_2O) with the labile hydrogen of malonic acid, followed by decarboxylation, yields tritiated acetic acid. In catalyzed exchanges, either tritiated water or glacial acetic acid (CH_3COO^3H) is the usual medium and platinum black the catalyst most commonly used. This method is suitable only for compounds that are stable in aqueous solution at temperatures up to 120°C. Garnett (10) has rather thoroughly reviewed catalytic tritium labeling. Evans (7) describes this technique as the most general one for tritium labeling.

Although high specific activities may be obtained by this method, some of the introduced tritium may be labile. Removal of this labile tritium and purification of the desired product are troublesome necessities. Exchange can take place into a labile position (uncatalyzed), or exchange can occur into a stable position, using catalysis. One gram of compound can be expected to be labeled with 40 to 400 mCi of nonlabile tritium, if the exposure is carried out with 10 Ci of tritium. In general, the labeled product is of higher purity and specific activity than that from the gas-exposure method described, and complete purification is easier. Using this technique, one can prevent hydrogenation of unsaturated compounds or dehalogenation.

c. ***By Gas Exposure.*** In the mid-1950s Wolfgang and Rowland's (25, 36) work involving tritium-recoil labeling of organic compounds laid the foundation for a new method of ^3H-labeling. Wilzbach (34), in 1957, first described the simplified approach to random labeling with tritium that has come to be called the *Wilzbach gas-exposure method.* In this method, the compound to be labeled is exposed to curie amounts of carrier-free tritium gas in a sealed reaction vessel for a period of a few days to several weeks. The energy released in the disintegration of the tritium and absorbed by the system provides the activation necessary to effect labeling. Compounds labeled by the Wilzbach method are "generally labeled (G)." Wilzbach reported that specific activities of 1 to 125 mCi/g of purified compound have been obtained (24, pp. 4–11, 28–31).

Unfortunately, Wilzbach labeling is often accompanied by the formation of *tritiated by-products* of high specific activity. As in the case of exchange labeling (previously described), a considerable portion of the tritium in the labeled compound is often labile. Nystrom and Dutton (24, pp. 18–27, 46–49) have shown that addition of tritium to unsaturated molecules accompanies the labeling process to a greater or lesser degree, depending on the nature of the double bond. This formation of labeled by-products is the major problem of the gas-exposure method, since the specific activity of the tritium addition product may be several orders of magnitude higher than the desired compound. Moreover, in a typical Wilzbach operation, although it is easy to remove contaminants with molecular weights different from that of the parent compound (such as those formed by fragmentation or polymerization), it is difficult to remove contaminants formed by hydrogenation, racemization, or isomerization. Thus rigid purification procedures must be followed.

Modifications of the Wilzbach method have used external energy to accelerate the labeling process. Electric discharge through the reaction vessel has been widely employed (5, 13, 15, 32, 35). Microwaves (4), ultraviolet irradiation, and gamma irradiation have also been used. All these techniques increase the rate of incorporation and allow faster labeling with smaller amounts of tritium gas, but in many instances they lead to an increase in labeled degradation products. Mottlau (16) suggests that the addition of argon or helium gas to the reaction vessel atomosphere will reduce this competing decomposition.

The chief advantage of the gas-exposure method is that it permits the labeling of compounds of complex or even unknown structure that cannot be readily labeled in any other way (24, pp. 32–41). As examples, insulin (31), digitoxin (27), and a variety of proteins (28) have been successfully tagged. At the end of their excellent review of the gas-exposure method, Whisman and Eccleston (33) list preparative data on 138 different tritium-labeled compounds. The scanty information available on the yield of ^{3}H-labeled compounds by means of a Wilzbach operation does not yet permit one to predict the extent of tritiation for a given compound, particularly compounds having unusual structures. *In general, because of the magnitude of the purification procedure required and the random nature of the labeling, it is suggested that all other synthetic routes be explored before the gas-exposure method is chosen.*

4. Radiolysis of Labeled Compounds

In many situations, the experimenter will prefer to secure labeled compounds from commercial sources rather than attempt to synthesize them himself. The radiochemical purity of such purchased compounds cannot be assumed. Radiation-induced self-decomposition (*radiolysis*) can result in the formation of a variety of labeled degradation products, which must be removed before experimental use of the compounds. The extent of radiolysis depends on the nature of the labeled compound, the length of storage time, and the manner in which the compound is stored. Radiolysis is most significant with low-energy β^--particle emitters (especially tritium), since the decay energy in this case is dissipated almost entirely within the compound itself. Furthermore, impurities involving other radionuclides may be present.

Evans and Stanford (6) have made a most exhaustive study of radiolysis in a large number of tritium-labeled compounds stored under different conditions. Although knowledge in this field is still largely empirical, they were able to draw several general conclusions. Radiolysis is minimized by storage below room temperature (0°C is usually adequate). Storage in solution, rather than in the dry state, definitely reduces radiolysis. Tritium-labeled compounds with specific activities below 500 mCi/mmole stored in solution at 0°C suffer little serious decomposition in one year. At higher specific activities, the extent of radiolysis increases proportionately, but varies considerably, depending on the particular compound in question. These conclusions can be seen from the general data for *tritium-labeled amino acids* in Table 15-2.

TABLE 15-2

Specific Activity (mCi/mmole)	Storage State	Temperature (°C)	Maximum Decomposition per Year (percent)
200	Dry, solid	0, −40	5
200	Aqueous solution	0, −40	5
500	Aqueous solution	−40	10
2000	Aqueous solution	Room	30
5000	Aqueous solution	0	30

Bayly and Weigel (1) have tabulated data on radiolysis in ^{14}C-labeled sucrose and glucose under various storage conditions. Evans (7) has made extensive tables of the percentage decomposition of various tritium-labeled compounds under different storage conditions.

The most complete guide to commercially available labeled compounds is the Isotope Index (26). It lists hundreds of ^{14}C-tagged compounds, as well as those labeled with ^{131}I, ^{32}P, ^{35}S, and ^{3}H. In addition, consult the various radiochemical supply house catalogs for the most up-to-date information on compound availability. Reputable radiochemical suppliers will furnish chromatographic evidence of compound purity; nevertheless, establishment of the purity of the compound should be the initial step in any radiotracer experiment. This point is particularly important when the compound has been stored for some time since purchase.

BIBLIOGRAPHY

1. Bayly, R. J., and W. Weigel. "Self-decomposition of compounds labeled with radioactive isotopes," *Nature* **188**, 384–387 (1960).
2. Calvin, Melvin, *et al. Isotopic Carbon. Techniques in Its Measurement and Chemical Manipulation.* New York: Wiley, 1949.
3. Catch, John R. *Carbon-14 Compounds.* London: Butterworths, 1961.
4. Chanem, N. A., and T. Westermark. "Extensions of the techniques for the accelerated unspecific isotopic labeling of organic compounds." In *Radioisotopes in the Physical Sciences and Industry.* Vienna: International Atomic Energy Agency, 1962, Vol. 3, pp. 43–67.
5. Dorfman, Leon M., and Kenneth E. Wilzbach. "Tritium labeling of organic compounds by means of electric discharge," *J. Phys. Chem.* **63**, 799–801 (1959).
6. Evans, E. A., and F. G. Stanford. "Decomposition of tritium-labeled organic compounds," *Nature* **197**, 551–555 (1963).
7. Evans, E. A. *Tritium and Its Compounds.* Princeton, N.J.: Van Nostrand, 1966. Highly recommended, an excellent, comprehensive treatment.
8. "Experiments in Nuclear Science." *Report AN-34.* ORTEC, Oak Ridge, 1971.
9. Friedlander, G., J. W. Kennedy, and J. M. Miller. *Nuclear and Radiochemistry.* New York: Wiley, 1964.
10. Garnet, John L. "Catalytic tritium labeling attractive for organics," *Nucleonics* **20**(12), 86–91 (1962).
11. Harvey, B. G. *Introduction to Nuclear Physics and Chemistry.* 2nd ed. Englewood Cliffs, N.J.: Prentice-Hall, 1969.
12. International Atomic Energy Agency. *Proceedings of the Symposium on Detection and Use of Tritium in the Physical and Biological Sciences Vienna, 1961,* 1962 Vol. 2.
13. Jackson, Frank L., George W. Kittinger, and Frank P. Krause. "Efficient tritium labeling with an electric discharge," *Nucleonics* **18**(8), 102–105 (1960).
14. Kamen, Martin D. *Isotopic Tracers in Biology. An Introduction to Tracer Methodology.* 3rd ed. New York: Academic, 1957.
15. Lemmon, Richard M., *et al.* "Ionizing energy as an aid in exchange tritium labeling," *Science* **129**, 1740–1741 (1959).

16. Mottlau, A. Y. "Effect of a noble gas on the labeling of n-hexane by exposure to tritium," *J. Phys. Chem.* **64**, 931–933 (1960).

17. Murray, Arthur, III, and D. Lloyd Williams. *Organic Syntheses with Isotopes.* New York: Interscience, 1958.

18. New England Nuclear Corp. *Proceedings of the Symposium on Tracer Applications of Tritium New York, 1957.* Boston, 1958.

19. New England Nuclear Corp. *Proceedings of the Symposium on Advances in Tracer Applications of Tritium, New York, 1958. Boston,* 1959.

20. Oak Ridge National Laboratory. *Catalog: Radio and Stable Isotopes.* 4th ed. Oak Ridge, Tenn. 1963.

21. Pearlman, William H. "A method for labeling C_{21} and C_{19} steroid hormones with tritium at C-7: progesterone-7-^{3}H and Δ4-androstene-3, 17-dione-7-H," *J. Biol. Chem.* **236**, 700–704 (1961).

22. The Radiochemical Centre. *The Radiochemical Manual.* B. J. Wilson (Ed.), 1966.

23. Rosenblum, Charles. "The chemistry and application of tritum labeling," *Nucleonics* **17**(12), 80–83 (1959).

24. Rothchild, Seymour (Ed.). *Advances in Tracer Methodology.* New York: Plenum Press, 1963, Vol. 1.

25. Rowland, F. S., and Richard Wolfgang. "Tritium-recoil labeling of organic compounds," *Nucleonics* **14**(8), 58–61 (1956).

26. Sommerville, J. L. (Ed.). *The Isotope Index.* Indianapolis: Scientific Equipment Co., 1967, Vol. 8. A very useful, somewhat out-of-date list of suppliers, prices, and available radioisotopes and labeled compounds.

27. Spratt, James L., George T. Okita, and E. M. Geiling. "In vivo radiotracer stability of a tritium 'self-radiation'-labeled compound," *Intern. J. Appl. Radiation Isotopes* **2**, 167–168 (1957).

28. Steinberg, D., *et al.* "Preparation of tritiated proteins by the Wilzbach method," *Science* **126**, 447–448 (1957).

29. Tolbert, Bert M. "Radiation self-decomposition of labeled compounds." In *Advances in Tracer Methodology.* Vol. 1. New York: Plenum Press, 1963, pp. 64–68.

30. "Tritium tracing—a rediscovery." *Nucleonics* **16**(3), 62–67 (1958).

31. Von Holt, C., I. Voelker, and L. Von Holt. "Markierung von insulin mit tritium," *Biochim. Biophys. Acta* **38**, 88–101 (1960).

32. Westermark, Torbjorn, Hans Linroth, and Bengt Enander. "Isotope labeling by means of electrical gaseous discharges," *Intern. J. Appl. Radiation Isotopes* **7**, 331–334 (1960).

33. Whisman, Marvin L., and Barton H. Eccleston. "Gas-exposure labeling of organics with tritium," *Nucleonics* **20**(6), 98–105 (1962).

34. Wilzbach, Kenneth E. "Tritium-labeling by exposure of organic compounds to tritium gas," *J. Am. Chem. Soc.* **79**, 1013 (1957).

35. Wilzbach, K. E., and L. M. Dorfman. "Labeling of organic compounds by electric discharge in tritium gas." In *Radioisotopes in the Physical Sciences and Industry.* Vienna: International Atomic Energy Agency, 1962, Vol. 3, pp. 3–11.

36. Wolfgang, Richard, F. S. Rowland, and C. Nigel Turton. "Production of radioactive organic compounds with recoil tritons," *Science* **121**, 715–717 (1955).

16

Nuclear Safety

Nuclear radiation interacts with matter (such as tissue) largely through the process of ionization. This ionization—that is, rupturing of chemical bonds within the cells—disturbs the normal function of the cell. The overall macroscopic effect will depend on the number and the nature of the cells whose function has been altered. In the case of acute exposures to radiation, these macroscopic effects can range from genetic mutation to diseases like leukemia or to death. Table 16-1 shows some effects associated with massive radiation dosage. It should be pointed out, however, that radiation dosage at the low levels encountered in tracer experiments does not appear to cause the harmful effects just cited. Still, there is no strong evidence for a "threshold dosage" of radiation, below which no biological effects are observable.

TABLE 16-1

Short-Term Effects of Acute Radiation Exposure

Dose (rems)[a]	Effect
25	Some alteration in blood white cell count.
100	Nausea in about half those exposed; fatigue noticed; marked effect on blood.
200	Nausea in all exposed individuals; fatigue; death possible; increased susceptibility to infection.
400	Death to half the exposed population.
600	Death to all the exposed population.

[a]See Section A for definition of rem.

The nature of the nuclear safety problem in most radiotracer experiments is not one of serious health hazards but one of minimizing radiation exposure and control of possible undesirable spreading of radioactive material (*contamination*). Typical radiation exposures encountered in radiotracer experiments are shown in Table 16-2, along with the radiation exposures given us by natural and man-made sources in everyday life. As can be seen from examining the table, health hazards due to external exposure to radiation are usually not a problem in radiotracer experiments. However, several detailed nuclear safety responsibilities are involved in radiotracer research. Some of these responsibilities stem from the fact that a large number of carefully worded international, national, state, and sometimes local regulations govern the use of radioactivity (15). Compliance with these regulations requires considerable vigilance on the part of the radioisotope user. Specifically, some of the detailed responsibilities associated with nuclear safety involve:

1. Design of buildings and interior equipment to facilitate safe handling of radioisotopes.
2. Measurement of radiation levels during experiments, measurement of radiation exposure of personnel involved in such experiments, and the keeping of proper records of same.
3. Periodic surveys of all radioisotope laboratories to detect presence of loose radioactivity.
4. Disposal of radioactive waste.
5. Calibration of radiation-measuring instruments.
6. Ability to cope with radiation emergencies, such as cuts contaminated by radioactivity.
7. Education of all personnel concerning proper techniques for handling radioactivity.
8. Designing better and safer procedures for executing radiotracer experiments.

TABLE 16-2

Comparison of Radiation Exposures

Exposure Source	Exposure Range (rem)
Natural background radiation (United States)	
Sea level, outdoors (varies with soil)	0.09–0.20/yr
5000-ft altitude, outdoors	0.11–0.22/yr
Inside a wooden house	0.09–0.20/yr
Inside a brick or concrete house	0.11–0.50/yr
Man-made exposures	
Luminous dial wrist watch (dose to small part of body)	up to 4/yr
Chest X-ray	0.01–1.0/film
Dental X-ray	0.2–1.0/film
Cancer therapy (dose delivered to small area)	up to 10,000 total
Typical exposure in radiotracer experiment	0.001–0.01/expt.

Frequently, the responsibility for performing many nuclear safety operations in an institution is delegated to a radiation health officer or health physicist. This individual has received special training in the various aspects of nuclear safety and generally supervises the nuclear safety program. However, even if a health physicist is available, a significant portion of the responsibility for safety in a radiotracer experiment must rest with the experimenter. It is probable that he is the only person familiar with the small details of an experiment, unusual occurrences, and so on. It is also probable that he is the person best qualified to judge the scientific worth of an experiment. Since most scientific experiments involve an element of risk, which must be balanced against potential gain in the experiment's execution, the knowledge of the scientific worth of an experiment is necessary when judging the risk involved due to radiation.

A. THE STANDARD UNITS OF RADIATION EXPOSURE AND DOSE

In order to discuss radiation hazards adequately, it is necessary to consider the units used to denote the energy dissipation involved. Unlike a unit of radioactivity, the radiation dose is concerned only with the portion of energy associated with radiation that has been absorbed by the object in question. It is the absorbed energy that determines the biological effect of the radiation. In practice, unequivocal statements of radiation dose must take into account the area of the irradiated object involved and the time factor, as well as the actual energy of the ionizing radiation that is dissipated in the object. Thus dose rate and exposure time must be stated and the portion of the body affected defined (i.e., whole body, hands only, and so on) before the extent of radiation exposure can be evaluated. Casarett (2) and Johns (8) give a good general survey of this whole topic of radiation biophysics. Henry (6) discusses current recommendations of the International Commission on Radiological Units (ICRU) regarding radiation quantities and units.

1. The Roentgen

The most widely used unit of radiation exposure, the roentgen (named after the discoverer of X-rays), is based on the ionization in air produced by X radiation. The roentgen (abbreviated R) was officially defined by the International Radiological Congress at Chicago in 1937 as the quantity of gamma or X radiation such that the associated corpuscular emission per 1 cc of dry air at standard temperature and pressure produces, in air, ions carrying one electrostatic unit of electricity of either sign. When we convert this unit of electrical energy to the conventional energy unit (ergs) and express the energy dissipation in grams of air, we find that the energy dissipated in the air is approximately 87.6 ergs per gram of dry air. In health physics operations, the exposure is normally expressed—taking the rate factor into consideration—in units of roentgens per hour (R/hr) or milliroentgens per hour (mR/hr). By definition, the roentgen is valid only for photon interaction with air and cannot be applied to particulate radiation or tissue absorption. Notice also, that, contrary to common misuse, the roentgen is a measure of energy dissipation in air, not dose.

2. The Rad

From the biological standpoint, interest centers primarily on the energy of ionizing radiation absorbed by tissues rather than by air. Although the roentgen unit was in universal use by medical workers, some workers felt that it would be more convenient to devise an arbitrary unit defined on the basis of energy dissipation in biological tissues. In 1953 the International Commission on Radiological Units and Measurements proposed such a new unit and called it the *rad.* It was defined as 100 ergs of energy imparted by any ionizing radiation that is dissipated in 1 g of irradiated material and is not, therefore, directly related to the unit R. Since the absorption of radiation by various materials differs, one must state the nature of the irradiated material involved when using the rad unit. The rad is now the unit of choice when tissue irradiation is considered. Note that the rad is a unit of absorbed dose, not merely of air exposure.

3. The Q Factor, rem, and LET

It is also known that radiation damage to biological tissue depends on both the type of tissue and the type of ionizing radiation and the energy associated therewith. Consequently, all these factors must be considered before a given radiation exposure can be evaluated and correlated to another radiation exposure; hence the quality factor (QF) of the radiation absorbed must be considered. In the older literature, the quality factor is referred to as the *RBE,* relative biological effectiveness. QF values (in reality, representing ratios) are stated as the ratio of the biological response derived (or damage inflicted) in the particular radiation exposure as compared to another radiation exposure. Generally the biological damage incurred by X-rays, γ-rays, or β^--particles is taken to be unity. Table 16-3 shows some quality factor values for various types of radiation.

The QF concept makes it possible to express the energy dissipation in a biological tissue system in a more meaningful unit. Such a unit is the *roentgen equivalent man* (or *mammal*), abbreviated *rem.* This unit reflects not only the amount of energy dissipated but also the amount of biological damage derived from such energy dissipation. It is defined as equal to the product of rads (as measured in soft tissue) and QF. Today statements of permissible exposure of humans to ionizing radiation are expressed in rem units.

TABLE 16-3

Quality Factor for Various Radiations

Radiation Type	Quality Factor (or RBE)	Radiation Type (in MeV)	Quality Factor (or RBE)
Gamma rays	1	Neutrons:	
X rays	1	Thermal	3
Beta particles:		0.005	2.5
<0.03 MeV	1.7	0.02	5
>0.03 MeV	1	0.10	8
Conversion electrons	1	0.50	10
Alpha particles	10	1.00	10.5
Recoil nuclei	20	5.0	7
		10	6.5

The term QF, although useful, does create a crucial problem in actually determining the values of QF in precise experimental work. The reason is that biological damage from ionizing radiation depends on many other factors, including tissue pH, tissue temperature, and oxygen content in tissue. A more meaningful treatment of this subject concerns the rate of energy transfer to the tissue as the ionizing radiation travels through the biological system. Such a concept is represented by the use of the unit, linear energy transfer (LET), which, in reality, is expressed in energy units or as keV/μ (micron) of path. When expressing the energy dissipation in this unit—that is, in the linear energy transfer to water (keV/μ)—we find that the values range from 3.5 or less for X rays and electrons to well over 100 for heavy ionizing particles. Table 16-4 shows QF values for various LET values.

TABLE 16-4

Quality Factor Versus Linear Energy Transfer

LET (keV/μ in water)	QF
3.5 or less	1
3.5–7.0	1–2
7.0–23	2–5
23–53	5–10
53–175	10–20

B. HAZARD FACTORS IN HANDLING RADIOISOTOPES

Many facets of radioactivity that relate to radiation safety have already been discussed in this book. Now we turn to the factors that determine the hazard involved in a given radiotracer experiment. These include (a) half-life of the radioisotope, (b) energy and type of radiation, (c) rate of elimination from the body, (d) selective deposition or localization in the body, and (e) quantity involved and mode of handling in experiment. The hazards from radioisotope use can be classified as internal or external, depending on the location of the activity with regard to the body. Taylor (13) has reviewed the development of radiation protection standards.

1. External Hazards

Radiation from external radioisotope sources poses the hazard of either whole-body irradiation or local irradiation. Since the range of α- and β^--particles in air is relatively short, it is most unlikely that whole-body irradiation could occur except from γ- or X-rays. Local irradiation from an external beta source normally involves only the most superficial layers of skin (up to a few millimeters in thickness).

Current federal regulations (15, part 20.101) specify, in general, that the *maximum permissible doses* to radiation workers may be as follows: whole body, head and trunk, active blood-forming organs, lens of eyes, or gonads, $1\frac{1}{4}$ rems in any 13 consecutive weeks; hands and forearms, or feet and ankles, $18\frac{3}{4}$ rems in any 13 consecutive weeks; skin of whole body, $7\frac{1}{2}$ rems in any 13 consecutive weeks. In addition, the accumulated radiation dose that the individual has received must be considered and must not exceed 5 (N-18) rems, where N is the age of the individual. However, at this time (1975), the federal guideline for the average dose to the general public that is permitted is only 170 mrem per year. It should be emphasized that the maximum permissible doses should not be casually accepted; it is advisable to strive for the lowest possible exposures at all times.

In order to provide a rough determination of the exposure from an external γ-ray-emitting source, the following empirical equation may be used:

$$R = 6CE \tag{16-1}$$

In this equation, R represents exposure rate in mR/hr at 1 ft, C is mCi of activity, and E is total γ-ray energy per disintegration in MeV. The relationship stated holds fairly well for γ-ray energies from 0.3 to 3.0 MeV.

Another important factor affecting external exposure is the *inverse square law*, which states that the radiation intensity varies inversely as the square of the distance. Thus an exposure rate of 100 mR/hr at 1 ft would be reduced to $100 \times 1^2/2^2 = 25$ mR/hr at 2 ft. The use of long-handled tongs and remote control pipetting devices often constitutes a necessary precaution in handling small amounts of γ-ray emitters.

The inverse square law has important consequences in relation to handling radioactive materials. For example, consider a situation in which the experimenter maintained an average distance of 1 m between a source and himself and a modest radiation dose of 1 mrem to his whole body resulted. The dose to his fingers, had he been handling the same source for this time period (at an average distance of 2 cm), would have been $100^2/2^2 = 2500$ mrem, a nontrivial dose. The message should be clear. Be very cautious when handling radionuclides with the hands. Seemingly small quantities of activity can lead to large hand and finger exposures if improperly handled.

On the other hand, if proper distances are maintained between radioactivity and experimenter, the small amount of γ-activity used in most tracer work precludes the possibility of serious external radiation hazard. For example, it can be calculated that one would have to be exposed at a distance of 1 ft to a 100-μCi amount of ^{131}I for over 4500 hr to receive a 1 R exposure. To receive similar exposures from the same levels of ^{60}Co and ^{24}Na activities would require over 700 hr and nearly 400 hr, respectively.

Where higher activity levels of γ-emitting isotopes are being used, *shielding* may be required, in addition to maintaining reasonable distance from the source. Gamma ray shielding has been considered in Chapter 3, and we need not discuss the matter further at this point except to stress again the value of using half-thickness calculations in determining the required shielding to reduce γ-ray intensity by a given factor. Glass or Lucite barriers are usually adequate shielding when working with relatively high levels of energetic β^--emitters.

2. Internal Hazards

Deposition of radioisotopes within the body as a result of ingestion, inhalation, or skin absorption poses an entirely different problem. In this case, isotopes whose radiations have very *short ranges* are the most hazardous, for they dissipate all their energy within a very restricted volume of tissue. Thus α-emitters and weak β^--emitters present the greatest hazard, followed by energetic β^--emitters and γ-ray emitters, respectively.

An added hazard from internal emitters exists where a given radioisotope is *selectively concentrated* in the body, rather than generally distributed, since it will produce a more intense local irradiation. Examples of this selective concentration are iodine in the thyroid gland, or Pu, Ra, Sr, P, or Ca in the bones. The radiation exposure from an internal radioactive source continues only as long as the isotope remains in the system. This factor is influenced by both the disintegration half-life of the particular radioisotope and its biological half-life—that is, the length of time until one-half of a given amount is excreted. Of course, the most serious internal hazard comes from ingestion of long-lived β^--emitting radioisotopes that are selectively concentrated and not readily excreted. ^{90}Sr is a prime example.

Selected radioisotopes are grouped according to their relative radiotoxicity as internal emitters in Table 16-5 (16). The activity levels regarded as low, intermediate, and high, according to the degree of laboratory safety precautions require, are also indicated. The activities stated are somewhat arbitrary and highly dependent on the nature of the experiment being performed, as well as the chemical and physical form of the radioisotope.

TABLE 16-5

Relative Internal Hazard of Selected Radioisotopes

Hazard Class	Activity Levels (in millicuries)		
	Low	Intermediate	High
Only slightly hazardous: ^{24}Na, ^{42}K, ^{64}Cu, ^{52}Mn, ^{76}As, ^{77}As, ^{85}Kr, ^{197}Hg	<1	1–10	>10
Moderately hazardous: ^{3}H, ^{14}C, ^{22}Na, ^{32}P, ^{35}S, ^{36}Cl, ^{54}Mn, ^{59}Fe, ^{60}Co, ^{89}Sr, ^{95}Nb, ^{103}Ru, ^{106}Ru, ^{127}Te, ^{129}Te, ^{131}I, ^{137}Cs, ^{140}Ba, ^{140}La, ^{141}Ce, ^{143}Pr, ^{147}Nd, ^{198}Au, ^{199}Au, ^{203}Hg, ^{205}Hg	<0.1	0.1–1	>1
Very hazardous: ^{45}Ca, ^{55}Fe, ^{90}Sr, ^{91}Y, ^{95}Zr, ^{144}Ce, ^{147}Pm, ^{210}Bi	<0.01	0.01–0.1	>0.1

Particular attention should be given to the relative health hazards from tritium and ^{14}C, the most commonly used radiotracers in biology. Tolbert *et al.* (14) and Skipper (12) have discussed the hazards involved in the use of ^{14}C. The latter concluded that ^{14}C is relatively nonhazardous in tracer amounts in most experiments. Inhaled $^{14}CO_2$ and ingested $H^{14}CO_3^-$ are rapidly excreted; inhaled particulate $Ba^{14}CO_3$ undergoes 95% turnover in about 2 hr. Thymidine-^{14}C has a long biological half-life, however, becoming incorporated into nucleic acids. Handloser (5, pp. 201–202) has considered the health hazards in working with tritium. Tritium gas is less toxic than tritiated water. The latter is readily absorbed by the skin by a factor of 10^4 greater than tritium gas, thus requiring one to wear gloves when handling it.

The *maximum permissible body burdens* and *maximum permissible concentrations* of radioisotopes in air and water for occupational exposure are clearly stated in federal regulations (15, 22). For example, the maximum permissible concentration of soluble ^{14}C in water is 0.02 μCi/ml and of soluble tritium-containing compounds in water is 0.1 μCi/ml. Maximum permissible body burdens of the two isotopes are 400 and 2000 μCi, respectively. In contrast, the maximum permissible body burden for a severely radiotoxic isotope, such as ^{90}Sr, is only 20 μCi; that for ^{239}Pu is a mere 0.04 μCi.

Morgan, in his excellent article (10), has described the effect that the quantity of radionuclide being used has on nuclear safety considerations. Figure 16-1 shows the levels of radioactivity likely to be encountered in various operations.

Morgan points out the hazards involved in each level of operation. At the nanocurie level, the problem is one of reducing the background and preventing contamination of glassware and counting equipment. At the microcurie level, where most radiotracer experiments operate, some care must be taken. and energetic β-sources should be handled with tongs, although soft β-sources need not be treated in this manner. Work with β- and γ-ray emitters should be done over blotting paper in ordinary chemical

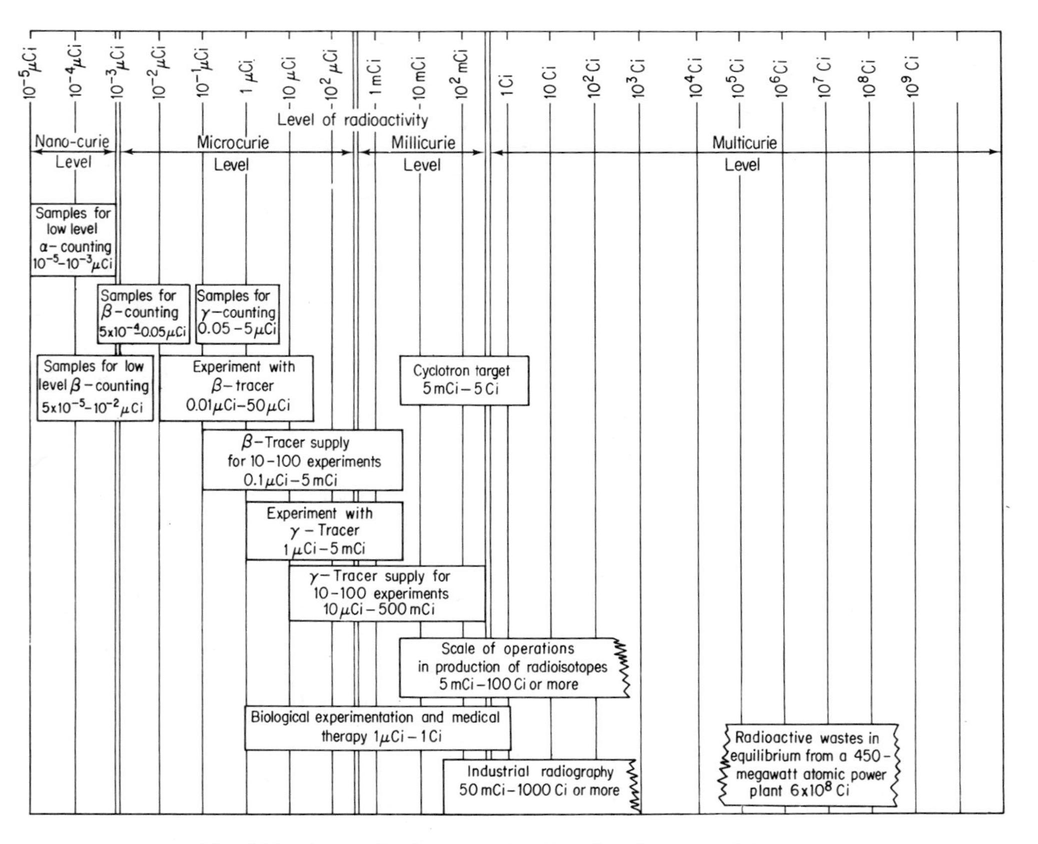

Fig. 16-1. Approximate amounts of radioactive material encountered during various research and industrial operations. From Morgan (10).

hoods. Very little shielding is necessary. All personnel should be equipped with personnel monitoring devices (see Section C). At the millicurie level of operation, sometimes encountered in the storage of primary radioisotopes for tracer experiments, special precautions are necessary. High-volume air-flow hoods, protective clothing, radiation shielding, and "dry runs" of all experimental procedures using inactive material are advised.

Detailed laboratory rules for use of tracer quantities of radioactivity are set forth at the end of this chapter. It will be seen that the fundamental purpose of these safety rules is to prevent ingestion, inhalation, or other entrance of radioisotopes into the body and to reduce the amounts of external irradiation to permissible levels. In general, it is a good principle to treat all radioactive materials as if they were pathogenic bacteria, even though they are being handled by purely chemical techniques. Thus the possibility of unnoticed contamination of the laboratory or one's person should always be considered. In many tracer experiments, however, chemical and fire hazards are far more significant than any hazard from the radiation involved.

C. RADIATION-MONITORING INSTRUMENTATION

In order to determine the extent of possible radiation hazard and/or to detect contamination, radiation-monitoring instruments should be used. Although the institutional radiation safety officer may conduct periodic radiation surveys in tracer laboratories, it is basically the responsibility of the individual isotope user to monitor his own laboratory for radiation hazards. In addition, where millicurie levels of γ-ray emitters are being used, each worker should have proper personnel-monitoring equipment. Among the many references that can be cited in the field of radiological monitoring, the article by Morgan (10) and Handloser's book (5) are perhaps the most suitable for radiotracer users.

1. Area Monitoring

Area monitoring of laboratory facilities is normally carried out by means of portable survey meters, although fixed monitors are used in certain cases. The most common portable monitor is a battery-operated G-M counter (see Chapter 5) with a rate meter. It is a very sensitive detector for relatively low levels of radiation, from either shielded isotopic sources or contamination. *G-M survey meters* are best used for the detection of radiation, not the measurement of radiation-dose rate. If a G-M counter is calibrated for a particular γ-ray energy, however, it can be used to measure the dose rate associated with that γ-ray energy. In general, readings in counts per minute will be more meaningful than readings in milliroentgens per hour.

A second type of survey meter is a portable *ionization chamber* (see

Chapter 5) that utilizes an electrometer. Ionization-chamber survey meters are reliable, hold their calibration well, and are relatively energy independent. Their major use is for measurement of γ-ray or X-ray dose rate at moderate-to-high radiation levels. They are commonly calibrated to read directly in mR/hr.

Various specialized area monitors are available, including such instruments as portable scintillation counters, designed for γ-ray measurement, and air monitors (as for tritium) that pull a volume of air over an ion chamber by means of an air pump or fan. Recently revised permissible concentrations of tritium in air make it necessary to recommend a sensitive electrometer, such as the vibrating-reed type, for tritium-monitoring devices. Photographs of a number of common radiation detection instruments used in area monitoring are shown in Figure 16-2.

Any monitoring instrument is only as good as its calibration, and usually that calibration can only be made for any average set of circumstances. Despite the existence of an mR/hr scale on an instrument, one still needs to establish experimentally the correlation between meter reading and true exposure rate. For γ-ray monitoring, the procedure is fairly simple. The instrument to be calibrated is placed in a known radiation field, and then use of the $R = 6CE$ rule and the inverse square law allows one to make a table of instrument reading versus true exposure rate.

Measurement of β^--radiation fields is quite difficult because of scattering effects. Most survey meters have "β-shields" on them, consisting of a cover that fits over the detector to prevent low-energy β^--particles from entering the gas volume and being counted. Survey meter readings taken with the shield in place and off ("window open" and "window closed" readings) allow one to measure the relative number of low-energy β^--particles present versus the number of high-energy β^--particles and γ-rays. The number of low-energy β^--particles is multiplied by a correction factor (between 1 and 50) for that instrument to yield an estimate of the β^--radiation field.

2. Personnel Monitoring

All personnel who work with radioactivity should use some sort of personnel monitor. A variety of such equipment is commercially available. The *film badge* (7) is one of the most widely used personnel-monitoring devices, for the developed film serves as a permanent record of the individual's cumulative exposure over a defined period and thus has some legal value. A typical film badge for γ-ray and β^--ray monitoring is shown in Figure 16-3. A typical film badge consists of two pieces of dental film covered by light-tight paper in a compact plastic container and is worn on the clothing. Such film badges are sensitive to γ-ray exposures from 10 to 1800 mR and β^--exposure from 50 to 1000 mR. Ionizing radiation darkens the film (see Chapter 10), and the degree of darkening is proportional to the radiation exposure. The film behind

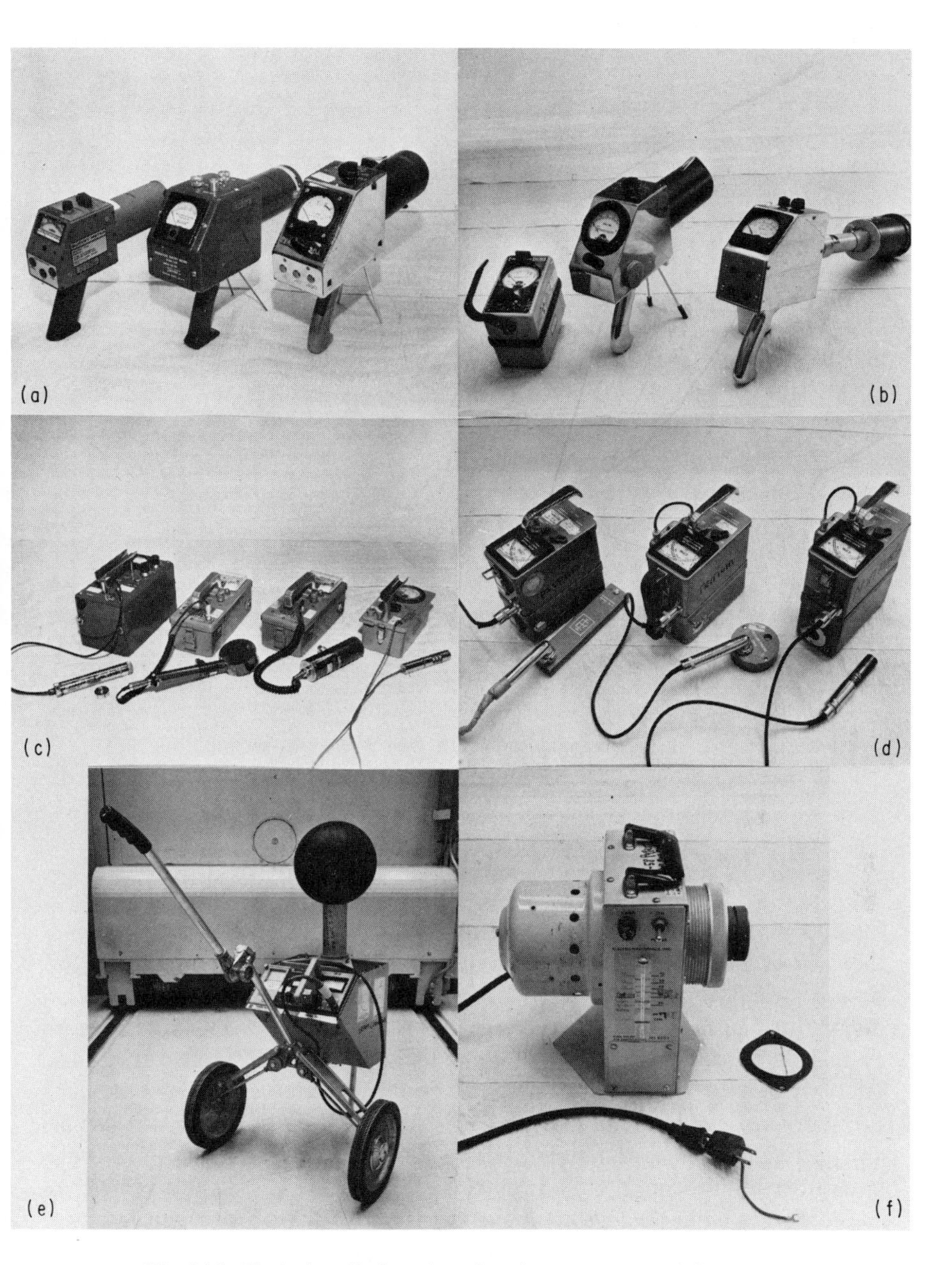

Fig. 16-2. Typical radiation detection instruments used in area monitoring. (a) Cutie-pie meter (high level β-γ meter). (b) Various high level β-γ survey meters. (c) Geiger-Müller meters (low level β-γ meters). (d) Low-level meters with special probes for alpha, tritium and neutron monitoring. (e) Neutron rem counter. (f) Portable, high-volume air sampler. Photo provided by Health Physics and Reactor Operations Staff at Oregon State University.

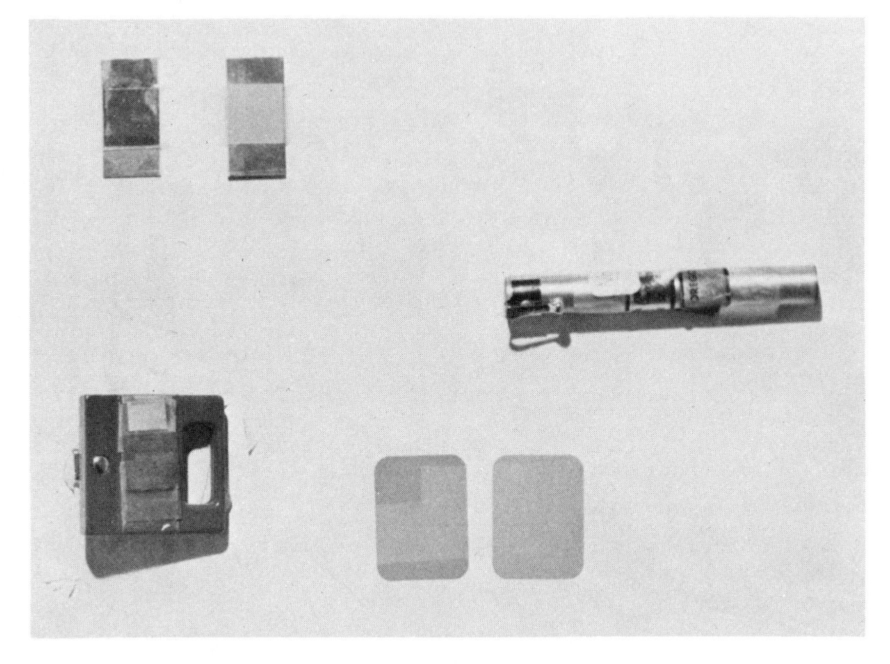

Fig. 16-3. Personnel monitors. The *film badge* shown with its holder, metal inserts and film packs monitors for β-, γ-, X-ray and neutron exposures. The *pocket ionization chamber* shown gives a direct reading of β-, γ-exposure at any time. Photo provided by Health Physics and Reactor Operations staff at Oregon State University.

the open area of the badge gives the total exposure, whereas the film behind the various metal filter discs, which have different absorption properties for β- and γ-radiation, is used to sort out the relative γ- and β-exposures. The major disadvantage of the film badge is the time lag between exposure and evaluation of the developed film. Films are changed and developed at regular intervals of 1 to 2 weeks. In no cases should the interval exceed a month. For workers dealing exclusively with ^{3}H and ^{14}C, the film badge will not be an effective personnel-monitoring device. In this case, urinalysis and routine laboratory monitoring should be employed.

Pocket ionization chambers (about the size of a fountain pen) are designed to be worn by the individual. These devices give a rough estimate of the total exposure over the period from time of charging to time of reading. In the form of dosimeters (electroscopes), the accumulated exposures may be read directly by the wearer at any time; pocket ionization chambers of a condenser type require a separate reading instrument to determine the exposure.

An important new development in personal dosimeters is that of the thermoluminescent dosimeter, or TLD. A typical TLD is made of LiF (a nearly tissue-equivalent material) or CaF_2 with a Mn impurity. Absorption of radiation energy by the dosimeter material excites electrons and holes in the crystal. These displaced electrons and holes remain trapped in the crystal. Heating the crystal releases the trapped energy in the form of light. The light output is then measured by using photomultiplier tubes and the appropriate electronic circuitry. TLDs are sensitive to doses ranging from 10 mrad to 100,000 mrad. The response is independent of the radiation energy for γ-rays in the range from 100 keV to 1.3 MeV. TLDs are very useful as finger dosimeters. They are worn on finger rings and measure the radiation dose given the hands during an experiment.

D. DECONTAMINATION

The use of rubber or disposable polyethylene gloves and protective laboratory coats will usually eliminate the necessity for rigorous skin decontamination. If all handling of open radioisotopic sources is restricted to areas over trays (preferably polished heavy stainless steel) that are lined with absorbent paper, preferably in hoods, laboratory work surface contamination can be kept to a minimum. Accidental contamination, however, will occur on some occasions despite the strictest precautions. A regular habit of monitoring the hands and all working surfaces at the end of each working period must be established in order to detect contamination. Again, it should be recalled that the hazard from tracer doses of isotopic material is primarily contamination of the laboratory, rather than a threat to the health of the radioisotope user. National Bureau of Standards Handbook 48 (17) will serve as a source of information on the control and removal of radioactive contamination in laboratories. Only general decontamination principles are considered in this chapter.

Routine *hand washing* for 2 to 3 min with mild soap and lukewarm water should always follow any handling of isotopic material. After rinsing, the hands should be monitored. If the first washing has not reduced contamination to an acceptable level, a repeated washing, using a soft brush and heavy lather, may be required. Be careful not to scratch or abrade the skin with the brush.

Laboratory clothing should be routinely monitored when high levels of activity and/or hazardous classes of isotopes are in use. Contaminated garments should be washed in the laboratory before being released. Rubber gloves are usually decontaminated easily by simple washing. Leather goods cannot be so readily decontaminated.

Laboratory glassware and implements that have become contaminated must be thoroughly washed so as to eliminate both the radiation hazard and the possibility of cross-contamination of successive experiments. There is a considerable tendency toward the absorption of radioactive materials onto glass. To reduce this problem, never allow radioactive solutions to dry on glass surfaces. In case of unavoidable contamination of glassware, treatment with warm cleaning solution ($K_2Cr_2O_7$ and concentrated H_2SO_4) or with detergent in a sonic bath has proved practical. A 10% solution of EDTA is especially effective in removing radioactive metal ion contaminants. Where the contamination is due to radiosulfate or radiophosphate, treatment with 6 *N* HCl is preferable.

There is a considerable difference in the ease of decontamination of various *laboratory surface materials* (9). Wood, concrete, and soapstone are particularly difficult to decontaminate because of their high porosity. Polished stainless steel, vinyl floor coverings, and strippable plastic coatings are generally the materials of choice with regard to ease of decontamination.

In the *routine cleaning* of a radiotracer laboratory, the possibility of spreading contamination into other areas must be considered. Depending on the activity level and the hazard rating of the isotopes used, it may be necessary to restrict the cleaning implements used to a single laboratory.

E. DISPOSAL OF RADIOACTIVE WASTES, LICENSING, AND TRANSPORTATION OF RADIOACTIVE MATERIALS

Laboratory use of radioisotopes inevitably results in radioactive wastes (i.e., paper wipes, disposable implements, carcasses, excreta, etc.) that must be disposed of without endangering the public. Possible routes of disposal are into sewers, by incineration, by venting directly into the atmosphere, by ground burial, or by disposal at sea. Hopefully, the means of disposal required will be reasonably convenient and economical.

Two general policies are usually followed in waste disposal: one seeks maximum dilution of the waste; the other obtains maximum concentration. Disposal by dilution generally involves release into the sewer, direct release into the air, or incineration of the wastes. Nuclear Regulatory Commission regulations (15, part 20.303) restrict disposal via *sewage* to material that is readily soluble or dispersible in water and to activities below specified levels (e.g., 100 μCi of ^{32}P per day). National Bureau of Standards Handbook 49 (18) discusses in detail such disposal of ^{32}P and ^{131}I wastes. For many radiotracer applications, this method offers both convenience and safety. There is always the possibility, however, that some aquatic organisms may reconcentrate these wastes to hazardous levels.

Direct release into the *air* via hood exhausts is feasible for $^{14}CO_2$ and other volatile or gaseous substances. But dilution by the air flow through the exhaust stack must be such that maximum permissible concentrations are not exceeded (e.g., for ^{14}C, this concentration is 1×10^{-6} μCi/ml air). Specific approval by the NRC is required to dispose of radioactive wastes by *incineration* (15, part 20.305).

Concentration and storage of radioactive wastes may be necessary for materials that cannot be disposed of by dilution. This method is more suitable for high levels of activity, such as reactor wastes (1). The storage may take the form of burial in the soil at a licensed site. *Soil burial* is a particularly attractive method for disposing of carcasses and excreta contaminated with isotopes that are noncombustible (e.g., iron and calcium). The quantity of wastes and the conditions of burial have been clearly specified (15, part 20.304). In a sense, burial amounts to a type of dilution because before the waste substances enter the biosphere, thus affecting man, they will, hopefully, have been diluted below the hazardous level.

The *physical half-life* of the isotopic material in the waste is an important factor in determining the most satisfactory method of disposal, especially for insoluble materials. For tracer levels of isotopes with half-lives of days to a few weeks, merely holding the wastes in storage will reduce the activity to tolerable levels. It can easily be calculated that after ten half-lives the original activity of a radioactive sample will be reduced by a factor of over 1000. Carcasses containing short-lived isotopic material can be either preserved in formalin or frozen for such a term of storage.

The biologist is perhaps most concerned with ^{14}C *waste* disposal. Here the long half-life precludes storage during decay. National Bureau of Standards Handbook 53 (20) describes various recommended methods of ^{14}C waste disposal. Isotopic dilution is a commonly applied method that is often used for ^{14}C disposal. Carbon-14 may be disposed of in any manner, provided that it is intimately mixed with stable carbon in the same chemical form, in a ratio that never exceeds 1 μCi of ^{14}C for every 10 g of stable carbon. Bureau of Standards Handbook 65 (21) describes precautions in handling human bodies containing isotopes; the same principles apply to animal carcasses.

In general, all people who possess or otherwise work with radioactivity in the United States must possess some sort of authorization or license issued by the state or the U.S. Nuclear Regulatory Commission. Exceptions are made for small quantities of certain radionuclides. For further information, one is advised to contact the U.S. Nuclear Regulatory Commission or one of its regional compliance offices. Transportation of radioactive material in the United States is now under the jurisdiction of the Department of Transportation. A very readable summary of the new and complex regulations concerning transportation of radioactive materials is found in Reference 23.

F. RADIOISOTOPE LABORATORY SAFETY RULES

It is not possible to specify working rules that will apply in all situations. As noted, the activity level and the hazard class of the isotope being used will determine the degree of precaution required. The following list of regulations, however, is intended to minimize internal and external hazards, to prevent contamination of the laboratory, and to comply with existing NRC or state (agreement states) rules (15).

1. Coats and other *personal belongings*, including books (except those required for work), should not be brought into a laboratory where they may become contaminated.
2. Eating, drinking, storing, or preparing *food*, smoking, or applying cosmetics are all forbidden or discouraged in any area where radioactive materials are stored or used.
3. *Direct contact* with radioactive materials must be avoided by using protective laboratory coats and by wearing rubber or disposable plastic gloves. Such protective clothing is not to be removed from the laboratory.
4. *Pipetting* liquids of any type by mouth or the performance of any similar operation by mouth suction is not permitted.
5. Complete *records* of receipts, transfers, and disposal of radioactive materials must be kept.
6. A *film badge* should be worn at all times when working with radioactivity, except for the case of ^{3}H, ^{14}C, or other low energy β^{-}-emitters where an alternative method of personal dosimetry should be used such as urinalysis.
7. Work should be carried out under a *hood* in all cases where radioactive material may be lost by volatilization, dispersion of dust, or by spraying or splattering. Wherever possible, work with closed containers.
8. All radioactive samples should be properly *labeled* with the isotope and activity indicated and should be covered.
9. *Liquid wastes* should not be poured into the drain or contaminated apparatus washed in the sink unless the levels of activity entering the sewer system have been calculated as permissible (15, part 20.303).
10. *Solid wastes* and contaminated articles (corks, paper wipes, and the like) should be disposed of in designated containers and should never be placed in ordinary trash receptacles.
11. The disposal of *gaseous waste* through the hood can be carried out only after careful examination of the air-dilution factor.
12. The *storage* of all radioactive material must be in properly designated locations.
13. At the close of a working period, the laboratory work surfaces should be carefully *monitored.*
14. Before leaving the laboratory after working with active materials, each person should *wash* his *hands* thoroughly and check them with a monitoring instrument.
15. All *laboratory glassware* and equipment should be properly decontaminated after use before being returned to general usage.
16. It is desirable to decontaminate one's hands and work surfaces com-

pletely, but the following arbitrary *surface contamination tolerances* (as measured by a G-M survey meter with a thin end-window) may be allowed after efforts at decontamination:

Hands	350 cpm
Working surface	250 cpm

The arbitrary nature of any such tolerances should be recognized, although, on the other hand, absolutely complete decontamination is not always feasible.

17. All *spills* of radioactive material must be reported to the person in charge of radiation safety immediately. In the event of a spill,
    a. any liquid should be blotted up immediately.
    b. attempts should be made to prevent spreading of the activity.
    c. the spill area should be isolated, identified as to the nature of the contaminant, and access to the area restricted.
    d. a radiation survey of the area and all involved personnel should be made immediately.
18. In the event of a *wound* that is possibly contaminated with radioactive material,
    a. flush wound in running water.
    b. immediately seek the aid of the person responsible for radiation safety.
19. Routine *urine analyses* should be carried out by means of liquid scintillation counting (3) whenever millicurie levels of carbon or tritium are handled.

BIBLIOGRAPHY

1. AMPHLETT, C. B. *Treatment and Disposal of Radioactive Wastes.* New York: Pergamon, 1961.

2. CASARETT, A. P. *Radiation Biology.* Englewood Cliffs, N.J.: Prentice-Hall, 1971.

3. CEMBER, H. *Introduction to Health Physics.* Oxford: Pergamon, 1969.

4. HANDLOSER, J. S. *Health physics instrumentation.* New York: Pergamon, 1959.

5. HANDLOSER, J. S. "Tritium health physics considerations." In S. Rothchild (Ed.), *Advances in Tracer Methodology.* New York: Plenum Press, 1963, Vol. 1.

6. HENRY, H. F. *Fundamentals of Radiation Protection.* New York: Wiley, 1969.

7. GETTER, E. S. and H. BLATZ. "Film measurement of radiation dose," *Nucleonics* **10**(10), 43 (1952).

8. JOHNS, H. E. *The Physics of Radiology.* 3rd ed. Springfield, Ill.: Charles C. Thomas, 1969.

9. LANE, W. *et al.* "Contamination and decontamination of laboratory bench-top materials," *Nucleonics* **11**(8), 49 (1953).

10. MORGAN, K. Z. "Techniques of personnel monitoring and radiation surveying." In A. H. Snell (Ed.), *Nuclear Instruments and Their Uses.* New York: Wiley, 1962, Vol. 1.

11. *Radiation Protection in Educational Institutions.* National Council on Radiation Protection and Measurement Report No. 32 (NCRP, Washington, D.C.), July 1966.

12. Skipper, Howard E. "The hazard involved in the use of carbon-14," *Nucleoneics* **10**(2), 40–44 (1952).
13. Taylor, Lauriston. "Radiation-protection standards," *Nucleonics* **21**(3), 58–60 (1963).
14. Tolbert, B. M., N. Garden, and P. T. Adams. "Special equipment for C^{14} work," *Nucleonics* **11**(3), 56–58 (1953).
15. U.S. Code of Federal Regulations, Title 10. Atomic energy, Part 20. Standards for protection against radiation.
16. U.S. Department of Commerce, National Bureau of Standards. *Safe Handling of Radioactive Isotopes.* Handbook No. 42. Washington, D.C., 1949.
17. U.S. Dept. of Commerce, National Bureau of Standards. *Control and Removal of Radioactive Contamination in Laboratories.* Handbook No. 48. Washington, D.C., 1951.
18. U.S. Dept. of Commerce, National Bureau of Standards. *Recommendations for Waste Disposal of Phosphorus-32 and Iodine-131 for Medical Users.* Handbook No. 49. Washington, D.C., 1951.
19. U.S. Dept. of Commerce, National Bureau of Standards. *Radiological Monitoring Methods and Instruments.* Handbook No. 51. Washington, D.C., 1952.
20. U.S. Dept. of Commerce, National Bureau of Standards. *Recommendations for the Disposal of Carbon-14 Wastes.* Handbook No. 53. Washington, D.C., 1953.
21. U.S. Dept. of Commerce, National Bureau of Standards. *Safe Handling of Bodies Containing Radioactive Isotopes.* Handbook No. 65. Washington, D.C., 1958.
22. U.S. Dept. of Commerce, National Bureau of Standards. *Maximum Permissible Body Burdens and Maximum Permissible Concentrations of Radionuclides in Air and Water for Occupational Exposure.* Handbook No. 69. Washington, D.C., 1959.
23. *Office of Hazardous Materials Newsletter* (Dept. of Transportation, Washington, D.C. 20590), Vol. II, No. 8, February 1972.

17

Radioanalytical Techniques

A. INTRODUCTION

In this chapter we wish to discuss some of the more popular analytical techniques that involve the use of radioactivity as an integral part of the technique. These radioanalytical techniques are not ultimate and all encompassing, but they are important weapons in the arsenal of an analyst. The general feature of radioanalytical techniques that is not always shared by ordinary analytical techniques is their high sensitivity. Since the experimenter is usually detecting sample activities of the order of 0.1 μCi, he may typically be tracing or measuring the behavior of a very small number of nuclei—that is, a very small mass of material.

In our discussion of radioanalytical techniques, we shall distinguish between those that involve the use of radiotracers, such as isotope dilution and the radiometric methods, and those that do not, such as activation analysis. We shall reserve until Chapter 19 the discussion of the numerous applications of radiotracers in testing and evaluating new chemical separation procedures. Here radiotracers provide an important way of determining the yield at each step of the separation and thus materially aid in the design of new analytical procedures.

B. ISOTOPE DILUTION ANALYSIS (IDA)

1. Direct IDA

The basic idea of isotope dilution analysis is to measure the changes in specific activity of a substance upon incorporation into a system containing an unknown amount of that substance. There are several types of isotope dilution analysis (IDA), and we start our discussion by considering *direct IDA*.

In direct IDA, we are faced with the problem of determining the amount of some inactive material A in a system. Let us define this unknown amount as x grams. To the system containing x grams of inactive A, one adds y grams of active material A^* of known activity D. Thus we know the specific activity of the added active material, S_1. That is,

$$S_1 = \frac{D}{y} \tag{17-1}$$

After thoroughly mixing the active material A^* with the inactive A in the system, one isolates, not necessarily quantitatively, and purifies a sample of the mixture of A and A^* and measures its specific activity, S_2. Clearly, conservation of material says that

$$S_2 = \frac{D}{x + y} \tag{17-2a}$$

Since, as Equation (17-1) tells us,

$$y = \frac{D}{S_1}$$

We can substitute for y in Equation (17-2a), obtaining

$$S_2 = \frac{D}{x + D/S_1} \tag{17-2b}$$

Rearranging, we get

$$x = \frac{D}{S_2} - \frac{D}{S_1} = \frac{D}{S_1}\left(\frac{S_1}{S_2} - 1\right) = y\left(\frac{S_1}{S_2} - 1\right) \tag{17-3}$$

This is the basic equation of direct isotope dilution analysis. The unknown amount x of material A is given in terms of the amount y of added labeled material A^* and the two measured specific activities S_1 and S_2.

Let us consider a practical problem to illustrate the use of this technique. A protein hydrolysate is to be assayed for aspartic acid. Exactly 5.0 mg of aspartic acid, having a specific activity of 0.46 μCi/mg, is added to the hydrolysate. From the hydrolysate, 0.21 mg of highly purified aspartic acid, having a specific activity of 0.01 μ Ci/mg, can be isolated. How much aspartic acid was in the original hydrolysate?

Using Equation (17-3), we say that

x = number of mg aspartic acid in original hydrolysate

$y = 5.0$ mg

$S_1 = 0.46\ \mu$Ci/mg

$S_2 = 0.01\ \mu$Ci/mg

$$x = y\left(\frac{S_1}{S_2} - 1\right) = (5.0)\left(\frac{0.46}{0.01} - 1\right)$$

$x = 225$ mg aspartic acid

Thus by isolating a small fraction of the added aspartic acid and measuring the diminution in its specific activity, the aspartic acid content of the original sample can be determined. Note that this example involved a large change in specific activity upon dilution. Poor experimental design or other circumstances may lead to a small change in specific activity upon dilution. In such cases, the results obtained from IDA involve a small difference between two large numbers and are quite uncertain.

2. Inverse IDA

Inverse IDA is a simple variant on the basic direct IDA. In inverse IDA, we measure the change in specific activity of an unknown radioactive material A^* after diluting it with inactive A. Specifically, let us assume that we have q mg (where q is unknown) of a radioactive substance A^* whose specific activity S_q is known (i.e., $S_q = D/q$). (S_q can be measured by isolating a small portion of A^*, weighing it, and measuring its activity.) Let us add r mg of inactive A to A^* and thoroughly mix the A and A^*. Suppose that we then isolate and purify some of the mixture and measure its specific activity S_r. Note that $S_r = D/q + r$. And so we have

$$S_r = \frac{D}{q + r} = \frac{qS_q}{q + r} \tag{17-4}$$

by substitution. Rearranging, we have

$$S_r(q + r) = qS_q \tag{17-5}$$

$$\frac{r}{q} = \frac{S_q}{S_r} - 1 \tag{17-6}$$

$$q = \frac{r}{[(S_q/S_r) - 1]} \tag{17-7}$$

Equation (17-7) is the basic equation of inverse isotope dilution analysis and says that the unknown amount q of active material A^* can be deduced by adding r g of inactive material A to A^* and measuring the specific activities before and after the addition S_q and S_r, respectively.

3. General Comments

Certain general comments can be made about the experimental techniques used in isotope dilution analysis. First, reagents and tracers of high purity are necessary. They must not contain any spurious radioactivity or any unknown compounds, for the presence of either could effect the specific activity of substances being analyzed. Although pure reagents and tracers are generally available commercially, it would be wise to check for contaminants before use.

One of the key steps in any isotope dilution analysis concerns the isolation and purification of the diluted activity, plus the measurement of its specific activity. Two techniques are usually preferred for the separation: precipitation, and solvent extraction. As a purification step, precipitation has the advantage that the precipitate can easily be weighed at the time of separation, thereby allowing a quick determination of the specific activity. The main problem with the use of precipitation techniques involves the occurrence of *coprecipitation* phenomena, in which unwanted materials are precipitated along with the desired substance, thus altering the sample specific activity. Precipitation techniques are used primarily for the isolation of inorganic components.

Solvent extraction is a frequently employed technique in isotope dilution analysis. It gives very clean separations, resulting in high-purity samples. It has the disadvantage of requiring further chemical processing in order to determine the mass of material isolated and the specific activity. Other separation techniques, such as electroplating, ion exchange, and paper chromatography, have also been used in IDA. In measuring the specific activity of the isolated diluted sample, the sample mass must be measured. This process is usually done by gravimetric, colorimetric, or spectrophotometric means.

One must be aware of the possible occurrence of certain problems in isotope dilution analysis. One of them is *incomplete isotopic exchange*, in which the active and inactive atoms do not mix. This lack of exchange can be due to differing physical and chemical states of tracer and inactive materials. Steps must be taken to ensure complete exchange, for it is a fundamental assumption of the method that the active and inactive material have been completely mixed before sampling. One should also be sure that the labeled position in any compound is relatively inert. If the atom in question is very labile, one can get a reduction in specific activity without any dilution having taken place. In order to compare specific activities, all samples must be counted under identical conditions with proper corrections being made for varying self-absorption in samples of different mass (see Chapter 13). Finally, to reduce uncertainty in the results, both diluted specific activities must be measured precisely so that the ratio of these numbers is well defined.

In summary, we can say that isotope dilution analysis is a highly sensitive, selective analytical method capable of high precision. It offers the opportunity to determine the amount of material present in a system without the need for a quantitative separation of the material from that system. This fact explains its popularity for analyzing organic materials. The conditions most favorable for the employment of isotope dilution analysis are as follows:

1. Quantitative isolation of unknown material not possible, or
2. Rapid analysis needed, or
3. Material being analyzed present in very low concentrations so that ordinary quantitative procedures will suffer by absorption loss of material on walls of beakers, etc., or
4. Only a part of the total amount of substance being determined is available

for use. For example, if one wishes to determine the amount of a metal ion in human blood, or a large lake, it is not feasible to isolate all the material in the system.

The applications of isotope dilution analysis cited in the literature are myriad. Perhaps the best, up-to-date summary of these applications is the book by Tölgyessy, Braun, and Kyrs (10). Table 17-1, taken from the excellent discussion of Duncan and Cook (5), shows some typical applications of IDA.

4. Special IDA Techniques

Substoichiometric isotope dilution analysis was first developed by Ruzicka and Stary (9) as a variant on the basic IDA technique. The basic idea of substoichiometric IDA is to isolate equal but substoichiometric amounts of both the diluted and the undiluted substance being analyzed and count these samples. Since the mass of both samples is the same, the specific activities in Equations (17-3) and (17-7) can be replaced by the activities. In this way, the sometimes tricky task of measuring specific activities is avoided. The key to the technique is obviously whether the analyst can isolate exactly equal quantities of both diluted and undiluted sample.

Another variant on the basic isotope dilution technique is that of *double isotope dilution*, or a variant thereof, as first proposed by Block and Anker (1). It is used in reverse IDA where the specific activity of the original unknown radioactive material A^* cannot be measured for some reason. Hence a second dilution is made to determine the specific activity of the original sample.

Consider a system containing an unknown amount q of some active substance A^* whose specific activity S_q cannot be measured. Take two equal aliquots of this unknown substance A^*. Add r mg of inactive A to one aliquot and p mg of inactive A to the other aliquot. Measure the specific activities of the two aliquots, S_r and S_p, respectively. For the first sample of specific activity S_r, we have, using Equation (17-7),

$$q = \frac{r}{[(S_q/S_r) - 1]} \qquad S_q = S_r + \frac{r}{q} S_r \tag{17-8}$$

For the second sample, we have

$$q = \frac{p}{[(S_q/S_p) - 1]} \qquad S_q = S_p + \frac{p}{q} S_p \tag{17-9}$$

Setting the two expressions for S_q equal and rearranging yield

$$q = \frac{rS_r - pS_p}{S_p - S_r} \tag{17-10}$$

One major difficulty with double isotope dilution analysis is that, because of

TABLE 17-1

Isotope Dilution Analysis*

Element or Compound	Label	Sample	Outline of Method	Results
Hydrogen	^{2}H	Alkali metals and alloys	Heat weighed metal sample in closed container with known quantity of $^{2}H_2$ gas to equilibrate with hydride, etc. in sample. Measure masses 2, 3, and 4 in gas with mass spectrometer.	On 2 g sample H-content down to ± 2 ppm can be determined.
Cobalt	^{60}Co	Steels and Ni alloys	Dissolve sample and add ^{60}Co solution. Extract iron as chloride with ether. Precipitate potassium cobaltinitrite, dissolve and electrodeposit hydrated cobaltic oxide. Dry anode, weigh and determine γ-activity.	On few g sample containing 0.14–0.154% Co maximum error was 0.02%.
γ-isomer of hexachlorobenzene	^{36}Cl or ^{14}C	Mixture with other isomers	Add approximately 120 mg labeled γ-hexachlorobenzene to 1 g crude mixture of isomers. Separate γ-isomer by multiple solvent extraction; dry, weigh, and determine β-activity.	Typically $13.9 \pm 0.15\%$ accuracy.
Penicillins	^{2}H, ^{35}S or ^{14}C	Production broths	Add labeled penicillin in known quantity and precipitate the *N*-ethyl-piperidine salt.	For single sample in duplicate $\pm 6\%$ at 95% confidence limit.
Gibberellic acid	^{2}H	Production broths	Add 100 mg labeled acid to broth containing approximately same amount. Isolate and purify by solvent extraction and recrystallization. Burn product and estimate deuterium content by infrared.	Overall accuracy $\pm 4\%$.
Vitamin B_{12} (cyano-cobaltamin)	^{60}Co	Fermentation broth, feed supplements, vitamin capsules, etc.	Add labeled vitamin to sample. Purify by solvent extraction, precipitation, or ion exchange. Measure yield by spectrophotometry and radioactivity by γ-counting.	For samples containing down to 100 μg at concentration of 0.1 μg/ml.

Element or Compound	Label	Sample	Outline of Method	Results
Corticosterone	^{14}C	Human plasma	Add labeled compound to plasma. Separate by solvent extraction with CCl_4/CH_2Cl_2 mixture and paper chromatography. Extract steroid from paper and estimate ^{14}C by β-counting and mass of steroid by fluorescence.	Accuracy for 0.5 μg/100 ml about $\pm 20\%$; >2 μg/100 ml, about $\pm 5\%$. For more than 100 μg, accuracy is typically $\pm 5\%$.
m-nitrotoluene	^{14}C	Products of nitration of toluene	Nitrate ^{14}C-labeled toluene; add inactive carriers of *o*, *m*, *p* nitrotoluene and separate by fractionation; determine mass and ^{14}C content of metafraction.	For more than 100 mg, accuracy is typically $\pm 5\%$.

*Reprinted from J. F. Duncan and G. B. Cook. *Isotopes in Chemistry*. © 1968. Oxford University Press. By permission of the Clarendon Press, Oxford.

the double dilution, the specific activities involved become low and therefore uncertain. As a result, there is usually considerable uncertainty attached to double IDA.

C. RADIOMETRIC TECHNIQUES

The basic idea behind all *radiometric* techniques of analysis is to have radioactive reagent R^* of known activity combine quantitatively with some unknown amount of material U to form a radioactive addition product R^*U. By measuring the activity of the product R^*U, the original amount of unknown material U is deduced. The advantages of such techniques are the high sensitivity due to the use of radioactivity and the requirement that the product R^*Q need not be chemically pure. All that is required is that R^*Q not contain any spurious radioactivity. The disadvantages of these techniques are that the reaction between R^* and U must be quantitative and there must not be another species present in the system besides U that will react with R^*. Several discussions of radiometric techniques are in the literature, including the monograph on radiometric titrations by Braun and Tölgyessy (3) and the survey of radiometric techniques by Bowen (2).

One of the radiometric techniques is that of *precipitation with a radioactive reagent*. For example, the halides can be precipitated with ^{110}Ag, silver precipitated with ^{131}I, the sulfates and chromates precipitated with ^{212}Pb or ^{131}Ba, while Al, Be, Bi, Ga, In, Th, U, Zr, and the rare earth elements can be precipitated with $^{32}PO_4^{\equiv}$. This method suffers from lack of selectivity and the need to make careful corrections for self-absorption in counting the precipitates.

Recently some interest has been shown in radiometric techniques that involve the *measurement of very low concentrations of substances in water* by causing these materials to react quantitatively with a radioactive solid and thus causing the solid to dissolve. The increase in solution activity is then measured. This technique has been used to measure the concentration of dissolved oxygen in water and dichromate in water (2). Dissolved oxygen can react with Tl metal labeled with ^{204}Tl as shown below:

$$2\,Tl(s) + \tfrac{1}{2}\,O_2 + H_2O \longrightarrow 2\,Tl^+ + 2\,OH^-$$

The increase in ^{204}Tl activity in the water is measured to determine the oxygen content of the water.

One of the most popular radiometric techniques is that of *radiometric titration*. In a radiometric titration, the unknown is titrated with a radioactive reagent and the radioactivity of product or supernate, etc. formed by the chemical reaction of the reagent and the unknown material is monitored as a function of titrant volume to determine the endpoint. This means that the radioactive component being followed must be isolated during the titration and its activity measured—that is, a discontinuous titration.

The classic example of a radiometric titration concerns the titration of some unknown material with a radioactive reagent to give a radioactive precipitate. In this case, the activity of the supernatant or the precipitate can be followed as a function of titrant volume, as shown in Figure 17-1. In this type of titrations, the tracer must have a long half-life and must emit high-energy β^-- or γ-rays so as to minimize self-absorption corrections (assuming, as is common practice, that the supernatant or precipitate is removed from the system and counted in an external sample counter after the addition of each volume of titrant).

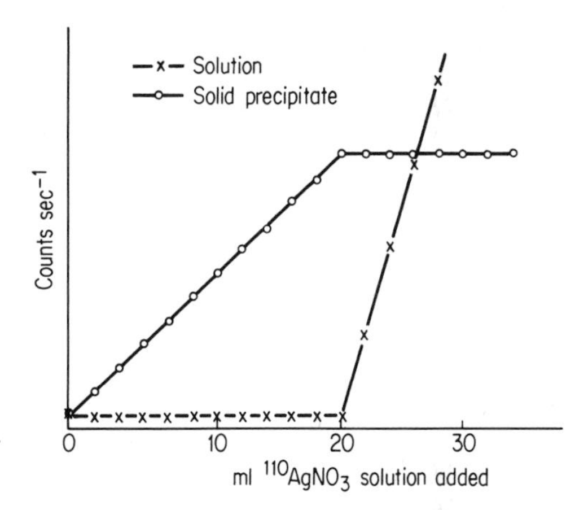

Fig. 17-1. Radiometric titration of $^{110}AgNO_3$ with NaCl.

D. ACTIVATION ANALYSIS

1. Basic Description of Method

Activation analysis is an analytical technique that allows one to measure the amount of a given element X contained in some material Y. The basic steps in the activation analysis technique are as follows:

1. Irradiate Y with a source of ionizing radiation so that X will change into X*, a radioactive isotope of X.
2. Using chemical or instrumental techniques, "isolate" X and X* from all other elements in Y (not necessarily quantitatively) and measure the activity of X*.†
3. Calculate the amount of X present.

†Chemical "isolation" of the activity of interest is performed by simply separating it chemically from all other activities. Instrumental "isolation" of the activity of interest involves the detection of radiation that can be uniquely identified with the activity of interest, such as γ-ray photopeak.

These basic steps are shown schematically for neutron activation analysis in Figure 17-2.

How does one calculate the amount of X present, knowing the activity X* produced in the irradiation? It can be shown that

$$D_{x^*} = N_x \sigma \phi (1 - e^{-\lambda_{x^*} t_i}) e^{-\lambda_{x^*} t_d} \tag{17-11}$$

where D_{x^*} is the activity of X* present at time t_d after the end of the bom-

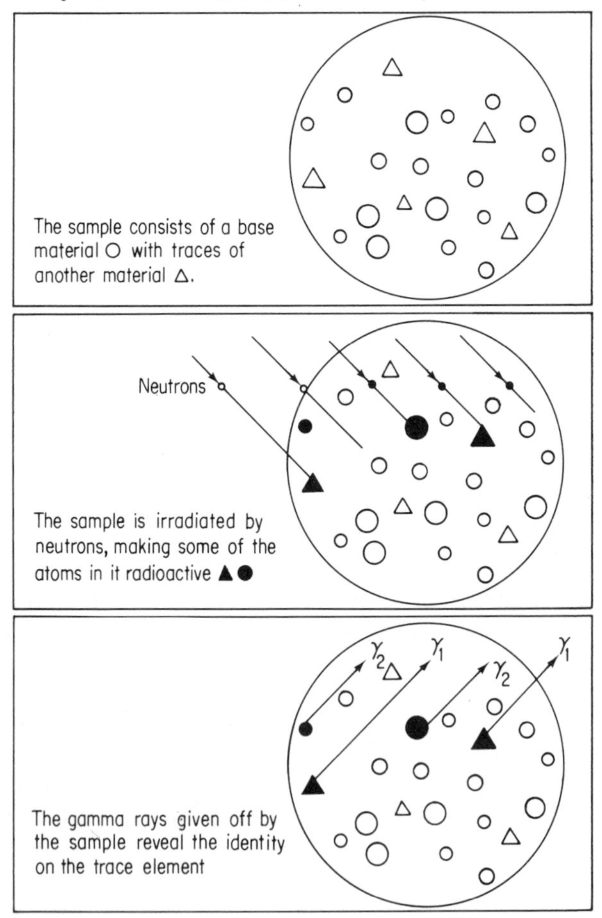

Fig. 17-2. A schematic view of activation analysis. From Corliss (4).

bardment, N_x is the number of X nuclei present initially, σ is the nuclear reaction cross section, ϕ is the flux of activating particles, t_i is the length of the irradiation, and λ_{x^*} is the decay constant of X*. From Equation (17-11), one could calculate N_x from D_{x^*}, knowing all the other variables. But a simpler technique is to irradiate and count a known amount of X under the same conditions used for Y. Then

$$\text{Mass X in Y} = (\text{known mass X})\left(\frac{\text{activity of X* in Y}}{\text{activity of X* in known}}\right)$$

2. Advantages and Disadvantages of Activation Analysis

Since we know that $A = C\lambda N$, where A is the measured radioactivity of a sample, λ is the decay constant, N is the number of radionuclei present, and C is a constant representing the detection efficiency, we know that just a few radioactive nuclei need to be present in order to give easily measurable activities. Use of activation analysis can therefore lead to measurement of elemental abundances of the order of 10^{-6} to 10^{-12} g. The actual detection sensitivities for various elements for activation analysis, as practiced by a commercial activation analysis service, are shown in Figure 17-3. The high sensitivity of activation analysis, in comparison to other analytical techniques, is emphasized by Table 17-2, which shows the activation analysis sensitivities compared to those of more conventional techniques. As shown in Figure 17-3 and Table 17-2, activation analysis is perhaps the most sensitive multielemental analytical technique.

Although the high sensitivity of activation analysis is perhaps its most striking advantage, there are other favorable aspects as well. Activation analysis is basically a multielemental technique, for many elements in the sample will become radioactive during irradiation; and if each of these elements can be "isolated" chemically or instrumentally, their abundances may be determined simultaneously. Since a great deal of activation analysis can be done instrumentally, it can be a nondestructive method of analysis. Numerous tests have shown that with careful experimental manipulation, activation analysis is an accurate ($\sim 1\%$ accuracy) and precise ($\sim 5\%$ precision) method of measuring elemental concentrations.

Activation analysis is not without its drawbacks, however. Among them are the need to use expensive equipment and irradiation facilities, the inability to determine the chemical state of the elements in question, the need to work with higher levels of radioactivity than ordinarily encountered in tracer applications, with their attendant radiation safety and legal problems, the long times needed to complete some analyses, and complex analysis sometimes needed to unscramble the γ-ray spectra in a given experiment.

Periodic Table of the Elements

Activation analysis sentivities

1 H NA																	2 He NA
3 Li 0.0008$_p$	4 Be 15.$_p$											5 B 1.1$_p$	6 C 0.1c	7 N 1.c	8 O 1.c	9 F 0.4	10 Ne 2.
11 Na 0.004	12 Mg 0.5											13 Al 0.004	14 Si 1. FS	15 P 0.2 b	16 S 4. bFS	17 Cl 0.05	18 Ar 0.002
19 K 0.2	20 Ca 4.	21 Sc 0.001	22 Ti 0.1	23 V 0.002	24 Cr 0.3	25 Mn 0.0001	26 Fe 2. FS	27 Co 0.01	28 Ni 0.7	29 Cu 0.002	30 Zn 0.1	31 Ga 0.002	32 Ge 0.1	33 As 0.005	34 Se 0.01	35 Br 0.003	36 Kr 0.01
37 Rb 0.02	38 Sr 0.005	39 Y 0.4	40 Zr 0.8	41 Nb 3.	42 Mo 0.1	43 Tc NA	44 Ru 0.04	45 Rh 0.005	46 Pd 0.03	47 Ag 0.004	48 Cd 0.005	49 In 0.00006	50 Sn 0.03	51 Sb 0.007	52 Te 0.03	53 I 0.002	54 Xe 0.1
55 Cs 0.001	56 Ba 0.02	57 ^{L}La 0.005	72 Hf 0.0006	73 Ta 0.1	74 W 0.004	75 Re 0.0008	76 Os 1.	77 Ir 0.0003	78 Pt 0.1	79 Au 0.0005	80 Hg 0.08	81 Tl 1.b	82 Pb 0.5$_p$	83 Bi 1.b	84 Po NA	85 At NA	86 Rn NA
87 Fr NA	88 Ra NA	89 ^{A}Ac NA															

Sensitivities are expressed as the micrograms of the naturally occurring element that must be present in the entire sample to be detected and determined by the Activation Analysis Service at General Atomic. By special arrangement, the sensitivities of many of the elements can be increased up to 100-fold. Sensitivities are for interference-free conditions.

L	58 Ce 0.2	59 Pr 0.03	60 Nd 0.03	61 Pm NA	62 Sm 0.001	63 Eu 0.0001	64 Gd 0.007	65 Tb 0.03	66 Dy 0.00003	67 Ho 0.003	68 Er 0.002	69 Tm 0.2	70 Yb 0.02	71 Lu 0.0003
A	90 Th 0.2	91 Pa NA	92 U 0.003	93 Np NA	94 Pu NA	95 Am NA	96 Cm NA	97 Bk NA	98 Cf NA	99 Es NA	100 Fm NA	101 Md NA	102 No NA	103(Lw) NA

Key

Atomic number → 33 As ← Symbol

0.005 ← Sensitivity in micrograms (interference free)

FS - Fast neutrons. Fission spectrum; b - Beta count. p - Reactor pulse;
c - Bremsstrahlung radiation required. NA - Analysis not normally performed at GGA

Fig. 17-3. Table of activation analysis sensitivities as offered by the General Atomic Company, San Diego, Calif.

TABLE 17-2

Comparison of Sensitivity by Neutron Activation Analysis and Spectrochemical Methods*

Element	Z	Neutron Activation Analysis	Spectrochemical Methods			
			Copper Spark	DC Arc	Flame Photometry	Colorimetry
Li	3		0.002		0.02	
Be	4		0.002		250	0.04
B	5		0.1		10	
Na	11	0.00035	0.1	20	0.0002	
Mg	12	0.03	0.01	0.1	1	0.06
Al	13	0.00005	0.1	0.2	20	0.002
Si	14	0.05	0.1	2		0.1
P	15	0.001	20	50		0.001
S	16	0.2				
Cl	17	0.0015				0.04
K	19	0.004	0.1		0.01	
Ca	20	0.19	0.1		0.03	
Sc	21	0.0001	0.005			
Ti	22		0.1		2	0.03
V	23	0.00005	0.05		2	0.2
Cr	24	0.01	0.05	2	1	0.02
Mn	25	0.00003	0.02	0.2	0.1	0.001
Fe	26	0.45	0.5	0.2	2	0.05
Co	27	0.001	0.5		10	0.025
Ni	28	0.0015	0.1	4	10	0.04
Cu	29	0.00035		0.2	0.1	0.03
Zn	30	0.000	2	20	2000	0.016
Ga	31	0.00235	1		1	
Ge	32	0.002				0.08
As	33	0.0001	5	10		0.1
Se	34	0.0025				
Br	35	0.00015				
Rb	37	0.0015	0.2		0.1	
Sr	38	0.03	0.5		0.1	
Y	39	0.0005	0.01		50	
Zr	40	0.015	0.1			0.13
Nb	41	0.5	0.2		20	50
Mo	42	0.005	0.05		30	0.1
Ru	44	0.005			10	0.2
Rh	45				1	0.2
Pd	46	0.00025	0.5		1	0.1
Ag	47	0.0055		0.1	0.5	0.1
Cd	48	0.0025	2	4	20	0.01
In	49	0.000005	1		1	0.2
Sn	50	0.01		0.2	10	
Sb	51	0.0002	5	4		0.03
Te	52	0.005	0.5		100	0.5
I	53	0.0001				
Cs	55	0.0015	0.5		1	
Ba	56	0.0025	0.1		3	
La	57	0.0001	0.05		5	
Ce	58	0.005	0.5		20	0.25
Pr	59	0.0001	0.2		100	
Nd	60	0.005	0.2		50	
Sm	62	0.00003	0.2		100	
Eu	63	0.0000015	0.02			

TABLE 17-2 (Continued)

Element	Z	Neutron Activation Analysis	Spectrochemical Methods: Copper Spark	DC Arc	Flame Photometry	Colorimetry
Gd	64	0.001	0.1		10	
Tb	65	0.0002				
Dy	66	0.0000015	0.5		10	
Ho	67	0.00002	0.2			
Er	68	0.001	0.5			
Tm	69	0.0001	0.05			
Yb	70	0.0001	0.1			
Lu	71	0.000015	2			
Hf	72	0.001	0.5			
Ta	73	0.00035	1			
W	74	0.00015	0.5			0.4
Re	75	0.00003	2			0.05
Os	76	0.001				1
Ir	77	0.000015	5			2
Pt	78	0.005	0.02			0.2
Au	79	0.00015	0.2		200	0.1
Hg	80	0.0065	5	2	100	0.08
Tl	81	0.03		0.2	1	
Pb	82	0.1	0.05	0.2	20	0.03
Bi	83	~0.02	0.2	0.2	300	1
U	92	0.0005	1		10	0.7

*From W. W. Meinke, "Trace-Element Sensitivity: Comparison of Activation Analysis with Other Methods," *Science* **121**, 177–184 (1955). Sensitivities in μg/ml based on gross normalizing factors introduced to adjust values to those for neutron activation analysis with a reactor flux of 10^{13} neutrons/cm^2/sec.

3. Practical Considerations in Activation Analysis

In order to better understand the practical details of how activation analysis may be applied to the solution of a given problem in elemental analysis, let us consider the various aspects of a typical activation analysis problem. To make our discussion more concrete, let us consider a specific problem, the measurement of the aluminum content of rocks and meteorites (7). The choice of this problem as an example was dictated by its pedagogic simplicity and the fact that conventional chemical analyses of aluminum in rocks are known to be inaccurate for low aluminum concentrations and, in general, not very precise.

The first step in an activation analysis procedure is *sample preparation.* The unknown and known samples (sometimes referred to as the unknown and standard samples) should have the same size, composition, and homogenity insofar as possible, to ensure that any attenuation of the initial activating radiation, or the sample radiation during counting, or any count-rate-dependent effects are exactly the same. In practice, this step is accomplished by making sure, that the unknown sample and known sample have the same physical volume, are irradiated in a homogeneous flux, and are counted under exactly the same conditions (geometry, detector, etc.). Preirradiation treatment of the sample should be kept to a minimum so as to lessen the possibility of sample contamination.

The second step in an activation analysis concerns the *choice of nuclear reaction* to change X into X*, plus the *irradiation facility* in which the reaction is to be carried out. In addition, the length of irradiation and decay must be chosen prior to counting so that the activity X* produced is enhanced relative to all other activities produced. Most activation analysis is done with thermal neutrons produced in the inside of nuclear reactors for the following reasons:

1. A great many elements have high cross sections for the absorption of thermal neutrons in the (n, γ) reaction.
2. Copious fluxes of thermal neutrons ($\Phi \sim 10^{12}$ neutrons/cm^2/sec) are available in nuclear reactors. Reactors are, in fact, overwhelmingly used as neutron sources.
3. Neutrons penetrate matter easily, and therefore there are few problems related to attenuation of the neutron flux in the sample.
4. The major elements, carbon, nitrogen and oxygen, are scarcely activated by thermal neutrons, thus making detection of other elements easier.

Although *most* activation analysis is done with reactor thermal neutrons, several other nuclear reactions and irradiation facilities can be used. Spontaneous fission of ^{252}Cf furnishes 3.8 neutrons per fission, and fluxes of up to 10^9 neutrons/cm^2/sec are available from ^{252}Cf isotopic neutron sources. Cockcroft-Walton accelerators can be used to accelerate deuterons to energies of ~ 150 keV, and then, using the ^{3}H(d, n) reaction, neutrons of ~ 14-MeV energy can be produced (*fast neutrons*). Typical neutron generators of this type give fluxes of $\sim 10^9$ neutrons/cm^2/sec of 14-MeV neutrons. These fast neutrons are useful for activating the light elements, such as silicon, nitrogen, fluorine, and oxygen, via (n, p) or (n, α) nuclear reactions (see Reference 6 for details). Charged particle or photon-induced reactions can also be used for activation purposes. For the sample problem of determining the Al content of rocks, the activating nuclear reaction was chosen to be ^{27}Al(n, γ) ^{28}Al, with the irradiation facility being a nuclear reactor.

Once a nuclear reaction and an irradiation facility have been selected, the possibility of *interfering reactions* must be carefully considered. This term means that quite often, although X will change to X* during irradiation, some other element Z may also change into X* during irradiation. Thus the activity of X* is proportional to the abundances of both Z and X in the sample, not just X. This effect is referred to as *an interfering reaction or interference*, and a correction must be made for it. In the case of the aluminum analysis, there is a very important interference—namely, the occurrence of the ^{28}Si

$(n, p)^{28}$Al reaction whereby silicon in the rock is converted into ^{28}Al by reactions involving the fast neutrons present in any reactor (along with the desired thermal neutrons). Thus the measured ^{28}Al activity will be due to the activation of ^{27}Al and ^{28}Si. Fortunately, the ^{28}Si$(n, p)^{28}$Al reaction contributed only $\sim 6\%$ of the total ^{28}Al activity; and by irradiating a known amount of Si and counting it, and from the well-known Si abundances of rocks, a correction for the ^{28}Al produced by the ^{28}Si$(n, p)^{28}$Al reaction could be calculated. Kruger (6) has an extensive table of potential interfering reactions.

The final decision concerning irradiation conditions involves the determination of the flux and irradiation duration. A rough rule is that the longer one irradiates and the longer one lets the sample decay before counting, the greater the activity of the long-lived species relative to shorter-lived species. One must keep in mind the saturation properties of nuclear irradiation are (see the Appendix) such that it rarely pays to irradiate any material for a time corresponding to more than one to two half-lives of the desired activity. The Appendix summarizes the basic principles of calculation of irradiation times and fluxes. In the Al analysis, a sample irradiation of 1.0 min in a neutron flux of $\sim 5 \times 10^{10}$ neutrons/cm^2/sec was used.

The next major step in any activation analysis procedure is the selection of a method of "isolating" the activity of interest, X*, in order to measure it. Two methods of "isolating" X* are commonly employed—*instrumental activation analysis* (IAA) and *radiochemical activation analysis*. In instrumental activation analysis, the energies of the γ-rays emitted by the radionuclides in question in the activated sample are used to identify them, and the corresponding photopeak areas give a measure of the activity. Instrumental activation is nondestructive, allowing further use of the sample. Furthermore, it permits the use of short-lived activities to identify various elements whose use might not be possible if a lengthy chemical separation were to precede the counting. Also, instrumental activation analysis (IAA) lends itself to automation procedures and generally reduces the time spent per sample on the analysis. The emergence of the Ge(Li) detector with its superior energy resolution has made IAA the preferred method of activation analysis.

Although most investigators prefer to use IAA, in some situations radiochemistry must be done prior to counting the sample, in order to isolate the activity of interest. An example of the need for this kind of activity is the determination of trace elements in biological materials, such as blood, that have very high sodium content. Copious quantities of ^{24}Na are produced via the ^{23}Na $(n, \gamma)^{24}$Na reaction, and they tend to "mask" the trace-element activities in the blood by creating a large Compton "background" in the region where the photopeaks of other trace-element activities are to be found. One solution to this problem is to separate the sodium chemically from the irradiated blood (using ion exchange with hydrated antimony pentoxide) and then instrumentally analyze the purified blood. This example does illustrate a trend in modern radiochemical activation analysis—that is, that of not completely separating the element of interest but of making a group separation of a relatively small number of activities and further resolving these activities by γ-ray spectroscopy.

4. Applications of Activation Analysis

The applications of activation analysis have been almost innumerable. Kruger (6) has an extensive bibliography of activation analysis applications. In the physical sciences, activation analysis has found use in trace-element analysis of semiconductor materials, metals, meteorites, lunar samples, and terrestial rocks. In most cases, the multielemental analysis feature of activation analysis has been used to measure the concentrations of several trace elements simultaneously. From detailed studies of these trace-element abundance patterns, one has been able to deduce information about the thermal and chemical history of the moon and meteorites, as well as identify the source or age of a given object.

The use of activation analysis in criminal investigations (*forensic activation analysis*) is growing. The basic idea here is to match the trace-element distributions found in bullets, paint, oil, and so on found at the scene of a crime with the trace-element distributions in objects found with criminal suspects. Such identification is rapid and nondestructive (allowing evidence to be presented in court). Moreover, the probability of its correctness can be ascertained quantitatively. Other prominent examples of the use of forensic activation analysis involve confirmation of the notion that Napoleon was poisoned (by finding significant amounts of arsenic in hair from his head) and the finding that activation analysis of the wiping taken from a suspect's hand can reveal not only whether he has fired a gun recently but also the type of gun and ammunition used.

Applications of activation analysis in the environmental sciences are just beginning. Determinations of the trace-element content of urban atmospheres, lakes, streams, and similar areas have allowed the flow of pollutants in various ecosystems to be traced. In addition, a few of the trace elements whose abundances have been measured by activation analysis have turned out to be biologically significant in themselves. The classic example is mercury and the large mercury concentrations in fish and other foodstuffs revealed by activation analysis. A particular combination of activation analysis and radiotracer methods is beginning to find important uses in the environmental sciences. This combination involves the use of stable isotopes instead of radioactive isotopes as tracers in various systems, with activation analysis being used to measure the tracer concentrations. Such a technique avoids the need to introduce radioactive materials into a system (such as the environment with its subsequent health and legal complications) and yet retains the selectivity and sensitivity of radiation measurement. The stable isotopes used as tracers are picked so as to have a negligibly low natural concentration in the system under study, a high-activation cross section, and convenient detectable radiation from the activated species. *Activable tracers*, as they are called, have been used to study atmospheric and hydrospheric particle transport, air pollution, and waste discharge.

In the applications of activation analysis, the biological sciences have not been neglected. Once again, Kruger (6) has an excellent summary of many of these uses. Many applications are summarized in the Proceedings of an IAEA Conference (8) on trace-element abundances in animal and plant systems, their toxicological effect, their use as tracers of the dynamics, and interactions of these materials, and in metabolism studies. In medical applications of activation analysis, normal and pathological trace-element distribution patterns are frequently measured and compared. This comparison can lead to the use of activation analysis measurements as diagnostic tools, a better understanding of disease mechanisms, and so forth. There has also been some use of *in vivo* activation analysis in which living human beings have been irradiated and then counted to determine whole-body trace-element distributions. Such studies are always carefully monitored to ensure the safety of the person being irradiated.

BIBLIOGRAPHY

1. BLOCH, K., and H. S. ANKER. "An extension of the isotope dilution method," *Science* **107**, 228 (1948).
2. BOWEN, H. J. M. *Chemical Applications of Radioisotopes*. London: Methuen, 1969. An excellent little monograph on chemical applications of tracers.
3. BRAUN, T. and J. TÖLGYESSY. *Radiometric Titrations*. Oxford: Pergamon, 1967. A comprehensive monograph on radiometric titrations with many examples of applications.
4. CORLISS, W. R. *Neutron Activation Analysis*. U.S. Atomic Energy Commission, 1963. A simple-minded discussion of the basic principles of activation analysis.
5. DUNCAN, J. F., and G. B. COOK. *Isotopes in Chemistry*. Oxford: Clarendon, 1968. An excellent survey with many examples of radioanalytical techniques.
6. KRUGER, P. *Principles of Activation Analysis*. New York: Wiley, 1971. The best available activation analysis textbook.
7. LOVELAND, W., R. A. SCHMITT, and D. E. FISHER. "Aluminum abundances in stony meteorites," *Geochim. et. Cosmochim. Acta* **33**, 375 (1969).
8. *Nuclear Activation Techniques in the Life Sciences*. IAEA, Vienna, 1967.
9. RUZICKA, J., and J. STARY. *Talanta* **11**, 697 (1964).
10. TÖLGYESSY, J., T. BRAUN, and T. KYRS. *Isotope Detection Analysis*. Oxford: Pergamon, 1972. A comprehensive monograph on isotope dilution analysis and its applications.

18

Environmental Applications of Radiotracers

A. INTRODUCTION

In recent years a great deal of applied research has centered on the study of problems related to the environment and environmental processes. In many of these studies, radiotracers have been used as primary tools to measure the dynamics of many physical and biological processes. In the best studies, the use of radiotracers to measure flow patterns, dispersion, and similar features is closely coupled to tests of theoretical models for the processes involved. This factor is important because in environmental studies the experimental conditions are difficult to control and, in general, only a few of the many possible conditions in a given experiment will be sampled. Therefore it is important to have some way (i.e., a model) to correlate experimental results measured under special conditions to general statements regarding an environmental process.

Radiotracers have been employed in studies of physical and biological processes in the *atmosphere* and the *hydrosphere*. Among the quantities that have been measured in atmospheric studies are the natural air-flow patterns in large- and small-scale investigations, the dispersion of atmospheric pollutants from various sources, and the identification of the source(s) of various pollution problems. In studies of the hydrosphere, radiotracers have been utilized to measure general water circulation patterns and various features of the hydrologic cycle, including precipitation, runoff and stream flow; total water inventories; infiltration; ground water problems, such as the origin and age of water, its flow velocity and direction; evaporative transport and aerosol production. Many studies of a biological nature, such as pollutant dispersal, uptake, and concentration in the ecosystem, have involved the use of radiotracers.

Three types of tracers have been used in these studies: (a) *artificial radiotracers,* (b) *naturally occurring radiotracers,* and (c) *stable-activable tracers.* Artificial radiotracers are those radionuclides released in the environment by the experimenter, whereas naturally occurring radiotracer studies involve "natural" radioactivity, cosmic-ray-produced radionuclides, and radioactivity introduced into the environment by human activity not controlled by the experimenter, such as bomb shot debris and nuclear power plant effluents. Stable-activable tracers are stable nuclides that are used as tracers in an experiment, collected, irradiated (usually with neutrons) to make them radioactive, and then counted as any other radiotracers. In this chapter we shall describe the techniques involved in using all these tracer methods in environmental research but oriented toward physical problems.

B. EXPERIMENTAL METHODS

1. Artificial Radiotracers

The controlled use of radioactivity to study processes occurring in our environment is a well-established and respected technique. Radioactive tracers have several significant advantages over conventional tracers for environmental studies. (a) The detectability of the tracer is not influenced by the physico-chemical nature of the environment (i.e., factors such as water color and pH). (b) Because nuclear radiation, particularly γ-radiation, is highly penetrating, the tracer can be detected while part of a living organism or when deeply buried in the ground, and so on. (c) Because only a small number of atoms are necessary to give a significant disintegration rate, there is a much better detection sensitivity in radiotracer experiments compared to conventional tracer experiments. This point is particularly important for environmental studies where high dilution factors are commonly encountered. (d) Because tracers of short half-life can be used and will rapidly disappear after the completion of an experiment, experiments can be repeated several times without damaging the environment or getting erroneous results due to persistence of tracers from previous experiments in the environment. (e) In many instances, radiotracers are the cheapest method of tracing pollutant flow. Channell and Kruger (22) have made an interesting comparison, shown in Table 18-1, of the relative costs of tracing pollutants in estuarine waters. As we can see from examining the table, radiotracers like ^{140}La and ^{3}H are the cheapest tracer method (when compared to a fluorescent dye, Rhodamine

TABLE 18-1

Tracing Costs for Model Field Studies*

	Scale of Study		
	Small	Intermediate	Large
Stable Lanthanum			
Weight needed (lb)	314	3,500	41,200
Tracer cost ($)	360	4,020	43,100
Analysis cost ($)	7,000	7,000	13,000
Total cost	7,360	11,020	56,100
Rhodamine WT			
Weight needed (lb)	945	10,550	124,000
Tracer cost ($)	1,965	22,000	258,000
Analysis cost ($)	100	100	200
Total cost	2,065	22,100	258,200
Tritium			
Activity needed (Ci)	280	2,830	28,300
Tracer cost ($)	560	4,750	34,000
Analysis cost ($)	3,000	4,500	6,000
	3,560	9,250	40,000
^{140}La			
Activity needed (Ci)	71	2,900	270,000
Tracer cost ($)	620	4,000	53,000
Analysis cost ($)	100	100	200
Total cost	720	4,100	53,200

*J. K. Channell and P. Kruger. "Activable rare earth elements as estuarine water tracers." In *Advances in Water Pollution Research*. New York: Pergamon Press, Vol. 2, pp. 111–16/1–20, 1970.

WT, and the stable activable tracer) because of the relatively low cost for the tracer itself.

The principal disadvantage of using radiotracers in environmental studies is the actual (or imagined) problem of nuclear safety. Public concern over possible harmful effects of ionizing radiation has increased in recent years. This fact, together with the lack of control over experimental conditions found in many environmental studies, has caused many regulatory agencies to establish extremely strict rules concerning the use of radiotracers in environmental research. Many research workers now find it difficult to demonstrate that the radiotracer concentrations will not exceed the regulatory agency's maximum permissible radionuclide concentration limits at *all* times and places. One must be especially aware, in this connection, of "concentration effects" present in food chains. At a minimum, the experimenter can usually look forward to a significant amount of red tape before performing environmental studies with radiotracers.

The criteria for choosing a particular radionuclide as a tracer for environmental studies are similar to those encountered in radiotracer laboratory experiments. Such items as the nature of the radiation emitted, the half-life of the radionuclide, the ease of obtaining it, the ease of detecting its radiations, and its cost play important roles. (Parts (a), (b), and (c) of Table 18-2 show some typical radiotracers that have been used in environmental studies (4).

Many examples of the application of radiotracers to environmental problems exist, and so we have selected only a few of the more interesting ones to discuss. Barry (3) has used ^{41}Ar, a short-lived gas found in reactor stack effluents, to trace the dispersion of stack effluent from the Chalk River nuclear

TABLE 18-2(a)

Radioactive Tracers for Gaseous Material*

Nuclide	Half-life	Radiation of Interest (MeV)	Chemical Form
^{35}S	87 days	β: 0.167 (100%)	H_2S
^{41}Ar	110 min	γ: 1.37	Gas
^{76}As	26.5 hr	γ: 0.55–2.02	AsH_3
^{82}Br	36 hr	γ: 0.55–1.32	CH_3Br
^{85}Kr	10 yr	β: 0.7 γ: 0.54	Gas
^{133}Xe	5.27 days	β: 0.34 γ: 0.03, 0.08	Gas

*From *Radioisotope Tracers in Industry and Geophysics.* Vienna: International Atomic Energy Agency, 1967.

TABLE 18-2(b)

Radioactive Tracers for Solid Material*

Nuclide	Half-life	Radiation of Interest	Chemical Form
^{24}Na	15 hr	γ: 1.37 (100%) γ: 2.75 (100%)	Na_2CO_3 in polypropylene balls
^{46}Sc	84 days	γ: 0.89 (100%) γ: 1.48 (100%)	Sc_2O_3
^{51}Cr	27.8 days	γ: 0.325 (9%)	Absorbed on quartz
^{64}Cu	12.8	β: 0.57 (38%) γ: 0.51 (19%)	CuO
^{65}Zn	274 days	γ: 1.11 (49%)	ZnO
^{82}Br	36 hr	γ: 0.55 (70%) γ: 1.32 (27%)	CaBr in polyethylene containers
^{110m}Ag	253 days	γ: 0.66 (100%) γ: 1.50 (13%)	Absorbed on solid grains
^{140}La	40 hr	γ: 0.33–2.54	La_2O_3 polypropylene balls
^{144}Ce	285 days	β: 3.1 (98%)	Ce_2O_3
^{182}Ta	115 days	γ: 1.19 γ: 1.12	Ta_2O_3
^{198}Au	2.7 days	γ: 0.41 (91%)	$AuCl_3$ absorbed on powder

*From *Radioisotope Tracers in Industry and Geophysics.* Vienna: International Atomic Energy Agency, 1967.

TABLE 18-2(c)

Radioactive Tracers for Organic Materials*

Nuclide	Half-life	Radiation of Interest	Chemical Form
^{3}H	12.26 yr	β: 0.018 (100%)	Various organic compounds
^{14}C	5568 yr	β: 0.155 (100%)	Various organic compounds
^{24}Na	15 hr	γ: 1.37 (100%) γ: 2.75 (100%)	Naphthenate Salicylate
^{35}S	97 days	β: 0.167 (100%)	Various organic compounds
^{38}Cl	37.3 min	γ: 1.60 (31%) γ: 2.15 (47%)	Chlorbenzene
^{59}Fe	44.5 days	γ: 1.1 (57%) γ: 1.29 (43%)	Ferrocene Dicyclopentadienyl-iron
^{60}Co	5.3 yr	γ: 1.17 (100%) γ: 1.33 (100%)	Naphthenate
^{64}Cu	12.8 hr	γ: 0.51 (19%)	Naphthenate
^{65}Ni	2.56 hr	γ: 0.37, 1.11, 1.49	Stearate Oxalate
^{77}Ge	11 hr	γ: 0.21–2.02	Various organic compounds
^{82}Br	36 hr	γ: 0.55–1.48	Bromobenzene Paradibromo-benzene
^{124}Sb	60 days	γ: 0.61 (99%) γ: 0.72 (14%)	Triphenylstibine
^{131}I	8.04 days	γ: 0.36 (80%) γ: 0.64 (9%)	I-Kerosene Iodobenzene
^{140}La	40 hr	γ: 0.33–2.54	Naphthenate
^{198}Au	2.7 days	γ: 0.41 (99%)	Sodium cyanide solution

*From *Radioisotope Tracers in Industry and Geophysics.* Vienna: International Atomic Energy Agency, 1967.

reactor and relate it to conventional dispersion models. The ^{41}Ar concentrations in the air at various distances from the emitting stack were measured by circulating the air through a counter consisting of layers of plastic scintillator. The β^--particles emitted in the decay of ^{41}Ar were detected by the plastic scintillators. The detectors were connected to automatic recording equipment that could monitor the stack effluent dispersion continuously.

A typical example of a radiotracer study in an aqueous system is the study of the concentration dynamics of soluble material in the Eshkol Reservoir of Israel by Gilath and Stuhl (18). The Eshkol Reservoir is a shallow lake (depth ~7 m, volume $\sim 3.5 \times 10^6$ m^3) with 10^6 m^3 of water flowing through the reservoir per day. As is typically done in studies of this type, a "spike" of radiotracer was injected into the reservoir and its dispersal was measured. ^{82}Br was chosen as the tracer because of its high maximum permissible concentration limit (100 μCi/m^3), its low detection limit ($\sim 2 \times 10^{-6}$ μCi/m^3), and its general solubility in water. A spike of ~15 μCi ^{82}Br was injected into the water over a 1-hr period, allowing the tracer concentration to be measured for ~3 to 4 times the typical residence time of the reservoir (~70 hr). The ^{82}Br activity was measured by using NaI detectors stationed at the reservoir outlet and in probes lowered into the water from boats.

An interesting set of radiotracer studies of the deposition of pesticides were made by Atkins and Eggleton (2). The pesticides were labeled with ^{14}C, which was assayed using liquid scintillation counting. Results of the study showed that direct absorption on soil or vegetation was a more effective way of removing the pesticides from the atmosphere than deposition in rain. Furthermore, the mean residence times of the pesticides in the atmosphere were shown to be long enough to allow global distribution of a pesticide from any given location.

2. Stable-Activable Tracers

The successful applications of radionuclides to trace the behavior of materials in the environment are almost countless. However, situations do arise where the use of a radionuclide as a tracer is inadvisable because of safety or public relations considerations, or where the element being traced has no radioisotope of suitable half-life, or where it emits no suitable radiation for detection. In recent years there has been increasing emphasis on using another nuclear tracer technique for such situations—the stable-activable tracer method. A stable nuclide is used as a tracer, and after the tracing has been completed (injection of tracer, dispersion, and sampling), the sample containing the stable nuclide is then activated, usually by irradiation with thermal neutrons in a nuclear reactor. Next, the radioactive nuclide is counted by using the same methods as for a conventional radiotracer. This method is especially useful in situations where the use of radiotracers is not—that is, *in vivo* studies in humans, tracing of potable water, air pollution studies, and so on.

Three types of stable-activable tracer may be employed: (a) an enriched stable isotope of an element already present in the system under study, (b) a stable nuclide that is chemically different from the material to be traced but whose physical properties and behavior are similar (presumably the tracer should also have superior nuclear characteristics, such as high activation cross section or appropriate decay properties.), and (c) a naturally occurring minor or trace elemental constituent of the material being traced may be used as a stable-activable tracer if its nuclear properties are suitable.

The criteria that a stable nuclide should meet in order to be used as a stable-activable tracer are as follows:

1. The thermal-neutron-capture cross section, $\sigma(n, \gamma)$, of the nuclide should be large, preferably greater than 0.1 b.
2. The radionuclide produced from irradiation of the stable nuclide should

have a half-life of days or less and should emit readily detected radiations, preferably γ-rays.
3. The stable nuclide should behave exactly as the element or material being studied. This feature is most easily ensured if the stable tracer is an isotope of the element of interest. However, in many studies, similar physical properties are all that is necessary to ensure proper tracer behavior.
4. The natural abundance of the stable tracer in the system under study should be low, preferably less than 10%.
5. The stable tracer should be commercially available, at moderate cost in high enrichments.
6. The stable tracer should not be toxic in the quantities used in the experiment.

Some of the most promising candidates for use as stable-activable tracers are listed in Table 18-3, together with their nuclear properties. Note the different types of tracer, such as (a) low-abundance isotopes of elements like ^{18}O and ^{48}Ca that can be used to trace the *chemical behavior* of that element and (b) other nuclides with high-capture cross sections, such as the rare earths and Au, that can be used as "*physical*" tracers. The rare earths are attractive for use as stable-activable tracers in studies of the physical behavior of pollutants on almost all grounds except cost. They have the advantage that a different rare earth can be used to spike each of several pollutant sources, and subsequent sampling and analysis can be done to study complex pollution problems.

Some of the advantages of the stable-activable tracer method are as follows:

1. Since no radioactivity is released in the system under study, no licensing, handling of radioactivity, employment of counters, and similar steps in the field are required.
2. Since the sample analysis is frequently nondestructive and does not involve radioactivity, other tests or procedures may be employed with the sample after analysis.
3. In some instances, the sensitivity is the same as radiotracer methods, although, in general, such is not the case.
4. Elements for which there are no suitable radiotracers, such as oxygen, can be traced.
5. The duration of the experiment can be quite long without loss of tracer due to decay.

Some of the disadvantages of the stable-activable tracer method are:

1. The analysis times are generally long; except for on-line systems, typical times of days, weeks, or months must elapse to complete the analysis. The

TABLE 18-3

Possible Stable-Activable Tracers

Tracer	Activation Product	Half-life[a]	Radiation(s) of Interest[a] Type, Energy (MeV) (Intensity)	Activation[a] Cross Section (barns)	Natural Isotopic[a] Abundance of Tracer (%)	Maximum Percent[b] in Enriched Form	Cost[b] ($/mg)
^{18}O	^{18}F	109.7 min	γ: 0.511 (194%)	—	0.204	97	0.47
^{26}Mg	^{27}Mg	9.46 min	γ: 0.84 (70%)	0.027	11.29	99.7	5.60
			γ: 1.013 (30%)				
^{30}Si	^{31}Si	2.62 hr	β^-: 1.48 (100%)	0.11	3.12	70	22.50
^{34}S	^{35}S	87.9 days	β^-: 0.167 (100%)	0.27	4.22	50	33.50
^{41}K	^{42}K	12.36 hr	β^-: 3.52 (100%)	1.2	6.77	99.5	23.50
			γ: 1.524 (18%)				
^{48}Ca	^{49}Ca	8.8 min	γ: 3.10 (89%)	1.1	0.185	>95	42.00
^{50}Ti	^{51}Ti	5.79 min	γ: 0.320	0.14	5.25	85	7.00
^{51}V	^{52}V	3.75 min	γ: 1.434 (100%)	4.9	99.75	100	0.95
^{50}Cr	^{51}Cr	27.8 days	γ: 0.320 (9%)	17	4.31	>95	10
^{54}Cr	^{55}Cr	3.52 min	β^-: 2.59 (100%)	0.38	2.38	>95	19.50
^{58}Fe	^{59}Fe	45.6 days	γ: 1.095 (56%)	1.1	0.31	85	56
			γ: 1.292 (44%)				
^{59}Co	^{60}Co	5.27 yr	γ: 1.173 (100%)	19	100	100	—
			γ: 1.332 (100%)				
^{64}Ni	^{65}Ni	2.56 hr	γ: 1.481 (25%)	1.5	1.16	99	28
^{70}Zn	^{71m}Zn	3.92 hr	γ: 0.385 (95%)	0.01	0.62	>75	21.50
			γ: 0.495 (75%)				
^{76}Ge	^{77}Ge	11.3 hr	γ: 0.21 (61%)	0.1	7.67	99	3.35
			γ: 0.263 (45%)				
^{75}As	^{76}As	26.4 hr	γ: 0.559 (43%)	4.5	100	100	—
^{82}Se	^{83m}Se	70 sec	γ: 1.01 (100%)	0.05	9.19	90	9.50
			γ: 2.02 (40%)				
^{81}Br	^{82}Br	35.34 hr	γ: 0.554 (66%)	0.2	49.48	97.5	4.20
			γ: 0.777 (83%)				
^{84}Sr	^{85}Sr	64.0 days	γ: 0.514 (100%)	0.8	0.56	60	28
^{96}Zr	^{97}Zr	17.0 hr	γ: 0.747 (92%)	0.05	2.80	90	14.50
^{100}Mo	^{101}Mo	14.6 min	γ: 0.191 (25%)	0.2	9.62	99	1.70
^{96}Ru	^{97}Ru	2.88 days	γ: 0.215 (91%)	0.2	5.46	90	17
^{108}Pd	^{109}Pd	13.47 hr	β^-: 1.028 (100%)	12	26.7	95	3
^{107}Ag	^{108}Ag	2.42 min	β^-: 1.64 (100%)	35	51.35	98.9	7.50
			γ: 0.632 (1.7%)				
^{116}Cd	^{117}Cd	2.4 hr	γ: 0.273 (31%)	1.4	7.58	95	2.50
^{115}In	$^{116m_1}In$	54.0 min	γ: 1.293 (80%)	154	95.8	100	0.85
^{124}Sn	^{125m}Sn	9.5 min	γ: 0.325 (97%)	0.1	5.98	96	1.95

Tracer	Activation Product	Half-life[a]	Radiation(s) of Interest[a] Type, Energy (MeV) (Intensity)	Activation[a] Cross Section (barns)	Natural Isotopic[a] Abundance of Tracer (%)	Maximum Percent[b] in Enriched Form	Cost[b] ($/mg)
^{121}Sb	^{122}Sb	2.80 days	γ: 0.564 (66%)	6	57.25	99.3	2.80
^{124}Te	^{125m}Te	58 days	X Te	5	4.61	85	3
^{129}I	^{130}I	12.3 hr	γ: 0.538 (99%) γ: 0.669 (100%) γ: 0.743 (87%)	28	0	100	5
^{136}Ba	^{137m}Ba	2.554 min	γ: 0.662 (89%)	0.01	7.81	>85	1.40
^{139}La	^{140}La	40.22 hr	γ: 0.487 (40%) γ: 1.596 (96%)	8.9	99.91	99.99	1.90
^{142}Ce	^{143}Ce	33 hr	γ: 0.293 (46%)	1	11.07	95	6.30
^{141}Pr	^{142}Pr	19.2 hr	γ: 1.57 (3.7%)	12	100	100	—
^{148}Nd	^{149}Nd	1.8 hr	γ: 0.210 (27%) γ: 0.27 (26%)	4	5.72	95	8.40
^{150}Nd	^{151}Nd	12 min	γ: 0.118 (40%)	1.5	5.60	97	8.40
^{152}Sm	^{153}Sm	46.8 hr	γ: 0.103 (28%)	210	26.63	99	1.40
^{154}Sm	^{155}Sm	23.5 min	γ: 0.104 (73%)	5	22.53	99	1.40
^{151}Eu	$^{152m_1}Eu$	9.3 hr	γ: 0.842 (13%) γ: 0.963 (12%)	2800	47.77	99.1	2.50
^{160}Gd	^{161}Gd	3.6 min	γ: 0.361 (66%)	0.8	21.9	—	—
^{159}Tb	^{160}Tb	72.1 days	γ: 0.299 (30%) γ: 0.966 (31%)	46	100	100	—
^{164}Dy	$^{165m_1}Dy$	1.26 min	γ: 0.514 (1.8%) γ: 3.62 (0.6%)	2000	28.18	95	3.60
^{165}Ho	^{166}Ho	26.9 hr	γ: 0.081 (5.4%)	64	100	100	—
^{170}Er	^{171}Er	7.52 hr	γ: 0.308 (63%)	9	14.88	99.7	15
^{168}Yb	^{169}Yb	31.8 days	γ: 0.063 (45%)	11,000	0.140	35	32
^{176}Lu	^{177}Lu	6.74 days	γ: 0.208 (6.1%)	2100	2.60	80	43
^{174}Hf	^{175}Hf	70 days	γ: 0.343 (85%)	400	0.163	20	89
^{181}Ta	^{182}Ta	115.1 days	γ: 1.122 (34%) γ: 1.222 (27%)	21	99.987	100	—
^{187}Re	^{188}Re	16.7 hr	γ: 0.155 (10%)	70	62.9	99.7	6.50
^{192}Os	^{193}Os	31.5 hr	γ: 0.139 (3%) γ: 0.460 (3.9%)	1.6	41.0	—	—
^{191}Ir	^{192}Ir	74.2 days	γ: 0.317 (81%)	750	38.5	99.7	90
^{198}Pt	^{199}Pt	31 min	γ: 0.540 (24%)	4	7.19	65	7
^{197}Au	^{198}Au	2.697 days	γ: 0.412 (95%)	98.8	100	100	—
^{196}Hg	^{197}Hg	65 hr	γ: 0.077 (18%)	880	0.146	8.9	200

[a] Nuclear data taken from *Table of Isotopes*. 6th ed. New York: Wiley, 1967.
[b] Cost data taken from J. H. Somerville (Ed.), *The Isotope Index, 1967*. Indianapolis, Ind.: Scientific Equipment Co., 1967.

use of X-ray fluorescence instead of activation analysis for sample analysis can obviate this problem somewhat.
2. Stable-activable tracers are usually fairly costly.
3. The analysis procedures are expensive and require special equipment, such as a nuclear reactor.
4. The tracer may persist in the environment for a long period of time, thus preventing further successful experimentation from taking place.

Although the technique of using stable-activable tracers in environmental research is fairly new, a number of examples of its use have appeared in the literature (1, 20, 22, 27, 28). Among the more interesting studies showing the potential of this tracer method is the work of Dahl and co-workers (10). Dahl *et al.*, used In_2O_3 particles to trace stack gases emitted from Norwegian aluminum factories. By carefully dispersing the In_2O_3 particles in the stack gas via an oil burner atomizer, they were able to achieve particle diameters of 0.005 to 0.07 μ, thus ensuring that the solid particles would behave the same as gases in the atmosphere. The fallout of the tracer-gas particles was measured by collecting them in several plastic trays filled with water. After sample collection, the indium was separated chemically from the water, irradiated, and analyzed by using the principles of activation analysis. Typical results obtained by Dahl *et al.*, for the tracer concentration as a function of distance from the source are shown in Figure 18-1. Here the dashed line

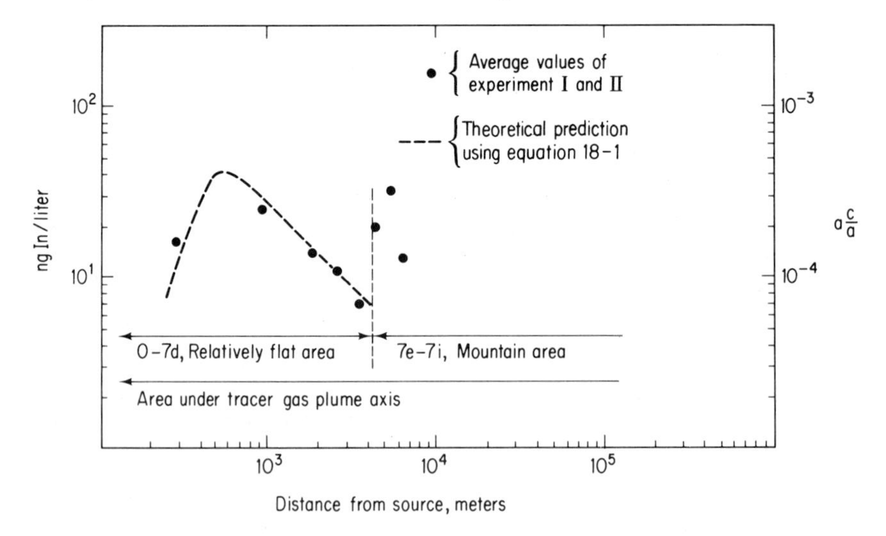

Fig. 18-1. Indium concentration at ground level along the tracer-gas plume axis (wind direction) in relation to theoretical values. From Dahl *et al. Nuclear Techniques in Environmental Pollution.* Vienna: International Atomic Energy Agency, 1971.

represents a theoretical value for the tracer concentration as predicted by the model of Pasquill (29) and Gifford (17); that is,

$$C = \frac{Q}{\pi \sigma_x \sigma_y \bar{\mu}} \exp\left[-\left(\frac{y^2}{2\sigma_y^2} + \frac{h^2}{2\sigma_z^2}\right)\right] \tag{18-1}$$

where C is the ground-level tracer concentration in grams per cubic meter, Q is the source strength in grams per second, $\bar{\mu}$ is the average wind velocity in meters per second, h is the source stack height (above the ground) in meters, x is the distance from the source along the wind direction in meters, y is lateral distance from the plume axis in meters, σ_y is the y dispersion of the plume, and σ_z is the z dispersion of the plume. The authors comment that, based on their results, it should be possible to trace the plume dispersion out to a distance of $\sim$100 km by using this technique. A similar study was carried out by Nakasa and Ohno (25) using ^{59}Co as a tracer. Further detailed studies comparing pollutant-dispersion patterns obtained with stable rare earth tracers and naturally occurring trace-element tracers have been carried out by Shum and Loveland (32).

Another novel application of the stable-activable tracer method was the work of Fendrik and Glubrecht (13). They labeled the pollen of a single tree, using inactive Mn tracer, and measured its dispersion into the surrounding area. To label the pollen, holes 8 cm in depth were drilled into the tree trunk and kept filled with 1 M $MnSO_4$ solution. The $MnSO_4$ uptake of the tree was 60 liters in 3 weeks. As a result of this procedure, the Mn content of the tree needles and pollen increased to $\sim$4 to 20 times the natural Mn concentration of 10^2 to 10^3 ppm. A typical single pollen grain had a mass of $\sim 6 \times 10^{-8}$ g with a Mn content of $\sim 10^{-10}$ g Mn. Thus the Mn present in one pollen grain could easily be detected by activation analysis. Furthermore, the experiment was allowed to continue for several years with no safety problems. The pollen grains were collected on a polyethylene foil, irradiated for 20 min, and counted by using autoradiography to determine the number of grains on a given collector foil.

Chatters and Peterson (8) demonstrated a novel way of tracing plant effluent from a paper mill by using stable-activable tracers. The wood fibers were treated with La salt prior to entry into the plant, and then all types of plant effluent were sampled to see where the La would appear. Normal plant operations were in no way altered during the experiment, as they might have been if radiotracers were employed. Lanthanum was picked as a tracer because of its favorable nuclear properties, its low abundance in wood, its ability to adhere to the wood fibers, and its relatively low cost. Moreover, the 40.2-hr half-life of the activation product ^{140}La is long enough to allow the decay of shorter-lived radionuclides, such as ^{38}Cl and ^{56}Mn, produced in the irradiation of the wood fiber.

An important demonstration of the applicability of stable-activable tracers to medicine was the experiment of Lowman and Krivit (24). Stable ^{58}Fe was used as a tracer for the iron in human plasma. ^{58}Fe concentrations of ~0.01 to 0.1 mg ^{58}Fe/ml of human heparinized plasma were used. After withdrawing a 1-ml aliquot of plasma from the body, it was dried by heating at 90°C for 20 min. The samples were then irradiated in a high flux reactor, allowed to decay for 2 weeks (for removal of ^{24}Na and ^{42}K), and then counted. Simultaneous experiments with stable ^{58}Fe tracer and radioactive ^{59}Fe tracer gave similar results, as shown in Figure 18-2. The stable ^{58}Fe tracer has obvious advantages in hematological studies on infants and newborn children.

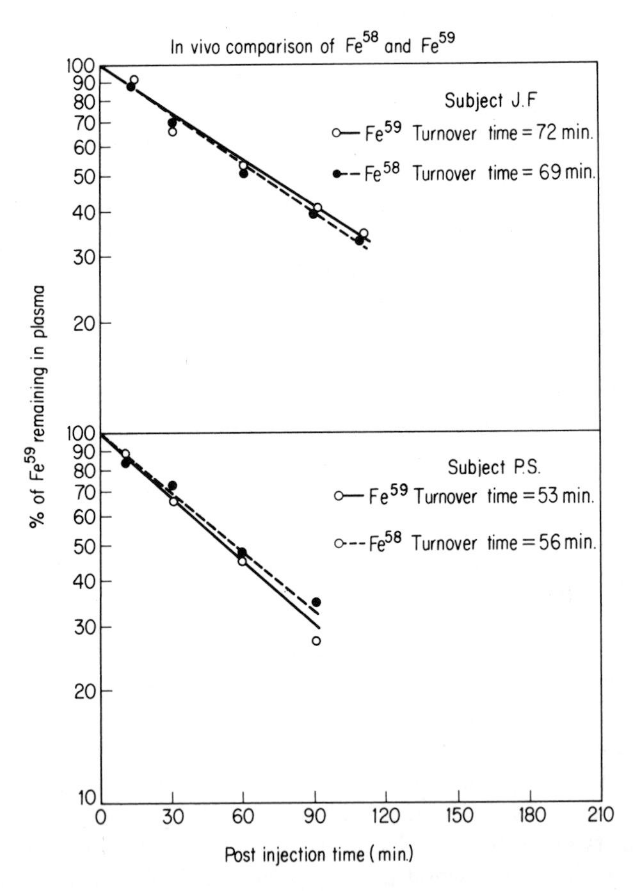

Fig. 18-2. In vivo plasma clearance rates for simultaneously injected stable and radioactive isotopes of iron in two human subjects. From J.T. Lowman and W. Krivit. New in vivo tracer method with the use of nonradioactive isotopes and activation analysis, "*J. Lab. Clin. Med.*" **61**, 1042–1048, (1963).

3. Naturally Occurring Radionuclides

A number of naturally occurring radionuclides can be used as tracers in various environmental studies. Such naturally occurring radionuclides can be divided into three classes: (a) *primary radionuclides*—that is, radionuclides formed in the synthesis of the elements that have survived to the present—(b) *cosmogenic radionuclides*—that is, radionuclides formed by the interaction of cosmic rays with matter—and (c) *artificial radionuclides*—radionuclides introduced into the environment by human activity over which the experimentor has no control. Let us briefly discuss each of these three types of natural radiotracers.

a. Primary Radionuclides. Of the approximately 340 nuclides found in nature, about 70 of them are radioactive. Most of these naturally occurring radionuclides are heavy-elements. For example, all isotopes of elements with $Z > 82$ are radioactive. Among the heavy-element radionuclides that have been used as tracers are

1. The isotopes of uranium, ^{234}U, ^{235}U, ^{238}U, which are found primarily in rocks and soils with a mean total U concentration in igneous rocks of ~4 ppm, as shown in Table 18-4.

TABLE 18-4

Average Radium, Uranium, Thorium, and Potassium Contents in Various Rocks*

Type of Rock	^{226}Ra (pCi/g)	^{238}U (pCi/g)	^{232}Th (pCi/g)	^{40}K (pCi/g)
Igneous	1.3	1.3	1.3	22.0
Sedimentary				
Sandstones	0.71	0.4	0.65	8.8
Shales	1.08	0.4	1.1	22.0
Limestones	0.42	0.4	0.14	2.2

*Adopted from UNSCEAR. United Nations Scientific Committee on the Effects of Atomic Radiation. 13th Session, Suppl. No. 17 (A/3838). New York: United Nations, 1958.

2. ^{226}Ra, which is found in rock and soils (see Table 18-4), in water ($\sim 0.4 \times 10^{-10}$ μCi/ml), in food ($\sim 10^{-9}$ μCi/g), and in human tissue ($\sim 10^{-4}$ μCi/person).
3. ^{232}Th, which is found mainly in rocks (see Table 18-4).
4. The decay products of (1), (2), and (3)—that is, ^{228}Ra, ^{222}Rn, ^{220}Rn, ^{210}Pb, etc. As geochemical tracers, these radionuclides are more important than the parent radionuclides (23). The Rn radionuclides are the only important tracers in this group.

Most of the preceding species are α-emitters and, as such, special care must be taken in assaying their activity (see Chapters 3 and 11). ^{40}K is the other naturally occurring radionuclide of importance. ^{40}K is a β^- and γ-emitter, comprising ~0.01% of all potassium and thus imparting to naturally occurring potassium a specific activity of 1900 dpm/g of K. A 70-kg man contains ~130 g of potassium, most of which is located in the muscle. Thus the ^{40}K content of the body is ~0.1 μCi, delivering a dose of ~19 mrem/year. ^{40}K is the predominant radioactive component in normal foods and human tissue.

The primary radionuclides are important tracers for use in the hydrosphere. The most important way for these radionuclides to enter the hydrosphere is by leaching from rocks and soils. Airborne dust that is leached in the ocean provides another important source of these radionuclides. Typical concentrations of geochemically important radionuclides are shown in Table 18-5.

TABLE 18-5

Concentrations and Activity Ratios in River and Ocean Waters of Some Radionuclides Belonging to U-Th Series*

Radionuclides	Rivers[a]	Oceans[a]
U^{238}	0.02–6.0	2.0–3.0
Th^{232}	0.01–0.10	5×10^{-5}–10^{-3}
Pa^{231}		$(1–2.5) \times 10^{-4}$
Ra^{226}	0.02–0.07	0.08–0.35
Pb^{210}	0.10–15	0.1–0.3
U^{234}/U^{238}	1.1–1.9	1.14 ± 0.02
Th^{230}/Th^{232}	0.6–1.3	1.5–10.5
Th^{228}/Th^{232}	1.2–1.4	12–25
Th^{234}/U^{238}	—	0.4–1.6

[a] All concentrations are given in units of dpm/liter of water.

*Reproduced, with permission, from "The radioactivity of the atmosphere and hydrosphere," *Annual Review of Nuclear Science*, Volume 18, page 411.

b. Cosmogenic Radionuclides. Nuclear reactions between cosmic rays and atmospheric nuclei can lead to radioactive products. Because there is a continuous generation of these radionuclides in our upper atmosphere, they are present in our environment in amounts sufficient to allow their use as tracers for geochemical and geophysical processes. A list of these nuclei and their typical abundances is shown in Table 18-6. Of these nuclei, ^{14}C, ^{3}H, and ^{7}Be are the most important. (^{14}C and ^{3}H are also produced in nuclear weapons tests and by other activities of man.) The literature concerning the use of ^{14}C and ^{3}H in environmental studies is so extensive as to almost preclude description. Certainly any detailed discussion is beyond the scope of this book;

TABLE 18-6

Cosmogenic Radionuclides*

Radionuclide	Half-life	Average Specific Activity[a]
^{10}Be	2.5×10^6 yr	10^{-3} oc
^{26}Al	7.4×10^5 yr	1.2×10^{-5} oc
^{36}Cl	3.1×10^5 yr	0.55
^{81}Kr	2.0×10^5 yr	—
^{14}C	5730 yr	2.4 a, 260 oc
^{32}Si	500 yr	2.4×10^{-2} oc
^{39}Ar	270 yr	2.9×10^{-3} oc
^{3}H	12.5 yr	7.0×10^{-2} a, 36 oc
^{22}Na	2.6 yr	6.7×10^{-5} a
^{35}S	87 days	7.8×10^{-3} a
^{7}Be	53 days	0.63 a
^{33}P	25 days	7.6×10^{-3} a
^{33}P	14.3 days	1.4×10^{-2} a
^{28}Mg	21.2 hr	—
^{24}Na	15.0 hr	—
^{38}S	2.9 hr	—
^{31}Si	2.6 hr	—
^{39}Cl	55.5 min	—
^{38}Cl	37.3 min	—
^{34m}Cl	32.0 min	—

[a] a = (dpm/kg) of air in the troposphere; oc = (dpm/ton) of water in the ocean.
*D. Lal and B. Peters. *Handbuch der Physik* **XLVI/2**, 551 (1967).

the reader is referred to the excellent review article by Lal and Suess (23) and references contained therein for further discussion.

c. Artificial Radionuclides. A dramatic increase in the levels of radioactivity in the atmosphere and hydrosphere occurred in 1954 as a result of atmospheric testing of thermonuclear weapons. Although such testing has decreased sharply since the 1963 moratorium, these bomb-produced radionuclides remain in concentrations large enough that many meteorological and hydrological phenomena can be studied. Figure 18-3 shows the atmospheric concentrations of some artificial and naturally occurring radionuclides in recent times. Note the evidence for seasonal transfer of radioactive material from the stratosphere to the troposphere.

Recently attention has focused on the artificial radionuclides released by nuclear power reactors and their fate in the environment. Typical radionuclide concentrations expected in river water, due to the discharge of a 1130-MWe-pressurized water reactor, are shown in Table 18-7(a), (b), and (c). As can be seen from studying the data, the expected radionuclide concentrations are extremely low and demand the ultimate in sampling and detection techniques. In fact, analysis of river-bottom sediments may be the only way of detecting the presence of some of these radionuclides.

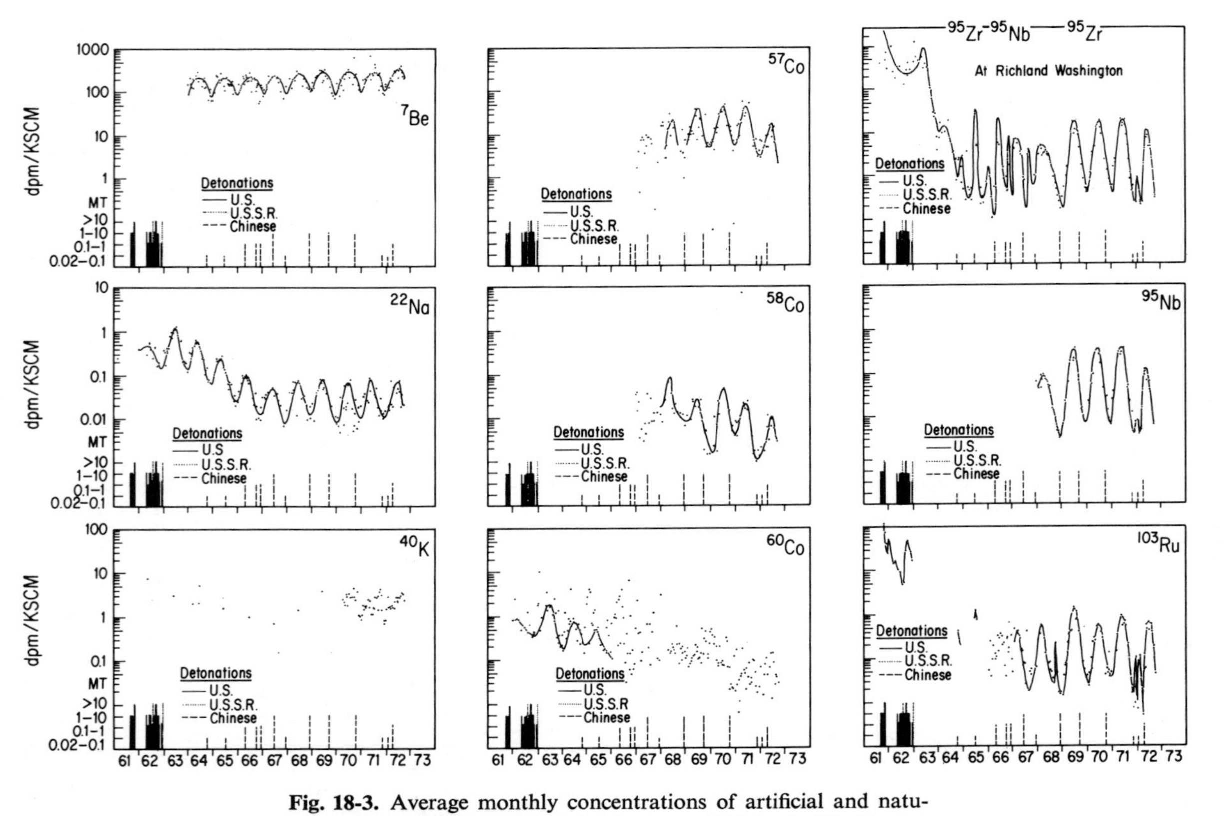

Fig. 18-3. Average monthly concentrations of artificial and natural radionuclides in ground level air at Richland, Washington (46°N) for the years 1961–73. (Data furnished by N. A. Wogman.)

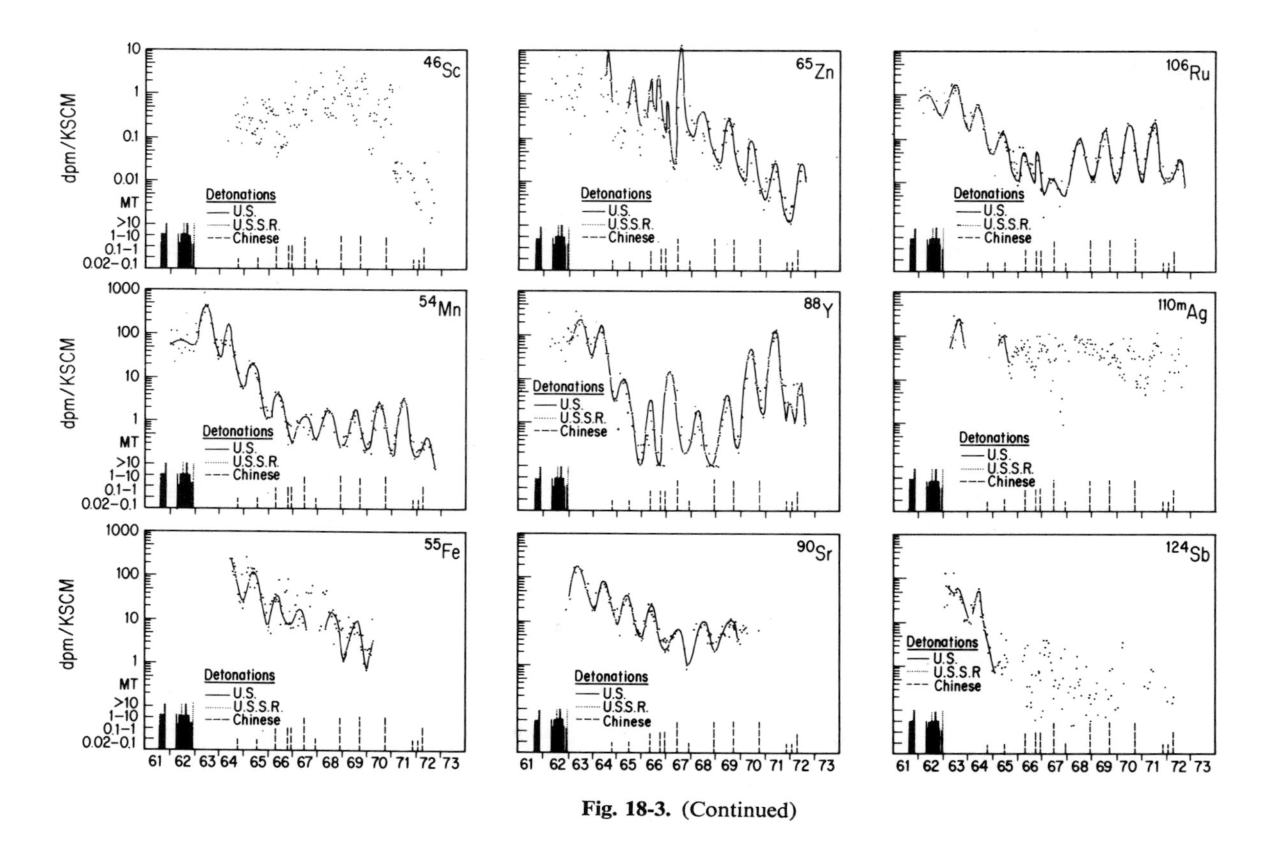

Fig. 18-3. (Continued)

TABLE 18-7(a)*

Projected Radionuclide Concentrations in Columbia River Water Resulting From Trojan Effluents[a]

Isotope	Normal Operation (μCi/cc)	Anticipated Operational Occurrences (μCi/cc)	Isotope	Normal Operation (μCi/cc)	Anticipated Operational Occurrences (μCi/cc)
H–3	3.4 E–09	3.4 E–09	Ru–103	1.6 E–16	1.8 E–15
Cr–51	1.3 E–15	2.2 E–14	Ru–106	1.5 E–17	1.6 E–16
Mn–54	1.1 E–15	1.8 E–14	Rh–103m	1.6 E–16	1.8 E–15
Mn–56	2.8 E–15	1.0 E–13	Ag–110m	5.3 E–19	4.4 E–16
Fe–59	1.1 E–14	2.2 E–13	Ag–110	6.9 E–21	5.7 E–18
Co–58	3.5 E–14	5.8 E–13	Te–132	7.0 E–14	1.1 E–12
Co–60	1.1 E–15	1.8 E–14	I–131	1.0 E–12	1.2 E–11
Zn–65	1.3 E–18	4.4 E–16	I–132	8.0 E–14	1.5 E–12
Br–84	1.7 E–16	6.7 E–15	I–133	3.8 E–13	1.1 E–11
Sr–89	1.9 E–15	2.2 E–14	I–134	3.8 E–15	6.3 E–14
Sr–90	7.3 E–17	8.4 E–16	I–135	8.6 E–14	2.9 E–12
Sr–91	1.1 E–16	3.6 E–15	Cs–134	6.4 E–13	1.2 E–11
Y–90	3.7 E–16	6.7 E–15	Cs–136	3.2 E–13	6.3 E–12
Y–91m	6.4 E–16	2.0 E–14	Cs–137	3.4 E–12	6.1 E–11
Y–91	1.6 E–15	3.7 E–15	Cs–138	4.0 E–14	1.6 E–12
Y–92	1.8 E–16	6.5 E–15	Ba–140	1.8 E–15	2.2 E–14
Zr–95	3.6 E–16	3.5 E–15	La–140	1.5 E–15	1.3 E–14
Nb–95	3.7 E–16	4.0 E–15	Ce–141	3.4 E–16	3.9 E–15
Mo–99	1.3 E–12	8.1 E–12	Ce–144	1.7 E–16	1.9 E–15
Tc–99m	1.0 E–12	1.3 E–11	—	—	—
Tc–99	5.7 E–20	2.4 E–19	—	—	—

[a]The exponential power of 10 in this table is expressed as 1.2 E–03 = 1.2×10^{-3} = 0.0012.
*Trojan Nuclear Power Plant, Rainier, Oregon, Docket No. 50–344, Final Safety Analysis Report.

C. METHODS OF SAMPLING

In experiments with naturally occurring radiotracers, much attention must be paid to the sampling techniques utilized. As a rule, one wishes to obtain a sample whose radionuclide concentration is representative of typical radionuclide concentrations in the system at the sampling point. Furthermore, because of the low radionuclide concentrations encountered in environmental studies, the sampling procedure must, in general, concentrate the radionuclides so as to allow their activity to be measured.

In atmospheric sampling, the normal procedure followed is to draw air through a filter at a known rate for a known period of time. The radioactivity of the filters may be counted and the activity per unit volume of air determined. When dealing with α-activity, one must be careful to delay measurement of the filter paper activity for a sufficient time ($\sim$6 to 8 hr), to allow short-lived α-emitting daughters to decay. Millipore and Nuclepore filter papers offer superior efficiency as collection media. Air samplers with flow rates up to several hundred cfm are available, as well as continuous air moni-

TABLE 18-7(b)

Projected Radionuclide Concentrations in Columbia River Biota Resulting from Trojan Liquid Effluents[a] (in μCi/cc)

	Fish Flesh		Invertebrate Flesh		Aquatic Plants	
Isotope	Normal Operation	Anticipated Operational Occurrences	Normal Operation	Anticipated Operational Occurrences	Normal Operation	Anticipated Operational Occurrences
H-3	3.4 E-09	3.4 E-09	3.4 E-09	3.4 E-09	3.4 E-09	3.4 E-09
Cr-51	2.5 E-13	4.4 E-12	2.5 E-12	4.4 E-11	5.0 E-12	8.7 E-11
Mn-54	9.0 E-14	1.5 E-12	3.3 E-10	5.5 E-09	1.7 E-10	2.8 E-09
Mn-56	2.3 E-13	8.5 E-12	8.5 E-10	3.1 E-08	4.3 E-10	1.6 E-08
Fe-59	2.2 E-12	4.3 E-11	2.8 E-10	5.6 E-09	7.5 E-11	1.5 E-09
Co-58	5.7 E-11	9.3 E-10	1.1 E-09	1.9 E-08	2.4 E-10	3.9 E-09
Co-60	1.7 E-12	2.9 E-11	3.5 E-11	5.8 E-10	7.3 E-12	1.3 E-10
Zn-65	2.2 E-15	7.6 E-13	4.2 E-14	1.5 E-11	4.0 E-15	1.4 E-12
Br-84	2.2 E-14	8.8 E-13	1.7 E-14	6.7 E-13	1.4 E-13	5.4 E-12
Sr-89	2.7 E-14	3.0 E-13	9.7 E-13	1.1 E-12	3.9 E-13	4.3 E-12
Sr-90	1.0 E-15	1.1 E-14	3.7 E-14	4.2 E-13	1.5 E-14	1.6 E-13
Sr-91	1.6 E-15	5.0 E-14	5.6 E-14	1.8 E-12	2.3 E-14	7.2 E-13
Y-90	3.7 E-14	6.7 E-13	3.7 E-13	6.7 E-12	3.7 E-12	6.7 E-11
Y-91m	6.4 E-14	2.0 E-12	6.4 E-13	2.0 E-11	6.4 E-12	2.0 E-10
Y-91	1.6 E-13	3.7 E-13	1.6 E-12	3.7 E-12	1.6 E-11	3.7 E-11
Y-92	1.8 E-14	6.5 E-13	1.8 E-13	6.5 E-12	1.8 E-12	6.5 E-11
Zr-95	3.6 E-14	3.5 E-13	3.6 E-13	3.5 E-12	3.6 E-12	3.5 E-11
Nb-95	1.1 E-11	1.2 E-10	3.7 E-14	4.0 E-13	3.7 E-13	4.0 E-12
Mo-99	1.3 E-10	8.1 E-10	1.3 E-10	8.1 E-10	1.3 E-10	8.1 E-10
Tc-99m	1.0 E-12	1.3 E-11	3.1 E-11	4.0 E-10	1.0 E-10	1.3 E-09
Tc-99	5.7 E-20	2.4 E-19	1.7 E-18	7.3 E-18	5.7 E-18	2.4 E-17
Ru-103	1.6 E-14	1.8 E-13	3.2 E-13	3.7 E-12	3.2 E-13	3.7 E-12
Ru-106	1.5 E-15	1.6 E-14	2.9 E-14	3.2 E-13	2.9 E-14	3.2 E-13
Rh-103m	1.6 E-14	1.8 E-13	3.1 E-13	3.6 E-12	3.1 E-13	3.6 E-12
Ag-110m	1.6 E-15	1.3 E-12	1.6 E-15	1.3 E-12	1.1 E-16	8.7 E-14
Ag-110	2.1 E-17	1.7 E-14	2.1 E-17	1.7 E-14	1.4 E-18	1.1 E-15
Te-132	7.0 E-14	1.1 E-12	1.4 E-11	2.2 E-10	1.4 E-11	2.2 E-10
I-131	9.3 E-12	1.1 E-10	3.3 E-10	4.0 E-09	7.2 E-11	8.5 E-10
I-132	7.2 E-13	1.3 E-11	2.6 E-11	4.7 E-10	5.5 E-12	1.0 E-11
I-133	3.4 E-12	9.7 E-11	1.2 E-10	3.3 E-09	2.6 E-11	7.1 E-10
I-134	3.4 E-14	5.7 E-13	1.2 E-12	2.0 E-11	2.6 E-13	4.4 E-12
I-135	7.7 E-13	2.6 E-11	2.8 E-11	9.3 E-10	5.9 E-12	2.0 E-10
Cs-134	3.2 E-09	5.8 E-08	6.4 E-10	1.2 E-08	5.8 E-10	1.1 E-08
Cs-136	1.6 E-09	3.1 E-08	3.2 E-10	6.3 E-09	2.9 E-10	5.7 E-09
Cs-137	1.7 E-08	3.1 E-07	3.4 E-09	6.1 E-08	3.1 E-09	5.5 E-08
Cs-138	2.0 E-10	7.9 E-09	4.0 E-11	1.6 E-09	3.7 E-11	1.4 E-09
Ba-140	1.8 E-14	2.2 E-13	3.6 E-13	4.3 E-12	9.1 E-13	1.1 E-12
La-140	1.5 E-13	1.3 E-12	1.5 E-12	1.3 E-11	1.5 E-11	1.3 E-10
Ce-141	2.8 E-14	3.2 E-13	3.7 E-13	4.3 E-12	1.1 E-12	1.3 E-11
Ce-144	1.4 E-14	1.6 E-13	1.9 E-13	2.1 E-12	5.5 E-13	6.1 E-12

[a]The exponential power of 10 in this table is expressed as 1.2 E-03 = 1.2×10^{-3} = 0.0012.

tors that draw samples through moving tapes and count the tapes later. Meteorological conditions should be continuously monitored to allow correlation of results with meteorological factors.

In sampling radionuclide concentrations in surface waters, one should sample both the water itself and the mud or sediment downstream from the

TABLE 18-7(c)

Projected Radionuclide Concentrations in Columbia River Bottom Sediment Resulting from Trojan Liquid Effluents[a]

Isotope	Normal Operation (mCi/km²)	Anticipated Operational Occurrence (mCi/km²)
H–3	0.0	0.0
Cr–51	3.60 E–05	6.20 E–04
Mn–54	1.40 E–03	2.30 E–02
Mn–56	1.20 E–06	4.60 E–05
Fe–59	2.10 E–03	4.10 E–02
Co–58	1.00 E–02	1.70 E–01
Co–60	8.50 E–03	1.40 E–01
Zn–65	1.30 E–06	4.40 E–04
Br–84	3.50 E–08	1.40 E–06
Sr–89	1.00 E–04	1.20 E–03
Sr–90	4.80 E–04	5.30 E–03
Sr–91	4.60 E–08	1.50 E–06
Y–90	4.00 E–06	7.40 E–05
Y–91m	9.20 E–08	2.80 E–06
Y–91	3.80 E–04	9.00 E–04
Y–92	1.10 E–07	4.00 E–06
Zr–95	9.60 E–05	9.40 E–04
Nb–95	5.30 E–05	5.80 E–04
Mo–99	1.50 E–02	9.30 E–02
Tc–99m	1.10 E–03	1.40 E–02
Tc–99	2.40 E–06	1.00 E–05
Ru–103	1.30 E–05	1.50 E–04
Ru–106	1.10 E–05	1.20 E–04
Rh–103m	2.60 E–08	2.90 E–07
Ag–110m	5.60 E–07	4.50 E–04
Ag–110	8.30 E–15	6.80 E–12
Te–132	9.30 E–04	1.40 E–02
I–131	3.40 E–02	4.10 E–01
I–132	3.20 E–05	5.80 E–04
I–133	1.40 E–03	3.70 E–02
I–134	5.60 E–07	9.40 E–06
I–135	9.90 E–05	3.30 E–03
Cs–134	2.00 E–00	3.60 E+01
Cs–136	1.70 E–02	3.40 E–01
Cs–137	9.20 E+01	1.70 E+03
Cs–138	3.50 E–06	1.40 E–04
Ba–140	4.80 E–05	5.70 E–04
La–140	1.00 E–05	8.90 E–05
Ce–141	4.60 E–05	5.40 E–04
Ce–144	2.00 E–04	2.20 E–03

[a]The exponential power of 10 in this table is expressed as 1.20 E–03 = 1.20×10^{-3} = 0.0012.

point of discharge. The exact sampling point should be selected after studying the characteristics of the water flow and sedimentation patterns. If radionuclide buildup is observed in sediments, the sampling program should be expanded to include plankton, crustaceans, bottom-feeding fish, and so forth. The importance of sampling locations is illustrated in Figure 18-4, which

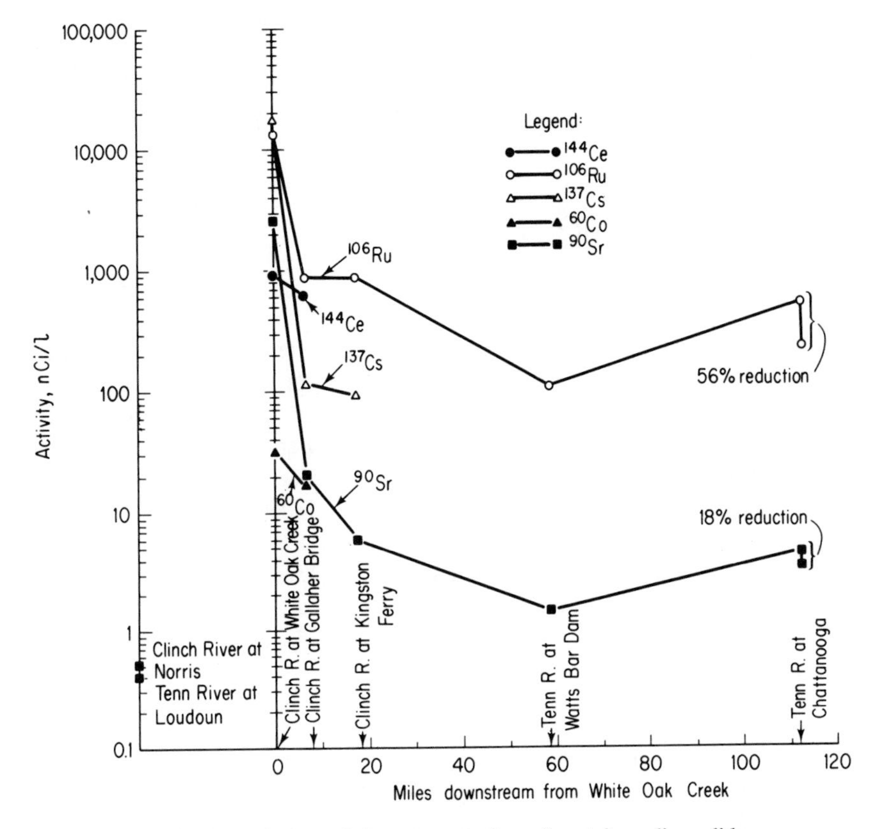

Fig. 18-4. Variation of the concentration of certain radionuclides in bottom muds from the Clinch and Tennessee rivers. The increase from Mile 58 to Mile 110 is due to differences in the adsorptive characteristics of the bottom muds (14).

shows the concentrations of radionuclides in bottom muds of some Tennessee rivers. Note that the radionuclide concentrations at 110 miles from the source are higher than those at 58 miles from the source. This unexpected result appears to be due to a high concentration of nonabsorbing quartz in the sediment at the 58-mile point and a very absorbing sediment and favorable river geometry at the 110-mile sampling point (14).

In sampling farming areas, one is usually concerned with monitoring soil, plants, and livestock so as to check on the buildup of various radionuclides. In general, a single monitoring station that consists of a small plot on which two or three animals are grazing located closest to the radionuclide source will be sufficient (12). In sampling foodstuffs for radioactivity, one must be concerned with such questions as what foods are to be collected, where they are to be collected, and how many samples are to be collected. As a rule, one should only sample those foodstuffs that might tend to concentrate the radionuclide. For example, studies of radioiodine can largely be limited to milk. If results from a few widely spaced sampling stations give similar results, one can limit the number of samples collected.

D. LOW-LEVEL TECHNIQUES

1. Introduction

Despite concentration of the radiotracers of interest during sampling procedures in environmental studies, quite often one is left with a sample containing a small quantity (<10 dpm) of radioactivity that must be assayed. Such assays are referred to as *low-level techniques.* Let us begin our discussion of low-level techniques by considering any chemical manipulations of the sample that must be made prior to counting it. Understandably, the fact of having activity levels <10 dpm puts severe restrictions on the nature of *low-level chemistry.* Among the requirements for low-level chemistry are a small constant blank, high chemical yields for all procedures, high radiochemical purity for all reagents employed, and the ability to place the sample in suitable chemical form for counting.

2. Blanks

The *blank* in low-level chemistry is defined as the contribution of the added reagents and other sample constituents to the activity being measured. Care must be taken to ensure that the blank is properly measured and that it includes all possible contributions to the activity that would be encountered in a real system. For example, in the determination of fission product ^{144}Ce in sea water, the blank must be determined for each new bottle of reagents used due to the high variability of the ^{144}Ce content in chemical reagents.

Clearly, one of the most effective ways of dealing with the blank correction is to reduce it to its lowest possible value. Among the factors contributing to the blank correction that can be reduced with care is *radiochemical contamination of analytical reagents used in chemical procedures.* DeVoe (11) has written an extensive review article on this subject, and his work should be consulted for detailed information. Typical contamination of most reagents is in the range of ~ 10 to 100 dpm/g reagent, although individual reagents may contain activity levels of >10000 dpm/g. Some especially troublesome reagents are rare earths (cerium salts in particular), Cl- or S-containing reagents (which may suffer ^{32}P contamination), Cs salts (which may contain ^{40}K or ^{87}Rb), and K salts or other obvious offenders. Precipitating agents, such as tetraphenylborates and chloroplatinates, are also particularly pernicious with regard to contamination problems.

Airborne contamination is another possible contribution to the blank correction. Here one is chiefly concerned with sample contamination with the daughters of ^{222}Rn, which have half-lives in the 30- to 40-min range. Steps that can be taken to avoid this problem include eschewing the use of suction filtration in chemical procedures, prefiltering of room air, and use of Rn traps.

Further lowering of the blank correction can be obtained by using *nonisotopic carriers* in chemical procedures to replace inert carriers of the element of interest when it is difficult to obtain the inert carrier in a contamination-free condition. Obviously, only clean glassware should be used, reagents should not be reused, and the laboratory should be kept in an immaculate condition.

3. Low-Level Counting: General Principles

Once the low-level radioactive sample has been collected and any chemical procedures performed prior to counting, it is ready for counting. Because of the extremely small disintegration rates encountered, special techniques, called *low-level counting*, must be used to assay the sample. We shall survey some of the highlights of this area, which has been the subject of many articles and monographs (5, 9, 21, 30, 35).

In beginning our discussion of low-level counting, let us consider the typical problem of having a fixed time T to assay a sample with time t_b used to measure the background count rate. Assume that S is the sample plus background count rate with a background count rate of B. Using the principles outlined in Chapter 12, we can say that the standard deviation of the source counting rate, when corrected for background, σ, is

$$\sigma = \left[\frac{S}{(T - t_b)} + \frac{B}{t_b}\right]^{1/2} \tag{18-2}$$

Minimizing σ, we have

$$\frac{d\sigma}{dt_b} = 0 = \frac{S/(T - t_b)^2 - B/t_b^2}{2[S/(T - t_b) + B/t_b]^{1/2}} \tag{18-3}$$

The right-hand side of Equation (18-3) is zero when

$$\frac{S}{(T - t_b)^2} = \frac{B}{t_b^2} \tag{18-4}$$

or, in other words,

$$t_b = \frac{T}{1 + (S/B)^{1/2}} \tag{18-5}$$

That is, the optimum apportionment of counting time is to count background for a time t_b given by Equation (18-5) and spend the rest of the time measuring the source plus background counting rate. [The reader should note the equivalence of Equation (18-5) and the equations in Chapter 12.]

Substituting for t_b the value given in Equation (18-5) back into Equation (18-2), we can calculate the minimum standard deviation of the source counting rate as

$$\sigma_{\min} = \left\{ \frac{S}{T\left[1 - \frac{1}{1 + (S/B)^{1/2}}\right]} + \frac{B/T}{1 + (S/B)^{1/2}} \right\}^{1/2} \tag{18-6}$$

To put things in a somewhat simpler form, let us consider the minimum *fractional* standard deviation of the source rate, C, as

$$C = \frac{\sigma_{\min}}{S - B} = \frac{1}{[(S/B)^{1/2} - 1]T^{1/2}B^{1/2}} \tag{18-7}$$

Clearly, in all low-level counting, we shall want to minimize C, the fractional uncertainty in the source counting rate.

Rearranging Equation (18-7), we can solve for S; that is,

$$S = \frac{1 + 2CT^{1/2}B^{1/2} + C^2TB}{CT^{1/2}} \tag{18-8}$$

If we denote the source count rate as D, we have

$$D = S - B \tag{18-9}$$

and using Equation (18-8)

$$D = \frac{1 + 2CT^{1/2}B^{1/2}}{C^2T} \tag{18-10}$$

The value of D given in Equation (18-10) is the *minimum observable counting rate* in a counter with background B, given time T for all counting and desiring a fractional uncertainty in the rate of C. Clearly, the most important features in low-level counting are to decrease the background and increase the sample activity. Equation (18-10) can be used to calculate the feasibility of various low-level assay procedures. Equations (18-5) and (18-10) can be combined to yield an equation for t, the time required to measure the sample plus background rate in order to achieve a percentage fractional standard deviation, $P = C \times 10^2$, in the source rate, D. That is,

$$t = \frac{(10^4)(D + B)}{D^2P^2 - (10^4B/t_b)} \tag{18-11}$$

where

$$t + t_b = T \tag{18-12}$$

Equation (18-11) is frequently useful in planning low-level counting procedures.

What are the most important characteristics that a detector must possess for low-level counting? The first general characteristic is one of stability. Low-level counting frequently requires counting periods of long duration; hence counter stability is quite important. As described in Chapter 12, if the sample count rate S (source plus background) is small compared to the background rate B, one's detector should be picked to maximize S^2/B—that is, low background and high efficiency. If the sample count rate is large with respect to background, one need only maximize S—that is, one chooses a high-efficiency detector.

4. Low-Level Counting: Details

For *low-level α-particle counting*, the choice is generally between a gas-filled ionization chamber and a semiconductor detector system. The former has a counting efficiency of ~50% and a background of ~3 to 4 cpm; the latter has a background rate of ~0.5 cph and detection efficiencies approaching 50%. The semiconductor detector is usually the detector of choice, although large sample sizes may be better assayed with gas-filled ionization chambers. Background radiation is primarily due to α-particle-emitting impurities in the counter, counter support material, and so on, plus the occurrence of cosmic-ray-induced (n, α) reactions.

Low-level "soft radiation" counting has its own techniques. The term "soft radiation counting" refers to detecting EC and low-energy β^--emitters where self-absorption of the radiation is important. In solving this problem, one typically tries to incorporate the radionuclide to be counted into the detector. One typical method of assay is liquid scintillation counting, which is used to assay samples whose activity is greater than 10 pCi. Typical liquid scintillation counter backgrounds can be as high as ~100 cpm, whereas special counters have been built with background count rates of ~10 cpm or less. Liquid scintillation counting is a speedy, simple method of low-level counting. Another technique that has been used to count low-level "soft radiation" samples involves the use of a gas-filled proportional counter. The sample to be counted is converted to gaseous form and added to the counter gas at a concentration of ~0.05 mole percent. This method of low-level counting, although tedious and time consuming, allows one to assay samples whose activity is less than 0.5 pCi. Typical counter backgrounds are ~1 to 2 cpm, with 100% counting efficiency down to radiation energies of ~100 eV.

The counting of tritium in water is a special problem about which much has been written (30). Current methods for assay of ^3H in water have a range of 0.1 to 5000 TU, where a tritium unit (TU) has a value of 7.2 dpm/liter. A comparison of different counting methods is shown in Table 18-8. The most desirable feature of a tritium measuring system is that it be capable of measuring a large number of samples as rapidly, simply, and cheaply as possible

TABLE 18-8

Comparison of Gas Counters and Liquid Scintillation Counters for Tritium Assay*

Precision Required	Counter	Counting Time t (min) — Sample Activity (TU) 100	500	1000	5000
0.03 (3%)	Proportional counter (IAEA)	860	120	60	10
	Geiger counter (Stockholm)	12,140	860	330	50
	Packard Tri-Carb (benzene)	—	750	240	28
	Large-volume scintillator (water-dioxane)	—	1150	330	25
	Packard Tri-Carb (water-Triton-X)	—	1150	360	39
	Packard Tri-Carb (water-dioxane)	—	2500	750	71
0.05 (5%)	Proportional counter (IAEA)	310	44	20	4
	Geiger counter (Stockholm)	4370	310	120	20
	Packard Tri-Carb (benzene)	—	270	90	10
	Large-volume scintillator (water-dioxane)	—	415	120	9
	Packard Tri-Carb (water-Triton-X)	—	415	130	14
	Packard Tri-Carb (water-dioxane)	—	900	270	26
0.10 (10%)	Proportional counter (IAEA)	78	11	5	(1)
	Geiger counter (Stockholm)	1092	77	30	5
	Packard Tri-Carb (benzene)	—	68	22	2
	Large-volume scintillator (water-dioxane)	—	105	30	2
	Packard Tri-Carb (water-Triton-X)	—	104	33	4
	Packard Tri-Carb (water-dioxane)	—	224	67	6

Counter	C	B
Proportional counter (IAEA)	C = 1 cpm/50 TU	B = 1.1 cpm
Geiger counter (Stockholm)	C = 1 cpm/210 TU	B = 2 cpm
Packard Tri-Carb (benzene)[a]	C = 1 cpm/100 TU	B = 12 cpm
Large-volume scintillator (water-dioxane)	C = 1 cpm/66 TU	B = 52 cpm
Packard Tri-Carb (water-Triton-X)	C = 1 cpm/138 TU	B = 10 cpm
Packard Tri-Carb (water-dioxane)	C = 1 cpm/230 TU	B = 8.4 cpm

[a]Estimated figures.

*From J. F. Cameron. *Radioactive Dating and Methods of Low Level Counting*. Vienna: International Atomic Energy Agency, 1967.

with an uncertainty of $\sim \pm 10\%$ or better. It is generally more important to assay 100 samples with an uncertainty of $\pm 10\%$ than to assay 10 samples with an uncertainty of $\pm 3\%$.

The basic feature of *low-level β^--counting* that distinguishes it from ordinary β^--counting is the use of an *anticoincidence guard ring* around the main β^--counter(s). An anticoincidence guard is a single detector, or array of detectors, that surrounds the primary detector. The output of the anticoincidence detector is fed to an anticoincidence circuit along with the output of the primary detector. When nuclear radiation passes through both detectors simultaneously, as in the case of a highly penetrating cosmic ray striking both detectors, no output results from the anticoincidence circuit. When the anticoincidence circuit receives a signal *only* from the primary detector, an output signal results. The net effect is that the anticoincidence guard ring detector "guards" or shields against exterior background radiation entering the primary detector. Typical ring assemblies reduce the background rate in the primary counter by a factor of ~ 50. A well-designed guard ring will allow several different types of central counter to be inserted into it. Low-background β^--counters constructed of especially pure materials with anti-coincidence shields have exhibited background count rates of ~ 1 cph with efficiencies of $\sim 50\%$.

Low-level counting of gamma ray emitters using solid scintillation counters is an extensively used technique. For a detailed discussion of this technique, the reader is referred to the excellent review article of Kreger and Mather (21). The most important aspect of low-level solid scintillation counting is to decrease the counter background. Typical contributions to a solid-scintillation-counter's background rate from various sources are shown in Table 18-9. Here, four factors are seen as the major contributors to the detector-

TABLE 18-9

Analysis of Background Contributions in the Region from 25 keV to 1.575 MeV in a 7-in. Diameter by $3\frac{1}{2}$-in. NaI Crystal at Argonne National Laboratory*

Source		Rate
Cosmic rays, charged particle interactions		55 cpm
Cosmic rays, soft or cascade component		120
Fe and Pb shield		110
5-in. phototube and preamplifier		100
^{40}K in crystal (2 ppm estimated)		30
Residual, unaccounted for		140
	Total	555 cpm

*From Reference 21.

background count rate. They are (a) the cosmic-ray shield, (b) the atmosphere surrounding the detector, (c) the detector itself, and (d) cosmic rays. For the cosmic-ray shield about the detector, it is advisable to use old or virgin lead—that is, lead that was purified over 100 years ago, thus allowing any ^{210}Pb to decay. One should expect ~ 1-cpm/g shield material. Iron can also be used in constructing the detector shield, but care must be taken to ensure that the Fe or steel is pre-1945 in origin. (Iron processed in the post-1945 period has ^{60}Co contamination due to the use of ^{60}Co in the blast furnace operation.) Mercury is a very good, easily purified shield material but is quite expensive. The main portion of the atmospheric contribution to the detector background is due to Rn and its daughters. Particularly troublesome in this regard is the fact that atmospheric Rn concentrations can fluctuate by a factor of 40 during the course of a day. Once again the problem is best handled by filtering room air, rapid air turnover, and use of inert atmospheres in counting assemblies. The NaI detector will contain ^{40}K impurity, which will contribute to the background. The detector can is also a potential contributor to the background. Copper appears to be the best material for a detector can, with Al being the least preferred, for it might be expected to contain $\sim 10^{-13}$ Ci Ra/g Al. In reducing the cosmic-ray component of the background, one tries to stop the "soft" cosmic rays (e^-, X, γ) in the detector shield, while using an anticoincidence system to stop the "hard" component of the cosmic radiation (mesons, etc.).

At first, one might think that Ge(Li) detectors with their low detection efficiencies would have little use in low-level counting. Such is not the case, however, and the reason can be understood by looking at Figure 8-13 in Chapter 8. Figure 8-13 shows the γ-ray spectrum from an air filter as measured by a NaI(Tl) detector. Note, because of the complexity of the radionuclide mixture, the virtual impossibility of drawing any significant conclusions from the NaI(Tl) detector spectrum due to the poor energy resolution. Clearly, there are many cases in which the very good energy resolution of the Ge(Li) detector is a necessity. Furthermore, recent developments in detector-fabrication techniques have allowed the production of Ge(Li) detectors with detection efficiencies that are one-fourth to one-third that of NaI(Tl) detectors.

A number of special techniques have evolved to increase the detection sensitivity in γ-ray counting (5, 9, 30). One of the most important is the suppression of the Compton scattering events in the γ-ray spectrum by use of an anticoincidence annulus about the central γ-ray detector. Sketches of one such typical assembly and its associated electronic circuitry are found in Figures 18-5 and 18-6. The idea behind a Compton-suppression spectrometer is that most events in which the incident photon was Compton-scattered out of the central detector will result in that event not being accepted as part of the spectrum. This result is accomplished by setting up an anticoincidence between the central detector signal and any signal coming from the annulus. A photon that is Compton-scattered from the central detector will probably give rise to a signal from the annulus (see Figure 18-5). Thus such events will not be accepted. Other events, such as photoelectric events in the central detector, will not produce signals from the annulus and will be counted. This reduction in the number of Compton scattering events in the γ-ray spectrum leads to a more easily interpreted spectrum, since the peak-to-Compton ratios are much higher. Typically, the use of such annuli reduces the number of Compton events tenfold. The effect on the detection sensitivity is shown in Table 18-10, which is adapted from Reference 26.

A very expensive but powerful refinement of the anticoincidence technique, one that involves the use of multidimensional γ-ray spectroscopy, is applicable only to radionuclides emitting two or more γ-rays in cascade and

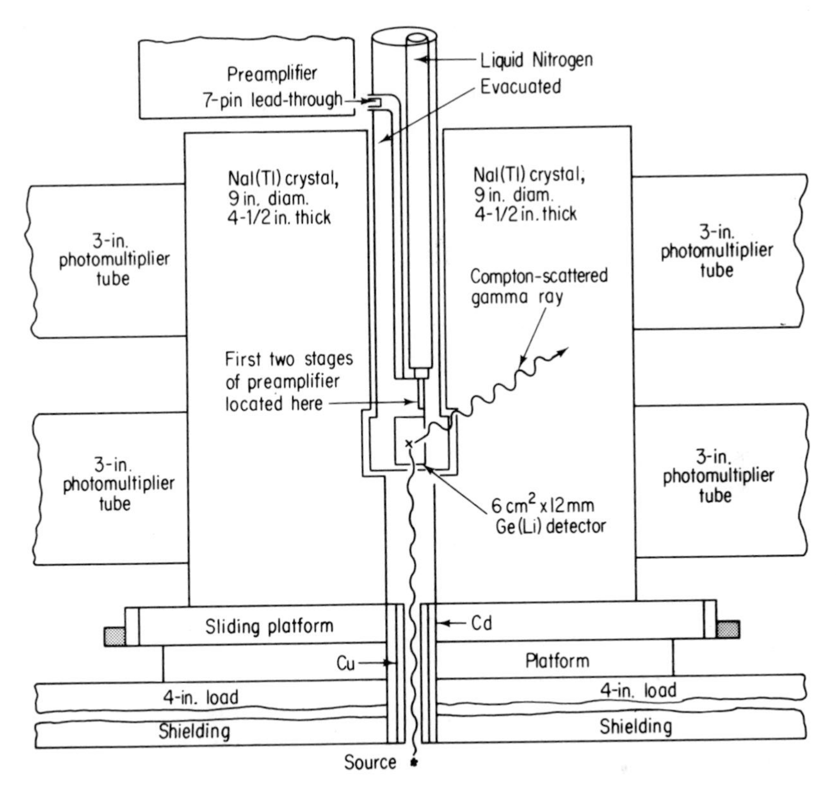

Fig. 18-5. A schematic drawing in cross section of the Compton-suppression spectrometer used at LRL (Livermore). The Ge(Li) detector is a 7-cm^3, 12-mm-depletion-depth planar detector.

consists of two detectors operated in coincidence that measure coincident γ-rays emitted by the sample. The entire two-detector assembly is surrounded by an anticoincidence annulus (see Figure 18-7). The energies of each of the two γ-rays detected are plotted as a point in matrix. Figure 18-8 shows what a multidimensional γ-ray spectrum of a mixture of ^{60}Co and ^{46}Sc looks like. Also shown is an ordinary γ-ray spectrum of the same mixture. Both ^{46}Sc and ^{60}Co emit two γ-rays, but the detector resolution in the ordinary γ-ray spectrum is not sufficient to separate the 1.118-MeV peak of ^{46}Sc from the 1.173-MeV peak of ^{60}Co. However, the multidimensional spectrum shows a clear separation of the ^{60}Co and ^{46}Sc peaks. Typical detection sensitivities obtained by using this method are given in Table 18-9. Through this technique, a factor of $\sim$10 improvement of detection sensitivity over γ-ray spectroscopy with anticoincidence is obtained, plus a factor of $\sim$50 improvement over ordinary γ-ray spectroscopy.

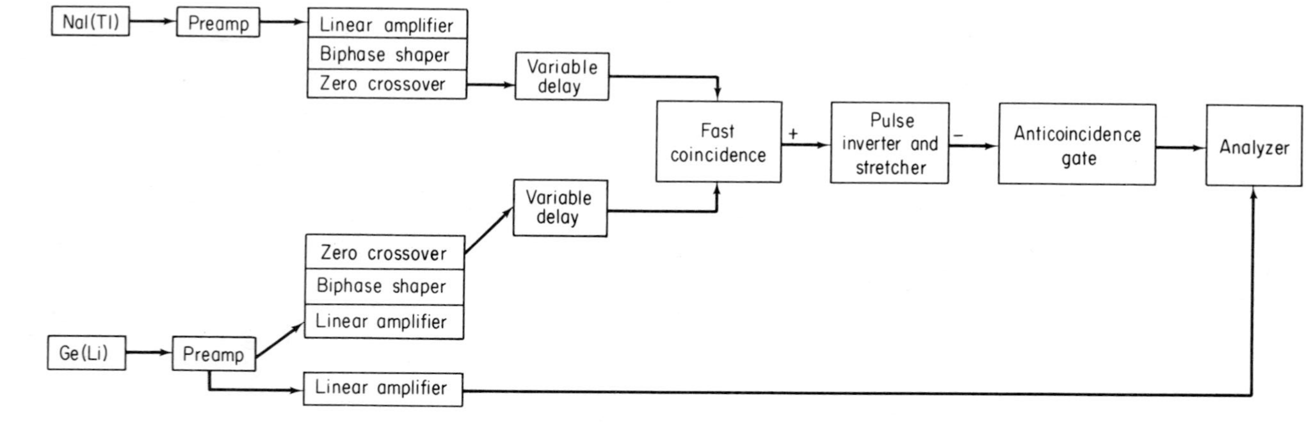

Fig. 18-6. Electronic block diagram for the Compton-suppression system. From Camp (5).

TABLE 18-10

Comparison of the Sensitivities for Some Radionuclides of the 11-in. Diameter by 6-in. Thick Dual Crystal System Using Three Different Counting Techniques*

Nuclide Being Assayed	Gamma Peak Energy (MeV)	Detection Limit (dpm) Gamma Spectrometry[a]	Gamma Spectrometry with Anticoincidence[a]	Multidimensional Gamma Ray[b] Spectroscopy
^{22}Na	0.51	45	13	0.20
	0.51 + 1.28	50	10	
^{24}Na	1.38	48	12	0.07
	2.75	23	3	
^{26}Al	0.51	43	12	0.11
	0.51 + 1.83	35	6	
^{39}Cl	0.25	72	25	0.67
	1.27	109	22	
^{44}Ti	0.51	47	13	0.14
	0.51 + 1.16	56	11	
^{60}Co	1.17 and 1.33	42	8	0.21

[a]Assume activity equivalent to 0.1 background rate measurable.
[b]Assume activity equivalent to 0.5 background rate measurable.
*From *Radioactive Dating and Methods of Low-Level Counting*. Vienna: International Atomic Energy Agency, 1967.

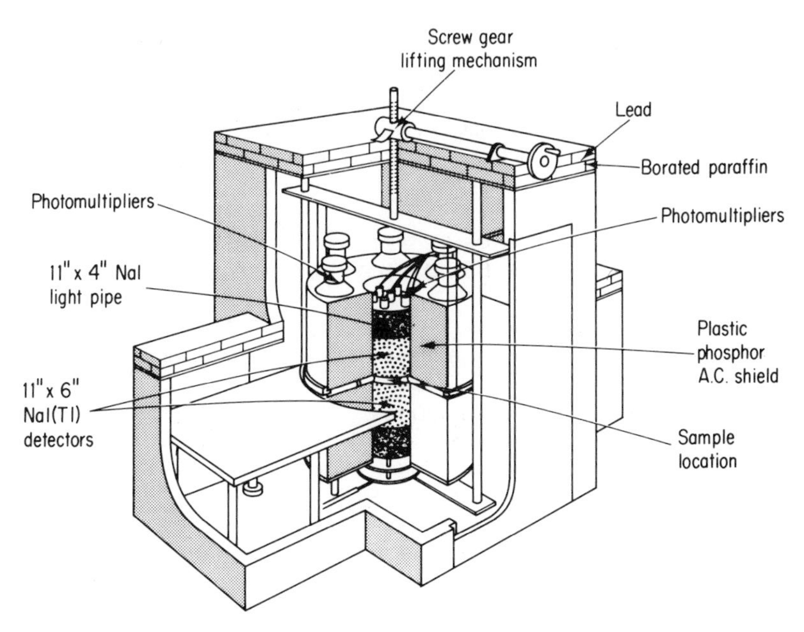

Fig. 18-7. Large multidimensional γ-ray spectrometer. From Neilsen and Perkins (26).

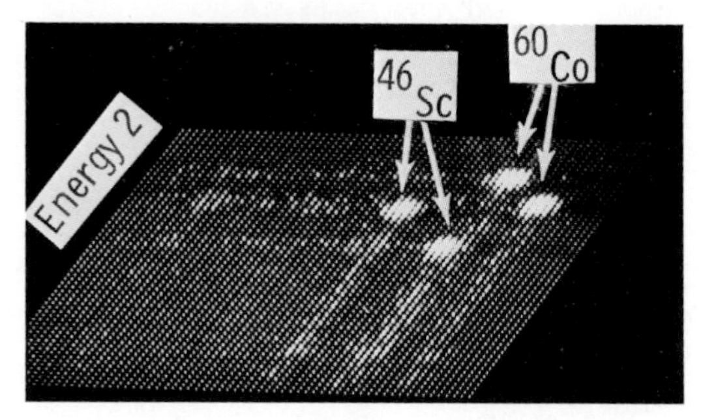

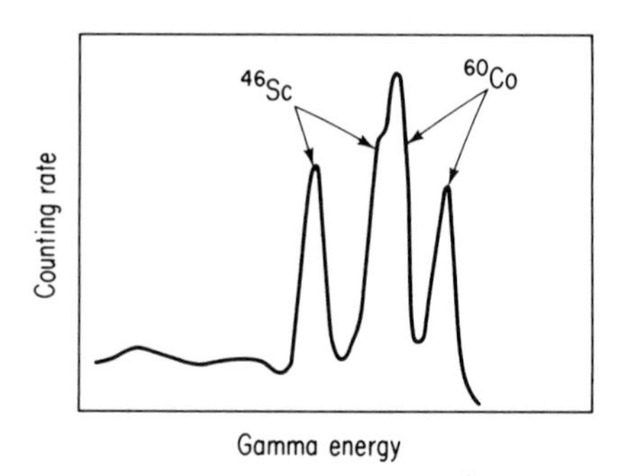

Fig. 18-8. Comparison of multidimensional γ-ray and single-crystal spectra. From Neilsen and Perkins (26). "Anticoincidence-shielded, multi dimensional gamma ray spectrometers for low-level counting." In *Radioactive Dating and Methods of Low Level Counting*, IAEA, Vienna, 1967

BIBLIOGRAPHY

1. *Assessment of Airborne Radioactivity.* Vienna: International Atomic Energy Agency, 1967. A useful survey of the instruments and techniques used to monitor atmospheric radioactivity.
2. ATKINS, D. H. F., and A. E. J. EGGLETON. "Studies of atmospheric washout and deposition of γ-BHC, dieldrin, and *p-p*DDT using radio-labelled pesticides." In *Nuclear Techniques in Environmental Pollution.* Vienna: International Atomic Energy Agency, 1971, pp. 521–534.
3. BARRY, P. J. "Use of argon-41 to study the dispersion of stack effluents." In *Nuclear Techniques in Environmental Pollution.* Vienna: International Atomic Energy Agency, 1971, pp. 241–255.
4. BESWICK, C. K. "Routine industrial uses of radioactive tracers." In *Radioisotope Tracers in Industry and Geophysics.* Vienna: International Atomic Energy Agency, 1967, pp. 32–35.

5. CAMP, D. C. "Applications and optimization of the Ge(Li) detector system," *UCRL-50156.*, March 1967.

6. CAPPADONA, C. "Measurements of movements of solid substances in water by means of stable tracers and activation analysis." In *Modern Trends in Activation Analysis.* Washington, D.C.: National Bureau of Standards, 1969, pp. 72–75.

7. CHANNELL, J. K., and P. KRUGER. "Post-sampling activation analysis of stable nuclides for estuary water tracing. In *Modern Trends in Activation Analysis.* Washington, D.C.: National Bureau of Standards, 1968, pp. 600–606.

8. CHATTERS, R. M., and R. L. PETERSON. "Use of tracers in in-plant evaluation of processes and in pollution control of effluent from pulp and paper mills into rivers." In *Radioisotope Tracers in Industry and Geophysics.* Vienna: International Atomic Energy Agency, 1967, pp. 251–258.

9. COOPER, J. A. "Applied Ge(Li) gamma ray spectroscopy," *BNWL-SA-3603*, January 1971.

10. DAHL, J. B., E. STEINNER, and J. THOMASSEN. "Air pollution studies by means of inactive indium tracer and activation analysis." In *Nuclear Techniques in Environmental Pollution.* Vienna: International Atomic Energy Agency, 1971, pp. 283–296.

11. DEVOE, J. R. *Radioactive Contamination of Materials Used in Scientific Research.* Publication 895, NAS-NRC, 1961.

12. EISENBUD, M. *Environmental Radioactivity*, 2nd Ed. New York: Academic Press, 1973. A useful survey of some techniques of detecting and measuring natural radioactivity, particularly in relation to nuclear power plants.

13. FENDRIK, I., and H. GLUBRECHT. "Investigations of the propagation of plant pollen by an indicator activator method." In *Nuclear Activation Techniques in the Life Sciences.* Vienna: International Atomic Energy Agency, 1967.

14. FRIEND, A. G. *Proceedings on the Effects of Radioactivity in Estuary Surveillance and Sampling Programs* (USAEC, 1961).

15. GASPAR, E., and M. ONCESCU. *Radioactive Tracers in Hydrology.* Amsterdam: Elsevier, 1972.

16. GATZ, D. F., A. N. DINGLE, and J. W. WINCHESTER. "Detection of indium as an atmospheric tracer by neutron activation," *J. Appl. Meteor.* **8**, 229–235 (1969).

17. GIFFORD, F. A. "Use of routine meteorological observations for estimating atmospheric dispersion," *Nucl. Sat.* **2**, 4 (1961).

18. GILATH, C., and Z. STUHL. "Concentration dynamics in a lake with a water current flowing through it." In *Nuclear Techniques in Environmental Pollution.* Vienna: International Atomic Energy Agency, 1971, pp. 483–496.

19. GIRARDI, F., M. MERLINI, C. BIGLIOCCA, G. POZZI, and A. BERG. "A radiotracer technique for the study *in vivo* of the biological pathway of heavy metals in aquatic organisms." In *Nuclear Techniques in Environmental Pollution.* Vienna: International Atomic Energy Agency, 1971, pp. 721–33.

20. GREENE, R. E. "Tracing with activable tracers. Part I and Part II." *Isotopes and Radiation Technology* **5**(4), 269 (1968) and **6**(1), 70 (1968).

21. KREGER, W. E., and R. L. MATHER. "Background, shielding and collimation." In S. Shafrath (Ed.), *Scintillation Spectroscopy of Gamma Radiation.* New York: Gordon and Breach, 1967, Vol. 1.

22. KRUGER, P. *Principles of Activation Analysis.* New York: Wiley, 1971. Chapter 9 contains a very complete discussion, with bibliography, of the use of stable-activable tracers.

23. LAL, D., and H. SUESS. "The radioactivity of the atmosphere and hydrosphere," *Ann. Rev. Nucl. Sci.* **18**, 407 (1968). An excellent review containing many references on the present state of knowledge of natural radioactivity.

24. LOWMAN, J. T., and W. KRIVIT. "New in vivo tracer method with the use of nonradioactive isotopes and activation analysis," *J. Lab. Clin. Med.* **61**, 1042 (1963).

25. NAKASA, H., and H. OHNO. "Application of neutron activation analysis to stack-gas tracing." In *Radioisotope Tracers in Industry and Geophysics.* Vienna: International Atomic Energy Agency, 1967, pp. 239–250.

26. NIELSEN, J. M., and R. W. PERKINS. "Anticoincidence shielded multidimensional gamma ray spectrometers for low-level counting." In *Radioactive Dating and Methods of Low-Level Counting.* Vienna: International Atomic Energy Agency, 1967, p. 687.

27. *Nuclear Activation Techniques in the Life Sciences.* Vienna: International Atomic Energy Agency, 1967. A useful survey of many applications of activation analysis in the life sciences, with some discussions of the use of stable-activable tracers.

28. *Nuclear Techniques in Environmental Pollution.* Vienna: International Atomic Energy Agency, 1971. An excellent volume dealing with many interesting applications of radiotracers and stable-activable tracers in solving environmental problems. "Must" reading for the environmental scientist.

29. PASQUILL, F. "The estimation of the dispersion of windborne material," *Meteorol. Mag.* **90**, 33 (1961).

30. *Radioactive Dating and Methods of Low-Level Counting.* Vienna: International Atomic Energy Agency, 1967.

31. *Radioisotope Tracers in Industry and Geophysics.* Vienna: International Atomic Energy Agency, 1967. A useful, comprehensive survey of practical physical applications of tracers, including many examples of environmental interest.

32. SHUM, Y. S., and W. LOVELAND, to be published.

33. SUGIHARA, T. T. *Low-level radiochemical separations.* NAS-NS-3103, April 1964.

34. THOMAS, C. W., and N. A. WOGMAN. "Atmospheric behavior of airborne radionuclides," *USAEC Res. and Develop. Report BNWL-281-2*, 4 (1967).

35. WATT, D. E., and D. RAMSDEN, *High Sensitivity Counting Techniques.* New York: Macmillan, 1964.

19

Tracer Applications in the Physical Sciences

Although the applications of radiotracers in the physical sciences are so numerous as to virtually defy description, it was believed worthwhile to discuss a few examples of some of the more important ones. Extensive bibliographies do exist (3, 4, 8, 9) that catalog many of these applications, and the reader is referred to them in any search for further information.

A. TRACING OF PHYSICAL PROCESSES

In many cases in which radiotracers are used, the chemical identity of the tracer is not important. These applications can be referred to *tracing physical processes*. For example, consider those experiments that seek to *locate an object in some system* by labeling it with radioactivity and then measuring the position of the radioactivity in the system. Quite often the tracer that decays by γ-ray emission is picked so that the radiation from the source will penetrate large masses of tissue, pipe, earth, and so on. *Mixing studies* are frequently carried out by using radiotracers. Here the objective is usually to see if proper mixing has taken place between two components of a system. Generally one of the components is labeled with a radiotracer and its distribution in the system is monitored as a function of time. A short-lived tracer is often used so that it can quickly decay away at the conclusion of the experiment and leave an essentially "nonradioactive" mixed system for further use.

A form of isotope dilution is frequently used to measure the *volume of an inaccessible container* (11). A small volume V_1 of tracer solution is assayed to give its activity A_1. The tracer is added to the liquid in the container and mixed, and a sample of size V_1 is removed and assayed to show its activity A_2. Then from Equation (18-3), the volume of the original container is given as

$$V = V_1\left(\frac{A_1}{A_2} - 1\right) \tag{19-1}$$

Leak testing can also be done by using radiotracers. Here the basic idea is simple—namely, to inject radiotracer into the pipe, flask, or whatever that is suspected to be leaking and look for activity that appears outside the container and so on.

One important application of tracers is the *measurement of the surface area of solids* by using tracers. One measures the exchange of active ions between a solution and the surface of a solid. Consider the case of a precipitate of $PbSO_4$ in a solution of $^{212}Pb(NO_3)_2$. At equilibrium, the following relation will hold; that is,

$$\frac{\text{Number Pb atoms in solution}}{\text{Number Pb atoms on surface}} = \frac{^{212}\text{Pb activity in solution}}{^{212}\text{Pb activity on surface}} \tag{19-2}$$

Since three of the four quantities in Equation (19-2) can be measured, the number of Pb atoms on the surface of the $PbSO_4$ precipitate can then be calculated. If one assumes (or knows) that the average area of a Pb atom in $PbSO_4$ is $\sim 18.6\text{Å}^2$, one can calculate the surface area of the solid. Related to this study are many studies of the absorption of gases, liquids, and solids using radiotracers.

The use of radiotracers can allow one to study diffusion that occurs without a concentration gradient in the system, inasmuch as only minute amounts of tracer are needed. The classic work of Groh and Hevesy (10) is an example of this type of experiment. Groh and Hevesy studied the self-diffusion of ^{210}Pb into a lead bar. After allowing the tracer to diffuse into the bar, the bar was cut into sections and counted. The diffusion coefficient $\mathfrak{D}$ was determined by the application of Fick's law

$$\frac{dc}{dt} = \mathfrak{D}\frac{d^2c}{dx^2} \tag{19-3}$$

where c is tracer specific activity, t the elapsed time, and x the distance the tracer has diffused.

B. CHEMICAL APPLICATIONS OF TRACERS

One of the most important uses of radiotracers in chemistry has been to *test separation procedures* in analytical chemistry. Tracers furnish a specific, easy-to-apply, quick method of following the path of a given material in a chemical separation. Physical chemical data or separation parameters can also be determined. An example of this type of application is the work of Sunderman and Meinke (15) in studying separations by precipitation. The scavenging efficiency of $Fe(OH)_3$ was evaluated by seeing how many other radiolabeled ions would coprecipitate with $Fe(OH)_3$ and the quantitative extent to which they were incorporated in the $Fe(OH)_3$ precipitate. A famous example of the use of radiotracers in evaluating separation procedures is the work of Kraus and Nelson (13), in which they studied the pH dependence, eluant volume, and similar factors in separations of metal ions by ion exchange. Tracers have also been used to locate the position of a particular fraction in column, thin layer (TLC), and paper chromatography. When used in TLC and paper chromatography, autoradiography is frequently utilized to locate the position of the activity.

The use of radiotracers is an excellent technique for *measuring the solubility product constant* of sparingly soluble salts or for making other studies of substances present in low concentrations. Another very important and classic example of the use of radiotracers is that of studying the occurrence and properties of *isotopic exchange reactions*—that is, reactions of the type

$$AX^* + BX \rightleftharpoons BX^* + AX \qquad (19\text{-}4)$$

where X, X* represent stable and radioactive atoms, respectively, of the same element.

Perhaps the most significant of the numerous applications of radiotracers in chemistry has been in the study of *chemical reaction mechanisms*. In fact, most of the proposed reaction mechanisms have been "verified" by means of a radiotracer study. However, it should be noted that it is impossible to use tracer methods intelligently to study reaction kinetics unless some preliminary measurements of other features (thermodynamics, possible intermediates, salt effects, etc.) have been made, using ordinary chemical techniques.

One of the simplest mechanistic experiments using radiotracers is to *test the equivalence of various atoms in molecules* in chemical reactions. An example of this type of study is the work of Volpin *et al.*, (16) on the equivalence of the seven carbon atoms in the tropylium ring. Volpin *et al.*, reacted labeled diazomethane with benzene and brominated the cyclohepatriene product to form a labeled tropylium bromide, as shown below.

$$C_6H_6 \xrightarrow{^{14}CH_2N_2} C_7H_8 \xrightarrow{Br_2} {}^{14}C_7H_7Br \qquad (19\text{-}5)$$

Then the tropylium bromide was subjected to a Grignard reaction and the product oxidized to give a labeled benzoic acid. Thus one had

$$^{14}C_7H_7Br \xrightarrow{C_6H_5MgBr} {}^{14}C_7H_7C_6H_5 \xrightarrow{HNO_3} C_6H_5{}^{14}COOH \qquad (19\text{-}6)$$

The specific activity of the labeled benzoic acid was found to be one-seventh that of the initial labeled diazomethane, thus showing the equivalence of the seven carbon atoms of the tropylium ring. A similar study was that of Downs and Johnson (7), which showed, in the reaction of labeled $*Cl_2$ with PCl_5, that three chlorine atoms in PCl_5 were quickly labeled, while the other two atoms were slow to become labeled, thus showing the non-equivalence of the equatorial and axial chlorine atoms in the trigonal bi-pyramidal PCl_5 structure.

Another popular use of radiotracers in studying chemical reaction mecha-nisms is the study of *molecular rearrangements*. A typical example is the study of the benzilic acid rearrangement by Collins and Neville (5).

$$C_6H_5CH_2\text{–}CO\text{–}{}^{14}CO\text{–}C_6H_5 \ \text{(I)} \xrightarrow{NaOH} HO\text{–}{}^{14}C(CH_2C_6H_5)(C_6H_5)\text{–}COOH \ \text{(II)} \qquad (19\text{-}7)$$

Oxidation of the product (**II**) left all the ^{14}C activity in $C_6H_5\text{–}{}^{14}C(=O)\text{–}CH_2\text{–}C_6H_5$ and none in the CO_2. Thus one concluded that the benzilic acid rearrange-ment proceeds by migration of the benzyl group rather than the phenyl group.

Another example of the study of this class of reactions that illustrates the use of radiotracers is the cyclization of ω-phenoxyacetophenone (**I**) to 2-phenylbenzofuran (**II**). Two possible mechanisms are shown below.

$$C_6H_5\text{–}O\text{–}CH_2\text{–}{}^{14}COC_6H_5 \ \text{(I)} \longrightarrow \text{3-}{}^{14}C\text{-3-phenylbenzofuran} \longrightarrow \text{2-phenylbenzofuran-3-}{}^{14}C \ \text{(II)} \qquad (19\text{-}8)$$

or

$$\text{(I)} \longrightarrow \left[\text{phenol (OH)} + \overset{+}{C}H_2-{}^{14}C(=O)-C_6H_5 \right] \xrightarrow{-H^+} \text{o-}(C_6H_5{}^{14}COCH_2)C_6H_4OH \longrightarrow \text{(II)}$$

(I): phenyl ether, $O-CH_2-{}^{14}COC_6H_5$; (II): 2-phenylbenzofuran, label at C-2 bearing C_6H_5 (14)

(II) (19-9)

Note that the two possible mechanisms, (19-8) and (19-9), can be distinguished by the position of the labeled atom in the 2-phenylbenzofuran.

Two problems that are common in tracer studies of reaction mechanism occurred in this study. The first was the synthesis of the labeled starting material (**I**) for the reaction. A chemical synthesis, shown below, was used to produce (**I**).

$$^{14}CH_3COBr \xrightarrow[\text{Craft Reaction}]{\text{Freidel}} C_6H_5{}^{14}COCH_3 \xrightarrow[\text{acetone}]{Br_2\text{ in}} C_6H_5{}^{14}COCH_2Br \xrightarrow[Na_2CO_3]{\text{phenol in acetone with}}$$

$$C_6H_5-{}^{14}\overset{\overset{O}{\|}}{C}-CH_2-O-C_6H_5 \quad \text{(I)} \qquad (19\text{-}10)$$

(**I**)

After obtaining the labeled starting material and carrying out the cyclization reaction, one had the problem of degrading the product (**II**) to reveal the position of the label. This was done by using the following steps:

$$\text{2-phenylbenzofuran } (C_6H_5) \xrightarrow[+CH_3I\text{ in ethanol}]{\text{metallic Na}} \text{o-}(CH_2-CH_2-C_6H_5)C_6H_4OCH_3 \xrightarrow[\text{and triethylamine}]{N\text{-bromosuccinimide}}$$

$$\text{o-}(CH{=}CH-C_6H_5)C_6H_4OCH_3 \xrightarrow{\text{oxidation with permanganate}} \text{o-}(COOH)C_6H_4OCH_3 + C_6H_5COOH$$

(19-11)

Note that in reaction (19-11), the two carbon atoms in the five-membered ring of **II** end up in different reaction products. Thus one can check to see which product contains the label and decide on the correct mechanism (19-8).

Studies of reaction intermediates can also be carried out using tracers to measure the *kinetics of slow reactions.* The easiest way to do so is to add a labeled version of a proposed reaction intermediate and observe its behavior and that of the reaction products as a function of time. Kondratyev (12) used radiotracers to study the oxidation of butane at 285°C.

$$C_4H_9 + O_2 \longrightarrow CO + CO_2 \tag{19-12}$$

It had been established prior to the tracer study that CO and CO_2 were both formed in the reaction. The question was whether the CO_2 was formed by oxidation of the CO or was itself a primary reaction product. Kondratyev added ^{14}CO to the reaction mixture and followed the activities of CO and CO_2 as a function of time (see Figure 19-1). From the fact that the specific activity of the CO_2 product did not change with time but that the CO activity decreased with time (due to dilution by CO_2), it was concluded that $<5\%$ of the CO_2 was formed from CO.

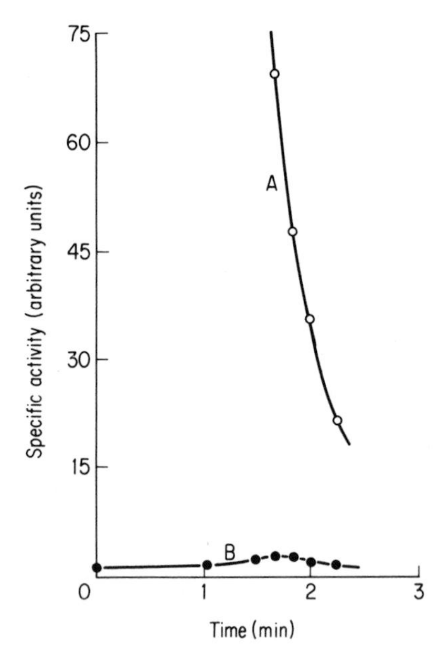

Fig. 19-1. Variation of specific activity of (*A*) carbon monoxide and (*B*) carbon dioxide with time, after the addition of ^{14}CO to butane, during oxidation at 285°.

C. ISOTOPE EFFECTS

Up to now in our discussion of tracer methodology, we have assumed that all isotopes of a given element, stable or radioactive, would behave alike chemically and physically. We should now examine this point more critically to see how different isotopes behave differently and how this difference in behavior (*the isotope effect*) can be detected and used to our advantage. (See Chapter 14 for a discussion of these effects in relation to experimental design.) The difference in behavior between different isotopes of the same element is due to the differing masses of the isotopes. This mass difference will affect the kinetic energy of the molecules (giving rise to *physical isotope effects*) or will change the vibrational or rotational properties of molecules (giving rise to chemical isotope effects).

Some examples of physical isotope effects are

1. *Mass spectrometry*—Trivially, one notes that a radioisotope-labeled molecule will have a different mass than a nonlabeled molecule and thus different mass spectrometric properties.
2. *Gaseous diffusion*—Graham's law states that

$$R = \sqrt{\frac{m_1}{m_2}} \tag{19-13}$$

   where R is the relative rate of diffusion of two isotopes of mass m_1 and m_2, respectively, through an orifice. Thus a radioisotope-labeled molecule will diffuse at a different rate than an unlabeled molecule, with the difference in rates being given by Equation (19-13).
3. *Distillation*—For a given temperature, the velocity of a light isotope will be greater than that of a heavy isotope, so that the lighter isotope will have a higher vapor pressure. It can be shown that the separation factor S for the distillation of two isotopes is given by

$$S = \frac{P_1^0}{P_2^0} \tag{19-14}$$

   where P^0 is the saturation vapor pressure of the ith isotope.

Chemical isotope effects can be divided into two classes—those affecting the position of the equilibrium in a chemical reaction and those affecting the rate or speed of a chemical reaction. *Equilibrium* isotope effects involve changes in the position of the equilibrium due to the substitution of a labeled atom in a molecule. These effects have their origin in the fact that the extent to which any chemical reaction "goes" is governed by the number of possible ways it can proceed. The more equally probable reaction paths available, the more likely the reaction will "go." To illustrate this point, consider the exchange reaction

$$AX + BX^* \rightleftharpoons AX^* + BX \tag{19-15}$$

The equilibrium constant K for the reaction is given by

$$K = \frac{f_{AX^*} f_{BX}}{f_{AX} f_{BX^*}} \tag{19-16}$$

where f_i is the partition function for the ith species.† The partition functions can be calculated by using statistical mechanics, and they depend in a complicated way on the relative masses of AX, AX*, and so on. Table 19-1 shows some typical values of K for various exchange reactions and illustrates the point that chemical equilibrium can be and is shifted when a radioisotope of given mass is substituted for another isotope. (Note that when $X = X^*$, $K = 1$.) Table 19-1 also shows the important fact that the greater the mass difference between isotopes, the larger such equilibrium isotope effects are. In general, it has been found (8) that such effects can be neglected when $Z > 10$.

TABLE 19-1

Typical Equilibrium Isotope Effects

Reacting System	K
$H^2H(g) + H_2O(l) \rightleftharpoons H_2(g) + {}^2HOH(l)$	3.2
${}^{14}CO_2(g) + {}^{12}COCl_2(g) \rightleftharpoons {}^{14}COCl_2(g) + {}^{12}CO_2(g)$	1.0884
$[Co(NH_3)_4{}^{12}CO_3]^+ + {}^{14}CO_3^= \rightleftharpoons {}^{12}CO_3^= + [Co(NH_3)_4{}^{14}CO_3]^+$	0.8933

Kinetic isotope effects are very important in the study of chemical reaction mechanisms. The substitution of a labeled atom for an unlabeled one in a molecule will cause a change in the reaction rate for $Z < 20$, and this change can be exploited to deduce the reaction mechanism. In 1970 an ACS monograph (9) dealing with this subject was published. The change in reaction rate with change in mass of reacting species is largely due to differences in vibrational frequency along the reaction coordinate in the transition state or activated complex reaction intermediate. (A detailed treatment of the theory of kinetic isotope effects is found in References 2, 6, and 14.)

Kinetic isotope effects are classified into two types—intramolecular and intermolecular. An example of the *intramolecular* kinetic isotope effect is the decarboxylation of malonic acid shown below:

$$H_2{}^{12}C\begin{cases} {}^{12}COOH \\ {}^{14}COOH \end{cases} \longrightarrow \begin{cases} \xrightarrow{k_1} {}^{12}CO_2 + CH_3{}^{14}COOH & \text{(a)} \\ \xrightarrow[k_2]{} {}^{14}CO_2 + CH_3{}^{12}COOH & \text{(b)} \end{cases} \qquad (19\text{-}17)$$

The insertion of a ${}^{14}C$ label into the system has caused a symmetrical molecule, malonic acid, to become unsymmetrical, and thus chemical bonds that were equivalent in the unlabeled molecule are now nonequivalent. This leads

†f_i reflects the probability of occurrence of the ith species and is formally the probability of occurrence of a set of vibrational, rotational, and translational energy levels for that species.

to the situation where the decarboxylation reaction can now take place in two different ways (a or b) and the rates for these two ways (k_1 and k_2) are different. The reason is that in (a) one breaks a $^{12}C—^{12}C$ bond, whereas in (b) one breaks a $^{14}C—^{12}C$ bond.

The decarboxylation of benzoic acid can furnish an example of an *intermolecular kinetic isotope effect*. Consider an attempt to study the decomposition of $C_6H_5—^{12}COOH$ by studying the decomposition of $C_6H_5—^{14}COOH$. The rates of the two decompositions will not be equal because of the presence of the ^{14}C label in the one compound. Thus labeling can affect the comparison between two separate molecules (*intermolecular effect*), whereas in the *intramolecular* effect bonds within a given molecule are made nonequivalent by the presence of a labeled atom. Note that (as indicated in Chapter 14) the intermolecular isotope effect can be avoided in ordinary tracer studies by driving the reactions to completion.

Experimentally, it is relatively straightforward to measure the existence and magnitude of kinetic isotope effects. Consider the reaction shown below, proceeding through a reaction intermediate, $AB^‡$, as

$$\mathrm{A} + \mathrm{B} \longrightarrow \mathrm{AB}^‡ \longrightarrow \text{products} \tag{19-18}$$

Let S_0 and S_γ be the specific activities of B at time $t = 0$ and after a fraction γ of the reaction has been completed, respectively. If B is the only labeled reactant, we have

$$\frac{d[B]}{dt} = k[A]^a[B]^b \tag{19-19}$$

and assuming B^* is present as a tracer

$$\frac{d[B^*]}{dt} = k^*[A]^a[B^*][B]^{b-1} \tag{19-20}$$

where a and b are the reaction order with respect to A and B. Dividing (19-19) by (19-20), we have

$$\frac{d[B]}{d[B^*]} = \frac{k[B]}{k^*[B^*]} \tag{19-21}$$

Integrating gives

$$\log_{10}\left(\frac{S_\gamma}{S_0}\right) = \left(\frac{k^*}{k} - 1\right)\log_{10}(1 - \gamma) \tag{19-22}$$

Thus plotting $\log_{10}(S_\gamma/S_0)$ versus $\log(1 - \gamma)$ gives a straight line of slope $(k^*/k - 1)$. If there was no isotope effect, $k = k^*$, then the slope will be zero. Any finite slope in the preceding plot will give k^*/k. This effect is shown in Figure 19-2 for the study of the Cannizzaro reaction by Downes and Harris (6).

As indicated earlier, kinetic isotope effects have important uses in establishing chemical reaction mechanisms. For example, consider the *decarboxy-*

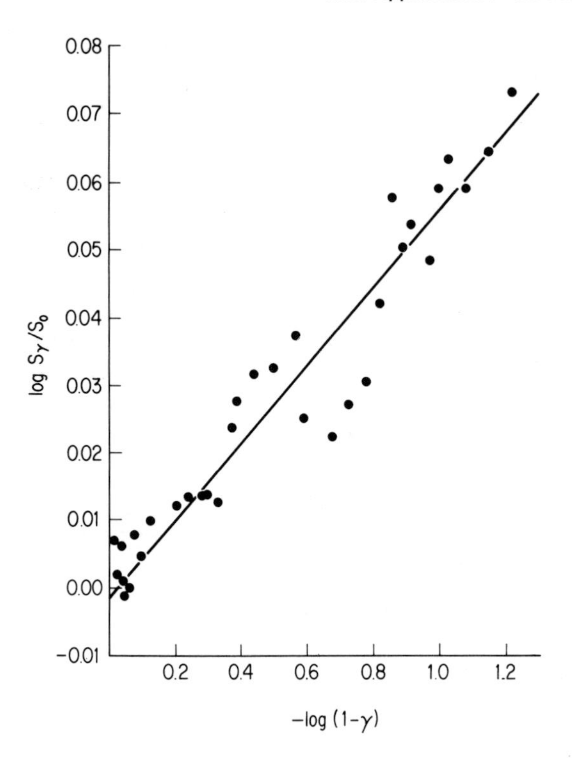

Fig. 19-2. Plot of log (S_γ/S_0) as a function of log $(1 - \gamma)$ for the Cannizzaro reaction, showing the isotope effect. From A.M. Downes and G. M. Harris (6).

lation of a dicarboxylic acid as shown below:

$$\mathrm{R{-}\overset{\alpha}{C}H(\overset{\beta}{C}({=}O){-}OH)(C({=}O){-}OH)} \longrightarrow \left[\mathrm{R\overset{\alpha}{C}H{=}C(HO){-}O\cdots H{-}O{-}\overset{\beta}{C}({=}O)}\right]^{\ddagger} \longrightarrow \mathrm{CO_2 + R{-}\overset{\alpha}{C}H{=}C(HO)(OH)} \qquad (19\text{-}23)$$

By doing successive labeling experiments, the first of which involved a ^{14}C on the $^{\alpha}C$ position and the second of which involved a ^{14}C at the $^{\beta}C$ position, one was able to demonstrate that the rate-determining step in the reaction was a C—C bond rupture. (The experimental observation was that, for the ^{14}C label in the α position, $k/k^* = 1.076$, whereas $k/k^* = 1.064$ for the ^{14}C label in the β position.)

The study of the *mechanism of elimination* reactions is another important application of radiotracers in chemistry. For example, consider the following

TABLE 19-2*

Reaction	Temperature (°C)	Experimental Isotope Effect	Theoretical Isotope Effect
	Intramolecular [Intermolecular marked†]		
$^{13}CH_2(COOH)_2 \xrightarrow{k_4} CO_2 + {}^{13}CH_3COOH$	137	$k_3/2k_2 = 1.046, 1.037, 1.036$	1.021, 1.020, 1.035
$CH_2(COOH)_2 \xrightarrow{k_3} CO_2 + CH_3COOH$			
$CH_2(^{13}COOH)(COOH) \xrightarrow{k_2} {}^{13}CO_2 + CH_3COOH$		$k_1/k_2 = 1.020, 1.024, 1.026$	1.020
$CH_2(^{13}COOH)(COOH) \xrightarrow{k_1} CO_2 + CH_3{}^{13}COOH$		$k_3/(k_1 + k_2) = 1.025$	
(ring) $CONH_2$, $CONH_2$ $\xrightarrow{k_1}$ (ring) $(CO)_2NH + NH_3$		$k_2/k_3 = 1.014$	1.016
	180	$k_1/2k_2 = 0.994$† $k_1/2k_3 = 1.006$†	>0.97
(ring) $CO^{15}NH_2$, $CONH_2$ $\xrightarrow{k_2}$ (ring) $(CO)_2{}^{15}NH + NH_3$		$k_1/(k_2 + k_3) = 1.000$†	
(ring) $CO^{15}NH_2$, $CONH_2$ $\xrightarrow{k_3}$ (ring) $(CO)_2NH + {}^{15}NH_3$			
$(COOH)_2 \xrightarrow{k_1} CO_2 + CO + H_2O$	100	$k_2/k_3 = 1.032$	
$^{13}COOH{-}COOH \xrightarrow{k_2} {}^{13}CO_2 + CO + H_2O$		$k_1/(k_2 + k_3) = 1.034$†	
$^{13}COOH{-}COOH \xrightarrow{k_3} CO_2 + {}^{13}CO + H_2O$			
Same reaction with ^{14}C-labeling	103 80	$k_3/k_2 = 1.055$ $k_3/k_2 = 1.067$	1.043 1.0435
	Intermolecular		
$H_2N.{}^{13}CONH_2 + H_2O \xrightarrow[\text{catalyst}]{\text{urease}} {}^{13}CO_2 + 2NH_3$	30	$k_{12}/k_{13} = 1.010$	

TABLE 19-2 (Continued)

Reaction	Temperature (°C)	Experimental Isotope Effect	Theoretical Isotope Effect
Same reaction with ^{14}C-labeling $2H^{14}CHO + OH^- = {}^{14}CH_3OH + H^{14}COO^-$	30 60	$k_{12}/k_{14} = 1.032, 1.011$ $k_{14}/k_{12} = 1.0, 0.943$	
$(CH_3)_3C_6H_2{}^{13}COOH$ (2,4,6-trimethyl ring with $^{13}COOH$) $\rightarrow (CH_3)_3C_6H_3 + {}^{13}CO_2$	60	$k_{13}/k_{12} = 1.038$	1.041
$H_2{}^{16}O^{18}O + 2Fe^{2+} + 2H^+ = H_2{}^{16}O + H_2{}^{18}O + 2Fe^{3+}$	25	$k_{16}/k_{18} = 0.94$	0.939
$(COOH)_2 \xrightarrow{k_1} CO_2 + CO + H_2O$	100	$k_2/k_3 = 1.032$	
$^{13}COOH-COOH \xrightarrow{k_2} {}^{13}CO_2 + CO + H_2O$ $^{13}COOH-COOH \xrightarrow{k_3} CO_2 + {}^{13}CO + H_2O$		$k_1/(k_2 + k_3) = 1.034†$	
Same reaction with ^{14}C-labeling	103 80 Intermolecular	$k_3/k_2 = 1.055$ $k_3/k_2 = 1.067$	1.043 1.0435
$H_2N\cdot{}^{13}CONH_2 + H_2O \xrightarrow[\text{catalyst}]{\text{urease}} {}^{13}CO_2 + 2NH_3$	30	$k_{12}/k_{13} = 1.010$	
Same reaction with ^{14}C-labeling $2H^{14}CHO + OH^- = {}^{14}CH_3OH + H^{14}COO^-$	30 60	$k_{12}/k_{14} = 1.032, 1.011$ $k_{14}/k_{12} = 1.0, 0.943$	
$(CH_3)_3C_6H_2{}^{13}COOH$ (2,4,6-trimethyl ring with $^{13}COOH$) $\rightarrow (CH_3)_3C_6H_3 + {}^{13}CO_2$	60	$k_{13}/k_{12} = 1.038$	1.041
$H_2{}^{16}O^{18}O + 2Fe^{2+} + 2H^+ = H_2{}^{16}O + H_2{}^{18}O + 2Fe^{3+}$	25	$k_{16}/k_{18} = 0.94$	0.939

*Reprinted from J. F. Duncan and G.B. Cook. *Isotopes in Chemistry*. ©1968. Oxford University Press by permission of The Clarendon Press, Oxford.

reaction mechanisms:

E1

$$\mathrm{H{-}\overset{\beta}{C}{-}\overset{\alpha}{C}{-}X \xrightarrow[\text{slow}]{} X^- + H{-}\overset{\beta}{C}{-}\overset{\alpha}{C} \xrightarrow[\text{slow}]{} XH + \overset{\beta}{C}{=}\overset{\alpha}{C}} \qquad (19\text{-}24)$$

E2

$$\mathrm{Y^- + H{-}\overset{\beta}{C}{-}\overset{\alpha}{C}{-}X \xrightarrow[\text{slow}]{} [\overset{\delta^-}{Y}\cdot\cdot H\cdot\cdot \overset{\beta}{C}\cdot\cdot\overset{\alpha}{C}\cdot\cdot \overset{\delta^-}{X}] \longrightarrow YH + \overset{\beta}{C}{=}\overset{\alpha}{C} + X^-} \qquad (19\text{-}25)$$

Carbanion

$$\mathrm{Y^- + H{-}\overset{\beta}{C}{-}\overset{\alpha}{C}{-}X \xrightarrow[\text{slow}]{} YH + {-}\underset{\beta}{\overset{\ddot{-}}{C}}{-}\underset{\alpha}{C}{-}X \xrightarrow[\text{fast}]{} \overset{\beta}{C}{=}\overset{\alpha}{C} + X^-} \qquad (19\text{-}26)$$

where the abbreviations *E1* and *E2* stand for elimination reaction, first order and second order respectively. For the *E1* mechanism, one expects an isotope effect for labeled X and $^{\alpha}$C and no effect for labeled $^{\beta}$C or the β-hydrogen. For the *E2* mechanism, one expects an isotope effect for labeled Y, the β-hydrogen, $^{\beta}$C, $^{\alpha}$C, and X. For the carbanion mechanism, if the first step is the rate-determining step, one expects an isotope effect for the labeled Y, β-hydrogen, and $^{\beta}$C. Thus a study of the kinetic isotope effect for elimination reactions can determine the mechanism by which the reaction proceeds. A summary of some observed kinetic isotope effects is shown in Table 19-2, taken from Duncan and Cook (8).

BIBLIOGRAPHY

1. Bigeleisen, J. *J. Chem. Phys.* **17**, 675 (1969).
2. Bigeleisen, J., and M. Wolfsberg. *Adv. Chem. Phys.* **1**, 15 (1958).
3. Bowen, H. J. M. *Chemical Applications of Radioisotopes.* London: Methuen, 1969. A fine, small monograph that surveys the application of radiotracer methods in chemistry.
4. Choppin, G. R. *Experimental Nuclear Chemistry.* Englewood Cliffs, N.J.: Prentice-Hall, 1961. This now out-of-print volume contains many succinct descriptions of the chemical applications of tracers and laboratory experiments that demonstrate such applications.
5. Collins, C. J., and O. K. Neville. *J. Am. Chem. Soc.* **73**, 2471 (1951).
6. Downes, A. M., and G. M. Harris. *J. Chem. Phys.* **20**, 196 (1952).
7. Downs, J., and R. E. Johnson. *J. Am Chem. Soc.* **77**, 2098 (1955).
8. Duncan, J. F., and G. B. Cook. *Isotopes in Chemistry.* Oxford: Clarendon, 1968.

9. FRY, A. "Heavy atom isotope effects in organic reaction mechanism studies." In C. J. Collins and N. S. Bowman, (Eds.), *Isotope Effects in Chemical Reactions*. ACS Monograph 167. New York: Van Nostrand Reinhold, 1970.

10. GROH, J., and G. HEVESY. *Ann. Phys.* **65**, 216 (1921).

11. HALL, D. E., B. A. FRIES, and J. T. GILMORE. *Intern. J. Appl. Radiation Isotopes* **16**, 19 (1965).

12. KONDRATYEV, V. N. *Proc. Int. Conf. on Peaceful Uses of Atomic Energy, Geneva* **15**, P708 (1956).

13. KRAUS, K. A., and F. NELSON. *Ann. Rev. Nucl. Sci.* **7**, 31 (1957).

14. STERNAND, M. J., and M. WOLFSBERG. *J. Chem. Phys.* **45**, 405 (1966).

15. SUNDERMAN, D. N., and W. W. MEINKE. *Anal. Chem.* **29**, 1578 (1957).

16. VOLPIN, M. E. *et al. Zh. Obshch. Khim.* **29**, 3711 (1959).

Problems

NOMENCLATURE (CHAPTER 1)

1. Briefly describe, in words, the meaning of the terms isotope, isobar, isotone, and isomer.

2. Which of the following nuclides are (a) isotopes, (b) isotones, (c) isobars, or (d) isomers?

$$^{60m}Co,\ ^{14}C,\ ^{15}O,\ ^{14}N,\ ^{12}C,\ ^{59}Co,\ ^{60}Co$$

3. Using the expression for the nuclear radius, R,

$$R = 1.20A^{1/3} \times 10^{-13}\ \text{cm}$$

where A is the mass number of the nucleus in question, calculate the approximate density in kg/m^3 and metric tons/mm^3 for nuclei (1 metric ton = 1000 kg).

ENERGETICS AND DECAY KINETICS (CHAPTER 2)

1. From the following information, construct a set of decay schemes:

Parent Nuclide	Daughter Nuclide	Particle or Photon Emission
a. 164 day ^{45}Ca	Stable ^{45}Sc	β^-, 0.256 MeV, 100%
b. 18.66 day ^{86}Rb	Stable ^{86}Sr	β_1^-, 0.698 MeV, 10%
		β_2^-, 1.776 MeV, 90%
		γ, 1.078 MeV
c. 2.696 day ^{198}Au	Stable ^{198}Hg	β_1^-, 0.282 MeV, 1%
		β_2^-, 0.959 MeV, 99%
		β_3^-, 1.371 MeV, <1%
		γ_1, 0.4118 MeV
		γ_2, 0.676 MeV
		γ_3, 1.089 MeV
d. 3.03×10^5 yr ^{36}Cl	Stable ^{36}A	β^-, 0.714 MeV, >98%
	Stable ^{36}S	EC (1.19 MeV), <2%

2. In order to determine the decay constant (λ) and half-life ($t_{1/2}$) of an unknown radionuclide, the following activity determinations were made:

Time of Determination (hr)	Corrected Net Count Rate (cpm)
0	8500
0.16	7050
0.30	6000
0.50	4756
0.65	3998
0.80	3370
1.06	2499
1.30	1880

a. Plot these results on both a linear and a semilogarithmic basis (logarithm of count rate), as shown in Figures 2-12 and 2-13. As illustrated in Figure 2-15, determine the half-life in hours of this radionuclide by the graphic method.

Answer: ~0.6 hr

b. Now, using the graphically determined half-life value ($t_{1/2}$) in hours, calculate the decay constant (λ) of this radioisotope from the equation

$$\lambda = \frac{0.693}{t_{1/2}}$$

Answer: $\lambda = 1.155\ \text{h}^{-1}$

3. An investigator is using colloidal ^{198}Au, which has a half-life of 2.696 days. He has on hand a shipment assayed at 50 μCi as of 12 noon Monday. He intends to use this radioactive substance on Thursday of the same week at 9 A.M. Determine the activity as of the time of use.

Answer: ~24 μCi

4. An investigator has received a shipment labeled only "30% ^{14}C enriched $BaCO_3$—weight 2285 mg." He realizes that this designation means that 30% (by weight) of the material is $Ba^{14}CO_3$, the rest being $Ba^{12}CO_3$ and $Ba^{13}CO_3$. He desires to know how much radioactivity the shipment contains.

Answer: ~220 mCi

5. Using the same general procedure followed in Problem 4, calculate the specific activity of carrier-free ^{32}P in the form of Na_3PO_4. Express your result as mCi/mg Na_3PO_4. Use a half-life value of 14.221 days for ^{32}P. It is understood that in a carrier-free compound all the atoms of the radioactive element in question are radioactive.

Answer: 5.55×10^4 mCi/mg $Na_3{}^{32}PO_4$

6. In a synthetic reaction, carrier-free tritium gas (T_2) is used to hydrogenate fumaric acid to succinic acid as follows:

$$\mathrm{HOOC{-}\overset{H}{\overset{|}{C}}{=}\overset{H}{\overset{|}{C}}{-}COOH} + T_2 \xrightarrow{\text{catalyst}} \mathrm{HOOC{-}\underset{T}{\underset{|}{\overset{H}{\overset{|}{C}}}}{-}\underset{T}{\underset{|}{\overset{H}{\overset{|}{C}}}}{-}COOH}$$

Using the same procedure as in Problems 4 and 5, calculate the specific activity in mCi/mg of the resulting succinic acid. Use a half-life value of 12.262 yr for tritium.

Answer: ~470 mCi/mg succinic acid-T_2

7. Predict the most probable mode of decay and draw a decay scheme for
a. ^{57}Co **b.** ^{32}P **c.** ^{40}K **d.** ^{41}Ca **e.** ^{141}Ce

8. Calculate the activity in microcuries of ^{40}K in a 160-lb man if 0.35% of the body weight is potassium.

$t_{1/2}(^{40}\mathrm{K}) = 1.2 \times 10^9$ yr. Isotopic Abundance of ^{40}K is 0.012% of all K.

9. If a sample has a decay rate of 10^5 dpm at time t, what is its decay rate in dpm at 6.93 half-lives *prior* to t?

Answer: 1.2×10^7 dpm

10. In the radioactive decay of 86.7-d ^{35}S to ^{35}Cl, a particle with $E_\beta^{max} = 0.168$ MeV is emitted. Write the equation for decay and the expression for the Q of this reaction.

11. A 10-mCi source of 36-h ^{82}Br is purchased for the study of Br uptake and distribution in a rat.

a. Calculate the disintegration rate of this source at time of purchase in dps.
b. Calculate the number of mCi of ^{82}Br after 7.0 days' decay.
c. Calculate the number of ^{82}Br *atoms* and the number of *grams* of ^{82}Br present at *time* of purchase.
d. The average Br content of an adult rat is about 1 mg. If the rat is injected with 10 mCi ^{82}Br, what fraction of the total Br atoms in the rat will be ^{82}Br atoms?

12. A small quantity of the tracer 163-d ^{45}Ca is injected into a bony biological specimen. Assuming no significant ^{45}Ca is lost via natural functions, how much ^{45}Ca is present after 6.5 yr?

13. A biochemist requiring 36-h ^{82}Br for a tracer experiment was given a solution containing 10^9 ^{82}Br atoms. After 3.0-d, how many microcuries of ^{82}Br activity are present in the solution?

14. What is the specific activity of carrier-free ^{3}H in the form of water? Express your answer in the form of curies per millimole. What amount of ordinary water would be needed to dilute 1 ml of carrier-free ^3H_2O to a specific activity of 220 dpm/ml. What is the cost of 1 ml ^3H_2O (carrier-free) at the current cost of $2 per curie?

DECAY AND INTERACTION OF RADIATION WITH MATTER (CHAPTER 3)

1. A β^-, γ-sensitive G-M detector with a wall thickness of 30 mg/cm^2 is used for assay of an nuclide emitting 1-MeV γ rays. Ignoring air absorption, what fraction of γ radiation of this energy would be transmitted through the detector wall? This wall is of aluminum, which has a mass absorption coefficient for 1-MeV γ-rays of 0.062 cm^2/g.

Answer: Virtually 100%

2. Could the same detector of Problem 1 now be used to assay a ^{14}C-containing sample emitting β^--particles with an R_{max} of about 28 mg/cm^2 in aluminum? Would any of these β^--particles be expected to penetrate the aluminum wall of the detector?

Answer: No

3. An investigator has on hand a 1-Ci source that emits 1.5-MeV γ-radiation. He desires to reduce the radiation level at the outside of his source shielding to the equivalent of that from a 5-μCi source. He plans to use steel plates to shield the source, which have a linear absorption coefficient (μ_1) of 0.4/cm for this γ-ray energy. How many inches of steel will be needed around his 1-Ci source to attain the desired shielding?

Answer: ~12.0 in. thickness

4. What is the minimum energy that an α-particle must have to be counted in a G-M counter with a Mylar window of thickness 1.0 mg/cm^2? Formula for Mylar is $C_8H_{10}O_4$.

5. The thickness of the aluminum shielding on a NaI crystal is ~0.05 in. If one arbitrarily says that any γ-ray that is attenuated by 50% by the shielding will not be counted, what is the minimum-energy γ-ray that can be detected in this detector?

6. ^{55}Fe decays by electron capture. How would you detect the occurrence of ^{55}Fe decay?

7. The metal $_{50}$Sn, as well as $_{13}$Al, may also be used to absorb γ rays.

a. For the photoelectric absorption effect, the cross section for 0.05-MeV γ-rays is 7.3 barns/atom of Al. What is the approximate photoelectric cross section per atom of Sn?

b. For the Compton effect, the cross section for 0.5-MeV γ-rays is 3.8 barns/atom of Al. What is the approximate CE cross section per atom of Sn?

8. Complete the following nuclear disintegration schemes.

$$^{86}_{40}\text{Zr} + e\text{-capture} \longrightarrow$$

$$^{245}_{97}\text{Bk} \xrightarrow{4.95\text{-d}} \alpha +$$

$$^{51}_{25}\text{Mn} \xrightarrow{45\text{-m}} \beta^{+} +$$

$$^{45}_{19}\text{K} \xrightarrow{20\text{-m}} \beta^{-} +$$

9. The absorption of γ rays by lead results in a 4-cm-thick absorber reducing the intensity of the γ-rays to 4.2% of the original γ-ray intensity. A 1-cm-thick absorber will reduce the intensity by how much?

10. Complete the following chart:

Type of Decay or Process	Change in Z	Change in N	Change in A	Examples
α-emission				
β^--emission				
β^+-emission				
γ-emission				
Electron capture				
Isomeric transition				
Internal conversion				

11. What is the difference between electrons emitted from beta emitters and those emitted during internal conversion?

NUCLEAR ELECTRONICS (CHAPTER 4)

1. The detector of a G-M counter consists of an ionization chamber that has an applied potential of 1100 to 1500 volts during operation. When a β-particle enters the chamber, a discharge occurs, giving a pulse of approximately 0.25 volt. Explain how one detects a 0.25-volt pulse superimposed on a 1100- to 1500-volt potential.

G-M DETECTORS (CHAPTER 5)

1. The counting efficiency for γ-rays in a G-M counter is ~1 to 2%. What causes this low efficiency?

2. Calculate the current flow (amperes) from a 1-μCi ^{14}C-source in a one-liter ion chamber that is measured by an electrometer. E_{max} for ^{14}C is 0.155 MeV.

3. The equation for the pulse size in a detector operating in the proportional region is

$$\Delta V = A\frac{ne}{C}$$

where A = gas amplification factor
n = number of primary ion pairs
e = coulomb charge on an electron (1.6×10^{-19})
C = capacitance of the scaler in farads

Assume reasonable values for A and n, set C equal to 50 picofarads, and calculate the ΔV for a 1-MeV beta particle.

GAMMA RAY COUNTING (CHAPTER 6)

1. Consider a counting system that uses scintillation detectors in which the resolving time is limited by the decay of the luminescence and is approximately equal to the decay time of the scintillator. Calculate the counting rate at which the counting losses are 1% for both anthracene and NaI scintillators.

SCINTILLATION SPECTROSCOPY (CHAPTER 7)

1. Consider an experiment in which one is measuring the γ-ray spectrum due to a radionuclide emitting ~1-MeV γ-rays. What will be the effect on the backscatter peak of increasing the lead shielding around the detector?

2. Gamma ray spectra, taken with NaI(Tl) detectors in a lead shield, present many photopeaks. Illustrate such a typical spectrum that would be expected from a radioisotope source, emitting two γ-rays of 0.7-MeV and 2.6-MeV energies and of about equal intensity. Identify all photopeaks.

3. Describe the change in pulse height resolution that occurs when the conversion efficiency in the photocathode changes so that 20 photoelectrons, instead of 10, reach the first dynode.

4. ^{137}Cs has a 0.662-MeV γ-ray associated with its decay scheme. What should the energy of the 180°-backscattered Compton photon be?

LIQUID SCINTILLATION COUNTING (CHAPTER 9)

1. Explain, in detail, why even the most costly liquid scintillation spectrometer will detect tritium with a maximum efficiency of only 50 to 60%. Tritium has an E_{max} of 18 keV and a half-life of 12.5 yr.

2. One ml of an aqueous solution containing 1 mg of glucose-U-^{14}C is assayed in a Thixcin gel preparation, using a liquid scintillation spectrometer. At the balance point, a net count rate of 65,000 cpm was registered. The counter had previously been calibrated at the same balance point settings, using two different standards as follows: 0.10 cc of $Na_2\,^{14}CO_3$ standardized at 4900 dpm gave a net count rate of 2200 cpm; 1.0 mg of toluene-^{14}C standardized at 3400 dpm gave a net count rate of

1870 cpm. Considering the nature of the sample, which one of the standards would be chosen for calibration purposes? Why? What was the activity of the glucose-U-^{14}C sample in microcuries?

Answer: ~0.006 μCi

3. a. Cite the advantages of counting a gamma emitter with external scintillation methods rather than by liquid scintillation techniques.
b. Compare the efficiencies of counting an alpha emitter with liquid scintillation and external scintillation methods.

AUTORADIOGRAPHY (CHAPTER 10)

1. a. Cite two advantages of using autoradiography rather than counting some particular specimens with other techniques.
b. In general, why do tritium-labeled compounds give higher resolution than ^{32}P-labeled compounds when subjected to the autoradiography method?

SAMPLE PREPARATION (CHAPTER 11)

1. In radiotracer applications, gaseous, liquid, and solid samples are counted by a variety of techniques. Discuss some of the factors involved in sample preparation and counting.

a. ^{14}C as $^{14}CO_2$ gas
b. ^{14}C by liquid scintillation methods
c. ^{90}Sr by liquid scintillation methods
d. ^{14}C and ^{3}H by solid sample counting

NUCLEAR STATISTICS (CHAPTER 12)

1. Collection of 10,000 counts from a radioactive sample, permits one to state that
(a) there is a 99% probability that the true total count was 10,000 ± 100 counts.
(b) there is a 90% probability that the true total count was 10,000 ± 196 counts.
(c) there is a 99.73% probability that the true total count was 10,000 ± 300 counts.
(d) none of the above estimations.

2. The radioisotope 45-d ^{59}Fe is injected into a rat, and after a few days the rat is sacrificed and various organs are transferred to a counting vial and counted in the well of a 3 × 3-in. NaI(Tl) detector. Ten-minute counts are taken for five organs, resulting in (a) 1000 cpm, (b) 950 cpm, (c) 600 cpm, (d) 1100 cpm, and (e) 200 cpm of ^{59}Fe. Show which of the above five organs have incorporated statistically different quantities of ^{59}Fe.

3. In a Geiger-Müller counter, a sample registered a total of 60 cpm with a background of 30 cpm. In a proportional counter, the same sample gave 95 cpm with a background of 60 cpm. If three such samples are to be counted, which system will give the same uncertainty in less time?

4. Six aliquots of a tritiated water sample were assayed, using a liquid scintillation counter. The following time intervals were required to accumulate a gross count of 10,000 counts for the six samples, respectively: 36.30, 35.51, 26.24, 36.01, 36.98, and 37.40 min. Subsequently, 1900 background counts were registered in 15 min. Calculate the average net count rate of these six samples and its uncertainty.

Answer: 148 ± 4 cpm

5. Assuming a 10-min background count is taken, how long should each of the following samples be counted for the most efficient utilization of counting time?

Sample Number	Sample Activity (cpm)	Background Activity (cpm)
1	110	110
2	50	25
3	500	250

6. A set of observations of the activity of a long-lived radioactive sample were made. The results are shown below.

Observation #	Activity (cpm)
1	280
2	267
3	277
4	299
5	269
6	260
7	290
8	285
9	318
10	259

Calculate

a. the average value of the activity.
b. the standard deviation of the set of measurements.
c. the standard deviation of the mean.
d. Should any of the data be rejected? If so, what is the new average value?

CORRECTION FACTORS (CHAPTER 13)

1. Draw simple, labeled diagrams that illustrate typical behavior for each of the following phenomena:

a. effect of sample thickness on counting rate for a given amount of tracer:

b. percent increase in counting rate versus thickness for a β^--source on top of a backscatterer.

2. For a fixed amount of ^{36}Cl tracer (β^-, 0.7 MeV) in the counting sample, counted in a fixed position near the counter window, would you expect it to make any difference in the observed counting rate if the sample was prepared in the form of NaCl, AgCl, or TlCl (of about the same total weight in each case)? Why or why not?

3. Estimate the counting rate at which the dead time loss becomes 5% for both a typical proportional counter and a G-M counter.

EXPERIMENTAL DESIGN (CHAPTER 14)

1. Compute the amount of a radionuclide necessary to perform an experiment with the following features:

a. Sample count rate	500 cpm
b. Detector efficiency	50%
c. Sample aliquot for counting	10% of total isolated sample
d. Percent incorporation of nuclide into total isolated sample	0.1%

2. A physiologist decides to tag molecule X with a phosphorus tracer in order to ascertain the uptake of this tagged molecule into various compartments of the system under study. Suppose that small quantities of fluid, say 1 cc, are withdrawn periodically for measurement of their X content via their radioactive P content. The available P tracers are 14-d ^{32}P (1.71-MeV β^-, no γ-ray) and 25-d ^{33}P (0.25-MeV β^-, no γ-ray). Both ^{32}P and ^{33}P are present in the purchased tracer; assume that the ^{32}P/^{33}P ratio is 100 at the initiation of the experiment. Assume a fully equipped counting laboratory and that 10 samples are to be counted each day, over a 100-day period. Also, $\pm 2\%$ accuracy is required in these measurements.

a. Discuss the best instrument to use.

b. Discuss the second best instrument you would use.

c. What means would be employed to check the stability of your counting system?

3. Complete the following table.

Radioisotope	Chemical Form	Best Choice of Detector	Estimated Efficiency
^{14}C	$^{14}CH_4$ in air		
^{3}H	$^{3}H_2$ in argon		
^{32}P	$H_3\,^{32}PO_4$ in aqueous solution		
^{55}Fe (EC decay)	$^{55}Fe^{++}$ in solution		
^{131}I	In vivo		

4. Isotope X, half-life of 5 days, is to be used in an experiment that includes the following factors.

a. Sample count rate of 1000 cpm.
b. Counter efficiency of 10%.
c. Duplicate samples to be taken each week for 3 weeks—a total of six samples.
d. Assume that the sample with the lowest count rate will represent a 0.1% incorporation.
e. All samples will represent only 5% of the total isotope administered.

What amount of isotope must be used?

5. Three tracers, 90-y ^{151}Sm (0.076 MeV β^-, 100% of disintegrations and 0.022 MeV X-ray, 4%), 245-d ^{65}Zn (0.33 MeV β^+, 1.7% of disintegrations and 0.511 MeV γ-rays, 3.4%), and 14-d ^{32}P (1.71 MeV β^-, 100% of disintegrations), will be used simultaneously in a multitracer experiment. (After each radiotracer, the principal β^--, β^+-, and γ-ray energies are given in parenthesis with the percentage for each disintegration.) You are given a well-equipped nuclear laboratory that has G-M counters, external scintillation, and liquid scintillation counters. Suppose that you wish to measure the uptake (a measure of blood volume) of these three elements in the blood of a rat and the loss of these elements to the rest of the rat's organs and tissue from the blood.

a. What levels of activity of each tracer would you inject into the rat? Why? Assume that you will withdraw 0.1-ml blood volumes every hour for 24 hr. (The total blood volume of an adult rat is $\approx$15 ml.)

b. What counting instrument or instruments would you utilize? Briefly describe the method.

c. If you decided on liquid scintillation counting, state qualitatively the factors involved in formulating the mathematical expressions for simultaneously counting all three radiotracers. It is not necessary to write down the mathematical expression.

6. List the instrument you would recommend for the quantitative assay of the following isotopes:

Isotope	Type of Decay	Instrument	Counting Efficiency and Other Justification
$Ba^{14}CO_3$	β^- (0.156 MeV)		
^3H_2O	β^- (0.018 MeV)		
$^3H_2$ (in air)	β^- (0.018 MeV)		
$^{14}CO_2$ (in air)	β^- (0.156 MeV)		
$^{55}Fe^{3+}$	Electron capture		
$^{13}NO_3^-$	β^+ (1.2 MeV)		
$^{137}Ba^{++}$	γ (0.662 MeV)		
$^{65}Zn^{++}$	β^+ (0.325 MeV–2.5%) γ (1.12 MeV–51%) Electron capture (97.5%)		
$^{211}Po^{4+}$	α (7.43 MeV)		
$^{211}Pb^{++}$ in a mixture of ^{211}Pb and ^{211}Po	β^- (1.39 MeV)		

7. This is an experiment in which isotope X is utilized and the following facts are known and the following assumptions are made. You are to calculate the amount of isotope required. State the additional assumptions that you make.

Facts: Isotope X Half-life: 10.0 days
Radiations: β^- 0.2 MeV—50%
gamma 4.0 MeV—50%
Electron capture—50%

Assumptions:

a. 10% of the isotope is incorporated.
b. 5% of the incorporated isotope is in the fraction to be analyzed.
c. 25% of the fraction to be analyzed is used for each sample to be counted.
d. The experiment will require 6 days.

8. Imagine that you wish to study the path of phosphate pollutants in a river. You propose to do so by injecting ^{32}P-labeled phosphate into the river at one point, followed by periodic sampling of the river water at points downstream from the injection point. A commercial supplier has furnished you with a Na_3PO_4 solution that contained 1 mCi of ^{32}P 300 days ago. The samples withdrawn from the river should have 10^3 times less activity per milliliter than the material injected. How long would you expect to have to count the samples you recover from the river in order to have $\pm 1\%$ uncertainty in the results if you dump a 1-ml solution containing all the ^{32}P into the river today? (Assume 100% counting efficiency, $t_{1/2}$ [^{32}P] = 14.3 days).

ISOTOPE PRODUCTION (CHAPTER 15)

1. In the following incomplete nuclear reactions, supply the missing items:

a. $^{27}Al(n, \alpha)$______ **b.** $^{23}Na(n,$ _____$)^{24}Na$
c. ______$(n, p)^{14}C$ **d.** $^{12}C(n, 2n)$______
e. $^{9}Be($_____$, n)^{12}C$ **f.** $^{58}Fe(n, \gamma)$______
g. ______$(n, p)^{27}Mg$ **h.** $^{32}S(n,$ _____$)^{32}P$
i. $^{35}Cl(n, p)$______ **j.** $^{6}Li($_____$, \alpha)^{3}H$
k. $^{44}Ca($_____$, \gamma)^{45}Ca$ **l.** ______$(n, \gamma)^{65}Zn$

2. In a particular synthetic reaction, carrier-free tritium gas (T_2) is used to hydrogenate ethylene to ethane as follows:

$$H_2C = CH_2 + T_2 \longrightarrow \underset{\displaystyle T}{\underset{|}{H_2C}} - \underset{\displaystyle T}{\underset{|}{CH_2}}$$

Calculate the specific activity in mCi/mg of the synthesized ethane.

NUCLEAR SAFETY (CHAPTER 16)

1. ^{24}Na emits two γ-rays of 1.37 and 2.75 MeV energy per disintegration. Recalling the equation $R = 6CE$, given in your text, calculate the approximate γ-radiation exposure in mR/hr at one-yard distance from an unshielded one-curie ^{24}Na source.

2. A 100-mCi standard of 30-y ^{137}Cs emits 0.66-MeV γ-rays in 85% of its disintegrations. The γ-ray exposure in mR/hr at 3 feet from this source is ________.

Answer: ~37 mR/hr

3. Radioisotopes, deposited within the body, present various radiation hazards. Compare the relative internal body hazards of these radionuclides, all having the same activity level.

163-d ^{45}Ca, 0.25-MeV β^-, no γ-ray
312-d ^{54}Mn, 0.84-MeV γ-ray (100% of disintegrations)
65-h ^{197}Hg, 0.077-MeV γ-ray (18% of disintegrations)
7650-y ^{243}Am, 5.2-MeV α-particle (100% of disintegrations)

Discuss the reasons for your evaluation in ranking these radionuclides, in say, their decreasing hazardous character. Assume that the ingested quantities of all elements are below the toxicity level.

4. Absorbed doses from internal emitters pose a serious radiation hazard. Ignoring excretion and loss of activity by radioactive decay, the approximate absorbed dose for a tissue can be calculated in rads/day (D) as follows:

$$D = \frac{(3.7 \times 10^4 \text{ dps}/\mu\text{Ci})(\text{mean } \beta^-\text{-energy in eV})(1.60207 \times 10^{-12} \text{ ergs/eV}) \\ (\text{tissue concentration of isotope in } \mu\text{Ci/g})(86{,}400 \text{ sec/day})}{100 \text{ ergs/rad}}$$

(Attempt to think your way through the derivation of this equation for dose calculation.) Using this equation, calculate the respective doses to liver tissue in rads/day where the liver tissue concentrations are maintained uniformly at 1 μCi/g of ^{32}P, ^{14}C, and ^{3}H, respectively. The results should show the relation of beta disintegration energy to internal dose.

Answers: ^{32}P ~36 rads/day, ^{14}C ~2.5 rads/day, ^{3}H ~0.3 rad/day

5. If no excretion or turnover occurred, what would be the approximate absorbed doses in the liver tissue of Problem 4 two months later from these three radioisotopes respectively? The results will reveal the relation of physical half-life to internal dose.

Answers: ^{32}P ~2 rads/day, ^{14}C ~2.5 rads/day, ^{3}H ~0.3 rad/day

6. Assume that the turnover time (biological half-life) of ^{14}C is 8 days and that of ^{3}H is 57 days in the situation discussed in Problem 4. What would be the approximate absorbed dose in rads/day to the liver tissue two months later from ^{14}C and ^{3}H, respectively? The decline in activity due to turnover can be treated mathematically in the same manner as physical decay. These results should illustrate the importance of biological half life to internal dose.

Answers: ^{14}C ~0.014 rad/day, ^{3}H ~0.15 rad/day

RADIOANALYTICAL TECHNIQUES (CHAPTER 17)

1. Isotope dilution analysis permits one to determine the purity of a radiochemical. Compound X, molecular weight of 100 (specific activity 1.0 mCi/mmole), was checked for purity by carefully weighing 1.0 mg of the radiochemical and mixing with 1000 mg of unlabeled compound X and recrystallizing until a constant specific activity. Radioassay gave a value of 2000 dpm/mg. What was the purity of the radiochemical in percent?

2. A 5-kg batch of crude penicillin was assayed by isotope dilution analysis: To a 1-g sample of the batch was added 10 mg of pure penicillin having an activity of 10,600 cpm; only 1.40 mg of pure penicillin having an activity of 280 cpm was recovered. What is the penicillin content of the batch?

3. Isotope dilution analysis is applied to the following analysis. Calculate the amount of the compound Y present in the sample and express your answer as percent by weight. A 1-g sample is analyzed for compound Y, molecular weight of 100. A derivative is formed of compound Y and the added radioactive Y (1.5 μCi at a specific activity of 3 mCi/mole). The derivative, molecular weight of 150 (1 mole of compound Y per mole of derivative), is recrystallized until pure. It has a specific activity of 4.44×10^3 dpm/mg.

4. A 10-ml sample of blood is withdrawn from a patient, and the red cells are labeled with ^{51}Cr, a 27-day γ-emitter. One ml of the labeled blood diluted to 10 ml with water gave a net counting rate of 33,000 cpm (background corrected). The remaining labeled blood is injected back into the patient, and after several hours 10 ml of blood is withdrawn and counted as before. The net counting rate (background corrected) was 500 cpm. What is the total volume of the patient's blood?

5. A sample of household Al foil is irradiated in a reactor along with a 100-μg Mn standard. One hour after the end of the irradiation, a γ-ray spectrum of the Mn standard showed 24,000 cpm in the 0.84-MeV ^{56}Mn photopeak. Two hours after the end of the irradiation, the γ-ray spectrum of the Al sample showed 12,000 cpm in the 0.84-MeV ^{56}Mn photopeak. How much ^{55}Mn was present in the original Al sample? $t_{1/2}$ ^{55}Mn $= 2.58$ hr.

6. It has been reported that the Hg concentration in the water of 14 states is above the maximum permissible level. Imagine that you have been given the assignment of measuring the Hg concentration in soda pop. You are to do so by irradiating the pop with neutrons which changes ^{202}Hg into ^{203}Hg via the ^{202}Hg (n, γ) ^{203}Hg reaction, and counting the γ-rays from decay of ^{203}Hg ($t_{1/2} = 46.6$ days). What is the minimum amount of ^{202}Hg you could detect, assuming a 1-hr irradiation of your sample in a reactor ($\Phi = 10^{12}$ neutrons/cm^2/sec), followed by a delay of 3 hours before you may begin counting? Assume that your detection system is 10% efficient, $\sigma(n, \gamma) = 4$ barns, and that the minimum detectable count rate is 100 cpm.

Appendix

Nuclear Reaction Calculations

A. ENERGETICS OF NUCLEAR REACTIONS

The act of writing down any possible nuclear reaction and successfully balancing it is not evidence that such a reaction will occur. One necessary, but not sufficient, condition for the occurrence of a reaction is that "the energetics must be favorable." What do we mean by this statement? First, the sum of the Q value of the reaction (the energy released in the reaction) and the incoming projectile energy must be positive. In other words, the reaction must be energetically possible. Formally, if

$$Q = (M_{\text{react}} - M_{\text{pdts}})c^2 \qquad \text{(A-1)}$$

where c is the speed of light, then if

$$\begin{aligned} &Q + E_p \text{ is positive, the reaction can occur.} \\ &Q + E_p \text{ is negative, no reaction will occur.} \end{aligned} \qquad \text{(A-2)}$$

M_{react}, M_{pdts} are the masses of the reactants and products, respectively, and E_p is the incident projectile energy. For example, consider the possible reaction

$$^{195}\text{Pt}(p, n)^{195}\text{Au}$$

to be carried out with 10.5-MeV protons. From a table of atomic masses, we have

$$M^{195}\text{Pt} = 194.964813$$
$$M^{195}\text{Au} = 194.965051$$
$$M_p = 1.007825$$
$$M_n = 1.008665$$

$$\begin{aligned} Q &= [194.964813 + 1.007825 - (194.965051 + 1.008665)]c^2 \\ &= -0.001078c^2 \\ &= -1.004157 \text{ MeV} \end{aligned}$$

Our condition is that $Q + E_p$ be positive, and it is as shown below.

$$Q = -1.004157 \text{ MeV}$$
$$E_p = 10.5$$
$$Q + E_p \simeq 9.5 \text{ MeV} \qquad \text{(reaction } can \text{ occur)}$$

Note that even though the basic reaction was endoergic, the additional energy furnished by the incoming projectile allowed the reaction to occur.

A second necessary, but not sufficient, condition that must pertain before one can say a given reaction will occur is that the quantity

$$Q + E_p - \frac{M_p}{M_p + M_T} E_p$$

must be positive. (M_p and M_T are the projectile and target mass, respectively.) This requirement derives from the need to conserve momentum in the reaction. Applying this condition to the ^{195}Pt$(p, n)^{195}$Au reaction, we have

$$\begin{aligned} Q + E_p - \frac{M_p}{M_p + M_T} E_p &= -1.004157 + 10.5 - \frac{1.007825}{1.007825 + 194.964813} 10.5 \\ &= 9.5 \text{ MeV} \qquad \text{(reaction can occur)} \end{aligned}$$

Note, from the preceding example, that this condition can, for all practical purposes, be neglected unless the sum $E_p + Q$ is very close to zero or the projectile and target are of similar mass.

A third energetic condition that pertains to reactions involving charged incident projectiles only is that the energy of the incident projectile be sufficient to overcome the electrostatic repulsion between the positively charged projectile and the positively charged nucleus. This requirement is to ensure that the incident projectile can get close enough to the target nucleus to react with it. The minimum projectile energy necessary to fulfill this condition for a given target nucleus is called the *Coulomb barrier* for that nucleus. Formally, it is given as

$$\text{Coulomb barrier} = 1.44 \frac{Z_p Z_T}{R_p + R_T} \text{ MeV} \qquad \text{(A-3)}$$

where Z_p, R_p, Z_T, R_T are the atomic number and radii of the incoming projectile and target, respectively. The radii in this formula are to be expressed in fermis (1 fermi $\equiv 10^{-13}$ cm) and can be calculated by using the equation

$$R = 1.4A^{1/3} \text{ fermi} \tag{A-4}$$

where A is the nuclear mass number.

Applying this condition to our example of the $^{195}\text{Pt}(p, n)^{195}\text{Au}$ reaction, we calculate that

$$R_{\text{pt}} = 1.4A^{1/3} = 1.4(195)^{1/3} = 8.0 \text{ fermi}$$

$$R_p \approx 1.4 \text{ fermi}$$

$$\text{Coulomb barrier} = 1.44\frac{(1)(78)}{(8.0 + 1.4)} = 11.9 \simeq 12 \text{ MeV}$$

Thus we would conclude that the reaction would not occur with appreciable probability with 10.5-MeV protons. With 15-MeV protons as the incident projectile, the reaction could occur.

B. THE PROBABILITY OF A NUCLEAR REACTION

If the energetic conditions outlined in Section A have been satisfied, we can still only say that the reaction *could* occur but not with what probability it will occur. In this section we wish to discuss how one determines the probability of a reaction taking place and how much product will be produced.

To begin, consider a slab of material that is so thin that when we look at it, no one nucleus lies "behind" another (see Figure A-1). Let the slab area be A, the total number of nuclei in the slab be N, and let σ denote the effective or apparent cross-sectional area covered by each nucleus ($\sigma \sim \pi R^2$, where R is the nuclear radius). Then the *fraction* of the total area of the slab covered by nuclei is $N\sigma/A$. Now imagine a group of projectiles incident on the slab,

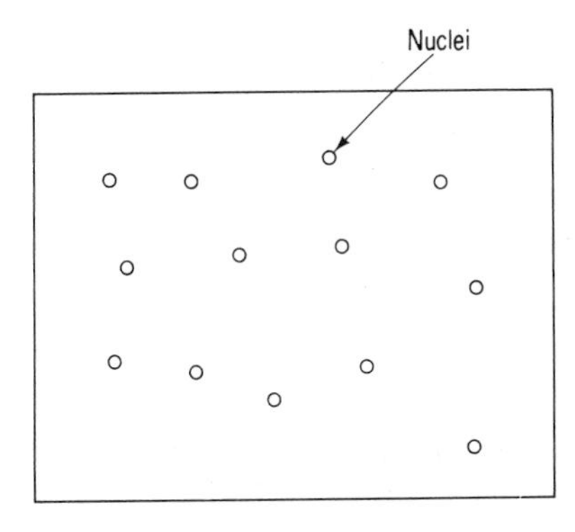

Fig. A-1. A schematic view of a section of thin slab of material such that no nucleus lies behind another nucleus.

where the number of particles hitting the front is n_1 and the number of particles coming out the back of the slab is n_2. Then the *fraction* of incident projectiles that do not emerge from the back of the slab is $(n_1 - n_2)/n_1$. The projectiles that are "lost" in the slab are those projectiles that strike slab nuclei and undergo nuclear reactions. Therefore we say that the fraction of incident projectile nuclei that do not emerge from the back of the slab is equal to the fraction of the slab area covered by target nuclei, $N\sigma/A$. Speaking formally, we say

$$\frac{n_1 - n_2}{n_1} = \frac{N\sigma}{A} \qquad \text{(A-5)}$$

Note that the larger the value of σ, the more reactions that take place. σ is then a measure of the probability that a given reaction will take place and is referred to as the *cross section* for the reaction. Formally, we can rearrange Equation (A-5) to yield

$$\sigma = \frac{A}{N}\frac{(n_1 - n_2)}{n_1} \qquad \text{(A-6)}$$

The units of σ are those of area, such as cm^2. Since σ represents something like the cross-sectional area of a nucleus, it will be a very small number, such as 10^{-24} cm^2. Rather than use such small numbers continually, cross sections are expressed in units of *barns* (b), where 1 barn $= 10^{-24}$ cm^2. The origin of the term barn to designate cross section is said to arise in the phrase "hitting the broad side of a barn," where one implies that nuclei with cross sections of 10^{-24} cm^2 are as easy to hit as "the broad side of a barn."

Returning to our discussion of reaction rate [as given by Equation (A-5)], we can say that if the fraction of incoming particles that react is $N\sigma/A$, then the number of reactions per second is $\phi N\sigma/A$, where ϕ is the total number of incident projectiles striking the target per second. Formally, we write

$$R_s \equiv \text{number of reactions per second} = \frac{\phi N\sigma}{A} = \phi\mathfrak{N}\sigma \qquad \text{(A-7)}$$

where $\qquad \mathfrak{N} \equiv \text{number of target nuclei per unit area} = \dfrac{N}{A}$

Equation (A-7) gives the general formula for the reaction rate in a thin target. In the special case of a nuclear reactor, the incident projectiles arrive at the target from all directions. Then we designate Φ as the number of incident particles per second per square centimeter of target area and N as the total number of nuclei in the target then we rewrite Equation (A-7) as

$$R_s = N\sigma\phi \qquad \text{(A-8)}$$

If the product nuclei are nonradioactive, then the number of product nuclei formed in a bombardment of t seconds duration is $R_s t$. However, it is exceedingly rare for the product nuclei to be nonradioactive, and our interest lies strictly in those reactions that produce radioactive nuclei for use as radiotracers. Because the product nuclei are radioactive, some of them may decay during irradiation. The equation that governs the production rate of radioactive nuclei in a nuclear reaction states that the number of radioactive product nuclei of decay constant λ present at some time t during an irradiation, N, is given as

$$N = \frac{R_s}{\lambda}(1 - e^{-\lambda t}) \tag{A-9}$$

In terms of the disintegration rate, $\mathfrak{D}$, of the product, we have

$$\mathfrak{D} = \lambda N = R_s(1 - e^{-\lambda t}) \tag{A-10}$$

Equations (A-9) and (A-10) are shown graphically in Figure A-2.

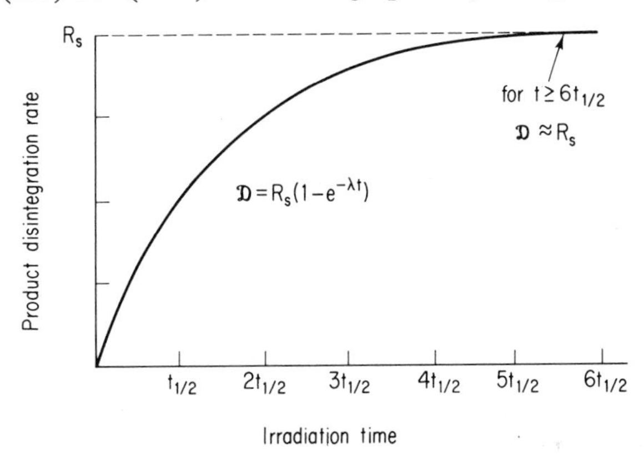

Fig. A-2. Buildup of daughter-product distintegration rate in a target subjected to constant irradiation, which produces a radioactive daughter nuclide.

Note, as shown in the figure, that the product disintegration rate does *not* increase linearly with bombardment time but asymptotically approaches some limiting value, *the saturation activity*, R_s. Thus, in producing a radionuclide, little advantage is gained by irradiating the target for more than a few half-lives of the product nucleus.

When the bombardment time is short compared to the product half-life ($t_{1/2} \geq 10t$), then Equation (A-10) can be written as

$$\mathfrak{D} \approx R_s \lambda t \tag{A-11}$$

Large compilations of cross-sectional data for various nuclear reactions are available (1, 2, 4, 5), thus allowing one to predict, using Equations (A-10), (A-9), and (A-8), just what is necessary, for example, to produce 20μCi of ^{59}Fe for use in a radiotracer experiment. Suppose, further, that we have the use of a nuclear reactor with a thermal neutron flux, Φ, of 10^{12} neutrons/cm^2/sec for a time period of 2 hours. How much natural iron must we irradiate in order to produce the desired 20 μCi of ^{59}Fe via the ^{58}Fe$(n, \gamma)^{59}$Fe reaction?

From the Table of Isotopes,

$$t_{1/2}(^{59}\text{Fe}) = 45 \text{ days}$$

$$\lambda^{59}\text{Fe} = \frac{0.693}{t_{1/2}} = \frac{0.693}{(45 \text{ days})(24 \text{ hr/days})} = 6.42 \times 10^{-4} \text{ hr}^{-1}$$

$$\Phi = 10^{12} \text{neutrons/cm}^2\text{/sec}$$

$$\mathfrak{D} = 20\ \mu\text{Ci} = (20)(3.7 \times 10^4 \text{ dis/sec-}\mu\text{Ci}) = 7.4 \times 10^5 \text{ dis./sec}$$

From the Chart of Nuclides,

$$\sigma[^{58}\text{Fe}(n, \gamma)^{59}\text{Fe}] = 1.2 \text{ barns} = 1.2 \times 10^{-24} \text{ cm}^2$$

Equation (A-11) states†

$$\mathfrak{D} \approx R_s\lambda t = N^*\sigma\phi\lambda t$$

Solving this equation for N^*, we obtain

$$N^* = \frac{\mathfrak{D}}{\sigma\phi\lambda t}$$

Entering the values from above, we have

$$N^* = \frac{(7.4)(10^5)}{(1.2)(10^{-24})(10^{12})(6.42)(10^{-4})(2)}$$

$$N^* = 4.80 \times 10^{20}\ {}^{58}\text{Fe nuclei}$$

How much natural iron do we need in order to provide 4.80×10^{20} ^{58}Fe nuclei? Natural iron is 0.33% ^{58}Fe. Therefore we can say that we need $(100/0.33)(4.80)(10^{20})$ Fe atoms or

$$\text{Mass of Fe required} = \frac{(100/0.33)(4.80)(10^{20})}{(6.02)(10^{23})}(55.8) = \underline{\underline{13.5 \text{ g Fe}}}$$

where 55.8 is the atomic weight of Fe and 6.02×10^{23} is Avogadro's number.

Although Equations (A-8), (A-9), (A-10), and (A-11) are applicable to any problem that involves calculations of activity found in an irradiation, it is frequently more convenient to use a nomograph to do such calculations in a specific irradiation situation. Figure A-3 shows such a nomograph for use in calculating the amount of activity formed in a neutron irradiation (3).

Figure A-3 shows the specific activity formed by irradiating the naturally occurring target material in a neutron flux of 10^{11} neutrons/cm²/sec. For example, if one wishes to determine the specific activity of ^{24}Na produced by irradiating natural sodium in a flux of 10^{11} neutrons/cm²/sec. for 1 hour, one performs the following steps:

1. Locate the ^{24}Na dot in Figure A-3 (coordinates 38 mCi/g, 21.7 hr). The abscissa value is equal to the mean life, $1.44t_{1/2}$.

†Equation (A-11) is used rather than Equation (A-10), since $t_{1/2} \geq 10t$.

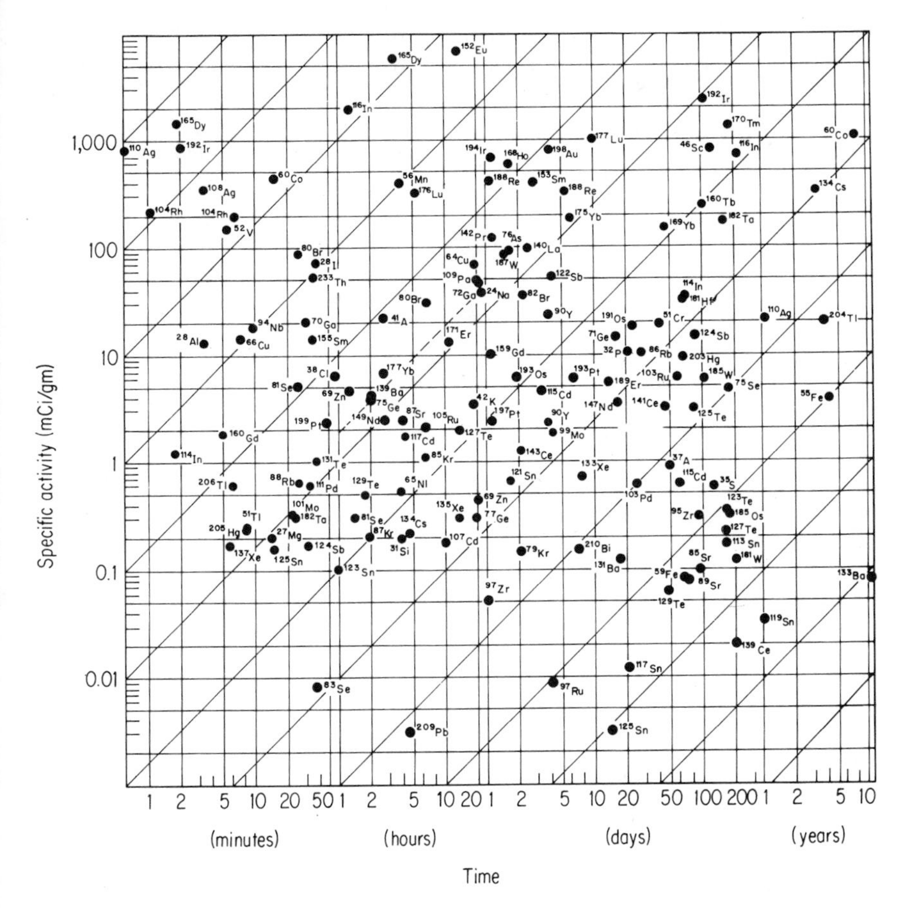

Fig. A-3. Nomograph for calculating activity formed in a neutron irradiation with $\Phi = 10^{11}$ neutrons/cm^2/sec. From (3).

2. Draw a line from this dot parallel to the 45° lines of the table until it intersects the vertical line at one hr (see dashed line).
3. Read off the specific activity produced as 1.5 mCi/g.

BIBLIOGRAPHY

1. ALLEY, W. E., R. W. GELL, and R. M. LESSLER. *Semiempirical Neutron-Induced Reaction Cross Sections*, USAEC Report UCRL-50484 (1968). A graphical display of cross sections from 0.025 eV to 15 MeV for 796 neutron-induced reactions in stable isotopes from $Z = 1$ to $Z = 83$.
2. JARMIE, N., and J. D. SEAGRAVE (Eds.). *Charged Particle Cross Sections*, USAEC Report LA-2014 (1957). Plots of charged particle induced reaction cross sections with targets up to F, projectiles up to 4 He, and energies up to 30 MeV.

3. MESLER, R. B. "Rapid assessment of neutron activation," *Nucleonics* **18**(1), 73 (1960).
4. SMITH, D. B., N. JARMIE, and J. D. SEAGRAVE (Eds.). *Charged Particle Cross Sections, Neon to Chromium*, USAEC Report LA-2424 (1961). Same as 2 except that region from Ne to Cr is covered.
5. STORM, E., and H. I. ISRAEL. *Photon Cross Sections from 0.001 to 100 MeV for Elements 1 Through 100*, USAEC Report LA-3753 (1967).

Index

Absolute counting, 96, 163, 323
Absorption, *see* types of radiation
Absorption coefficient
  linear (μ_l), 63-65, 68
  mass (μ_m), 65-66, 69-71
Abundance of stable isotopes, 9-10
Actinium series, 11
Activation analysis, *see* Neutron activation analysis
Alpha particles, 22, 39-46
  absorption, 41-46
  detection, 45, 111-114, 117, 180, 187, 195, 225, 250, 412, 416, 423
  emission, 22
  energy characteristics, 39-41
  energy dissipation, 41
  half-life and energy relationships, 40-41
  health hazard, 46, 368
  interaction with matter, 41-42, 99-100
  nature of, 22, 39
  range, 42-45
  range-energy relations, 43-44
  specific ionization, 42
Alpha plateau, 113-114
Alternate decay paths, 19, 21-24, 36
Amplifier, 75-82, 91, 105-106, 136-137, 157
  gain, 79
  linear, 79, 215
  logarithmic, 79, 200, 203, 206-207, 215
  pulse pileup, 79, 81, 158
Amplifier (*cont.*)
  pulse shaping, 75, 79-81, 94, 158
  signal-to-noise ratio, 78-81
Analog-to-digital converter (ADC), 87-89
Anticoincidence circuitry, 117, 424-430
Areal density, 44, 53-54
Artificially produced radioisotopes, 12, 344-352, 400-404, 413-418, 463-467
Ashing methods of sample preparation, 253-254
Atomic number (*Z* value), 7
Atoms, 5-7
Attenuation coefficient, *see* Absorption coefficient
Auger electrons, 23-24
Automated combustion for liquid scintillation samples
  Kaartinen apparatus, 257-260
  Peterson apparatus, 257-259
Automatic Quench Calibration (AQC), 204
Autoradiography, 99, 233-241, 435
  artifacts in, 237-238, 342
  cellular, 234, 236, 239-241
  general principles
    exposure time, 237
    resolution, 235-236
    sensitivity, 236-237
  gross specimens, 236, 238-239
  molecular, 241
  specific techniques, 238-241

Average counting rate, 302-303
Average life (T_A), 33-34

Background
 effect on choice of detector, 225, 305-306, 423-430
 effect on counting time, 303-305, 421-423
 effect on detector sensitivity, 165-166
 error, 135-136, 216-217, 300-305, 310-311
 reduction of, 113-114, 117, 129, 156-157, 175, 225, 310-311
 sources of, 113-114, 310,363
Backing thickness, 262, 317-318
Backscattering, 148-152, 175, 262, 317-318, 321
Backscatter peaks in gamma-ray spectra, 147, 149-154
Band theory of solids, 125-127, 168-173
 activator centers, 126-127
 conduction band, 125-126, 169-170, 172
 excitons, 126-127
 forbidden gap (band), 125-126, 169, 172-173
 holes, 126-127, 168, 170-173
 valence band, 125-126, 169-172
Barn, 464
Beta energy spectrometry, 47-48, 187, 204, 215-217
Beta particles
 detection of, 55-56, 101, 105-106, 113-114, 116-117, 185-193, 195, 212-225, 236-237, 249, 251, 372, 423-425
 emission of, 18-20, 49-51
 energy characteristics, 47-51
 energy spectra of, 47-50, 215, 219
 Feather analysis, 54
 health hazard, 56-57, 368-371
 interaction with matter, 51-53
 maximum energy (E_{max}), 47-49
 mean energy (E_{mean}), 48-49
 nature of, 18-20, 46-47
 negatrons, 18-19, 46-47
 positrons, 19-20, 46-51
 range, 53-55
 relation to neutrino, 49-51
 self-absorption, 320-322
 specific ionization, 52-53
 transmission curves for mica windows, 56
Beta plateau, 114
Billion electron volt (GeV), 6
Binding energy, *see* Nuclear binding energy
Bragg scattering, 60
Branching ratios, 139-140
Bremsstrahlung, 51-52, 54, 57, 154-155

Calibration of detectors, 160-163, 314
Carrier free radioisotopes, 37, 349-352
Cascade of gamma rays, 59
Cerenkov radiation, 52, 187-189
Channels ratio method in liquid scintillation counting, 200, 203, 205, 209, 219-221
 using external standard, 222-224
Chauvenet's criterion, 307
Chemiluminescence, 269, 279-280
Chromatogram scanning counter, 266-268
Chromatography, *see* Paper chromatography, and Thin-layer chromatography
Cloud chamber, 42, 47, 52
Coincidence circuitry, 90-94, 190, 200, 202-203, 206-207, 209, 311
Coincidence loss corrections, 314-317
Combustion methods, 254-260
Composite decay, 32-33
Compton edge, 147, 149-154
 technique in liquid scintillation counting, 222-223
Compton effect, 61-63, 67-72, 125, 143-144, 146-149
Compton scattering, *see* Compton effect
Compton "smear" in gamma-ray spectra, 146-147, 149-154
Confidence limits, 297
Contamination hazards, 363
 removal of, 375-376, 378-379
Continuous-flow scintillation detectors, 212-214
Conversion electrons, 22, 47
Correction factors in radiotracer assay
 absorption
  self-absorption, 45, 264, 277, 311, 319-323
  window and air, 45, 55-56, 319
 background, 113-114, 117, 129, 135-136 156-157, 175, 216-217, 225, 300-305, 310-311
 backscattering, 148-152, 175, 262, 317-318, 321
 detector efficiency, 97, 120, 225, 305-306, 313-314, 335
 geometry, 131, 137-139, 181, 193, 266, 312-313
 resolving time, 109-110, 314-317
Coulomb barrier, 40-41, 462-463
Counters, *see* Detectors
 comparison of, 225
Counting sample preparation, 248-280
 beta emitters, 249-280
 centrifugation techniques, 264
 choice of sample form, 249-252
 conversion to suitable form, 253-261
  ashing methods, 253-254
  combustion methods, 254-260
  miscellaneous methods, 260-261
 electroplating, 264-265

Counting sample preparation (*cont.*)
gamma emitters, 249, 261-262
gaseous form, 261, 271, 273-274
liquid, assay of, 261, 268-280
for liquid scintillation counting, 189, 253-261, 268-280
composition of solutions, 270, 272, 274-275
direct solution, 269-271
emulsion, 278-280
filter paper, 280
fluorescence quenching problems, 192-194, 268-270
gel, 277-278
indirect solution, 271-276
for inorganic ions, 273-275, 278
particle suspension, 277
Schöniger combustion, 256-260
solid form, 262-266
Counting time requirements, 302-305, 308, 421-425
Count rate meter, 84-85, 89, 266-267
Curie (unit), 34-35
Cyclotron, 350-352

Dark current, *see* Noise, electrical and thermal
Dating, age, 1-2, 12, 117, 269-271
Dead time, *see* Resolving time
Decay constant, 25-31, 34, 465, 466
for selected radioisotopes, 27
Decay correction methods
by log-log slide rule, 28
by log tables, 27-28
by prepared tables, 28-29
for ^{32}P, 29
for ^{131}I, 29
Decay scheme, 18-23
Decay series, 11, 22, 411-412
Decontamination, 375-379
Detectors
continuous-flow, 106, 212-214
electroscope, 103-104
gas flow, 112, 116-117
Geiger-Müller
characteristics of, 109-111, 118-120
operation of, 118-119, 261-262
types of, 114-117
internal gas type, 113
ion chamber, 100-106, 120, 261-262, 371-374
large-volume scintillation, 210-212
liquid scintillation (internal-sample), 98, 123, 181-225, 261-262
nuclear track, 242-245
optimum choice of, 305-306
proportional, 108, 111-115, 120
characteristics of, 108, 113-115
operation of, 113-115, 261-262

Detectors (*cont.*)
selection of, 224-225, 249-251
semiconductor, 98-99, 168-180, 261-262, 426-430
solid scintillation (external-sample), 98, 122-167, 261-262
well scintillation, 131, 139, 261-262, 313
Deviation, standard, 296, 299-301, 305-306
fractional, 303, 422
of the distribution (σ), 296, 299-301, 305-306
of the mean ($\sigma_{\bar{x}}$), 299-300, 421-422
Diffused *p-n* junction detector, 173-174
Discriminator, pulse, 76, 81-83, 86, 91-94, 113, 134, 141, 190, 200, 204, 214-217, 219-221
Disintegration rate, *see* Radioactive decay, rate of
Distribution, Gaussian or Normal, 295-300, 307-308
Distribution of label in compounds, terminology for, 353
Dosimeter, 104, 374-375
Double-labeled sample, 187, 191, 201, 214-217, 221
Dynamic capacitor electrometer, *see* Vibrating-reed electrometers
Dynodes of photomultiplier, 124, 128, 132-134

Efficiency, detection, 97, 305-306, 335
comparison of, 120, 225, 313-314, 423-430
electrometer-ion chamber, 105-106, 119-120, 423
Geiger-Müller, 117-120, 310-311, 335
liquid scintillation, 181, 185-190, 192-199, 203-204, 211-224, 269-280, 423-424
proportional, 112, 114, 119-120, 423
semiconductor, 174-180
solid scintillation, 128-131, 133-141, 163-166, 335, 425-430
Electrical noise, *see* Noise, electrical and thermal
Electromagnetic radiation, 10, 57-58
Electrometers, vibrating-reed (dynamic capacitor)
operating characteristics, 104-106, 119-120
Electron capture (EC), 20-21, 423
Electrons, 5-8
internal conversion (IC), 22, 47
Electron volt (eV), 6
Electroplating of counting samples, 264-265
Electroscopes, 103-104
Elements, 7
Emulsion, photographic
in autoradiography, 99, 233- 241

Emulsion sample in liquid scintillation counting, 278-280
Energy, *see* various types of radiation
Energy dissipation
excitation, 41, 51, 122-129, 183-186
ionization, 41-42, 51-53, 60-62, 97 ff, 168 ff, 183
Energy transfer
in liquid scintillation counting, 182-186
in semiconductor counting, 168-175
in solid scintillation counting, 98, 122-129, 132-134
"Energy Well" concept, 40-41
Environmental applications of radiotracer methods
artificial radiotracers, 400-404
low-level counting, 420-430
naturally occurring radionuclides, 411-418
relative costs, 400-402
sampling methods, 416-420
stable-activable tracers, 398, 400-401, 404-410
Equivalent thickness, *see* Areal density
Escape peaks in gamma-ray spectra, 145, 148-154
Excitation, *see* Energy dissipation
Experimental design
basic features of, 333-338
data analysis in, 339-342
evaluation of feasibility, 331-333
execution of, 338-339
preliminary considerations in, 327-331
External health hazards, 46, 56-57, 71-72, 362-363, 367-371
External-sample scintillation counters, *see* Solid scintillation counters
External standard method in liquid scintillation counting, 200-210, 221-224

Factors in relative counting, *see* Backscattering, Coincidence loss, Self-absorption
Fast neutrons, 347
Feather analysis, 54
"Figure of merit," 217, 305-306
Film badge, 234, 372, 374, 378
Filter paper sample in liquid scintillation counting, 280
Fission products, 349-351
Fission reactions, 347-351, 461 ff
Fluorescence decay times, 125, 130, 187
Fluorescence quenching, 192-196, 203-204, 217-224, 268-276
Fluorescence yield, 24
Fluors
energy transfer in, 98, 122-129, 182-186
Fluors (*cont.*)
liquid, 123, 130, 181-187, 197-199, 210-214, 268-280
plastic, 123, 130, 212-213
solid crystal, 98, 123-132
Four-π geometry, 181, 193, 266, 312-313
Fractional standard deviation, 303, 422

Gamma rays
absorption
dependence on energy and density, 67-71
exponential nature of, 63-67
half-thickness ($X_{1/2}$, $d_{1/2}$), 66-69
linear coefficient (μ_l), 63-65, 67-68
mass coefficient (μ_m), 65-67, 69-72
branching ratios, 139-140
detection of, 72, 74, 96-97, 113-114, 117-118, 122 ff, 142 ff, 175 ff, 187, 210, 225, 249, 425-430
electromagnetic nature of, 10, 57-58
energy characteristics, 59
health hazards of, 71-72, 367-368
interaction with matter, 60-74
isomeric transition, 22, 59
secondary energy dissipation
Bragg scattering (diffraction), 60
Compton effect, 61-63, 67-72, 125, 143-144, 146-149
nuclear transformation, 60
pair production, 62-63, 67-72, 125, 143-144, 148, 154
photoelectric effect, 60-63, 67-72, 125, 143-144, 146
shielding, 66-71, 368
spectrometry, 142-167, 176-180
Gamma ray spectra, characteristics of
backscatter peaks, 147, 149-154
Compton edge, 147, 149-154
Compton "smear," 146-147, 149-154
escape peaks, 145, 148-154
photopeaks, 148-154, 158-166, 176-178
sum peaks, 150-154, 156
X-ray peaks, 150-154, 176
Gamma ray spectrometry, 142-167
components of spectrometers, 156-158
energy resolution, 158-160, 172-179, 425-430
γ-ray absorption characteristics, 142-148
identification of radionuclides, 160-163, 176-179
quantitative analysis of γ-ray spectra, 163-166
photofraction in, 164
single channel counting, 166-167
special interaction effects, 148-156
backscattering, 148-152

Gamma ray spectrometry (*cont.*)
special interaction effects (*cont.*)
bremsstrahlung, 154-155
lead X-rays, 152-153
peaks at 0.511 MeV, 153-154
summation effects, 155-156
Gas ionization mechanism, 98-120
Geiger-Müller region, 107, 109-111
ionization potential, 99-100
limited proportional region, 107-109
potential gradient, 100-103, 106-110
proportional region, 107-108
saturation current, 103
Townsend avalanche, 107
with gas amplification, 106-111
without gas amplification, 100-106
Gaussian distribution, *see* Distribution, Gaussian
Geiger-Müller detectors, 109-111, 114-120
characteristics of, 109-111, 118-120
coincidence loss, 314-317
quenching mechanisms, 110-111, 314
types of, 114-117, 371-373
typical preamplifier output signals, 78
Geiger-Müller region, 107, 109-111
Gel sample in liquid scintillation counting, 277-278

Half-life
biological, 32, 368-369
determination of, 30-35
radioactive decay ($t_{1/2}$), 28-32, 34-37, 465-467
relation to decay constant, 30
relation to radioactive emission mechanism, alpha decay, 40-41
tables of common radioisotopes, 18, 31, 350-351
Half-thickness ($x_{1/2}$ and $d_{1/2}$)
relation to absorption coefficient, 66-67
relation to shielding requirements, 66-69
Half-value layer (HVL), *see* Half-thickness
Halogen quenching in G-M detectors, 110-111
Health hazards from handling radioactive materials
external, 46, 56-57, 71-72, 362-363, 367-371
internal, 46, 56-57, 72, 368-371
High-resistance leak method for ionization chambers, *see* Voltage-drop method
High voltage supply, *see* Power supply
Hoods, use of with radioactivity, 369, 371, 375-376, 378

Infinite thickness, *see* Saturation thickness
Integral counting, 134-141
Internal conversion, 22, 47
Internal health hazards, 46, 56-57, 72, 368-371
Internal-sample scintillation counters, *see* Liquid scintillation counters
Internal standard method in liquid scintillation counting, 217-218
with identical system properties (ISISP), 218
Interval distribution, 297-298
Inverse square law, 367
Ion chamber, 100-106, 120, 261-262, 371-374, 423
Ionization, 41-42, 99-100
gas, *see* Gas ionization
specific, 42, 52-53
Ionization chamber, *see* Ion chamber
Ionization potential, 99
Ion pair formation, 41-42, 99-100
IR drop method for ionization chambers, *see* Voltage-drop method
Isobars, 8
Isomeric transition (IT), 22, 59
Isotones, 8
Isotope dilution analysis (IDA), 381-388
direct, 381-383
double, 385, 388
general comments on, 383-385
inverse, 383
substoichiometric, 385
Isotope effect, 327-329, 438-445
Isotopes
defined, 8
number of, 8
stable, 9-10
natural abundance of, 10

Kaartinen automated combustion for liquid scintillation samples, 257-260
K-capture, *see* Electron capture
Kilo electron volt (keV), 6

Labeled compounds, availability of, 331, 344-360
Laboratory safety rules, 377-379
Large-volume external-sample detectors, 210-212
Lauritsen electroscope, 103-104
Limited proportional region, 107-109
Linear absorption coefficient (μ_l), 63-65, 67-68
Linear energy transfer (LET), 366
Line of stability, 14-15, 19
Liquid counting samples, 261-262
Liquid (internal-sample) scintillation counters
age dating, 269-271

Liquid scintillation counters (*cont.*)
beta particle spectrometry, 187-204, 215-217
chemiluminescence, 269, 279-280
components of
coincidence circuitry, 190, 200, 202-203, 206-207, 209, 311
detector assembly, 194-214
logarithmic amplifier, 200, 203, 206-207, 215
power supply, 200, 202, 206
preamplifier, 185
pulse discriminator, 190, 200, 204, 214-217, 219-221
pulse height analyzer, 200, 202, 206, 208-209, 214-223
scaler, 200, 202, 206, 209
summing amplifier, 200, 202, 206, 209
timer, 209
conversion of specimen to counting sample, 189, 253-261, 268-280
Kaartinen automated combustion, 257-260
Parr oxygen bomb combustion, 256
Peterson automated combustion, 257-259
Schöniger flask combustion, 256-260
sealed tube combustion, 255-256
wet oxidation, 257
counting sample types
accommodating aqueous solutions, 212-214, 271-272
double radioisotope labeled, 187, 191, 201, 214-217, 221
filter paper, 280
gases (absorbed), 197, 213, 271, 273-274
gel, 277-278
metallic ions, 197, 273-275, 278
minimum quenching, 268-270
detection of
alpha particles, 187, 195, 225, 250
beta particles, 185-193, 195, 212-225, 249, 251, 423-425
gamma rays, 187, 210-211, 249, 255
neutrinos, 210
neutrons, 210-212
protons, 210
detector assembly components, 194-199
photomultipliers, 182, 184-191, 194-195, 199-217
primary solutes, 184, 191, 197-199, 270-279
primary solvents, 182-185, 195-196, 269-280
sample vials, 185, 193-195

Liquid scintillation counters (*cont.*)
detector assembly components (*cont.*)
secondary solutes (wave shifters), 184, 198-199, 270, 272, 274-275, 277-279
secondary solvents, 196-197, 270-280
determination of counting efficiency
by channels ratio (pulse height shift) method, 200, 203, 205, 209, 219-221
by dilution method, 218-219
by external standard method, 200-210, 221-224
by internal standard method, 217-218
efficiency of, 181, 185-190, 192-199, 203-204, 211-224, 269-280
energy transfer in, 182-186
fluorescence quenching, 192-196, 203-204, 217-224, 268-276
chemical, 192-193, 268-270
color, 192-193, 268, 276
corrections for, 200-209, 217-224
dilution, 192
effect on counting, 192, 204, 217-225, 276
optical, 193-194
oxygen, 193
fluors
characteristics and varieties of, 181-187, 197-199, 210-214, 268-280
energy transfer in, 182-186
four-π geometry of, 181, 193, 313
operating characteristics of
effect of amplifier gain, 204, 214-217
effect of photomultiplier potential, 221
effect of window setting, 214-217, 219-221
flat spectrum counting, 216-217
special detector types
continuous-flow, 212-214
large volume, 187, 210-212
whole-body counter, 187, 210-211
thermal noise problems, 186, 189-192, 207, 209, 216-217
reduction of, 186, 190-192, 207, 209, 216-217
Lithium-drifted detectors, 173-179
germanium [Ge(Li)], 175-179, 426-430
silicon [Si(Li)], 175-176
Logarithmic amplifier, 79, 200, 203, 206-207, 215
Low background counter, 117, 129, 156-157, 175, 311, 423-430
Low-level counting techniques, *see also* Low background counter
blanks, use of, 420-421

Low-level counting techniques (*cont.*)
counting time, determination of, 421-423
techniques for various types of radiation, 157, 211, 423-430

Mass absorption coefficient (μ_m), 65-66, 69-71
Mass number (A value), 7
Maximum permissible body burden, 46, 369
Maximum permissible concentration, 46, 369
Maximum permissible dose, 367
Mechanical register, 83-84, 314-315
Microcurie (μCi), 35
Millicurie (mCi), 35
Million electron volt (MeV), 6, 15-16
Multichannel pulse height analyzer, 87-89, 157-158, 167, 222
analog-to-digital converter (ADC), 87-89

Nanocurie (nCi), 35
Naturally occurring radioisotopes, 11-12, 411-415
Negatron (β^-) emission, 18-19
Negatrons (β^-), 46-47
Neutrino, 49-51, 210
Neutron activation analysis, 381, 389-398, 400-401, 404-410
advantages and disadvantages of, 391-394
applications of, 397-398, 404-410
basic description of, 389-391
comparative cost of, 400-401
practical considerations in, 394-397
using stable-activable tracers, 404-410
Neutron-capture cross section, 390-391, 398, 404-407, 463-466
Neutron flux, 347-349, 390-391, 395-396, 464-467
Neutron generators, 395
Neutrons
defined, 6-7
detection of, 210-212
quality factors for, 366
reactions induced by, 345-352, 390-398, 404-410, 464-467
thermal, 347-349, 366, 395, 404, 465
Neutron sources, 60, 395
Noise, electrical and thermal, 77-81, 83, 141, 166
from photomultipliers, 135-137, 186, 189-192, 207, 209, 216-217, 219, 425
signal-to-cable noise ratio, 78-81
Nomographs for
absorption of X-rays and γ-rays by lead, 70, 73
neutron irradiation products, 466-467
optimum counting time, 303-304
Normal distribution, *see* Distribution, Gaussian
Nuclear binding energy (BE), 16
Nuclear emulsion, 99, 236-241
Nuclear instrumentation, 75-95
amplifiers, 75-82, 91, 105-106, 136-137, 157
linear, 79
logarithmic, 79, 200, 203, 206-207
coincidence circuitry, 90-94, 190, 200, 202-203, 206-207, 209, 311
count rate meters, 84-85, 89, 266-267
discriminators, 76, 81-83, 86, 91-94, 113, 134, 141, 190, 200, 204, 214-217, 219-221
mechanical registers, 83-84, 314-315
output devices, 89-90
power supplies, 100, 117, 134-135, 157, 200, 202, 206
preamplifiers, 76-79, 91, 113, 128, 135, 157, 166, 185
pulse height analyzers, 75-77, 86-89, 160, 166-167, 200-209, 214-217, 219-224
multichannel analyzer, 87-89, 157-158, 167, 222
single channel analyzer (SCA), 86-87, 91, 166-167, 200-204, 214-217, 219-224
scalers, 76-77, 81-84, 89, 113, 117, 200, 202, 206, 209
binary, 82-83
decade, 83-84
timers, 82, 84, 209
Nuclear potential barrier, 40-41, 462-463
Nuclear reactions, 15-17, 344-352, 390-396, 404, 461-467
energetics and probability of, 15-17, 346, 461-467
isotope production by, 347-352
terminology used, 344-346
Nuclear reactors, 347-352, 395, 404, 408, 413, 416-418, 465
Nuclear safety
decontamination procedures, 375-379
hazard factors
external, 46, 56-57, 71-72, 362-363, 367-371
internal, 46, 56-57, 368-371
instrumentation, 104, 106, 117, 371-375
laboratory safety rules, 377-379

Nuclear safety (*cont.*)
radiation dose and exposure, 96-97, 362-366
units of, 97, 364-368
Nuclear stability, 14-15, 19
Nuclear statistics
application to radiotracer experiments, 308
average counting rate, 302-303
background error, 300-305, 421-423
confidence limits, 297
counting time requirements, 302-305, 308
interval distribution, 297-298
normal distribution, 295-300, 307-308
optimum choice of detector, 305-306, 421-423
rejection of abnormal data, 307
significant figures, 300-302
standard deviations
fractional, 303, 422
of the distribution (σ), 296, 299-301, 305-306
of the mean (σ_x), 299-300, 421-422
statistical tests, 307-308
weighted averages, 306-307
Nuclear track detectors, 233, 242-245
Nuclear transformation, 60
Nucleons, 6
Nucleus of atom, 5-7
Nuclides, 8-12
defined, 8
number of, 8
radioactive, *see* Radioisotopes
stable, 9-10
symbolic expression of, 7

Oak Ridge National Laboratory's Graphite Reactor, 348
Operating potential for counters, 100-102, 107, 113-114, 118, 135-136, 214
Optimum choice of detector, 305-306
Output devices, 89-90

Paired sources for determination of resolving time, 316-317
Pair production, 62-63, 67-72, 125, 143-144, 148, 154
Paper chromatography
autoradiography of, 238, 435
counting of sections, 280
scanning of radiochromatograms, 266-268
Paralysis time, *see* Resolving time
"Parent-daughter" radionuclides, 33
Parr oxygen bomb, 256
Particle accelerators, 350-352
Persulfate combustion method, 255
Peterson automated combustion apparatus for liquid scintillation samples, 257-259
Phosphors, *see* Fluors
Photocathode, 124, 127-128, 132-133, 172-173, 182, 184-186, 188-190, 194
Photocoupling, 132, 194
Photoelectric effect, 60-63, 67-72, 125, 143-144, 146
Photofraction in gamma-ray spectrometry, 164
Photomultipliers
effect of potential applied to, 135-136, 166-167, 214
energy transfer in, 98, 122-123, 127-129, 172-173, 182, 184-186, 194
nature of, 124, 132-134
refrigeration of, 189-192
thermal noise, 135-137, 141, 166, 186, 189-192, 207, 209, 216-217
Photons, 10, 57-59
Photopeaks in gamma-ray spectra, 148-154, 158-166
Physical science applications of radiotracer methods
in chemistry
equivalence of various atoms in molecules, 435-436
kinetics of slow reactions, 438
molecular rearrangements, 436-438
separation procedures, 434-435
solubility product constants, 435
isotope effects
on equilibria, 439-440
on kinetics, 440-445
tracing physical processes
leak testing, 434
mixing studies, 433-434
surface area of solids, 434
Picocurie (pCi), 35
Pileup, pulse, 79, 81, 158
Pipetting radioactive solutions, 250, 252, 378
Planchets for counting sample, 262-264, 318
Plateau
for G-M detector, 118
for proportional detector, 113-114
for solid scintillation, 135-137
Pocket dosimeter, *see* Dosimeter
Positron, 19-20, 46-51
annihilation of, 47
emission of, 19-20, 48-51, 148
Power supply, 110, 117, 134-135, 157, 200, 202, 206
Preamplifier, 76-79, 91
in proportional counting assemblies, 113
in scintillation counting assemblies, 128, 135, 157, 166, 185
Primary solute in liquid scintillation counting, 184, 191, 197-199, 270-279

Primary solvent in liquid scintillation counting, 182-185, 195-196, 269-280
Proportional detectors
  alpha plateau, 113-114
  beta plateau, 114
  operating characteristics, 108, 113-114, 120, 225
  resolving time of, 108, 120, 225
  types of, 111-113, 423
  typical preamplifier output signals, 78
Proportional region, 107-108
Protons, 6-8, 461
Pulse height analyzer, 75-77, 86-89, 160, 166-167, 200-209, 214-217, 219-224
  multichannel analyzer, 87-89, 157-158, 167, 222
    analog-to-digital converter (ADC), 87-89
  single channel analyzer (SCA), 86-87, 91, 166-167, 200-209, 214-217, 219-224
Pulse height selector, *see* Discriminator, pulse
Pulse height shift method in liquid scintillation counting, *see* Channels ratio method
Pulse shaping, 75, 79-81, 94, 158
Purity, radiochemical, 330, 359-360

Quality factor (QF), 365-366
Quenching (fluorescence) in liquid scintillation counting, 192-196, 203-204, 217-224, 268-276
  chemical, 192-193, 268-270
  color, 192-193, 268-276
  corrections for, 200-209, 217-224
  dilution, 192
  effect on counting, 192, 204, 217-225, 276
  optical, 193-194
  oxygen, 193
Quenching mechanisms in G-M detectors
  external, 110
  internal, 110-111, 115-116, 314
Q value, 17-18, 461-462

Rad, 365
Radiation, *see* individual types of radioactive emission
Radiation dose, 96-97
  comparison of sources of, 363, 412
  effects of, 362
  factors in
    linear energy transfer (LET), 366
    quality factor (QF), 365-366
  units of
    rad, 97, 365
    roentgen (R), 97, 364-365, 367-368
Radiation dose (*cont.*)
  units of (*cont.*)
    roentgen equivalent man (rem), 365-368
Radiation dosimetry, 363-375
Radiation monitoring instruments, 371-375
Radioactive decay
  by alpha particle emission, 22
  average life (T_A), 33-34
  composite, 32-33
  constant, 25-31, 34
  correction tables for, 28
    ^{32}P, 29
    ^{131}I, 29
  by electron capture (EC), 20-21
  half-life ($t_{1/2}$), 28-32, 34-37, 40-41
    determination of, 30-35
    tables of, 18, 31, 350-351
  by internal conversion, 22, 47
  by isomeric transition (IT), 22, 59
  nature of, 14-38
  by negative beta (negatron) emission, 18-19
  by positive beta (positron) emission, 19-20
  rate of, 24-34
  schemes, 18-23
  types of, 18-24
Radioactive isotopes, *see* Radioisotopes
Radioactive wastes
  disposal of, 332, 376-378
Radioactivity
  absolute counting, 96, 163, 323
  detection by
    autoradiography, 99, 233-241, 435
    gas ionization, 98-120, 261-262
    liquid (internal-sample) scintillation method, 98, 123, 181-225, 261-262
    nuclear track method, 233, 242-245
    semiconductors, 98-99, 168-180, 261-262
    solid (external-sample) scintillation method, 98, 122-167, 261-262
  measurements of
    basic principles of, 97-99
    correction factors in, 310-325
    general considerations, 96-99
    radiation dose and, 96-97, 364-366
    statistical considerations, 294-309
    types of, 96-97
  nature of, 6, 10-11, 14-38
  relative counting, 96, 163, 323
  standard unit of, 34-37
Radioanalytical techniques
  activation analysis, 381, 389-398, 400-401, 404-410
    advantages and disadvantages of, 391-394

Radioanalytical techniques (*cont.*)
activation analysis (*cont.*)
applications of, 397-398, 404-410
basic description of, 389-391
comparative cost of, 400-401
practical considerations in, 394-397
using stable-activable tracers, 404-410
isotope dilution analysis (IDA), 381-388
direct, 381-383
double, 385, 388
general comments on, 383-385
inverse, 383
substoichiometric, 385
radiometric techniques, 388-389
increased solubility of a solid, 388
precipitation with a radioactive reagent, 388
titration, 388-389
Radioautography, *see* Autoradiography
Radiocarbon dating, 2, 12, 117, 269-271
Radiochemical purity, *see* Purity, radiochemical
Radiochromatograms
autoradiography of, 238, 435
counting of sections, 280
scanning of, 266-268
thin-layer, 268, 277, 435
Radiochromatogram scanning counter, 266-268
Radiocolloid, 330
Radioisotope labeled compounds
preparation by
biosynthesis, 356
chemical synthesis, 353-355
tritium exposure, 356-359
radiolysis of, 330, 359-360
terminology, 353
Radioisotopes
artificially produced, 12, 344-352, 400-404, 413-418, 463-467,
by nuclear reactors, 347-352, 395, 404, 408, 413-418, 463-467
by other neutron sources, 395
by particle accelerators, 350-352
availability of, 344-361
average life values for, 33-34
carrier-free, 37, 349-352
decay constants for, 25-31, 34
hazard factors in handling
external, 46, 56-57, 71-72, 362-363, 367-371
internal, 46, 56-57, 72, 368-371
naturally occurring, 11-12, 411-415
preparation of labeled compounds from, 352-359
safe handling of, 362-380
specific activity of, 37-38, 331, 335-340, 349-352, 356-358
Radiolysis of labeled compounds, 330, 359-360
Radiometric techniques, 388-389
increased solubility of a solid, 388
precipitation with a radioactive reagent, 388
titration, 388-389
Radionuclides, *see* Radioisotopes
Radiotracer experiments
anticipated findings, 338
basic assumptions, 327-331
data analysis, 339-342
expression of, 339-340
interpretation of, 340-342
design of, 38, 326-338
basic features of, 333-338
example of, 336-338
preliminary factors in, 327-333
execution of, 338-342
feasibility of, 331-333
hazards, 332
isotope effects, 328-329
scale of operation, 333-335
specific activity requirement, 38, 335-338
unique advantages of, 326-327
Range
of alpha particles, 42-45
of beta particles, 53-55
Range-energy relationships
for alpha particles, 43-44
for beta particles, 54-55
Rate meter, *see* Count rate meter
Rate-of-charge (-drift) method for ionization chambers, 102, 104-105
Recovery time, *see* Resolving time
Regulations, laboratory, 377-379
Rejection of abnormal data, 307
Relative biological effectiveness (RBE), 365-366
Relative counting, 96, 163, 323
Resolution in autoradiography, 235-241
Resolution in gamma-ray spectrometry, 158-160
Resolving (coincidence, dead, paralysis) time
correction for, 314-317
dead time component of (τ_d), 109-110, 315
determination of, 314-317
of detectors, 97, 108-110, 119-120, 129-130, 180, 190, 225, 314-317
of instrument components, 89-92
recovery time component (τ_r), 315
Roentgen (R), 97, 364-365, 367-368
Roentgen absorbed dose (*rad*), 97, 365
Roentgen equivalent man (*rem*), 365-368

Safety measures, *see* Nuclear safety
Sample geometry, 131, 137-139, 181, 193, 312-313
four-π, 181, 193, 266, 312-313
two-π, 131, 313

Sample mounting, 262-266, 317-318
Sample preparation, *see* Counting sample preparation
Sampling methods in environmental applications, 416-420
Saturation current, 103
Saturation thickness
in backscattering, 317-318
in self-absorption, 320-322
Scalers, 76-77, 81-84, 89, 113, 117, 200, 202, 206, 209
binary, 82-83
decade, 83-84
Scattering, *see* Backscattering, Self-scattering, Side scattering
Schöniger combustion method, 256-260
Scintillation counters, *see* Liquid scintillation counters, Solid scintillation counters
Scintillation mechanism, 98, 122-129, 142-148, 182-186
Scintillators, *see* Fluors
Secondary solute in liquid scintillation counting, 184, 198-199, 270, 272, 274-275, 277-279
Secondary solvent in liquid scintillation counting, 196-197, 270-280
Self absorption, 45, 264, 277, 311, 319-323
Self scattering, 321
Semiconductor detectors
advantages and disadvantages, 78, 81, 179-180, 261-262, 423, 426-430
energy resolution, 172-179
nature of semiconductors, *see also* Band theory of solids, 168-171
germanium, 173
silicon, 171-173
operating principles, 171-173
types of
diffused *p-n* junction, 173-174
lithium-drifted, 173-179
surface barrier, 174
Sensitivity of detection methods, 97, 313, 326, 331, 381, 400, 423-430
autoradiography, 236-237
Geiger-Müller, 117-118, 313-314
scintillation, 133, 165-166, 181, 186, 190
Shelf ratios, 313
Shielding
from beta particles, 56-57, 368
from gamma radiation, 66-71, 368
of detectors, 117, 129, 148-149, 152-154, 156-157, 311, 425-426
Side scattering, 318
Signal-to-noise ratio, 78-81
Significant figures, 300-302
Single channel pulse height analyzer (SCA), 86-87, 91, 166-167, 200-209, 214-217, 219-224
Solid angle, 179, 312-313
Solid counting samples, 262-266, 277-278
Solid (external-sample) scintillation counters,
components of
amplifier, 135-137, 157
detector assembly, 124, 129-135, 156-157
discriminators, 134-135, 141
multichannel analyzer, 157-158, 167
power supply, 135, 157
preamplifier, 78, 128, 135, 157, 166
single channel analyzer (SCA), 166-167
counting sample types
liquid, 261-262
solid, 262
detection efficiency
for beta particles, 129-131
for gamma rays, 71-74, 128-131, 134-135, 137-141, 163-166, 261-262, 313, 335, 425-430
detector assembly components
detector housing, 129
fluors, 98, 123-132
photocoupling, 132
photomultiplier, 98, 122-124, 127-129, 132-136
energy transfer in, 98, 122-129, 132-134
fluors
characteristics and varieties of, 122-132
energy transfer in, 98, 122-129
gamma ray spectrometry, 142-167, 176-180, 425-430
operating characteristics,
effect of amplifier gain, 136-137
effect of photomultiplier potential, 135-136
integral counting mode, 134-141
plateau, 135-137
resolving time, 129-130, 225, 315
sample preparation for, 131, 137-139, 249-251, 261-265
thermal noise, 135-137, 141, 166, 425
well detector, 131, 139, 261-262, 313
Specific activity, 37-38, 331, 335-338, 349-352, 356-358
expression of results as, 339-340
in isotope dilution analysis, 382-385
Specific ionization
of alpha particles, 42
of beta particles, 52-53
Spectra
of beta particles, 47-50, 215, 219
of gamma rays, 142-167, 176-180, 430
Spectrometry
of beta particles
by magnetic spectrometer, 47-49
by liquid scintillation spectrometer, 187-204, 215-217
of gamma rays, 142-167, 176-180, 425-430

Stable-activable tracers, 398, 400-401, 404-410
Stable isotopes (nuclides), 9-10, 400-401, 404-410
Standard deviation, *see* Deviation, standard
Standard source, 160, 218, 314
Starting potential of G-M detectors, 118
Statistical considerations, *see* Nuclear statistics
Straggling
  of alpha particles, 43
  of beta particles, 54
Stripping film, 237, 240-241
Sum peaks in gamma-ray spectra, 150-154, 156
Surface barrier detectors, 174
Synthesis of labeled compounds, 331, 352-359

Thermal neutrons, 347-349, 366, 395, 404, 465
Thermal noise, *see* Noise, electrical and thermal
Thermoluminescent dosimeters (TLD), 374-375
Thin-layer chromatography (TLC), 268, 277, 435
Thorium series, 11, 411-412
Threshold
  alpha, 113-114
  beta, 114
Time constant, 85, 267
Timers, 82, 84, 209
Townsend avalanche, 107
Tritium and tritium labeled compounds, 8, 12, 37, 47, 50, 328, 346, 349, 351, 412-418
  detection of, 3, 55-56, 105, 185-187, 190-191, 193, 195, 213, 215, 218-219, 221-222, 224, 234, 236, 240-241, 321, 423-424
  health hazards from, 57, 369-370, 374
  labeling methods, 353, 355-360
  sample preparation, 249, 254-261, 266, 271, 277-279, 424

Units
  of radiation absorption, 63-71
  of radiation dose and exposure, 97, 364-368
  of radiation energy, 6, 15-16
  of radioactive decay rate, 24-34
  of radioactivity, 34-37
Uranium-radium series, 11, 411-412

Van Slyke-Folch combustion, 254-255
Vibrating-reed electrometers, *see* Electrometers, vibrating-reed
Voltage-drop method for ionization chambers, 102, 104

Waste disposal of radioactive material, *see* Radioactive wastes
Wave shifter, *see* Secondary solute in liquid scintillation counting
Weighted averages, 306-307
Well scintillation detector, 131, 139, 261-262, 313
Wet ashing method, 253-254
Whole-body counters, 210-211
Wilzbach gas-exposure, 358-359
Windowless gas flow counter, 112, 117
Window of pulse height analyzer, 86, 166, 215-217, 220

X-ray film, 236-239
X-ray peaks in gamma spectra, 150-154
X-rays
  compared to gamma rays, 10, 57-58
  detection of, 175-176, 222, 236
  emission in electron capture decay, 21, 23-24, 39, 57, 59-60, 62, 236
  in gamma spectrometry, 144-146, 152-153, 159

Yield, expression of results as, 339

Z value, 7, 155